Green's Function Integral Equation Methods in Nano-Optics

Green's Function Integral Equation Methods in Nano-Optics

Thomas M. Søndergaard
Aalborg University, Aalborg, Denmark

CRC Press is an imprint of the
Taylor & Francis Group, an **informa** business

CRC Press
Taylor & Francis Group
6000 Broken Sound Parkway NW, Suite 300
Boca Raton, FL 33487-2742

First issued in paperback 2023

Version Date: 20190107

ISBN 13: 978-1-03-265312-9 (pbk)
ISBN 13: 978-0-8153-6596-9 (hbk)
ISBN 13: 978-1-351-26020-6 (ebk)

DOI: 10.1201/9781351260206

Publisher's Note
The publisher has gone to great lengths to ensure the quality of this reprint but points out that some imperfections in the original copies may be apparent.

Library of Congress Cataloging-in-Publication Data

Names: Søndergaard, Thomas, author.
Title: Green's function integral equation methods in nano-optics / Thomas Søndergaard.
Description: First edition. | Boca Raton, FL : CRC Press/Taylor & Francis Group, 2019. | Includes bibliographical references and index.
Identifiers: LCCN 2018026592| ISBN 9780815365969 (hardback : acid-free paper) | ISBN 9781351260206 (ebook)
Subjects: LCSH: Micro-optics--Mathematics. | Green's functions.
Classification: LCC TA1659.5 .S66 2018 | DDC 621.36--dc23
LC record available at https://lccn.loc.gov/2018026592

Visit the Taylor & Francis Web site at
http://www.taylorandfrancis.com

and the CRC Press Web site at
http://www.crcpress.com

Contents

Preface

The purpose of the book is to give a comprehensive introduction to using Green's function integral equation methods (GFIEMs) for solving scattering problems in nano-optics. The cases of interest from the area of nano-optics include scattering from metal nanoparticles, theoretical studies of the optics of nanostructured surfaces, and studies of scattering from objects placed on or near planar layered structures including optical waveguides. At optical wavelengths we will take into account the penetration of light into metals, the possibility of surface plasmon resonances of metal particles, and the possibility of excitation of surface plasmon polaritons propagating along metal surfaces or in narrow gaps in metals. The book covers different types of integral equation methods for 1D, 2D, and 3D scattering problems in nano-optics, how the integral equations can be discretized and solved numerically, and how this can be done efficiently.

Chapter 1 gives a brief introduction, without a lot of equations, to the types of GFIEMs and the scattering problems of interest in the book. The GFIEMs and subsequent extraction of data rely heavily on a theoretical foundation of electromagnetic theory, optics of planar layered structures including waveguides, and scattering theory. For this reason the most important parts of the theoretical foundation are reviewed in Chapter 2, and in appendices. It is recommended that readers should have knowledge of optics and electromagnetics corresponding to the theoretical foundation presented in Chapter 2. It is also recommended that the reader should have a good grasp of vector calculus, linear algebra, differential equations, complex numbers, and time-harmonic fields in a complex representation.

The subsequent Chapters 3-9 consider GFIEMs of different types for increasing complexity and dimensionality. After presenting each method there are sections with examples of applying the GFIEM to different problems in nano-optics. In addition, there is a section with guidelines for software implementation and another section with exercises. The book is accompanied by a separate electronic file with examples of numerical code implementations of GFIEMs in MATLAB®.

The book is sufficiently detailed such that it can be used for a self-study of GFIEMs. The book can be used in a lecture course focusing on GFIEMs, or alternatively it is also possible to select a few of the chapters and use these as part of a lecture course on nano-optics, nano-photonics, or numerical modeling methods. The book can also be used in student projects, where the aim is to develop a numerical program with an implementation of one of the GFIEMs and to use this to model the optics of a specific nanostructure. The book can also be used by researchers entering the area of GFIEMs either as a user of the methods or for developing such methods. It is my hope that the readers will implement and use the GFIEMs presented in the book for studying many new and interesting nano-optics problems.

Finally, I would like to mention that I have had a great time working in the field of nano-optics together with good colleagues. In particular I have had a long and fruitful cooperation with Sergey Bozhevolnyi and his group within the field of plasmonics. I have also had good cooperation in the fields of plasmonics and nano-optics with Kjeld Pedersen, Thomas G. Pedersen, Jesper Jung, Esben Skovsen, Peter K. Kristensen, Yao-Chung Tsao, Jonas Beermann, Niels Asger Mortensen, Alexandra Boltasseva, Tobias Holmgaard, Vahid Siahpoush, Alexander Roberts, Paw Simesen, Enok Skjølstrup, Andrei Lavrinenko, Niels Gregersen, and many others. I got introduced to some advanced topics of Green's functions by Bjarne Tromborg that I have used ever since, and spent a few years in the field of photonic crystals working together with Anders Bjarklev, Jes Broeng, Martin Kristensen, and Stig. E. Barkou. Our cooperation and discussions most certainly influenced the choice of examples of nano-optics that are considered in this book.

Thomas M. Søndergaard, May 30, 2018, Aalborg University, Denmark

1 Introduction

The Green's Function Integral Equation Methods (GFIEMs) considered in this book are numerical methods for studying theoretically what happens when light is incident on a scatterer placed in a reference structure. The GFIEMs give solutions to Maxwell's equations (see Section 2.1) for electric and magnetic fields. Rather than solving Maxwell's equations directly in differential form, the electromagnetic fields are obtained by solving integral equations that involve a Green's function [1, 2, 3], e.g., $g(\mathbf{r},\mathbf{r}';\omega)$, which describes the electromagnetic fields generated at an observation point $\mathbf{r}$ by a point source at a position $\mathbf{r}'$ in the reference structure.

1.1 OVERVIEW OF METHODS AND SCATTERING PROBLEMS

Two examples of the types of structures of interest in this book are illustrated in Fig. 1.1, namely the case of a scatterer placed in free space [Fig. 1.1(a)], and a scatterer placed on a planar layered structure [Fig. 1.1(b)]. Certain layered refer-

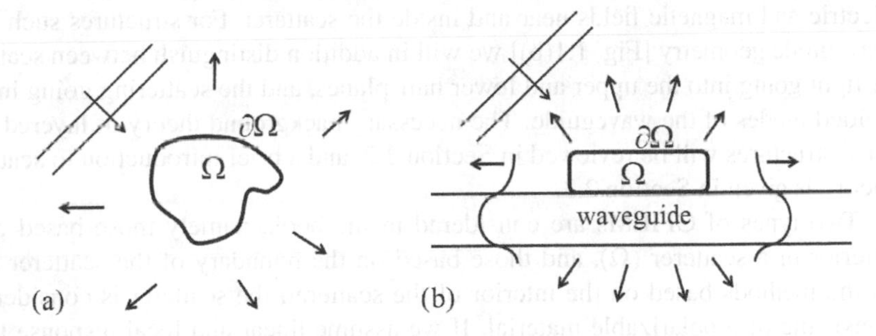

Figure 1.1 Schematic of (a) a scatterer placed in free space, and (b) a scatterer placed on a planar waveguide.

ence structures may support guided modes and are thus waveguides. Here, the free space or planar layered structure serve as a reference structure with a known Green's function. The total structure, including also the scatterer, is illuminated by an electromagnetic field, which can, for example, be a plane wave, a Gaussian beam, or any other beam profile. The structure can also be illuminated by the field generated by any given source distribution.

In the absence of the scatterer, the incident field will lead to a well-known resulting field. In the free-space case, this is identical to the incident field, while in the case of the waveguide shown in Fig. 1.1(b) the resulting field at, for example, positions above the waveguide, will also contain a contribution due to reflection from the layered waveguide geometry. In both cases we will refer to the total field in the absence of the scatterer as the reference field. Since the properties of the reference structure are already governed by the Green's function, then the numerical problem can be reduced at first to considering only the region of the scatterer, which is one of the main advantages of GFIEMs. The Green's function also automatically takes care of appropriate boundary conditions such as, for example, the radiating boundary condition in open geometries, which basically means that outside the scatterer, the total field must be the sum of the reference field and a field that propagates away from the scatterer. It should be emphasized that in the case of other popular methods, such as, for example, the Finite Element Method (FEM) [4, 5, 6, 7], it is necessary to include in the numerical problem a larger region than just the scatterer, and it is necessary to truncate the computational region by using, for example, absorbing boundary conditions or perfectly matched layers. When using GFIEMs, the region that must be discretized and solved for can thus be much smaller, and the boundary conditions are automatically taken care of via the Green's function.

For the situation in Fig. 1.1(a) the GFIEMs can be used to calculate the total scattered power, the angular distribution of the scattering, and the power absorbed in the particle. When the incident light is a plane wave, we may also be interested in expressing this in terms of the corresponding scattering and absorption cross sections. In addition, the GFIEMs can be used to calculate detailed maps of the distribution of electric and magnetic fields near and inside the scatterer. For structures such as the waveguide geometry [Fig. 1.1(b)] we will in addition distinguish between scattering of light going into the upper and lower half-planes, and the scattering going into the guided modes of the waveguide. The necessary background theory of layered reference structures will be reviewed in Section 2.2, and a brief introduction to scattering theory is given in Section 2.3.

Two types of GFIEMs are considered in the book, namely those based on the interior of a scatterer (Ω), and those based on the boundary of the scatterer ($\partial\Omega$). In the methods based on the interior of the scatterer, the scatterer is considered as consisting of a polarizable material. If we assume linear and local response theory the local polarization density at a position $\mathbf{r}$ inside the scatterer depends on the total field at $\mathbf{r}$, which can be divided into the reference field and the fields due to polarized material throughout the scatterer. In this way the polarization density (or the field) at one part of the scatterer will be linked to that in all other parts of the scatterer, which can be expressed in the form of an integral equation. In the case of large scatterers, or many scatterers, where the numerical problem is large, we will consider in this book different approaches to reduce the computational problem. In many cases the integrals in GFIEMs are convolution integrals, which (in discrete form) can be calculated fast by using the Fast Fourier Transform (FFT) algorithm. In addition, when the integral equation is discretized and converted to a matrix equation, the convolution property means that only a small part of the matrix needs to be stored. This greatly reduces memory requirements which will scale approximately linearly with the number of discretization elements.

In the methods based on the boundary of the scatterer ($\partial\Omega$) one often considers a scatterer made of a single material. The GFIEMs that will be considered for this case will be based on expressing the fields at any position **r** inside and outside the scatterer in terms of overlap integrals between the total fields at the boundary of the scatterer and a Green's function. Self-consistent equations for the surface fields can be obtained by using in the integral equations a position **r** that approaches infinitesimally the scatterer surface from either side. The numerical problem is thus reduced to considering only the boundary of the scatterer leading to a computational domain being one dimension smaller than for methods based on Ω. One of the great advantages of the surface integral equation methods is that the surface of the scatterer is often treated with high precision.

In the case of introducing many scatterers into the reference structure it is sometimes possible to obtain almost linear scaling of the numerical problem with the number of scatterers, and thus to be able to consider large numerical problems. In Sec. 4.3.8 it is discussed how concepts from the Fourier-Modal-Method (FMM), also known as rigorous coupled-wave analysis [8, 9, 10, 11], can be used to speed up calculations for large structures. This amounts to sub-dividing a structure with many scatterers into layers with an interface in-between neighboring layers. The response of a layer given from integral equations relates fields propagating out of the layer through the boundaries in terms of fields propagating into the layer. The response of a structure with several layers, as can be the case for a photonic crystal [12, 13, 14, 15], can then be determined iteratively in terms of single-layer responses in the same manner as in the FMM.

For three-dimensional structures with cylindrical symmetry we will also consider in this book GFIEMs formulated to take advantage of the symmetry. Here, the incident field will be decomposed into a number of components with different angular momentum, and the response of the structure to each component can be obtained by considering an integral equation where the region that must be discretized is reduced by one dimension. Especially for the surface-based GFIEM, it will be discussed in Chapter 9 how a three-dimensional structure with cylindrical symmetry can be modeled by discretizing in only one dimension. As examples of cylindrically symmetric structures we will consider a cylindrical scatterer on a dielectric substrate and on a planar waveguide, a microstructured gradient-index lens, and metal nano-discs.

1.2 OPTICS VERSUS MICROWAVES

A major difference between optics and radiation regimes with much longer wavelengths, such as, for example, microwaves, is that the material parameters can be quite different. In the case of microwaves it would be quite normal to model a metal as a perfect conductor, which is equivalent to using the approximation that the dielectric constant $\varepsilon(\omega) = -\infty$. A waveguide for microwaves could be constructed in the form of a metal tube, and, for example, a dipole antenna in free space would be resonant at a wavelength equal to two times the dipole antenna length.

For optics, and thus nano-optics, the perfect conductor approximation for metals is often not acceptable. In fact, much of the most interesting physics at optical wavelengths is a consequence of metals having a finite and highly wavelength-dependent dielectric constant. Here, metal surfaces may support waves known as surface plasmon polaritons (SPPs) [16, 17], and metal nanostructures may support optical resonances, which make them interesting as optical resonators and optical antennas [18]. Note that contrary to the case of microwaves a metal dipole antenna at optical wavelengths can be significantly shorter than half of the resonance wavelength. Optical resonators and antennas can be further used as building blocks in so-called metamaterials that are composite materials engineered at a subwavelength-scale leading to new materials with effective optical constants [19]. Strong near-fields in resonant metal nanostructures can also be exploited for strong enhancement of Raman scattering, and for enhancing non-linear optical effects. In optics, waveguides will often be constructed using dielectric materials, and in the case of using metals, the ohmic losses are significant, and contrary to microwaves, the light penetration into the metal cannot be ignored. In principle the methods considered in this book can be applied for both optical and microwave wavelengths, and GFIEMs are also popular in the microwave community [20, 21, 22, 23, 24, 25, 26], where they are also referred to as the method of moments. The structures that will be of interest in this book are those found in nano-optics and nano-photonics, such as metal nano-particles, nanostructured metal surfaces, optical waveguides, photonic crystals, grating couplers, microstructured lens geometries, anti-reflective microstructured surfaces, thin-film solar cells with nano- and microstructures, etc.

1.3 EXAMPLES OF NANO-OPTICS

The methods presented in each of the Chapters 3-9 are accompanied by examples from nano-optics. In the simplest one-dimensional case we will consider in Chapter 3 the reflection and transmission for a multi-layer structure with alternating layers of high and low refractive index of thicknesses of 68 and 114 nanometers (nm), which can be used for efficient reflection of red light in, for example, a helium-neon laser.

In Section 4.1 we start out for the two-dimensional case with an example of the optics of metal nanostrip resonators in a homogeneous medium. Metal nanostrips may support surface waves known as surface plasmon polaritons that can propagate back and forth along the strips and be efficiently reflected at strip terminations. For specific resonance wavelengths, where the phase change of the light during one round-trip back and forth along the strip matches an integer number of 2π, the strip will both strongly scatter and absorb light, and the nanostrip will perform as an optical resonator. The GFSIEM will be used to calculate the optical cross sections (extinction, scattering, and absorption) of metal nanostrips, which will exemplify their resonance properties. In addition, local field distributions or field enhancements compared with the incident field will be calculated. At resonance, the local fields can be very large at certain locations, also termed hot spots. Hot spots are often found in nanometer-size gaps between metal nanostructures. One application of strong local fields is in Raman scattering and for other non-linear phenomena, where the interaction between light and a molecule placed at a hot spot will be strongly increased. This type of enhancement of Raman scattering is responsible for the phenomenon of Surface-Enhanced-Raman-Scattering (SERS) that can be achieved with surfaces being covered with metal nanostructures and overlayered with the molecules of interest [27, 28]. Resonances exist for many different metal nanoparticles, and the resonance wavelengths are highly geometry and material dependent, which is responsible for metal nanoparticles of the same material but different shapes having different colors [29, 30].

In Section 4.2 we will consider examples of nanostructures on surfaces. Here, an example of a nanostrip on a dielectric substrate will be presented. Similar phenomena of local field enhancements and resonances will be observed as in Section 4.1 but the presence of the surface shifts the resonance wavelengths. In addition the substrate will affect the angular profile of the scattered light with most of the scattering now going into the high-refractive-index substrate. We will also consider the case of a nanostrip placed on a dielectric waveguide, in which case it will be shown that many large scattering resonances are related to the properties of the waveguide rather than to the properties of the metal nanostrip. Efficient coupling of light into waveguide modes seem to occur at wavelengths near the cut-off wavelengths of guided modes. This phenomenon can be of interest for thin-film solar cells and light couplers. In addition we will consider a metal nanostrip separated from a metal surface with a small gap. The gap will support another type of surface plasmon polariton known as gap-plasmon polariton. In this case the gap between metal surfaces rather than the strip itself will perform as a resonator. Another example which will be considered is a nano-groove in a metal surface, which depending on its shape can also perform as an optical resonator or as a broadband scatterer and absorber.

We will also consider an example of a dielectric microstructured lens on a dielectric slab. Here, the idea is that instead of varying the thickness of the lens to obtain a lens effect, we may instead change the effective refractive index by etching a microstructure, where the fraction of high-refractive index material will vary along the lens geometry. Since the structure is on a scale that is small compared with the wavelength of the light, this will result in an effectively varying refractive index giving a lens effect. This type of lens can be very small and may be monolithically integrated in optical components, and it is a convenient geometry from a fabrication point of view.

In Section 4.3 we will consider examples of different types of periodic structures. This will include an example of a dielectric surface nanostructure giving an antireflection effect, and an example of a periodic array of grooves in a metal leading to a shiny and highly reflecting metal being transformed into a broadband absorber that will look black, and thus the surface nanostructure has completely changed the appearance of the metal surface. Another example that will be considered is a photonic crystal. Photonic crystals [12] are wavelength-scale periodic structures in which light cannot propagate for certain wavelengths that are in an optical bandgap of the structure. This phenomenon, which is highly geometry and material dependent, is due to optical interference effects. By making, for example, point defects or line defects in the crystal, it is possible to create optical cavities and optical waveguides that confine light at wavelengths in the bandgap of the surrounding crystal. In particular, contrary to usual optical waveguides, it is possible to guide light in a defect channel with an on-average lower refractive index than the surrounding crystal structure.

In Chapter 5 we will consider an example of the Purcell factor of an emitter placed inside a photonic crystal and a photonic crystal cavity. The rate of spontaneous emission from an emitter (atom, molecule, or a quantum dot) is not just a property of the emitter itself but also depends on the close environment that the emitter is placed in [31, 32, 33, 34]. If an emitter is placed in a cavity with an optical cavity-mode which is strong at the emitter position, and at the emission wavelength, then emission can be strongly enhanced. If, on the other hand, the emitter is placed inside a photonic crystal supporting no propagating modes at the emission wavelength, because the wavelength is in the optical bandgap of the photonic crystal, then the emission can be strongly suppressed. The enhancement factor of spontaneous emission is known as the Purcell factor. The factor can be obtained classically (in the weak-coupling regime) by observing the enhancement in the emission from a dipole. In the example we will obtain the emission by both integrating the radiated power going through a circle in the far field, and from a component of the near-field at just the dipole location. The emission is proportional to the optical density of states, which is proportional to a component of the field at the emitter position, and can also be obtained from the Green's function, which basically governs the emission from a dipole emitter. In this chapter we will also consider an example for the case of excitation of surface plasmon polaritons by second harmonic generation in an organic nanofiber placed on a metal film on a substrate. In this case the fundamental harmonic incident field creates a polarization in the fiber at the 2nd harmonic frequency, which then at this frequency causes the nanofiber to act like a distribution of local emitters placed directly on a metal surface. Here, the SPP excitation can be observed via leakage-radiation into the substrate.

In the subsequent Chapters 6-9 we will consider in three dimensions the scattering from a nanorod on a thin metal film, a 2D periodic antireflective surface microstructure, a microstructured gradient-index lens, scattering into guided modes of a waveguide by a surface nanoparticle, and resonant metal nanoparticles. The physics is similar to examples considered with 2D modeling in previous chapters. However, since these are 3D geometries the modeling is much more complicated. Light can be scattered in more directions, two polarizations of the light must be considered, and we have to use a fully vectorial model of the electromagnetic fields. In order to reduce the computational task for some cases we will develop modeling approaches that can exploit cylindrical symmetry.

1.4 NOTATION, ABBREVIATIONS AND SYMBOLS

The notation, abbreviations and symbols used frequently throughout the book are listed in the following tables:

Object	Notation	Example
Vector in three-dimensional space	Bold	$\mathbf{r}$, $\mathbf{E}$
Unit vector	Bold with hat	$\hat{\mathbf{x}}$, $\hat{\mathbf{y}}$, $\hat{\mathbf{z}}$
Vector of expansion coefficients	Plain, single bar	$\overline{c}$
Matrix	Plain, double bar	$\overline{\overline{O}}$
Dyadic tensor	Bold	$\mathbf{G}$
Scalar (dot, inner) product	Single dot	$\hat{\mathbf{x}} \cdot \mathbf{E}$
Vector (cross) product	Single cross	$\hat{\mathbf{x}} \times \mathbf{E}$
Tensor (dyadic, outer) product or dyad	No dot	$\mathbf{RR}$

Abbreviation	Meaning
GFIEM	Green's function integral equation method
GFSIEM	Green's function surface integral equation method
GFAIEM	Green's function area integral equation method
GFVIEM	Green's function volume integral equation method
SPP	Surface Plasmon Polariton
FMM	Fourier Modal Method
c.c.	Complex conjugate

Symbol	Meaning	Symbol	Meaning
E or $\mathbf{E}$	scalar or vector electric field	H or $\mathbf{H}$	scalar or vector magnetic field
J or $\mathbf{J}$	scalar or vector current density	ε_0	vacuum dielectric constant
ε	relative dielectric constant	μ_0	vacuum permeability
χ	susceptibility	∇	gradient vector
$\nabla\cdot$	divergence	$\nabla\times$	rotation or curl
$\hat{\mathbf{n}}$	surface normal unit vector	$P\int$	principal value integral
t	time	ω	angular frequency
$\mathbf{r}$	position coordinate	i	$\sqrt{-1}$
k_0	free-space wave number (ω/c)	c	vacuum speed of light $(1/\sqrt{\varepsilon_0\mu_0})$
λ	free-space wavelength	∞	infinity
$\hat{\mathbf{r}}$, $\hat{\theta}$, $\hat{\phi}$	spherical coordinate unit vectors	r, θ, ϕ	spherical coordinates
$\hat{\rho}$, $\hat{\phi}$, $\hat{\mathbf{z}}$	cylindrical coordinate unit vectors (3D)	ρ, ϕ, z	cylindrical coordinates (3D)
$\hat{\mathbf{r}}$, $\hat{\theta}$	cylindrical coordinate unit vectors (2D)	r, θ	cylindrical coordinates (2D)
Real	Real part	Imag	Imaginary part
$*$	complex conjugation	$\delta(x)$	Dirac delta function
$\delta_{i,j}$	Kronecker delta function		

2 Theoretical foundation

This chapter reviews some of the most important theoretical foundation of GFIEMs and the scattering problems of interest. Maxwell's equations are reviewed in Sec. 2.1 along with the electromagnetics boundary conditions, wave equations, and the Poynting vector. The planar layered reference structures of interest in the book require knowledge of Fresnel reflection and transmission, and of waveguide modes, which are reviewed in Sec. 2.2. Finally, the optical cross sections that are of interest in many scattering calculations are derived and explained in detail in Sec. 2.3.

2.1 MAXWELL'S EQUATIONS

In this section we briefly review electromagnetic theory, which is the theoretical foundation for the GFIEMs of this book. The macroscopic Maxwell's equations in differential form are given by (see e.g., [35, 36, 37]):

$$\nabla \times \mathbf{E}(\mathbf{r},\mathbf{t}) = -\frac{\partial \mathbf{B}(\mathbf{r},t)}{\partial t}, \tag{2.1}$$

$$\nabla \times \mathbf{H}(\mathbf{r},\mathbf{t}) = \mathbf{J}(\mathbf{r},t) + \frac{\partial \mathbf{D}(\mathbf{r},t)}{\partial t}, \tag{2.2}$$

$$\nabla \cdot \mathbf{D}(\mathbf{r},t) = \rho(\mathbf{r},t), \tag{2.3}$$

$$\nabla \cdot \mathbf{B}(\mathbf{r},t) = 0, \tag{2.4}$$

where $\mathbf{E}$ is the electric field, $\mathbf{H}$ is the magnetic field, $\mathbf{D}$ is the electric displacement field, $\mathbf{B}$ is the magnetic induction field, ρ is the density of free charges, and $\mathbf{J}$ is the density of free currents.

The fields $\mathbf{D}$ and $\mathbf{B}$ are defined in terms of the macroscopic polarization density $\mathbf{P}$ and magnetization density $\mathbf{M}$ as follows:

$$\mathbf{D}(\mathbf{r},t) = \varepsilon_0 \mathbf{E}(\mathbf{r},t) + \mathbf{P}(\mathbf{r},t), \tag{2.5}$$

$$\mathbf{B}(\mathbf{r},t) = \mu_0 \left(\mathbf{H}(\mathbf{r},t) + \mathbf{M}(\mathbf{r},t)\right). \tag{2.6}$$

We will in this book assume linear, local, and isotropic media, which implies the following relation for the induced polarization density

$$\mathbf{P}(\mathbf{r},t) = \varepsilon_0 \int_{t'} \chi(\mathbf{r},t-t')\mathbf{E}(\mathbf{r},t')dt', \tag{2.7}$$

where $\chi(\mathbf{r},t-t') = 0$ for $t' > t$ (causality). Here, the susceptibility χ is specific for a material. A similar relation exists between $\mathbf{B}$ and $\mathbf{H}$. However, throughout the book we will assume that the magnetization density can be neglected, which is a reasonable approximation in optics for most materials of contemporary interest, and thus $\mathbf{B} = \mu_0\mathbf{H}$.

The free currents can be divided into a given source current density $\mathbf{J}_s$ and currents induced by the electric field (Ohms law):

$$\mathbf{J}(\mathbf{r},t) = \mathbf{J}_s(\mathbf{r},t) + \int_{t'} \sigma(\mathbf{r},t-t')\mathbf{E}(\mathbf{r},t')dt'. \tag{2.8}$$

The expressions (2.7) and (2.8) take into account dispersion in materials.

We now introduce the temporal Fourier-transform pair

$$\mathbf{P}(\mathbf{r},t) = \int_{\omega} \mathbf{P}(\mathbf{r};\omega)e^{-i\omega t}d\omega, \tag{2.9}$$

$$\mathbf{P}(\mathbf{r},\omega) = \frac{1}{2\pi}\int_{t} \mathbf{P}(\mathbf{r},t)e^{i\omega t}dt. \tag{2.10}$$

Thus $\mathbf{P}(\mathbf{r},t)$ is related to $\mathbf{P}(\mathbf{r},\omega)$ and vice versa. Similar relations relate $\mathbf{E}(\mathbf{r},t)$, $\mathbf{B}(\mathbf{r},t)$, $\mathbf{J}(\mathbf{r},t)$, $\chi(\mathbf{r},t)$ and $\sigma(\mathbf{r},t)$ to $\mathbf{E}(\mathbf{r},\omega)$, $\mathbf{B}(\mathbf{r},\omega)$, $\mathbf{J}(\mathbf{r},\omega)$, $\chi(\mathbf{r},\omega)$ and $\sigma(\mathbf{r},\omega)$.

In the frequency domain the constitutive relations (2.7) and (2.8) are much simpler because the convolution integrals in the time domain become simple products in the frequency domain:

$$\mathbf{P}(\mathbf{r};\omega) = \varepsilon_0\chi(\mathbf{r};\omega)\mathbf{E}(\mathbf{r};\omega), \tag{2.11}$$

$$\mathbf{J}(\mathbf{r};\omega) = \mathbf{J}_s(\mathbf{r};\omega) + \sigma(\mathbf{r};\omega)\mathbf{E}(\mathbf{r};\omega). \tag{2.12}$$

Consider now the Fourier transform of Eq. (2.2):

$$\nabla\times\mathbf{H}(\mathbf{r};\omega) = \mathbf{J}_s(\mathbf{r};\omega) + \sigma(\mathbf{r};\omega)\mathbf{E}(\mathbf{r};\omega) - i\omega\varepsilon_0\left(1+\chi(\mathbf{r};\omega)\right)\mathbf{E}(\mathbf{r};\omega). \tag{2.13}$$

Here both induced polarization and conduction currents can be treated on an equal footing by redefining the meaning of $\mathbf{D}(\mathbf{r};\omega)$ as follows

$$\mathbf{D}(\mathbf{r};\omega) = \varepsilon_0\left(1+\chi(\mathbf{r};\omega)+\frac{i\sigma(\mathbf{r};\omega)}{\omega\varepsilon_0}\right)\mathbf{E}(\mathbf{r};\omega) \equiv \varepsilon_0\varepsilon(\mathbf{r};\omega)\mathbf{E}(\mathbf{r};\omega), \tag{2.14}$$

where the complex relative dielectric constant is given by

$$\varepsilon(\mathbf{r};\omega) \equiv \left(1+\chi(\mathbf{r};\omega)+\frac{i\sigma(\mathbf{r};\omega)}{\omega\varepsilon_0}\right). \tag{2.15}$$

Maxwell's equations can now be formulated in the following way in the frequency domain:

$$\nabla\times\mathbf{E}(\mathbf{r};\omega) = i\omega\mu_0\mathbf{H}(\mathbf{r};\omega), \tag{2.16}$$

$$\nabla\times\mathbf{H}(\mathbf{r};\omega) = \mathbf{J}_s(\mathbf{r};\omega) - i\omega\varepsilon_0\varepsilon(\mathbf{r},\omega)\mathbf{E}(\mathbf{r};\omega), \tag{2.17}$$

$$\nabla\cdot\mathbf{D}(\mathbf{r};\omega) = \rho_s(\mathbf{r};\omega), \tag{2.18}$$

$$\nabla\cdot\mathbf{B}(\mathbf{r};\omega) = 0. \tag{2.19}$$

Here $\rho_s(\mathbf{r},\omega) = \nabla\cdot\mathbf{J}_s(\mathbf{r},\omega)/i\omega$, which follows from the continuity equation

$$\nabla\cdot\mathbf{J}(\mathbf{r},t) = -\frac{\partial\rho(\mathbf{r},t)}{\partial t}. \tag{2.20}$$

In this book we will be mainly concerned with solving Maxwell's equations in the frequency domain (2.16)-(2.19), and we will often drop the argument ω and, for example, replace $\mathbf{E}(\mathbf{r},\omega)$ with $\mathbf{E}(\mathbf{r})$ where the dependence on ω is implicit.

2.1.1 BOUNDARY CONDITIONS

The electric and magnetic fields on each side of an interface between different materials are related by boundary conditions. Consider the interface between two dielectric materials with dielectric constants ε_1 and ε_2 illustrated in Fig. 2.1. We will assume that fields and material parameters are all in the frequency domain and that their dependence on ω is implicit. The orientation of the interface is given by the surface

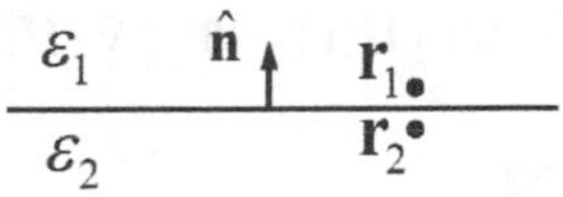

ε_1 $\hat{\mathbf{n}}$ $\mathbf{r}_1$

ε_2 $\mathbf{r}_2$

Figure 2.1 Illustration of interface between two dielectric media with dielectric constants ε_1 and ε_2, respectively.

normal vector $\hat{\mathbf{n}}$. Two positions $\mathbf{r}_1$ and $\mathbf{r}_2$ are considered that are separated by an infinitesimal distance and placed on opposite sides of the boundary in medium 1 and 2, respectively. Boundary conditions relating the fields at $\mathbf{r}_1$ and $\mathbf{r}_2$ follow directly from Maxwell's equations (2.16)-(2.19) by integration and applying Stokes' law or Gauss's law [37], which leads to

$$\hat{\mathbf{n}} \times \mathbf{E}(\mathbf{r}_1) = \hat{\mathbf{n}} \times \mathbf{E}(\mathbf{r}_2), \tag{2.21}$$

$$\hat{\mathbf{n}} \times \mathbf{H}(\mathbf{r}_1) = \hat{\mathbf{n}} \times \mathbf{H}(\mathbf{r}_2), \tag{2.22}$$

where it has been assumed that there is no given surface current density. These two boundary conditions for the tangential components of the electric and magnetic fields are usually sufficient, but sometimes it is also convenient to have boundary conditions for the normal component. For the case with no given surface charge density and non-magnetic materials it can be shown that [37]

$$\hat{\mathbf{n}} \cdot (\varepsilon_1 \mathbf{E}(\mathbf{r}_1)) = \hat{\mathbf{n}} \cdot (\varepsilon_2 \mathbf{E}(\mathbf{r}_2)), \tag{2.23}$$

$$\hat{\mathbf{n}} \cdot \mathbf{H}(\mathbf{r}_1) = \hat{\mathbf{n}} \cdot \mathbf{H}(\mathbf{r}_2). \tag{2.24}$$

2.1.2 WAVE EQUATIONS

Based on the frequency domain Maxwell's equations it is possible to formulate the following wave equations containing either only the electric field or the magnetic field:

$$-\nabla \times \nabla \times \mathbf{E}(\mathbf{r}) + k_0^2 \varepsilon(\mathbf{r}) \mathbf{E}(\mathbf{r}) = -i\omega\mu_0 \mathbf{J}_s(\mathbf{r}), \tag{2.25}$$

$$-\nabla \times \frac{1}{\varepsilon(\mathbf{r})} \nabla \times \mathbf{H}(\mathbf{r}) + k_0^2 \mathbf{H}(\mathbf{r}) = -\nabla \times (\mathbf{J}_s(\mathbf{r})/\varepsilon(\mathbf{r})), \tag{2.26}$$

where $k_0 = \omega/c$ is the free-space wave number, and $c = 1/\sqrt{\varepsilon_0\mu_0}$ is the vacuum speed of light. These equations must be supplemented by appropriate boundary conditions. For example, in the case of the structure in Fig. 1.1(a), an appropriate boundary condition would be that the total field is of the form $\mathbf{E}(\mathbf{r}) = \mathbf{E}_0(\mathbf{r}) + \mathbf{E}_s(\mathbf{r})$, where

$\mathbf{E}_0(\mathbf{r})$ is a given incident field, and $\mathbf{E}_s(\mathbf{r})$ is the scattered field, which is required to propagate away from the scatterer.

In regions where the given source current density vanishes, and where the dielectric constant does not depend on position, i.e., $\varepsilon(\mathbf{r}) = \varepsilon$, the wave equations simplify to

$$(\nabla^2 + k_0^2\varepsilon)\,\mathbf{E}(\mathbf{r}) = \mathbf{0}, \quad \nabla\cdot\mathbf{E}(\mathbf{r}) = 0, \tag{2.27}$$

and

$$(\nabla^2 + k_0^2\varepsilon)\,\mathbf{H}(\mathbf{r}) = \mathbf{0}, \quad \nabla\cdot\mathbf{H}(\mathbf{r}) = 0. \tag{2.28}$$

2.1.3 POYNTING VECTOR

The Poynting vector is defined as

$$\mathbf{S}(\mathbf{r},t) \equiv \mathbf{E}(\mathbf{r},t)\times\mathbf{H}(\mathbf{r},t). \tag{2.29}$$

It has the property that the outward flux of the Poynting vector through a closed surface $\partial\Omega$ surrounding a volume Ω as illustrated in Fig. 2.2 represents the net elec-

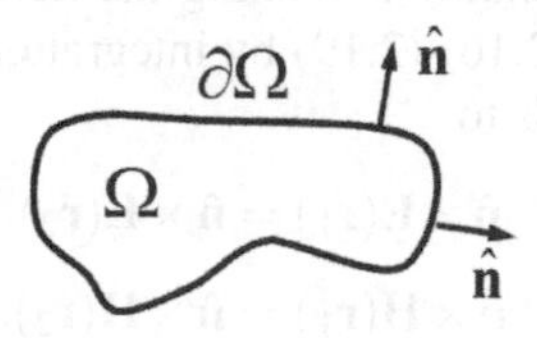

Figure 2.2 Illustration of volume Ω with boundary $\partial\Omega$ and outward surface normal vector $\hat{\mathbf{n}}$.

tromagnetic power P_{out} propagating out through the surface

$$P_{\text{out}} = \oint_{\partial\Omega} \mathbf{S}(\mathbf{r},t)\cdot\hat{\mathbf{n}}dA. \tag{2.30}$$

Here, $\hat{\mathbf{n}}$ is the outward surface normal vector. The Poynting vector is often given the interpretation that it points in the direction of the flow of electromagnetic radiation and that the magnitude of the vector represents the power flow per unit area. The Poynting vector is central in scattering calculations for calculating scattered and absorbed power.

The interpretation of the Poynting vector as describing the flow of electromagnetic power can be seen by first using Gauss's law on the flux of the Poynting vector

$$\oint_{\partial\Omega} \mathbf{S}(\mathbf{r},t)\cdot\hat{\mathbf{n}}dA = \int_{\Omega} \nabla\cdot\mathbf{S}(\mathbf{r},t)d\Omega. \tag{2.31}$$

We may then use that

$$\nabla\cdot\mathbf{E}(\mathbf{r},t)\times\mathbf{H}(\mathbf{r},t) = \mathbf{H}(\mathbf{r},t)\cdot\nabla\times\mathbf{E}(\mathbf{r},t) - \mathbf{E}(\mathbf{r},t)\cdot\nabla\times\mathbf{H}(\mathbf{r},t), \tag{2.32}$$

and the equations (2.1) and (2.2) to obtain

$$\oint_{\partial\Omega} \mathbf{S}(\mathbf{r},t)\cdot\hat{\mathbf{n}}dA + \frac{1}{2}\frac{d}{dt}\int_{\Omega}(\mathbf{B}(\mathbf{r},t)\cdot\mathbf{H}(\mathbf{r},t)+\mathbf{D}(\mathbf{r},t)\cdot\mathbf{E}(\mathbf{r},t))\,d\Omega =$$

$$-\int_{\Omega}\mathbf{E}(\mathbf{r},t)\cdot\mathbf{J}(\mathbf{r},t)d\Omega - \frac{1}{2}\int_{\Omega}\left\{\mathbf{E}(\mathbf{r},t)\cdot\frac{\partial\mathbf{P}(\mathbf{r},t)}{\partial t} - \mathbf{P}(\mathbf{r},t)\cdot\frac{\partial\mathbf{E}(\mathbf{r},t)}{\partial t}\right\}d\Omega. \quad (2.33)$$

Here we have already used the approximation $\mathbf{B}(\mathbf{r},t) = \mu_0\mathbf{H}(\mathbf{r},t)$.

Consider now linear and non-dispersive materials, i.e.,

$$\mathbf{P}(\mathbf{r},t) = \chi(\mathbf{r})\mathbf{E}(\mathbf{r},t), \quad (2.34)$$

where χ is a constant with respect to time. In this case the last term on the right-hand side of (2.33) vanishes, and (2.33) can be expressed as

$$\oint_{\partial\Omega} \mathbf{S}(\mathbf{r},t)\cdot\hat{\mathbf{n}}dA + \frac{dU}{dt} = -\int_{\Omega}\mathbf{E}(\mathbf{r},t)\cdot\mathbf{J}(\mathbf{r},t)d\Omega, \quad (2.35)$$

where the energy stored in the built-up electric and magnetic fields is given by

$$U = \frac{1}{2}\int_{\Omega}\{\mathbf{B}\cdot\mathbf{H}+\mathbf{D}\cdot\mathbf{E}\}\,d\Omega. \quad (2.36)$$

The expression (2.36) for the electromagnetic energy applies to the static limit and linear and non-dispersive materials [35]. The integral $\int_{\Omega}\mathbf{E}(\mathbf{r},t)\cdot\mathbf{J}(\mathbf{r},t)d\Omega$ represents dissipation inside Ω. Due to the sign, the right-hand side of (2.35) thus equals the rate of electromagnetic energy generation inside Ω. On the left-hand side of (2.35) the term dU/dt represents the time rate of increasing the built-up electromagnetic energy, and the term $\oint_{\partial\Omega}\mathbf{S}(\mathbf{r},t)\cdot\hat{\mathbf{n}}dA$ represents the electromagnetic power propagating out through the surface of Ω. With this interpretation, (2.35) is a continuity equation for electromagnetic energy or an energy conservation law for linear and non-dispersive materials.

In the case of linear and dispersive materials, and assuming fields with a narrow spectral width around $\omega = \omega_0$, it can be shown that Poynting's theorem instead becomes [35, 18]

$$\int_{\partial\Omega}\langle\mathbf{S}(\mathbf{r},t)\rangle\cdot\hat{\mathbf{n}}dA + \frac{dU_{\mathrm{eff}}}{dt} =$$

$$-\int_{\Omega}\langle\mathbf{E}(\mathbf{r},t)\cdot\mathbf{J}_s(\mathbf{r},t)\rangle\,d\Omega - \omega_0\varepsilon_0\int_{\Omega}\mathrm{Imag}\{\varepsilon(\mathbf{r},\omega_0)\}\langle\mathbf{E}(\mathbf{r},t)\cdot\mathbf{E}(\mathbf{r},t)\rangle\,d\Omega, \quad (2.37)$$

where here the effective electromagnetic energy built up in the electromagnetic fields is given by

$$U_{\mathrm{eff}} = \frac{1}{2}\varepsilon_0\int_{\Omega}\mathrm{Real}\left\{\frac{d\omega\varepsilon(\mathbf{r},\omega)}{d\omega}\right\}_{\omega=\omega_0}\langle\mathbf{E}(\mathbf{r},t)\cdot\mathbf{E}(\mathbf{r},t)\rangle\,d\Omega. \quad (2.38)$$

The brackets $\langle\ldots\rangle$ refer to the time average over one cycle of the carrier frequency ω_0. The current density $\mathbf{J}$ has been split into a given source current density $\mathbf{J}_s$

and conduction current density $\mathbf{J}_c$. In the frequency domain we have further used $\mathbf{J}_c(\mathbf{r},\omega) = \sigma(\mathbf{r},\omega)\mathbf{E}(\mathbf{r},\omega)$ and included the effect of the conduction currents using the effective dielectric constant (2.15). The right-hand side again expresses the rate of net energy generation inside Ω due to source currents, and conduction and polarization currents.

In the limit where the spectral width of the fields is so narrow that we are in the limit of a monochromatic field, it is convenient to calculate the time average of the Poynting-vector flow. The first step is to express the fields in the form

$$\mathbf{E}(\mathbf{r},t) = \text{Real}\left\{\mathbf{E}(\mathbf{r},\omega)e^{-i\omega t}\right\} = \frac{1}{2}\left(\mathbf{E}(\mathbf{r},\omega)e^{-i\omega t} + [\mathbf{E}(\mathbf{r},\omega)]^* e^{+i\omega t}\right), \quad (2.39)$$

and similar expressions apply for **H, D, B** and **J**. We further assume that $\mathbf{B}(\mathbf{r},\omega) = \mu_0\mathbf{H}(\mathbf{r},\omega)$ and $\mathbf{D}(\mathbf{r},\omega) = \varepsilon_0\varepsilon(\mathbf{r},\omega)\mathbf{E}(\mathbf{r},\omega)$ and apply the equations (2.16) and (2.17). The time-average of the Poynting vector (2.29) may in that case be expressed in terms of complex fields as follows

$$\mathbf{S}_{\text{av}}(\mathbf{r}) = \frac{1}{T}\int_{t=0}^{T}\mathbf{S}(\mathbf{r},t)dt = \frac{1}{2}\text{Real}\left\{\mathbf{E}(\mathbf{r},\omega)\times(\mathbf{H}(\mathbf{r},\omega))^*\right\}, \quad (2.40)$$

where $T = 2\pi/\omega$ is the period of one oscillation. We note that for monochromatic fields, the electromagnetic energy U is the same at the start and end of a period, and thus the time average of dU_{eff}/dt over one period vanishes. The time average of the power propagating out of the closed surface can thus be expressed as

$$P_{\text{out, av}} = \oint_{\partial\Omega}\mathbf{S}_{\text{av}}(\mathbf{r})\cdot\hat{\mathbf{n}}dA =$$

$$-\frac{1}{2}\int_{\Omega}\text{Real}\left\{\mathbf{E}(\mathbf{r},\omega)\cdot\mathbf{J}_s^*(\mathbf{r},\omega)\right\}d\Omega - \frac{\omega}{2}\int_{\Omega}\left(\varepsilon_0\text{Imag}\left\{\varepsilon(\mathbf{r},\omega)\right\}|\mathbf{E}(\mathbf{r},\omega)|^2\right)d\Omega. \quad (2.41)$$

Throughout the rest of the book we shall only use the time-averaged Poynting vector of monochromatic fields given by (2.40). The calculation of absorption, scattering, and extinction cross sections relies on the Poynting vector and will be presented in detail in Sec. 2.3.

2.2 PLANAR LAYERED STRUCTURES

In this section we review the properties of planar layered reference structures. The Fresnel reflection and transmission coefficients of planar layered structures (Sec. 2.2.1) enter into expressions for Green's functions and reference fields. In some cases layered reference structures support guided modes, and it may be of interest to calculate the excitation of guided modes in a scattering problem. The guided modes of layered reference structures are reviewed in Sec. 2.2.2.

2.2.1 FRESNEL REFLECTION AND TRANSMISSION

We start by considering the reflection and transmission for a single interface, which is illustrated for s and p-polarized light in Fig. 2.3. The dielectric constant of the

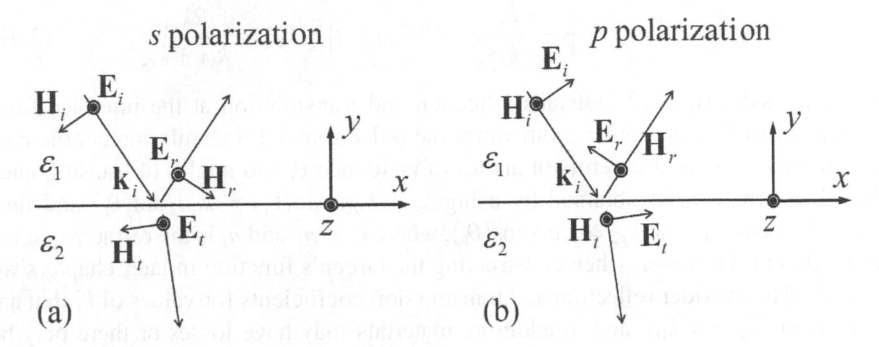

Figure 2.3 Illustration of Fresnel reflection and transmission of (a) s-polarized light, and (b) p-polarized light.

structure being considered is given by

$$\varepsilon(\mathbf{r}) = \begin{cases} \varepsilon_1, & y > 0 \\ \varepsilon_2, & y < 0 \end{cases}. \tag{2.42}$$

In the case of s-polarized light and an incident plane wave, the electric field can be expressed as

$$\mathbf{E}(\mathbf{r}) = \hat{\mathbf{z}} E(\mathbf{r}), \tag{2.43}$$

where $\mathbf{r} = \hat{\mathbf{x}}x + \hat{\mathbf{y}}y$, and

$$E(\mathbf{r}) = \begin{cases} E_0 \left(e^{-ik_{y1}y} + r^{(s)}(k_x) e^{ik_{y1}y} \right) e^{ik_x x}, & y > 0 \\ E_0 t^{(s)}(k_x) e^{ik_x x} e^{-ik_{y2}y}, & y < 0 \end{cases}. \tag{2.44}$$

Here $r^{(s)}(k_x)$ and $t^{(s)}(k_x)$ are the Fresnel reflection and transmission coefficients for s-polarized light, and $k_{yi} = \sqrt{k_0^2 \varepsilon_i - k_x^2}$ with $\text{Imag}\{k_{yi}\} \geq 0$ such that the wave equation (2.27) is satisfied for $y > 0$ and $y < 0$. For $y > 0$ this describes the total field as the sum of a downward propagating incident plane wave and an upward propagating reflected plane wave, and for $y < 0$ the field is a downward propagating transmitted plane wave. k_x is the same for all the plane wave components in accordance with the electromagnetics boundary conditions. The corresponding magnetic field can be obtained using Eq. (2.16).

The electromagnetics boundary conditions for the tangential electric and magnetic field components across the interface at $y = 0$ furthermore dictates that

$$1 + r^{(s)}(k_x) = t^{(s)}(k_x), \tag{2.45}$$

and

$$-ik_{y1}\left(1-r^{(s)}(k_x)\right)=-ik_{y2}t^{(s)}(k_x). \tag{2.46}$$

This can be solved for the single-interface reflection and transmission coefficients:

$$r^{(s)}(k_x)=r^{(s)}_{1,2}(k_x)=\frac{k_{y1}-k_{y2}}{k_{y1}+k_{y2}},\quad t^{(s)}(k_x)=t^{(s)}_{1,2}(k_x)=\frac{2k_{y1}}{k_{y1}+k_{y2}}. \tag{2.47}$$

Here, the subscript $1,2$ indicates reflection and transmission at the interface from medium 1 to 2 in that order. Sometimes the reflection and transmission coefficients are instead expressed in terms of angles of incidence θ_i and angles of transmittance θ_t. This form can be obtained by using $k_x = k_0 n_1 \sin(\theta_i) = k_0 n_2 \sin(\theta_t)$ and thus $k_{y1} = k_0 n_1 \cos(\theta_i)$ and $k_{y2} = k_0 n_2 \cos(\theta_t)$, where $\varepsilon_i = n_i^2$ and n_i is the refractive index of medium i. However, when constructing the Green's function in later chapters we will need to consider reflection and transmission coefficients for values of k_x that are very large ($k_x >> k_0$), and in addition, materials may have losses or there may be evanescent waves in one of the media, and for those cases the form (2.47) means that we avoid having to consider complex angles θ_i and θ_t.

In the case of p-polarized light, the reflection and transmission of a plane wave leads instead to the field

$$\mathbf{H}(\mathbf{r})=\hat{\mathbf{z}}H(\mathbf{r}), \tag{2.48}$$

where

$$H(\mathbf{r})=\begin{cases}H_0\left(e^{-ik_{y1}y}+r^{(p)}(k_x)e^{ik_{y1}y}\right)e^{ik_xx}, & y>0\\ H_0t^{(p)}(k_x)e^{ik_xx}e^{-ik_{y2}y}, & y<0\end{cases}. \tag{2.49}$$

In this case the electromagnetics boundary conditions lead to the single-interface reflection and transmission coefficients

$$r^{(p)}(k_x)=r^{(p)}_{1,2}(k_x)=\frac{\varepsilon_2k_{y1}-\varepsilon_1k_{y2}}{\varepsilon_2k_{y1}+\varepsilon_1k_{y2}},\quad t^{(p)}(k_x)=t^{(p)}_{1,2}(k_x)=\frac{2\varepsilon_2k_{y1}}{\varepsilon_2k_{y1}+\varepsilon_1k_{y2}}. \tag{2.50}$$

We now turn to the case of reflection and transmission for structures with two interfaces illustrated in Fig. 2.4(a), where a medium of dielectric constant ε_2 and thickness d is sandwiched between media with dielectric constants ε_1 and ε_3. Here, for s-polarized light the electric field (z-component) can be expressed as

$$E(\mathbf{r})=\begin{cases}E_0\left(e^{-ik_{y1}y}+r^{(s)}(k_x)e^{ik_{y1}y}\right)e^{ik_xx}, & y>0\\ E_0\left(Ae^{-ik_{y2}y}+Be^{ik_{y2}y}\right)e^{ik_xx}, & -d<y<0\\ E_0t^{(s)}(k_x)e^{ik_xx}e^{-ik_{y3}(y+d)}, & y<-d\end{cases}. \tag{2.51}$$

The electromagnetics boundary conditions at the first ($y=0$) and second ($y=-d$) interface can be expressed as

$$1+r^{(s)}(k_x)=A+B,\quad -ik_{y1}\left(1-r^{(s)}(k_x)\right)=-ik_{y2}(A-B), \tag{2.52}$$

$$t^{(s)}(k_x)=Ae^{ik_{y2}d}+Be^{-ik_{y2}d},\quad -ik_{y3}t^{(s)}(k_x)=-ik_{y2}\left(Ae^{ik_{y2}d}-Be^{-ik_{y2}d}\right). \tag{2.53}$$

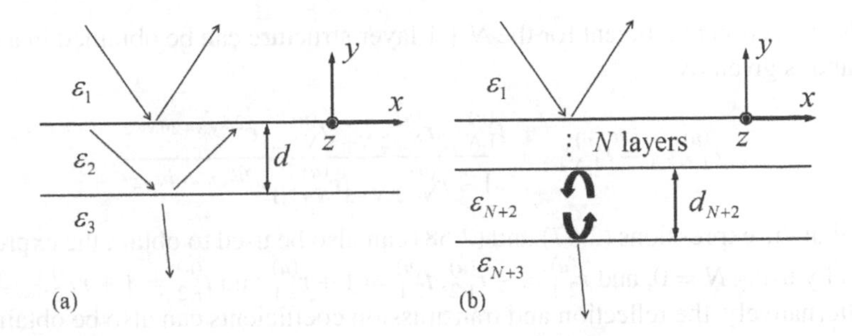

Figure 2.4 Illustration of Fresnel reflection and transmission in the case of (a) two interfaces or 1 layer, and (b) N layers.

By solving these equations the reflection and transmission coefficients can be obtained. A similar procedure applies for p-polarized light. In both cases it can be shown that the reflection and transmission coefficients for the two-interface reference structure are given by

$$r_{1,3}^{(u)} = \frac{r_{1,2}^{(u)} + r_{2,3}^{(u)} e^{2ik_{y2}d}}{1 + r_{1,2}^{(u)} r_{2,3}^{(u)} e^{2ik_{y2}d}}, \quad t_{1,3}^{(u)} = \frac{t_{1,2}^{(u)} t_{2,3}^{(u)} e^{ik_{y2}d}}{1 + r_{1,2}^{(u)} r_{2,3}^{(u)} e^{2ik_{y2}d}}, \tag{2.54}$$

where u can be either s or p, and the dependence on k_x is implicit.

The case of $N+1$ layers is illustrated in Fig. 2.4(b). The reflection and transmission coefficients for this case can be obtained iteratively by following all reflections back and forth between the N layer structure and the last interface. This can be expressed as

$$t_{1,N+3}^{(u)} = t_{N+2,N+3}^{(u)} e^{ik_{y,N+2}d_{N+2}} \left(\sum_{n=0}^{\infty} \left(r_{N+2,1}^{(u)} r_{N+2,N+3}^{(u)} e^{2ik_{y,N+2}d_{N+2}} \right)^n \right) t_{1,N+2}^{(u)}. \tag{2.55}$$

Here, the term with $n=0$ first corresponds to transmission through the N layer structure (from medium 1 to medium $N+2$), propagation through layer $N+1$ (medium $N+2$), and then direct transmission through the interface between medium $N+2$ and $N+3$. The next term with $n=1$ corresponds to light being not directly transmitted through the last interface but instead reflected back and forth once inside the layer $N+1$ (medium $N+2$) before being transmitted, and so on.

By using the relation

$$\sum_{n=0}^{\infty} x^n = \frac{1}{1-x}, \quad |x| < 1, \tag{2.56}$$

the expression for the transmission coefficient can be simplified to

$$t_{1,N+3}^{(u)} = \frac{t_{1,N+2}^{(u)} t_{N+2,N+3}^{(u)} e^{ik_{y,N+2}d_{N+2}}}{1 - r_{N+2,N+3}^{(u)} r_{N+2,1}^{(u)} e^{2ik_{y,N+2}d_{N+2}}}. \tag{2.57}$$

The reflection coefficient for the $N+1$ layer structure can be obtained in a similar way and is given by

$$r_{1,N+3}^{(u)} = r_{1,N+2}^{(u)} + \frac{t_{1,N+2}^{(u)} r_{N+2,N+3}^{(u)} t_{N+2,1}^{(u)} e^{2ik_{y,N+2}d_{N+2}}}{1 - r_{N+2,N+3}^{(u)} r_{N+2,1}^{(u)} e^{2ik_{y,N+2}d_{N+2}}}. \tag{2.58}$$

Note that the expressions (2.57) and (2.58) can also be used to obtain the expressions (2.54) by using $N=0$, and $r_{2,1}^{(u)} = -r_{1,2}^{(u)}$, $t_{2,1}^{(u)} = 1 + r_{2,1}^{(u)}$ and $t_{1,2}^{(u)} = 1 + r_{1,2}^{(u)}$.

Alternatively, the reflection and transmission coefficients can also be obtained using matrix methods where a 2×2 system matrix is obtained that relates forward and backward propagating waves in medium $N+3$ and 1 by multiplying together matrices with such relations for single interfaces and for propagation through layers [38]. Another matrix approach relates the tangential electric and magnetic field components at the first and last interface [39]. The reflection and transmission coefficients can be obtained from the matrix elements of the system matrix. In the case of adding a thick layer $N+1$ of metal, in which case the transmitted field would be negligible, the matrix methods may become numerically unstable, and in such cases the iterative approach in (2.57) and (2.58) is preferable.

2.2.2 PLANAR WAVEGUIDES AND GUIDED MODES

Scattering of light by particles on or near waveguides as illustrated in Fig. 1.1(b) can result in the excitation of waveguide modes. In this case part of the light being incident on the scatterer is transferred to the waveguide. Later in the book we will introduce Green's function integral equation techniques for calculating the scattering into guided modes. Here, it is necessary to know the field profile and the wavenumber, or alternatively the mode index, of the modes. The mode wave number can be identified from poles of the Fresnel reflection coefficient, and these poles can later be used to construct the part of a Green's function that governs the excitation of guided modes. In this section the foundation is laid for calculating the scattering into guided modes by introducing planar waveguides, guided modes, and the relation between mode wave number and poles of reflection coefficients.

The simplest planar structure that supports guided modes is a single interface between a metal and a dielectric. The guided modes of this structure are p polarized and known as surface plasmon polaritons (SPPs). The magnetic field (H) of the SPP wave is illustrated in Fig. 2.5(a). The magnetic field of an SPP propagating in the xy-plane is of the form

$$\mathbf{H}(\mathbf{r}) = \hat{\mathbf{z}} H(\mathbf{r}), \tag{2.59}$$

where $\mathbf{r} = \hat{\mathbf{x}}x + \hat{\mathbf{y}}y$, and

$$H(\mathbf{r}) = \begin{cases} A e^{ik_x x} e^{ik_{y1} y}, & y > 0 \\ B e^{ik_x x} e^{-ik_{y2} y}, & y < 0 \end{cases}. \tag{2.60}$$

By defining $k_{yi} \equiv \sqrt{k_0^2 \varepsilon_i - k_x^2}$ with $\mathrm{Imag}\{k_{yi}\} \geq 0$ the wave equation (2.26) or (2.28) is satisfied for $y > 0$ and for $y < 0$. The boundary condition of continuity of the

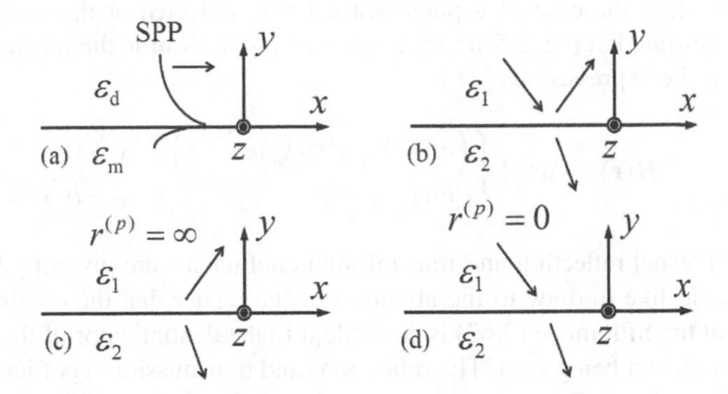

Figure 2.5 Schematic of: (a) magnetic field of an SPP wave bound to and propagating along a metal-dielectric interface, (b) reflection of a plane wave at a single planar interface between two media, (c) limiting case of an infinitely large reflection coefficient in the case of a pole, (d) case of vanishing reflection coefficient which occurs at the Brewster angle.

tangential component of the magnetic field across the interface at $y = 0$ leads to the requirement $A = B$. Furthermore, the requirement of continuity of the tangential (x) component of the electric field leads to

$$\frac{1}{\varepsilon_1}\left(\frac{\partial H}{\partial y}\right)_{y=0+} = \frac{1}{\varepsilon_2}\left(\frac{\partial H}{\partial y}\right)_{y=0-}, \tag{2.61}$$

where $\varepsilon_1 = \varepsilon_d$ (dielectric) and $\varepsilon_2 = \varepsilon_m$ (metal). This condition leads further to the requirement

$$k_{y1}\varepsilon_2 + k_{y2}\varepsilon_1 = 0. \tag{2.62}$$

For real-valued dielectric constants this equation can only be solved if one of the dielectric constants is positive and the other is negative. We note that the real part of the dielectric constant of metals can be negative [40] depending on the frequency or wavelength, while the (real part of the) dielectric constant of dielectrics is usually positive. By noting that k_{yi} is a function of k_x, this equation can be solved for k_x, in which case the SPP wave number is obtained [16]

$$k_{x,\mathrm{SPP}} = k_0\sqrt{\frac{\varepsilon_1\varepsilon_2}{\varepsilon_1+\varepsilon_2}}. \tag{2.63}$$

A solution for the field of the form (2.60) that satisfies the wave equation and electromagnetics boundary conditions thus only exists for $k_x = \pm k_{x,\mathrm{SPP}}$. For the case of real-valued dielectric constants we must also require that $\varepsilon_1 + \varepsilon_2 < 0$ for the wavenumber $k_{x,\mathrm{SPP}}$ to be real-valued. A dominating real part of $k_{x,\mathrm{SPP}}$ compared with the corresponding imaginary part is necessary for the mode to be considered propagating. In the case of real metals the dielectric constant of the metal will have an imaginary part, and the SPP wave number will also have an imaginary part. The SPP wave is thus exponentially damped as it propagates along the metal-dielectric interface.

Now consider the case of a plane wave being reflected at the metal-dielectric interface illustrated in Fig. 2.5(b). In the case of p polarization the magnetic field for this case can be expressed as

$$H(\mathbf{r}) = Ae^{ik_x x}\begin{cases}\left(e^{-ik_{y1}y} + r^{(p)}(k_x)e^{+ik_{y1}y}\right), & y > 0 \\ t^{(p)}(k_x)e^{-ik_{y2}y}, & y < 0\end{cases}, \tag{2.64}$$

where the Fresnel reflection and transmission coefficients are given by Eqs. (2.50). Here, I would like to draw to the attention of the reader that the condition for the existence of the SPP mode (2.62) is equivalent to the denominator of the Fresnel reflection coefficient being zero. The reflection (and transmission) coefficient thus has a pole at $k_x = k_{x,\mathrm{SPP}}$. By finding the pole of the reflection coefficient we thus find the wave number of the guided mode. From Eq. (2.64) it can be seen that the incident wave can be ignored if the reflection coefficient becomes infinitely large, which explains why the pole is equivalent to the existence of a guided mode. The case of ignoring the incident wave is shown as a schematic in Fig. 2.5(c). This is another type of schematic of the SPP wave, where the other one in Fig. 2.5(a) illustrates that the field is evanescent (exponentially decreasing) away from the interface into both the dielectric and the metal. This happens because (for real-valued dielectric constants) $k_x = k_{x,\mathrm{SPP}} > k_0 n_1$, where $n_1^2 = \varepsilon_1$, and the resulting $k_{yi} = k_{yi,\mathrm{SPP}}$ are imaginary. For p polarization it can also occur that the reflection coefficient vanishes instead, which is illustrated in Fig. 2.5(d). This can happen for a specific angle of light incidence known as the Brewster angle [38] if both materials are dielectrics ($\varepsilon_1 > 0$ and $\varepsilon_2 > 0$), since in that case

$$k_{y1}\varepsilon_2 - k_{y2}\varepsilon_1 = 0 \tag{2.65}$$

can be solved for a (predominantly) real k_x. At the Brewster angle, the in-plane wave number k_x is still given by the expression

$$k_x = k_0\sqrt{\frac{\varepsilon_1\varepsilon_2}{\varepsilon_1 + \varepsilon_2}} \tag{2.66}$$

but now $k_x < k_0 n_1$, and the corresponding k_{yi} are real-valued.

The case of a plane wave being incident on a three-layer geometry is illustrated in Fig. 2.6(a). The incident wave is shown as the dashed arrow. Here, in the case of p polarization, the magnetic field will be of the form

$$H(\mathbf{r}) = H_0 e^{ik_x x}\begin{cases} e^{-ik_{y1}y} + r^{(p)}(k_x)e^{ik_{y1}y}, & y > 0 \\ Be^{-ik_{y2}y} + Ce^{ik_{y2}y}, & 0 > y > -d \\ t^{(p)}(k_x)e^{-ik_{y3}(y+d)}, & y < -d \end{cases}, \tag{2.67}$$

where the Fresnel reflection and transmission coefficients are given by Eqs. (2.54), and

$$B = \frac{1}{2}\left[\left(1 + \frac{\varepsilon_2 k_{y1}}{\varepsilon_1 k_{y2}}\right) + r_{1,3}^{(p)}(k_x)\left(1 - \frac{\varepsilon_2 k_{y1}}{\varepsilon_1 k_{y2}}\right)\right], \tag{2.68}$$

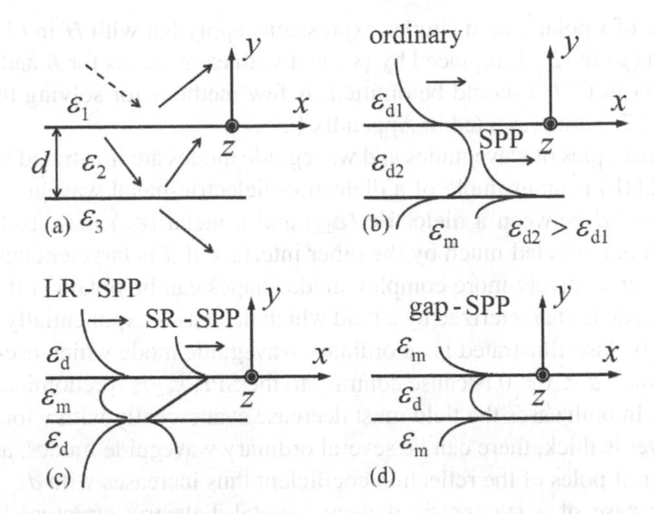

Figure 2.6 Schematic of: (a) reflection of a plane wave by a three-layer structure, or if the dashed arrow is absent it is a wave picture of a guided mode, (b) ordinary and SPP modes in a dielectric-dielectric-metal waveguide, (c) long-range SPP and short-range SPP in a dielectric-metal-dielectric waveguide, (d) gap SPP in a metal-dielectric-metal waveguide.

$$C = \frac{1}{2}\left[\left(1-\frac{\varepsilon_2 k_{y1}}{\varepsilon_1 k_{y2}}\right) + r_{1,3}^{(p)}(k_x)\left(1+\frac{\varepsilon_2 k_{y1}}{\varepsilon_1 k_{y2}}\right)\right], \tag{2.69}$$

follows from the boundary conditions at $y = 0$.

Depending on the dielectric constants and the thickness d of the waveguide, guided modes may exist for this structure. The condition for the existence of a guided mode is that there is a pole $k_x = k_{x,\mathrm{p}}$ where the denominator in the reflection coefficient vanishes. When a guided mode exists, the field solution of the guided mode will be of the form

$$H(\mathbf{r}) = Ae^{ik_{x,\mathrm{p}}x}\begin{cases} e^{ik_{y1,\mathrm{p}}y}, & y>0 \\ Be^{-ik_{y2,\mathrm{p}}y} + Ce^{ik_{y2,\mathrm{p}}y}, & 0>y>-d\,, \\ De^{-ik_{y3,\mathrm{p}}(y+d)}, & y<-d \end{cases} \tag{2.70}$$

where here

$$B = \frac{1}{2}\left(1-\frac{\varepsilon_2}{\varepsilon_1}\frac{k_{y1,\mathrm{p}}}{k_{y2,\mathrm{p}}}\right), \quad C = \frac{1}{2}\left(1+\frac{\varepsilon_2}{\varepsilon_1}\frac{k_{y1,\mathrm{p}}}{k_{y2,\mathrm{p}}}\right), \tag{2.71}$$

and

$$D = Be^{ik_{y2,\mathrm{p}}d} + Ce^{-ik_{y2,\mathrm{p}}d}. \tag{2.72}$$

In this case there is no simple closed-form expression like (2.63) for the guided-mode wave number. Here, for the three-layer structure, the guided-mode wave number $k_x = k_{x,\mathrm{p}}$ can be found by solving the equation

$$1 + r_{1,2}^{(p)}(k_x)r_{2,3}^{(p)}(k_x)e^{2ik_{y2}d} = 0. \tag{2.73}$$

In the case of s polarization similar expressions apply but with H in (2.70) replaced by E, with (p) in (2.73) replaced by (s), and with expressions for B and C where the factor $\varepsilon_2/\varepsilon_1$ in (2.71) should be omitted. A few methods for solving the dispersion equation (2.73) are presented in Appendix F.

Three examples of waveguides and waveguide modes are illustrated in Fig. 2.6(b)-(d). Fig. 2.6(b) is an example of a dielectric-dielectric-metal waveguide. The interface at $y = -d$ between a dielectric (ε_{d2}) and a metal (ε_m) may itself support an SPP that is not affected much by the other interface if d is large enough (case illustrated), or alternatively more complex mode shapes can be obtained if d is smaller. The SPP mode is characterized by a field which decreases exponentially in the region $-d < y < 0$. Also illustrated is an ordinary waveguide mode which oscillates with y in the region $-d < y < 0$ because contrary to the SPP, k_{y2} is (predominantly) real for this mode. In both cases the field must decrease evanescently with y for $y > 0$. If the middle layer is thick, there can be several ordinary waveguide modes, and generally the number of poles of the reflection coefficient thus increases with d.

For the case of a symmetric dielectric-metal-dielectric structure [Fig. 2.6(c)], where the metal layer is thin, e.g., a thickness of 10 nm, the metal-film geometry supports two types of SPP waves known as short-range SPPs and long-range SPPs. One is weakly bound to the metal film with most of the field outside the film, and this mode thus experiences a relatively small absorption in the metal, and thus propagates relatively long distances. The other mode has opposite symmetry in the field profile along y and is strongly bound with comparatively high absorption in the metal, and this mode thus propagates only relatively small distances. Finally, a metal-dielectric-metal waveguide with a nanometer-thin gap between two metal surfaces [Fig. 2.6(d)] supports so-called gap-SPP modes even for thicknesses that are much smaller than the wavelength, where k_{y2} is imaginary. For the case of a thick dielectric layer the structure may also support ordinary waveguide modes, where k_{y2} is real-valued. This case would be equivalent to waveguides of a type used for microwaves. A common photonics waveguide would be the one where all three media are dielectrics with the middle layer having the highest dielectric constant.

In this section we have focused on the case of p polarization due to the existence of SPP waves for this case. SPP waves do not exist for s polarization, where it is assumed that $\mathbf{E}(\mathbf{r}) = \hat{\mathbf{z}}E(x,y)$. In that case there are no guided modes bound to a single interface. However, a similar analysis to that presented above will show that there can be ordinary waveguide modes for a three-layer geometry with a dielectric layer in the middle.

2.3 SCATTERING THEORY

This section introduces the scattering, absorption and extinction cross sections that are the main quantities of interest in scattering calculations. Expressions will be obtained for these quantities for both two- and three-dimensional scattering problems, and for scatterers in homogeneous media and on layered reference structures supporting guided modes. The cross sections are meaningful in the limit where the scatterer is illuminated by a wide beam being much wider than the scatterer and much wider than the optical wavelength. In this limit the incident field at the site of the scatterer can be described as a plane wave carrying a certain power per unit area. By normalizing, for example, the scattered power by the incident power per unit area, or intensity, we obtain the scattering cross section. The optical cross sections are convenient parameters describing the properties of a scatterer since their values do not depend on the incident power per unit area. For example, the absorption cross section can be interpreted in the way that the absorbed power is equivalent to the power of the incident beam that falls within an area equivalent to the cross section. Note that the optical cross sections can be much larger than the physical cross section of the scatterer in the case of, for example, resonant metal nanoparticles.

2.3.1 SCATTERER IN HOMOGENEOUS MATERIAL (2D)

The first case we will consider is a two-dimensional scattering problem, where a scatterer is placed in a homogeneous material with refractive index n_1. The scatterer is illuminated by a Gaussian beam (Fig. 2.7). Furthermore, the scatterer is located near the beam waist, and the beam resembles a plane wave across the scatterer due to its large assumed width. The total field outside the scatterer will be a superposition of the incident and the scattered fields.

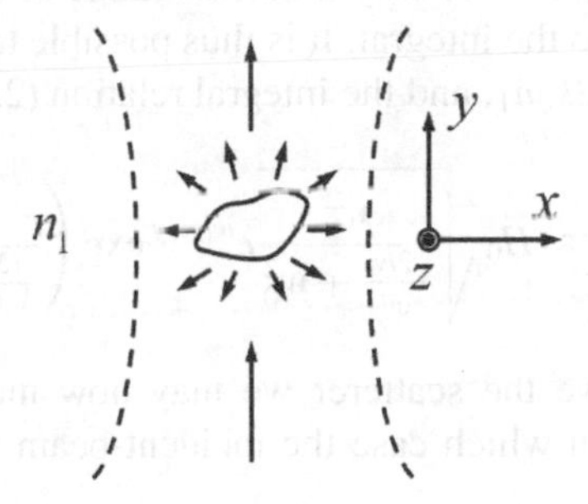

Figure 2.7 Schematic of a scatterer placed in a homogeneous material with refractive index n_1 being illuminated with a Gaussian beam from below.

We assume that the incident field is p polarized with a magnetic field along the z-axis and propagating mainly along the positive y-axis in the xy-plane. In the plane of the beam waist ($y = 0$) the incident field can be expressed as

$$\mathbf{H}_{0,\mathrm{i}}(x, y = 0) = \hat{z}H_{0,\mathrm{i}}(x, y = 0) = \hat{z}H_0 e^{-x^2/w_0^2}, \tag{2.74}$$

where w_0 is the beam waist radius. The dependence on ω is implicit.

In order to calculate the magnetic field at other values of y the spatial Fourier-transform pair may be applied:

$$H_{0,\mathrm{i}}(x, y) = \int_{k_x} \tilde{H}_{0,\mathrm{i}}(k_x; y) e^{ik_x x} dk_x, \tag{2.75}$$

$$\tilde{H}_{0,\mathrm{i}}(k_x;y) = \frac{1}{2\pi}\int_x H_{0,\mathrm{i}}(x,y)e^{-ik_x x}dx. \tag{2.76}$$

The requirement that $H_{0,\mathrm{i}}$ must satisfy the wave equation and propagate along the positive y-axis leads to

$$\tilde{H}_{0,\mathrm{i}}(k_x;y) = \tilde{H}_{0,\mathrm{i}}(k_x;0)e^{ik_{y1}y}, \tag{2.77}$$

where $k_{y1} = \sqrt{k_0^2 n_1^2 - k_x^2}$ and $\mathrm{Imag}\{k_{y1}\} \geq 0$.

By using the integral relation

$$\int_{-\infty}^{\infty} e^{-ax^2+bx}dx = e^{b^2/4a}\sqrt{\frac{\pi}{a}} \tag{2.78}$$

it can be shown that the angular spectrum (2.76) for $y=0$ is given by

$$\tilde{H}_{0,\mathrm{i}}(k_x;y=0) = H_0\frac{w_0}{2\sqrt{\pi}}e^{-\frac{1}{4}w_0^2 k_x^2}. \tag{2.79}$$

When inserting the expression for $\tilde{H}_{0,\mathrm{i}}(k_x;y=0)$ into Eq. (2.75) we note that for a wide beam ($k_0 n_1 w_0 >> 1$) only a narrow range of k_x centered on $k_x = 0$ will contribute significantly to the integral. It is thus possible to use the paraxial approximation $k_{y1} \approx k_0 n_1 - k_x^2/2k_0 n_1$, and the integral relation (2.78) once more to obtain

$$H_{0,\mathrm{i}}(x,y) \approx H_0\sqrt{\frac{w_0^2}{\frac{2iy}{k_0 n_1}+w_0^2}}e^{ik_0 n_1 y}\exp\left(\frac{-x^2}{\frac{i2y}{k_0 n_1}+w_0^2}\right). \tag{2.80}$$

For large distances above the scatterer we may now make the replacements $y = r\cos\theta$ and $x = r\sin\theta$, in which case the incident-beam far field can be expressed as

$$H_{0,\mathrm{i}}^{(\mathrm{ff})}(r,\theta) \approx H_0 w_0\sqrt{\frac{k_0 n_1}{2r}}e^{-i\pi/4}e^{ik_0 n_1 r}e^{-\frac{\theta^2}{4}(k_0 n_1 w_0)^2}, \tag{2.81}$$

where it was exploited that only small values of θ are of practical relevance as the field is negligible for all but very small θ when the beam waist radius is large.

When the scatterer is present the total field will also contain the scattered field:

$$H(r,\theta) = H_{0,\mathrm{i}}(r,\theta) + H_{\mathrm{scat}}(r,\theta). \tag{2.82}$$

The transmitted beam power (per unit length along z) may now be obtained by integrating the time-averaged Poynting vector flux over the beam cross section in the far field ($y > 0$, $k_0 n_1 r >> 1$):

$$\begin{aligned} P_{\mathrm{beam}} &= \int_{\theta=-\theta_b}^{+\theta_b}\frac{1}{2}\mathrm{Real}\left\{\mathbf{E}^{(\mathrm{ff})}(r,\theta)\times\left[\mathbf{H}^{(\mathrm{ff})}(r,\theta)\right]^*\right\}\cdot\hat{\mathbf{r}}rd\theta \\ &= \frac{1}{2}\sqrt{\frac{\mu_0}{\varepsilon_0}}\frac{1}{n_1}\int_{\theta=-\theta_b}^{\theta_b}|H_i^{(\mathrm{ff})}(r,\theta)+H_{\mathrm{scat}}^{(\mathrm{ff})}(r,\theta)|^2 rd\theta, \end{aligned} \tag{2.83}$$

where $\hat{\mathbf{r}} = \hat{\mathbf{x}}x/r + \hat{\mathbf{y}}y/r$, (ff) means far field, and θ_b is a small angle that is large enough to include the beam, but also small enough to avoid including any scattered power of significance. In the limit of plane-wave incidence ($w_0 \to \infty$) the angle θ_b will approach zero ($\theta_b \to 0^+$). In this limit the total beam power can be divided into two terms

$$P_{\text{beam}} = P_i - P_{\text{ext}}, \tag{2.84}$$

namely the beam power in the absence of scattering

$$P_i = \frac{1}{2}\sqrt{\frac{\mu_0}{\varepsilon_0}}\frac{1}{n_1}\int_{\theta=-\theta_b}^{+\theta_b} |H_i^{(\text{ff})}(r,\theta)|^2 r d\theta, \tag{2.85}$$

and minus the extinction power

$$-P_{\text{ext}} = \sqrt{\frac{\mu_0}{\varepsilon_0}}\frac{1}{n_1}\int_{\theta=-\theta_b}^{\theta_b} \text{Re}\left\{H_i^{(\text{ff})}(r,\theta)\left[H_{\text{scat}}^{(\text{ff})}(r,\theta)\right]^*\right\} r d\theta. \tag{2.86}$$

Note that we have ignored the term

$$\frac{1}{2}\sqrt{\frac{\mu_0}{\varepsilon_0}}\frac{1}{n_1}\int_{\theta=-\theta_b}^{+\theta_b} |H_{\text{scat}}^{(\text{ff})}(r,\theta)|^2 r d\theta, \tag{2.87}$$

since the scattered field is much smaller in magnitude than the incident-beam field across the beam, and the angle $\theta_b \approx 0^+$.

The power per unit area of the incident beam I_i at the location of the scatterer is given by the y-component of the Poynting vector for the incident field there given by

$$I_i = \frac{1}{2}\sqrt{\frac{\mu_0}{\varepsilon_0}}\frac{1}{n_1}|H_0|^2. \tag{2.88}$$

The extinction power represents the power that has been removed from the reference beam due to scattering and absorption. By inserting the analytic expression for $H_i^{(\text{ff})}(r,\theta)$, and using that $H_{\text{sc}}^{(\text{ff})}(r,\theta) \approx H_{\text{sc}}^{(\text{ff})}(r,\theta=0)$ across the beam, we obtain the following simple expression for the extinction cross section

$$\sigma_{\text{ext}} = \frac{P_{\text{ext}}}{I_i} = -\frac{2\sqrt{\frac{2\pi}{k_0 n_1}}}{|H_0|^2}\text{Real}\left\{\left[H_{\text{scat}}^{(\text{ff})}(r,\theta=0)\right]^* \sqrt{r}e^{ik_0 n_1 r}e^{-i\pi/4}H_0\right\}. \tag{2.89}$$

The scattering cross section is defined as the total scattered power normalized by I_i, and is given by

$$\sigma_{\text{scat}} = \frac{P_{\text{scat}}}{I_i} = \frac{1}{|H_0|^2}\int_\theta |H_{\text{scat}}^{(\text{ff})}(r,\theta)|^2 r d\theta. \tag{2.90}$$

Here, the integral is over all angles θ.

The power absorbed in the scatterer is given from the total Poynting-vector flux into the scatterer and is given by

$$P_{\text{abs}} = -\frac{1}{2}\oint \text{Real}\left\{\mathbf{E}\times\mathbf{H}^*\right\}\cdot\hat{\mathbf{n}}dl, \tag{2.91}$$

where the integral is over the surface of the scatterer, and $\hat{\mathbf{n}}$ is the outward surface normal vector.

In the Green's Function Surface Integral Equation Method (GFSIEM) one finds the field and its normal derivative at the surface of the scatterer. It is thus convenient to cast the expression for the absorption cross section as

$$\sigma_{\text{abs}} = \frac{P_{\text{abs}}}{I_i} = \frac{-1}{k_0 n_1} \frac{1}{|H_0|^2} \oint \text{Imag}\{(\hat{\mathbf{n}} \cdot \nabla H) H^*\} \, dl. \tag{2.92}$$

If we carry out the same type of calculations for the case of s polarization but otherwise the same situation, then we will be considering an incident Gaussian-beam *electric* field of the form

$$\mathbf{E}_{0,\text{i}}(x, y=0) = \hat{z} E_{0,\text{i}} = \hat{z} E_0 e^{-x^2/w_0^2}, \tag{2.93}$$

and similar expressions for the extinction, scattering, and absorption cross sections, as those just obtained for p polarization, can be obtained, with the only difference that H must be replaced by E in the final expressions.

Note that on physical grounds the extinction must equal the sum of scattering and absorption, and thus

$$\sigma_{\text{ext}} = \sigma_{\text{scat}} + \sigma_{\text{abs}}. \tag{2.94}$$

This will serve as a useful check of numerical calculations.

2.3.2 SCATTERER ON A LAYERED STRUCTURE (2D)

We are now going to consider a scattering problem where the scatterer is placed on a layered reference structure (Fig. 2.8). Here, the scatterer is illuminated by a Gaussian beam from above. In the absence of the scatterer, the total field will consist on the upper side ($y > 0$) of the incident Gaussian beam and a reflected Gaussian beam, and on the lower side ($y < -d$) it will consist of a transmitted Gaussian beam. In the presence of the scatterer there will also be a scattered field component.

The scatterer can both lead to extinction of the reflected and transmitted beams, and scattering can be divided into out-of-plane scattering going into the upper and lower half-plane, respectively, and scattering into guided modes bound to and propagating along the film geometry. The power carried by the guided modes normalized with the incident power per unit area gives yet another scattering cross section, that we shall refer to as the guided-mode scattering cross section.

We shall again start with considering p polarization, and fields and a geometry that are invariant along the z-axis (2D calculations). The incident magnetic field and its corresponding Fourier transform in the plane $y = 0^+$ are identical to expressions (2.74) and (2.79) found in the previous section. We are now going to find the reference field corresponding to the case where the scatterer is absent. The reflection and transmission of each Fourier component of the incident field follows from standard Fresnel reflection. This leads to the reflected part of the reference field ($y > 0$)

$$H_{0,\text{r}}(x, y) = \int_{k_x} \tilde{H}_{0,\text{i}}(k_x; 0) r^{(p)}(k_x) e^{ik_x x} e^{ik_{y1} y} \, dk_x, \tag{2.95}$$

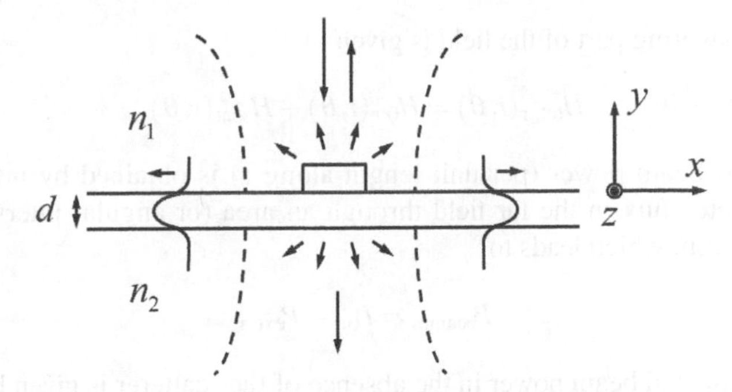

Figure 2.8 Schematic of a scatterer placed on a film of thickness d being illuminated with a wide Gaussian beam from above. On the upper side of the film ($y > 0$) the refractive index is n_1, and below the film ($y < -d$) the refractive index is n_2. It is illustrated that reflected and transmitted light consists of the reflection and transmission that would occur in the absence of the scatterer superposed with scattered light, where the latter component can both be propagating out of the plane, but can also propagate in the form of guided modes in the film geometry.

where $r^{(p)}(k_x)$ is the Fresnel reflection coefficient for the layered structure. When inserting $\tilde{H}_{0,\mathrm{i}}(k_x; y=0)$ we note again that for a wide beam ($k_0 n_1 w_0 >> 1$) only a narrow range of k_x centered on $k_x = 0$ will contribute significantly to the integral. It is thus possible to move $r^{(p)}(k_x)$ outside the integral and replace it with $r^{(p)}(k_x = 0)$, and to use again the approximation $k_{y1} \approx k_0 n_1 - k_x^2/2k_0 n_1$. We can use once more the integral relation (2.78) to obtain

$$H_{0,\mathrm{r}}(x,y) \approx \left(r^{(p)}(k_x = 0) H_0\right) \sqrt{\frac{w_0^2}{\frac{2iy}{k_0 n_1} + w_0^2}}\, e^{ik_0 n_1 y} \exp\left(-x^2 \frac{1}{\frac{i2y}{k_0 n_1} + w_0^2}\right). \tag{2.96}$$

For large distances r we may now again make the replacements $y = r\cos\theta$ and $x = r\sin\theta$, in which case the reflected far field can be expressed as

$$H_{0,\mathrm{r}}^{(\mathrm{ff})}(r,\theta) = \left(r^{(p)}(k_x = 0) H_0\right) w_0 \sqrt{\frac{k_0 n_1}{2r}}\, e^{-i\pi/4} e^{ik_0 n_1 r} e^{-\frac{\theta^2}{4}(k_0 n_1 w_0)^2}. \tag{2.97}$$

For the transmitted field we may instead use $x = r\sin\tilde{\theta}$ and $y = -r\cos\tilde{\theta}$, in which case a similar derivation based on the angular spectrum approach leads to the transmitted far field given by

$$H_{0,\mathrm{t}}^{(\mathrm{ff})}(r,\tilde{\theta}) = \left(t^{(p)}(k_x = 0) H_0\right) w_0 \sqrt{\frac{k_0 n_2}{2r}}\, e^{-i\pi/4} e^{-ik_0 n_2 d} e^{ik_0 n_2 r} e^{-\frac{\tilde{\theta}^2}{4}(k_0 n_2 w_0)^2}. \tag{2.98}$$

When the scatterer is present the total reflected field will be modified by the scattering from the scattering object, and thus in the upper half-plane ($y > 0$) the total upward propagating part of the field is given by

$$H_{\text{tot, r}}(r,\theta) = H_{0,\text{r}}(r,\theta) + H_{\text{scat}}(r,\theta). \tag{2.99}$$

The reflected beam power (per unit length along z) is obtained by integrating the Poynting vector flux in the far field through an area (or angular interval) that just covers the beam, which leads to

$$P_{\text{beam},r} = P_{0,\text{r}} - P_{\text{ext, r}} \tag{2.100}$$

where the reflected beam power in the absence of the scatterer is given by

$$P_{0,\text{r}} = \frac{1}{2}\sqrt{\frac{\mu_0}{\varepsilon_0}}\frac{1}{n_1}\int_{\theta=-\theta_b}^{\theta_b} |H_{0,\text{r}}^{(\text{ff})}(r,\theta)|^2 r d\theta, \tag{2.101}$$

and minus the extinction power for the reflected beam is given by

$$-P_{\text{ext, r}} = \sqrt{\frac{\mu_0}{\varepsilon_0}}\frac{1}{n_1}\int_{\theta=-\theta_b}^{+\theta_b} \text{Real}\left\{H_{0,\text{r}}^{(\text{ff})}(r,\theta)\left[H_{\text{scat}}^{(\text{ff})}(r,\theta)\right]^*\right\} r d\theta. \tag{2.102}$$

By normalizing with the incident power per unit area, we find the extinction cross section for the reflected beam:

$$\sigma_{\text{ext},r} = -\frac{2\sqrt{\frac{2\pi}{k_0 n_1}}}{|H_0|^2}\text{Real}\left\{r^{(p)}(k_x=0)\left(H_{\text{scat}}^{(\text{ff})}(r,\theta=0)\right)^* \sqrt{r}e^{ik_0 n_1 r}e^{-i\frac{\pi}{4}}H_0\right\}. \tag{2.103}$$

The extinction cross section for the transmitted beam is obtained by a similar procedure and is given by

$$\sigma_{\text{ext},t} = -2\frac{n_1}{n_2}\frac{\sqrt{\frac{2\pi}{k_0 n_2}}}{|H_0|^2}\text{Real}\left\{t^{(p)}(k_x=0)\left(H_{\text{scat}}^{(\text{ff})}(r,\tilde{\theta}=0)\right)^* \sqrt{r}e^{ik_0 n_2 r}e^{-i\frac{\pi}{4}}H_0 e^{-ik_0 n_2 d}\right\}. \tag{2.104}$$

We shall now consider a position in the upper half-plane which is both in the far-field but also not so close to the film geometry that guided modes will have an effect on the scattered field. In that case the scattered field will be of the form

$$H_{\text{scat}}^{(\text{ff})}(r,\theta) = A(\theta)\frac{e^{ik_0 n_1 r}}{\sqrt{r}}. \tag{2.105}$$

Note that near the surface ($\theta \approx \pm\pi/2$) the magnitude of this scattered field component can be reduced arbitrarily toward 0 by increasing r. This will not be the case for guided modes, which (in loss-less media) will have unchanged magnitude with increasing distance to the scatterer, and unchanged magnitude of the confined field distribution along y. The out-of-plane scattering cross section governing scattering of light into the upper half-plane is now given by

$$\sigma_{\text{scat},r} = \frac{1}{|H_0|^2} \int_{\theta=-\pi/2^+}^{\theta=\pi/2^-} |H_{\text{scat}}^{(\text{ff})}(r,\theta)|^2 r d\theta, \tag{2.106}$$

and the scattering cross section for light going into the lower half-plane is correspondingly given by

$$\sigma_{\text{scat},t} = \frac{n_1}{n_2} \frac{1}{|H_{0,i}|^2} \int_{\tilde{\theta}=-\pi/2^+}^{\tilde{\theta}=+\pi/2^-} |H_{\text{scat}}^{(\text{ff})}(r,\tilde{\theta})|^2 r d\tilde{\theta}. \tag{2.107}$$

The factor n_1/n_2 is due to the Poynting vector of the considered field components being proportional to $|H|^2/n$, where n is the refractive index.

We are now going to consider the power scattered into guided modes. We are thus going to consider values of y such that the position is in the vicinity of the waveguide or inside it. We are also going to consider large values of $|x|$, meaning that $k_0|n_i x| >> 1$ for any refractive index of the layered geometry n_i. Here, the scattered field is dominated by the guided modes and is of the form

$$H_{\text{g}\pm}(x,y) = \sum_j H_{\text{g},j}^{(\pm)}(y) e^{ik_0 n_j |x|}, \tag{2.108}$$

where $H_{\text{g},j}$ is the field distribution along y for the j'th guided mode, n_j is the mode-index of mode j, and $+$ is used when x is to the right of the scatterer, and $-$ when x is to the left. The field distribution $H_{\text{g},j}$ will be evanescent in the upper and lower half-planes above and below the film. Naturally, if there are propagation losses, meaning that n_j has an imaginary part, then the magnitude of the guided modes will vanish for large $|x|$. However, for loss-less materials the magnitude of fields will be independent of $|x|$ (in the far field). Furthermore, for loss-less materials the guided modes will be orthogonal in the sense that

$$P_{\text{g}\pm} = \frac{1}{2} \int_y \text{Re}\left\{\mathbf{E}_{\text{g}\pm}(x,y) \times \left[\mathbf{H}_{\text{g}\pm}(x,y)\right]^*\right\} \cdot (\pm\hat{\mathbf{x}})\, dy = \sum_j P_{\text{g},j\pm}, \tag{2.109}$$

where

$$P_{\text{g},j\pm} = \frac{1}{2} \int_y \text{Re}\left\{\mathbf{E}_{\text{g},j\pm}(x,y) \times \left[\mathbf{H}_{\text{g},j\pm}(x,y)\right]^*\right\} \cdot (\pm\hat{\mathbf{x}})\, dy. \tag{2.110}$$

It is thus meaningful to assign an amount of power to each mode separately. This orthogonality property may still hold when there are propagation losses but this cannot generally be ensured [41]. If we now normalize the right- and left-propagating guided mode powers $P_{\text{g}\pm}$ with the power per unit area of the incident beam, we find right-and-left propagating guided-mode scattering cross sections

$$\sigma_{\text{g}\pm} = \frac{P_{\text{g}\pm}}{I_i}. \tag{2.111}$$

In addition there can be absorption in the scatterer itself. The absorption cross section can, for p polarization, and for the scatterer placed in the upper medium with refractive index n_1, be calculated using again (2.92).

If the scatterer is the only part of the structure with absorption then the requirement of energy conservation leads to

$$\sigma_{\text{ext},r} + \sigma_{\text{ext},t} = \sigma_{\text{scat},r} + \sigma_{\text{scat},t} + \sigma_{\text{abs}} + \sigma_{g+} + \sigma_{g-}. \tag{2.112}$$

Thus, the total power removed from transmitted and reflected beams must equal the total scattered and absorbed power. The expression (2.112) allows you to check the numerical calculation by comparing scattering and extinction cross sections for a lossless layered geometry.

When using the Green's function area integral equation method (GFAIEM) it may be preferable to express the optical cross sections in terms of the electric field. The absorption cross section (still p polarization) can be alternatively written

$$\sigma_{\text{abs}} = \frac{k_0/n_1}{|\mathbf{E}_0|^2} \int \text{Imag}\left\{\varepsilon(\mathbf{r})\right\} |\mathbf{E}(\mathbf{r})|^2 dA. \tag{2.113}$$

In order to obtain the other cross sections in terms of the electric field we note that the incident electric field at the beam waist ($y = 0$) can be written as

$$\mathbf{E}_{0,\text{i}}(x, y = 0) \approx \hat{x} E_0 e^{-x^2/w_0^2}. \tag{2.114}$$

Strictly speaking we can choose an x-component in this way but the field must also have a y-component in order to be divergence-free. However, for a wide beam waist ($k_0 n_1 w_0 >> 1$) this can be ignored. The field at other values of y can be expressed as

$$\mathbf{E}_{0,\text{i}}(x, y) = \int \tilde{\mathbf{E}}_{0,\text{i}}(k_x; y = 0) e^{ik_x x} e^{-ik_{y1} y} dk_x. \tag{2.115}$$

A derivation similar to the one carried out for the magnetic field now leads to the reflected far field

$$E_{0,\text{r},x}(r, \theta) \approx \left(-r^{(p)}(k_x = 0) E_0\right) w_0 \sqrt{\frac{k_0 n_1}{2r}} e^{-i\pi/4} e^{ik_0 n_1 r} e^{-\frac{\theta^2}{4}(k_0 n_1 w_0)^2}. \tag{2.116}$$

It is important here to notice the sign change which is because the reflection coefficient $r^{(p)}$ is defined relative to the convention that the in-plane component of the reflected electric field has the opposite sign compared with the incident field.

For the transmitted electric far field we find

$$E_{0,\text{t},x}(r, \tilde{\theta}) \approx \left(\frac{n_1}{n_2} t^{(p)}(k_x = 0) E_0\right) w_0 \sqrt{\frac{k_0 n_2}{2r}} e^{-i\pi/4} e^{-ik_0 n_2 d} e^{ik_0 n_2 r} e^{-\frac{\tilde{\theta}^2}{4}(k_0 n_2 w_0)^2}. \tag{2.117}$$

The additional factor n_1/n_2 is because the transmission coefficient is the one for the magnetic field. Thus, here for normal propagation the transmitted field just below the layered structure (at $z = -d$) is given by $E_{0,\text{t},x} = E_0 t^{(p)}(k_x = 0) n_1/n_2$.

We can now express the total reflected beam power as

$$P_{\text{beam},r} = \frac{1}{2}\sqrt{\frac{\varepsilon_0}{\mu_0}}n_1 \int_{\theta=-\theta_b}^{\theta=\theta_b} |\mathbf{E}_r^{(\text{ff})}(r,\theta) + \mathbf{E}_{\text{scat}}^{(\text{ff})}(r,\theta)| r d\theta, \tag{2.118}$$

which, as before, can be divided into the reflected beam power in the absence of the scatterer, and minus the extinction power. Similar expressions can be constructed for the transmitted beam power. By normalizing these with the incident power per unit area

$$I_i = \frac{1}{2}\sqrt{\frac{\varepsilon_0}{\mu_0}}n_1|E_0|^2, \tag{2.119}$$

we finally obtain expressions for the extinction cross sections related to transmitted and reflected beams:

$$\sigma_{\text{ext},r} = +\frac{2\sqrt{\frac{2\pi}{k_0 n_1}}}{|E_0|^2}\text{Real}\left\{ r^{(p)}(k_x=0)\left(E_{\text{scat},x}^{(\text{ff})}(r,\theta=0)\right)^* \sqrt{r}e^{ik_0 n_1 r}e^{-i\pi/4}E_0 \right\}, \tag{2.120}$$

$$\sigma_{\text{ext},t} = -2\frac{\sqrt{\frac{2\pi}{k_0 n_2}}}{|E_0|^2}\text{Real}\left\{ t^{(p)}(k_x=0)\left(E_{\text{scat},x}^{(\text{ff})}(r,\tilde{\theta}=0)\right)^* \sqrt{r}e^{ik_0 n_2 r}e^{-i\pi/4}E_0 e^{-ik_0 n_2 d} \right\}. \tag{2.121}$$

In this case, the Poynting vector of the considered field components is proportional to $|\mathbf{E}^{(\text{ff})}|^2 n$, where n is the refractive index.

The out-of-plane scattering cross sections for light scattered into the upper and lower half-plane are given by

$$\sigma_{\text{scat},r} = \frac{1}{|E_0|^2}\int_{\theta=-\pi/2^+}^{\pi/2^-} |\mathbf{E}_{\text{scat}}^{(\text{ff})}(r,\theta)|^2 r d\theta, \tag{2.122}$$

$$\sigma_{\text{scat},t} = \frac{n_2}{n_1}\frac{1}{|E_0|^2}\int_{\tilde{\theta}=-\pi/2^+}^{\pi/2^-} |\mathbf{E}_{\text{scat}}^{(\text{ff})}(r,\tilde{\theta})|^2 r d\tilde{\theta}. \tag{2.123}$$

We will finish this subsection by presenting the extinction and scattering cross sections for s polarization. In this case the incident field has the same form as for p polarization except that H must be replaced by E, and reflection and transmission coefficients $r^{(p)}$ and $t^{(p)}$ must be replaced by the corresponding coefficients for s polarization, namely $r^{(s)}$ and $t^{(s)}$. These reflection and transmission coefficients are defined for the electric field directly, such that here the transmitted field at $y=0$ equals the incident field times the transmission coefficient, and the same for the reflected field (no sign change). In this case, the Poynting vector in the far field must be proportional to $|H^{(\text{ff})}(\mathbf{r})|^2/n$ or $|E^{(\text{ff})}(\mathbf{r})|^2 n$, which means that some refractive index ratios will be replaced by their inverse ratios when expressions are cast in terms of the electric field. This leads to

$$\sigma_{\text{ext},r}^{(s\text{ pol})} = -\frac{2\sqrt{\frac{2\pi}{k_0 n_1}}}{|E_0|^2}\text{Real}\left\{r^{(s)}(k_x=0)\left(E_{\text{scat}}^{(\text{ff})}(r,\theta=0)\right)^* \sqrt{r}e^{ik_0 n_1 r}e^{-i\pi/4}E_0\right\}, \tag{2.124}$$

$$\sigma_{\text{ext},t}^{(s\text{ pol})} = -2\frac{n_2}{n_1}\frac{\sqrt{\frac{2\pi}{k_0 n_2}}}{|E_0|^2}\text{Real}\left\{t^{(s)}(k_x=0)\left(E_{\text{scat}}^{(\text{ff})}(r,\tilde{\theta}=0)\right)^* \sqrt{r}e^{ik_0 n_2 r}e^{-i\pi/4}E_0 e^{-ik_0 n_2 d}\right\}, \tag{2.125}$$

$$\sigma_{\text{scat},r}^{(s\text{ pol})} = \frac{1}{|E_0|^2}\int_{\theta=-\pi/2^+}^{\theta=\pi/2^-} |E_{\text{scat}}(r,\theta)|^2 r d\theta, \tag{2.126}$$

$$\sigma_{\text{scat},t}^{(s\text{ pol})} = \frac{n_2}{n_1}\frac{1}{|E_0|^2}\int_{\tilde{\theta}=-\pi/2^+}^{\tilde{\theta}=+\pi/2^-} |E_{\text{scat}}(r,\tilde{\theta})|^2 r d\tilde{\theta}. \tag{2.127}$$

$$\sigma_{\text{abs}}^{(s\text{ pol})} = \frac{k_0/n_1}{|E_0|^2}\int \text{Imag}\{\varepsilon(\mathbf{r})\}|E(\mathbf{r})|^2 dA. \tag{2.128}$$

The expressions for the guided-mode scattering cross sections given previously can be used for both s and p polarization.

2.3.3 SCATTERER IN HOMOGENEOUS MEDIA (3D)

In this section we extend the scattering theory to *three* dimensions for a wide Gaussian beam incident on a scatterer in a homogeneous material with refractive index n_1. We will consider the scatterer to be placed near the plane $z=0$ and consider a Gaussian beam that propagates along the positive z-axis. We will assume that the electric field of the incident Gaussian beam is predominantly polarized along the x-axis.

Assuming a Gaussian spot profile at the beam waist, the x-component of the incident electric field can be written as

$$E_{xi}(x,y,z=0) = E_0 e^{-(x^2+y^2)/w_0^2}. \tag{2.129}$$

The incident field will also generally have a z-component, which can be constructed from the x-component by requiring the incident field to be divergence-free. This can be ignored for a sufficiently large beam waist radius w_0. The incident field can be expressed as a Fourier integral

$$E_{xi}(x,y,z) = \int_{k_x}\int_{k_y} \tilde{E}_{xi}(k_x,k_y;z)e^{ik_x x}e^{ik_y y}dk_x dk_y, \tag{2.130}$$

and the requirement that the field must satisfy the scalar Helmholtz equation and propagate along the positive z-axis leads to

$$\tilde{E}_{xi}(k_x,k_y;z) = \tilde{E}_{xi}(k_x,k_y;0)e^{ik_z z}, \tag{2.131}$$

where $k_z = \sqrt{k_0^2 n_1^2 - k_x^2 - k_y^2}$.

The two-dimensional Fourier transform can be calculated leading to

$$\tilde{E}_{xi}(k_x, k_y; 0) = \frac{1}{(2\pi)^2} \int_x \int_y E_{xi}(x, y, 0) e^{-ik_x x} e^{-ik_y y} dx dy = \frac{E_0 w_0^2}{4\pi} e^{-\frac{1}{4}(k_x^2 + k_y^2) w_0^2}. \tag{2.132}$$

By inserting (2.131) and (2.132) into Eq. (2.130) and carrying out the integrations it follows that

$$E_{xi}(x, y, z) = E_0 e^{ik_0 n_1 z} \frac{w_0^2}{w_0^2 + \frac{2iz}{k_0 n_1}} \exp\left(-\frac{x^2 + y^2}{w_0^2 + \frac{2iz}{k_0 n_1}} \right). \tag{2.133}$$

Clearly, this reduces to Eq. (2.129) in the case $z = 0$.

Now we shall consider the far field, that is the case of $k_0 n_1 r >> 1$ and $k_0 n_1 z >> 1$, with spherical coordinates such that $z = r\cos\theta$, $x = r\sin\theta\cos\phi$, and $y = r\sin\theta\sin\phi$. In this limit the field assumes the simpler form

$$E_{xi}^{(\mathrm{ff})}(r, \theta, \phi) \approx E_0 \frac{k_0 n_1 w_0^2}{2r} e^{-i\pi/2} e^{ik_0 n_1 r} e^{-\frac{1}{4}\theta^2 (k_0 n_1 w_0)^2}, \tag{2.134}$$

where it has been assumed that w_0 is sufficiently large that it is acceptable to use $\sin\theta \approx \theta$ and $\cos\theta \approx 1$.

The total field can be written as the sum of incident and scattered fields:

$$\mathbf{E}(\mathbf{r}) = \mathbf{E}_i(\mathbf{r}) + \mathbf{E}_{\mathrm{scat}}(\mathbf{r}). \tag{2.135}$$

The power carried by the beam can now be obtained from the Poynting vector flux through the beam cross section in the far field, thus

$$P_{\mathrm{beam}} = \int_{\theta=0}^{\theta_b} \int_{\phi=0}^{2\pi} \frac{1}{2} \mathrm{Re}\left\{ \mathbf{E}^{(\mathrm{ff})}(\mathbf{r}) \times [\mathbf{H}^{(\mathrm{ff})}(\mathbf{r})]^* \right\} \cdot \hat{\mathbf{r}} r^2 \sin\theta d\theta d\phi, \tag{2.136}$$

where θ_b is large enough to include the beam but not much more in order to avoid the inclusion of scattered power. And, similar to the 2D case, in the limit of $w_0 \to \infty$, the angle $\theta_b \to 0^+$. Rewriting the expression using Maxwell's equations we find

$$P_{\mathrm{beam}} = \frac{1}{2} \sqrt{\frac{\varepsilon_0 n_1^2}{\mu_0}} \int_{\theta=0}^{\theta_b} \int_{\phi=0}^{2\pi} |\mathbf{E}_i^{(\mathrm{ff})}(\mathbf{r}) + \mathbf{E}_{\mathrm{scat}}^{(\mathrm{ff})}(\mathbf{r})|^2 r^2 \sin\theta d\theta d\phi. \tag{2.137}$$

This can be divided into two terms

$$P_{\mathrm{beam}} = P_i - P_{\mathrm{ext}}, \tag{2.138}$$

where

$$P_i = \frac{1}{2} \sqrt{\frac{\varepsilon_0 n_1^2}{\mu_0}} \int_{\theta=0}^{\theta_b} \int_{\phi=0}^{2\pi} |\mathbf{E}_i^{(\mathrm{ff})}(\mathbf{r})|^2 r^2 \sin\theta d\theta d\phi \tag{2.139}$$

is the power of the incident beam corresponding to the case of no scattering, and

$$-P_{\text{ext}} = \sqrt{\frac{\varepsilon_0 n_1^2}{\mu_0}} \int_{\theta=0}^{\theta_b} \int_{\phi=0}^{2\pi} \text{Real}\left(\mathbf{E}_i^{(\text{ff})}(\mathbf{r}) \cdot \left[\mathbf{E}_{\text{scat}}^{(\text{ff})}(\mathbf{r})\right]^*\right) r^2 \sin\theta d\theta d\phi, \tag{2.140}$$

is minus the extinction power. It is reasonable to assume that the scattered field will not vary with θ within the small angular range considered, and by inserting the far-field expression for $\mathbf{E}_{\text{i}}(\mathbf{r})$ and carrying out the integration, the extinction power can be finally expressed as

$$P_{\text{ext}} = -\frac{2\pi}{k_0}\sqrt{\frac{\varepsilon_0}{\mu_0}} \text{Real}\left\{E_0 e^{-i\pi/2} r e^{ik_0 n_1 r} \left[\hat{\mathbf{x}} \cdot \mathbf{E}_{\text{scat}}^{(\text{ff})}(r, \theta = 0, \phi)\right]^*\right\}. \tag{2.141}$$

Note that the value of ϕ has no importance when $\theta = 0$.

The total scattered power can be written

$$P_{\text{scat}} = \frac{1}{2}\sqrt{\frac{\varepsilon_0 n_1^2}{\mu_0}} \int_{\theta=0}^{\pi} \int_{\phi=0}^{2\pi} |\mathbf{E}_{\text{scat}}^{(\text{ff})}(\mathbf{r})|^2 r^2 \sin\theta d\theta d\phi. \tag{2.142}$$

Absorption in the scatterer is given by

$$P_{\text{abs}} = -\oint \frac{1}{2} \text{Real}\left(\mathbf{E}(\mathbf{r}) \times [\mathbf{H}(\mathbf{r})]^*\right) \cdot \hat{\mathbf{n}} dA, \tag{2.143}$$

where the surface integral is over the surface of the scatterer, and $\hat{\mathbf{n}}$ is the outward surface normal vector. By applying Gauss law to rewrite this into a volume integral, and using Maxwell's equations, this can also be written

$$P_{\text{abs}} = \frac{\omega\varepsilon_0}{2} \int \text{Imag}\{\varepsilon\} |\mathbf{E}|^2 dV, \tag{2.144}$$

which is convenient when working with volume integral equation methods.

By normalizing with the incident power per unit area we obtain the extinction, scattering, and absorption cross sections:

$$\sigma_{\text{ext}} = -\frac{4\pi}{k_0 n_1 |E_0|^2} \text{Real}\left\{E_0 e^{-i\pi/2} r e^{ik_0 n_1 r} \left[\hat{\mathbf{x}} \cdot \mathbf{E}_{\text{scat}}^{(\text{ff})}(r, \theta = 0, \phi)\right]^*\right\}, \tag{2.145}$$

$$\sigma_{\text{scat}} = \frac{1}{|\mathbf{E}_0|^2} \int_{\theta=0}^{\pi} \int_{\phi=0}^{2\pi} |\mathbf{E}_{\text{scat}}^{(\text{ff})}(\mathbf{r})|^2 r^2 \sin\theta d\theta d\phi, \tag{2.146}$$

$$\sigma_{\text{abs}} = \frac{k_0/n_1}{|\mathbf{E}_0|^2} \int \text{Imag}\{\varepsilon\} |\mathbf{E}|^2 dV. \tag{2.147}$$

2.3.4 SCATTERER ON A LAYERED STRUCTURE (3D)

In this section we shall consider again the situation in Fig. 2.8 but extended to three dimensions. Thus, now the incident Gaussian beam will propagate along the negative

z-axis in a medium with refractive index n_1 and be incident on a scatterer placed on a film of thickness d with another material with refractive index n_2 below the film. This corresponds to interchanging the y and z axes in Fig. 2.8. For the incident Gaussian beam in the plane $z=0$ we will consider the same three-dimensional Gaussian field distribution as in the previous section. Thus, if we think of the incident beam as being p polarized, then the reflected beam propagating along the positive z-axis in the medium with refractive index n_1 can be expressed in the following way in the far field:

$$E_{\mathrm{r},x}^{(\mathrm{ff})}(r,\theta,\phi) \approx \left(-r^{(p)}(k_x=0)E_0\right)\frac{k_0 n_1 w_0^2}{2r}e^{-i\pi/2}e^{ik_0 n_1 r}e^{-\frac{1}{4}\theta^2(k_0 n_1 w_0)^2}, \quad (2.148)$$

and the corresponding transmitted beam propagating in the medium with refractive index n_2 is given by

$$E_{\mathrm{t},x}^{(\mathrm{ff})}(r,\theta,\phi) \approx \left(\frac{n_1}{n_2}t^{(p)}(k_x=0)E_0\right)\frac{k_0 n_2 w_0^2}{2r}e^{-i\pi/2}e^{ik_0 n_2 r}e^{-\frac{1}{4}\theta^2(k_0 n_2 w_0)^2}e^{-ik_0 n_2 d}. \quad (2.149)$$

Note the factor n_1/n_2 which is needed since $t^{(p)}$ is really defined with respect to transmission of the magnetic field.

Since many examples of how to calculate extinction, scattering, and absorption cross sections have already been given, and calculations here follow closely those for the 2D case, the final expressions are simply presented here:

$$\sigma_{\mathrm{ext},r} = -\frac{4\pi r}{k_0 n_1 |E_0|^2}\mathrm{Real}\left\{e^{-i\pi/2}\left(-r^{(p)}E_0\right)e^{ik_0 n_1 r}\left[E_{\mathrm{scat},x}^{(\mathrm{ff})}(r,\theta=0,\phi)\right]^*\right\}, \quad (2.150)$$

$$\sigma_{\mathrm{ext},t} = -\frac{4\pi r}{k_0 n_1 |E_0|^2}\mathrm{Real}\left\{e^{-i\pi/2}\frac{n_1}{n_2}t^{(p)}e^{-ik_0 n_2 d}E_0 e^{ik_0 n_2 r}\left[E_{\mathrm{scat},x}^{(\mathrm{ff})}(r,\theta=\pi,\phi)\right]^*\right\}, \quad (2.151)$$

$$\sigma_{\mathrm{scat},r} = \frac{1}{|E_0|^2}\int_{\theta=0}^{\pi/2^-}\int_{\phi=0}^{2\pi}|\mathbf{E}_{\mathrm{scat}}^{(\mathrm{ff})}(\mathbf{r})|^2 r^2 \sin\theta d\theta d\phi, \quad (2.152)$$

$$\sigma_{\mathrm{scat},t} = \frac{n_2}{n_1}\frac{1}{|\mathbf{E}_{0,i}|^2}\int_{\theta=\pi/2^+}^{\pi}\int_{\phi=0}^{2\pi}|\mathbf{E}_{\mathrm{scat}}^{(\mathrm{ff})}(\mathbf{r})|^2 r^2 \sin\theta d\theta d\phi, \quad (2.153)$$

$$\sigma_{\mathrm{abs}} = \frac{k_0/n_1}{|\mathbf{E}_{0,i}|^2}\int \mathrm{Imag}\{\varepsilon\}|\mathbf{E}|^2 dV. \quad (2.154)$$

Here, it is implicit that the reflection and transmission coefficients are those corresponding to normally incident light.

A few more words are needed to deal with the guided modes. At a large distance from the scatterer, but inside or in the near vicinity of the film, the contribution to the electric field due to guided modes can be expressed as

$$\mathbf{E}_{\mathrm{g}}^{(\mathrm{ff})}(\rho,\phi,z) = \sum_{\alpha=s,p}\sum_{j}\mathbf{E}_{\mathrm{g}\alpha j}^{(\mathrm{ff})}(\rho,\phi,z), \tag{2.155}$$

where here α represents the polarization of the guided mode, which can be either s or p polarized, and the index j is used to sum over different modes. Here we now use cylindrical coordinates.

The corresponding magnetic field of guided modes is given by

$$\mathbf{H}_{\mathrm{g}}^{(\mathrm{ff})}(\rho,\phi,z) = \frac{-i}{\omega\mu_0}\nabla\times\mathbf{E}_{\mathrm{g}}^{(\mathrm{ff})}(\rho,\phi,z). \tag{2.156}$$

The p-polarized modes will generally have an electric field of the form

$$\mathbf{E}_{\mathrm{g}pj}^{(\mathrm{ff})}(\rho,\phi,z) = (\hat{\rho}A_{pj}(z)+\hat{\mathbf{z}}B_{pj}(z))\,C_{pj}(\phi)\frac{e^{ik_0 n_{\mathrm{g}pj}\rho}}{\sqrt{\rho}}. \tag{2.157}$$

Here $n_{\mathrm{g}pj}$ is the mode index of the j'th p-polarized guided mode. The s polarized modes will have electric fields of the form

$$\mathbf{E}_{\mathrm{g}sj}^{(\mathrm{ff})}(\rho,\phi,z) = \hat{\phi}A_{sj}(z)C_{sj}(\phi)\frac{e^{ik_0 n_{\mathrm{g}sj}\rho}}{\sqrt{\rho}}, \tag{2.158}$$

where $n_{\mathrm{g}sj}$ is the mode index of the j'th s-polarized guided mode. The functions $A_{pj}(z)$, $B_{pj}(z)$, and $A_{sj}(z)$ will decay exponentially into the media on either side of the film, such that this corresponds to modes that are bound to and propagating along the film. While the magnitude of these fields will decrease as $1/\sqrt{\rho}$, then the modes that are not bound to the film will decrease instead as $1/\rho$, and thus for large enough ρ (in the far field), only the contribution from guided modes will be of significance inside or near the film, if the film geometry is not absorbing.

If there are no propagation losses in the film geometry, then the total power carried by the guided modes can be calculated as

$$P_{\mathrm{g}} = \frac{1}{2}\int_{\phi=0}^{2\pi}\int_{z}\mathrm{Re}\left\{\mathbf{E}_{\mathrm{g}}^{(\mathrm{ff})}(\rho,\phi,z)\times\left[\mathbf{H}_{\mathrm{g}}^{(\mathrm{ff})}(\rho,\phi,z)\right]^{*}\right\}\cdot\hat{\rho}\rho d\phi dz, \tag{2.159}$$

where orthogonality between modes in the case of absence of losses in the film geometry means that

$$P_{\mathrm{g}} = \sum_{\alpha=s,p}\sum_{j}P_{\mathrm{g}\alpha j}, \tag{2.160}$$

where

$$P_{\mathrm{g}\alpha j} = \frac{1}{2}\int_{\phi=0}^{2\pi}\int_{z}\mathrm{Re}\left\{\mathbf{E}_{\mathrm{g}\alpha j}^{(\mathrm{ff})}(\rho,\phi,z)\times\left[\mathbf{H}_{\mathrm{g}\alpha j}^{(\mathrm{ff})}(\rho,\phi,z)\right]^{*}\right\}\cdot\hat{\rho}\rho d\phi dz, \tag{2.161}$$

such that a certain amount of power can be assigned to each guided mode. We can now define the scattering cross section related to guided modes as

$$\sigma_g = \frac{P_g}{I_i}, \tag{2.162}$$

where

$$I_i = \frac{1}{2}\sqrt{\frac{\varepsilon_0}{\mu_0}} n_1 |E_0|^2. \tag{2.163}$$

2.4 EXERCISES

1. Derive the wave equations (2.25)-(2.28) by using Maxwell's equations in the frequency domain (2.16)-(2.19).

2. Derive the expression for the time-averaged Poynting vector flux (2.41) by starting from (2.33) and using fields of the form (2.39).

3. Derive the Fresnel reflection and transmission coefficients (2.54) from the boundary conditions (2.52) and (2.53).

4. Show that there are no *s*-polarized guided modes for a single metal-dielectric interface.

5. Show that both (2.62) and (2.65) lead to (2.66) governing either the Brewster angle or the SPP dispersion relation.

6. Study the methods of Appendix F and construct numerical programs for obtaining field distributions and plots of dispersion relations for the guided modes of the structures shown in Fig. 2.6.

7. Derive the far-field expression (2.81) from (2.80). Use this expression to evaluate the extinction power and finally obtain the extinction cross section (2.89).

8. Derive the expression for the absorption cross section (2.92).

9. Derive the optical cross sections for the 3D case (2.150)-(2.154).

3 One-dimensional scattering problems

In this chapter we will consider a Green's function integral equation method for one-dimensional scattering problems with a scattering object placed in a homogeneous medium (dielectric constant $\varepsilon_{\mathrm{ref}} = n_{\mathrm{ref}}^2$). The scattering object is illuminated by a given plane wave being incident from the left (E_{in}), and as a result of scattering there will be a reflected wave (E_{r}), and a transmitted wave (E_{t}). The situation is illustrated in Fig. 3.1.

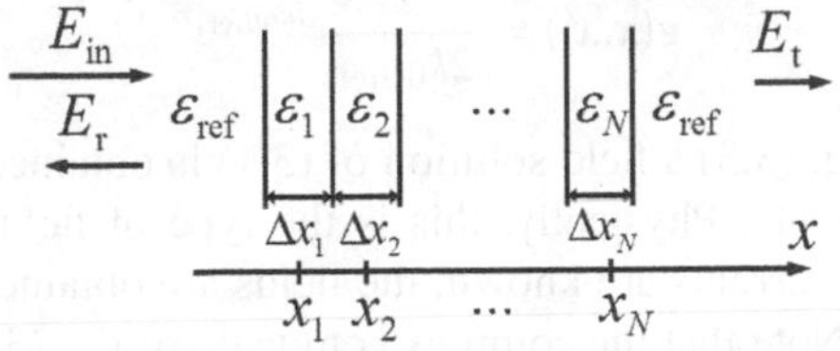

Figure 3.1 Schematic of one-dimensional scattering problem.

While the dielectric constant inside the scattering object may in principle vary continuously, in a numerical calculation, we shall discretize the scattering object into N layers or elements of widths Δx_i and dielectric constants ε_i, where $i = 1, 2, ..., N$. The physics of the situation is that the incident field will cause the scattering object to be polarized, which is equivalent to setting up polarization currents in the scatterer that generate new radiation being superposed onto the incident field. The current distribution in the scatterer can only generate radiation that propagates away from the scatterer. If the scatterer is divided into a large number of polarizable regions, then the polarization of each region will be driven by the external field at its position. This external field will consist of both the given incident field but also of the field generated at this position coming from all other polarizable elements.

3.1 GREEN'S FUNCTION INTEGRAL EQUATIONS

Imagine for a moment that the current distribution being set up in the scatterer is given by $J(x)$. The field E_{scat} generated by this current distribution must satisfy the following wave equation

$$\left(\frac{\partial^2}{\partial x^2} + k_0^2 \varepsilon_{\mathrm{ref}}\right) E_{\mathrm{scat}}(x) = -i\omega\mu_0 J(x). \tag{3.1}$$

This equation can readily be solved by using a solution $g(x,x')$ to the following equation

$$\left(\frac{\partial^2}{\partial x^2}+k_0^2\varepsilon_{\mathrm{ref}}\right)g(x,x')=-\delta(x-x'). \tag{3.2}$$

Such a solution is known as a *Green's function*. In terms of $g(x,x')$, a particular solution of Eq. (3.1) is given by the integral

$$E_{\mathrm{scat}}(x)=i\omega\mu_0\int g(x,x')J(x')dx'. \tag{3.3}$$

Now there are an infinite number of solutions $g(x,x')$ to Eq. (3.2) since any homogeneous solutions of Eq. (3.2) can be added to obtain another solution. However, only one of the solutions satisfies the radiating boundary condition, meaning that $g(x,x')$ will describe fields propagating away from x'. This solution is given by

$$g(x,x')=\frac{i}{2k_0n_{\mathrm{ref}}}e^{ik_0n_{\mathrm{ref}}|x-x'|}. \tag{3.4}$$

By using (3.4) in Eq. (3.3) a field solution of (3.1) is obtained which is propagating away from the currents. Physically, this is the type of field that the currents will generate, and if the currents are known, the fields are obtained straightforwardly by simple integration. Note that the complex conjugate of Eq. (3.4) is also a solution of Eq. (3.2) but inserted in Eq. (3.3) this would result in a solution of $E_{\mathrm{scat}}(x)$ with fields propagating toward the currents, which is not physically possible. We can now add the homogeneous solution of Eq. (3.1) that corresponds to the incident field resulting in the total field

$$E(x)=E_0(x)+E_{\mathrm{scat}}(x). \tag{3.5}$$

This procedure, however, only works if the currents are given, and at the moment we do not know the currents being set up in the scattering object. Therefore, we will return to the actual scattering problem.

The total field solution of the scattering problem must satisfy the wave equation

$$\left(\frac{\partial^2}{\partial x^2}+k_0^2\varepsilon(x)\right)E(x)=0, \tag{3.6}$$

where $\varepsilon(x)$ is the dielectric constant of the total structure including the scatterer.

The incident field E_0 is, on the other hand, a solution for the case where the scatterer is absent, i.e.,

$$\left(\frac{\partial^2}{\partial x^2}+k_0^2\varepsilon_{\mathrm{ref}}\right)E_0(x)=0. \tag{3.7}$$

The incident field solution corresponding to a field propagating along the x-axis and thus being incident on the scatterer from the left can be expressed as

$$E_0(x)=Ae^{ik_0n_{\mathrm{ref}}x}, \tag{3.8}$$

where A is a constant.

The expressions (3.6) and (3.7) can be rearranged into

$$\left(\frac{\partial^2}{\partial x^2}+k_0^2\varepsilon_{\text{ref}}\right)(E(x)-E_0(x))=-k_0^2\left(\varepsilon(x)-\varepsilon_{\text{ref}}\right)E(x). \tag{3.9}$$

Clearly, the right-hand side is now equivalent to $-i\omega\mu_0 J(x)$, and the result for the scattered field in terms of $J(x)$ can be used to finally arrive at the integral equation

$$E(x)=E_0(x)+\int g(x,x')k_0^2\left(\varepsilon(x')-\varepsilon_{\text{ref}}\right)E(x')dx'. \tag{3.10}$$

The total field is now expressed in a form satisfying the boundary condition that it should be in the form of the sum of the incident field and a scattered field propagating away from the scatterer. Here, however, we note that the field is itself inside the integral. The integral will only include the field at positions x where $(\varepsilon(x)-\varepsilon_{\text{ref}})\neq 0$. Thus, it is only necessary to know the field inside the scattering object (equivalent to knowing the currents J being set up in the scatterer). By inserting those fields in Eq. (3.10) the field at any other position x can be obtained.

3.2 NUMERICAL APPROACH

In order to solve Eq. (3.10) numerically we need to discretize the equation. Here, a simple strategy is chosen where the scatterer is discretized into N elements with the approximation that the field is constant within each element. This is a reasonable approximation if the cell width is small enough compared with the wavelength in the medium. The ith element has center position x_i, width Δx_i, and dielectric constant ε_i, and the field in element i is denoted E_i. We apply point matching at positions x_i. The integral equation (3.10) may now be approximated with the discretized equation

$$E_i=E_{0,i}+\sum_{j=1}^{N}A_{ij}E_j, \tag{3.11}$$

where $E_{0,i}=E_0(x_i)$, and

$$A_{ij}=g_{ij}k_0^2\left(\varepsilon_j-\varepsilon_{\text{ref}}\right)\Delta x_j \tag{3.12}$$

with

$$g_{ij}=g(x_i,x_j). \tag{3.13}$$

Since the Green's function does not have any singularities in the 1D case, this is an acceptable approximation.

The discretized integral equation may now be arranged in matrix form. First, the discrete field values are arranged in vectors

$$\overline{E}=\begin{bmatrix}E_1 & E_2 & \dots & E_N\end{bmatrix}^T, \tag{3.14}$$

$$\overline{E}_0=\begin{bmatrix}E_{0,1} & E_{0,2} & \dots & E_{0,N}\end{bmatrix}^T. \tag{3.15}$$

The unit matrix is introduced

$$\bar{\bar{I}} = \begin{bmatrix} 1 & 0 & 0 & \dots & 0 \\ 0 & 1 & 0 & \dots & 0 \\ \vdots & & & & \vdots \\ \vdots & \dots & \dots & \dots & \vdots \\ 0 & 0 & \dots & 0 & 1 \end{bmatrix}, \tag{3.16}$$

and the matrix with the coefficients A_{ij} is given by

$$\bar{\bar{A}} = \begin{bmatrix} A_{1,1} & A_{1,2} & \dots & A_{1,N} \\ A_{2,1} & A_{2,2} & \dots & A_{2,n} \\ \vdots & & & \vdots \\ \vdots & \dots & \dots & \vdots \\ A_{N,1} & A_{N,2} & \dots & A_{N,N} \end{bmatrix}. \tag{3.17}$$

The discretized integral equation can now be written as

$$\left(\bar{\bar{I}} - \bar{\bar{A}}\right)\bar{E} = \bar{E}_0, \tag{3.18}$$

and the discrete field values inside the scatterer are thus readily given by

$$\bar{E} = \left(\bar{\bar{I}} - \bar{\bar{A}}\right)^{-1}\bar{E}_0. \tag{3.19}$$

At this stage, the field inside the scatterer can already be plotted by simply plotting the values E_i as a function of the values x_i. Furthermore, the total field at any other position x outside the scatterer can be obtained from

$$E(x) \approx E_0(x) + \sum_{j=1}^{N} g(x,x_j)k_0^2(\varepsilon_j - \varepsilon_{\text{ref}})E_j\Delta x_j. \tag{3.20}$$

The reflectance can be calculated from

$$R = \left|\frac{E(x_-) - E_0(x_-)}{A}\right|^2, \tag{3.21}$$

where x_- is a position to the left of the scatterer, and the transmittance can be calculated from

$$T = \left|\frac{E(x_+)}{A}\right|^2, \tag{3.22}$$

where x_+ is a position to the right of the scatterer.

3.3 EXAMPLE OF A SIMPLE BARRIER

As a simple example, a barrier is chosen with dielectric constant $\varepsilon_b = n_b^2 = 2.32^2$ and width W = 0.5 μm in free space ($\varepsilon_{\text{ref}} = 1$). The problem is of the same type as one that was treated in Ref. [42]. The barrier is illuminated from the left with light of the wavelength $\lambda = 633$ nm.

The barrier is discretized into N elements of the same width $\Delta x_i = W/N$. The coefficients A_{ij} then assume the simple form

$$A_{ij} = g_{ij} k_0^2 (\varepsilon_b - 1) W/N. \tag{3.23}$$

A calculation of the field inside the barrier is shown in Figure 3.2. In Fig. 3.2(a) a

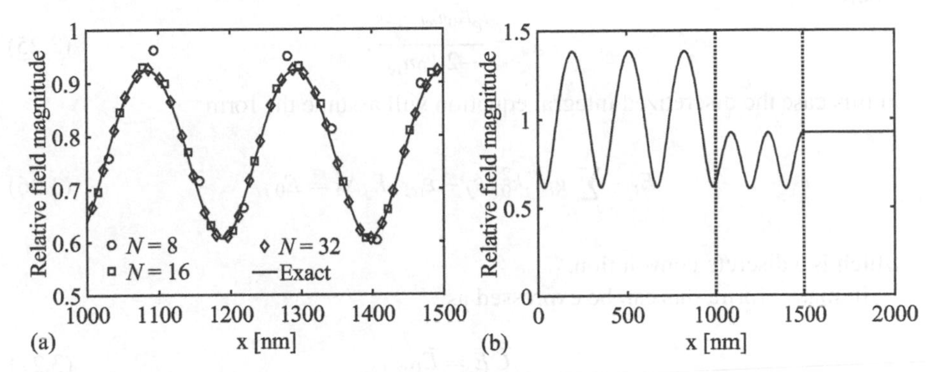

Figure 3.2 Magnitude of electric field inside and outside a barrier of refractive index $n_b = 2.32$ being illuminated from the left with light of wavelength 633 nm. (a) Field inside the barrier calculated with the GFIEM using 8, 16, and 32 elements. Also shown is the exact result. (b) Field both inside and outside barrier.

calculation of the field magnitude inside the barrier is shown using $N = 8$, 16 and 32. The ratio of barrier width to $\lambda/n_b \approx 1.8$. As a rule of thumb, at least 10 elements or sampling points are needed per wavelength in the medium. Here, this is equivalent to $N \geq 18$. It is seen that $N = 16$ already results in a reasonably good calculation of the fields. When using $N = 8$ there is a clear deviation from the exact result. In Fig. 3.2(b) the calculation of the field outside the barrier has been added. On the left-hand side of the barrier, interference fringes are seen as a result of interference between counter-propagating waves, namely between the forward propagating (along the positive x-axis) incident field, and the backward propagating scattered (or reflected) field. On the right-hand side, both incident field and scattered fields are forward propagating, and consequently no interference fringes are seen.

3.4 ITERATIVE FFT-BASED APPROACH FOR LARGE STRUCTURES

For small structures the procedure of solving Eq. (3.18) by matrix inversion is straightforward. However, in the case of a very large N then, while the number of unknowns in $\overline{E}$ is also N, the matrix $\overline{\overline{A}}$ will contain N^2 elements. Furthermore, the number of operations for inversion of the left-hand-side matrix in Eq. (3.18) by standard methods will scale as N^3. Thus, if N becomes very large, e.g., $N = 10^6$, then you may run out of computer memory, and the calculation time also becomes prohibitive.

Consider the situation where $\Delta x_i = \Delta x$ is the same for all elements. Then the discrete Green's function values will be of the form

$$g_{i,j} = g_{i-j}, \tag{3.24}$$

where

$$g_u = \frac{e^{ik_0 n_{\text{ref}}|u|\Delta x}}{-2ik_0 n_{\text{ref}}}. \tag{3.25}$$

In this case the discretized integral equation will assume the form

$$E_i - \sum_{j=1}^{N} g_{i-j} k_0^2 (\varepsilon_j - \varepsilon_{\text{ref}}) E_j \Delta x = E_{0,i}, \tag{3.26}$$

which is a discrete convolution.

In matrix form, this can be expressed as

$$\overline{\overline{C}}\,\overline{E} = \overline{E}_0, \tag{3.27}$$

where

$$\overline{\overline{C}} = \left(\overline{\overline{I}} - \overline{\overline{g}} k_0^2 \left(\overline{\overline{\varepsilon}} - \overline{\overline{I}} \varepsilon_{\text{ref}}\right) \Delta x\right) \tag{3.28}$$

with the matrices $\overline{\overline{g}}$ and $\overline{\overline{\varepsilon}}$ given by

$$\overline{\overline{g}} = \begin{bmatrix} g_0 & g_{-1} & \cdots & g_{1-N} \\ g_1 & g_0 & \cdots & g_{2-N} \\ \vdots & \cdots & \cdots & \vdots \\ g_{N-1} & g_{N-2} & \cdots & g_0 \end{bmatrix}, \quad \overline{\overline{\varepsilon}} = \begin{bmatrix} \varepsilon_1 & 0 & 0 & \cdots & 0 \\ 0 & \varepsilon_2 & 0 & \cdots & 0 \\ \vdots & \cdots & \cdots & \cdots & \vdots \\ 0 & 0 & \cdots & 0 & \varepsilon_N \end{bmatrix}. \tag{3.29}$$

Notice that as a result of the equidistant spacing, most of the elements of $\overline{\overline{g}}$ are redundant information. The matrix is a Toeplitz matrix, which is characterized by having the same element along diagonals, and the whole matrix can be reconstructed from only the upper row and the left column. This information can also be placed in the vector

$$\overline{g} = \begin{bmatrix} g_{1-N} & g_{2-N} & \cdots & g_{N-1} \end{bmatrix}^T. \tag{3.30}$$

The matrix problem can be solved by a number of iterative algorithms that avoid matrix inversion, but do require the calculation of many matrix-vector products. These methods also do not require the storage of full matrices. The conjugate gradient algorithm presented in Appendix B requires the calculation of both matrix-vector products such as $\overline{\overline{C}}\,\overline{y}$ and $\overline{\overline{C}}^\dagger\,\overline{y}$, where $\dagger$ represents Hermitian conjugation which is both conjugation and transposition. These matrix-vector products can be carried out with computer memory storage requirements scaling only as N and calculation times scaling only as $N \log N$. Clearly, the operation $\bar{y}_1 = \Delta x k_0^2 \bar{\bar{\varepsilon}} \bar{y}_0$ can be carried out in N steps with storage of only the diagonal elements of $\bar{\bar{\varepsilon}}$.

The calculation of $\bar{y}_2 = \bar{\bar{g}} \bar{y}_1$ can be carried out efficiently using the following procedure:

$$\bar{y}_2 = \bar{\bar{g}} \bar{y}_1 = \text{IFFT}_N \left(\text{FFT}\left(\bar{g}\right) .* \text{FFT}\left(\bar{y}_1\right) \right), \tag{3.31}$$

where FFT and IFFT are the Fast Fourier Transform and the corresponding inverse transform algorithm. We thus do not need to store $\bar{\bar{g}}$. It is only needed to store the vector $\bar{g}$. The N in IFFT_N represents that only N of the elements in the vector obtained after the IFFT should be used. In addition, prior to applying the FFT, the vector $\bar{y}_1$ must be extended by a number of zeroes (zero-padded) to assume the same length as $\bar{g}$. The operation .* represents element-wise multiplication such that

$$\bar{x} = \bar{a} .* \bar{b} \tag{3.32}$$

means that the ith element of vector $\bar{x}$ is given by $x_i = a_i b_i$. The calculation time of this procedure scales as $N \log N$, and the computer-memory requirements scale as N.

The other matrix-vector product can be expressed as

$$\bar{\bar{C}}^\dagger \bar{y} = \left(\bar{\bar{I}} - \left(\bar{\bar{\varepsilon}}^\dagger - \bar{\bar{I}} \varepsilon_{\text{ref}}^* \right) \bar{\bar{g}}^\dagger k_0^2 \Delta_x \right) \bar{y}. \tag{3.33}$$

Here * refers to complex conjugation. The matrix-vector product involving $\bar{\bar{g}}^\dagger$ can also be carried out using the FFT approach, and for this particular case it can be used that $\bar{\bar{g}}^\dagger = \bar{\bar{g}}^*$.

3.5 GUIDELINES FOR SOFTWARE IMPLEMENTATION

The structure of a GFIEM (1D) program may be based on carrying out the following steps:

Structure of GFIEM (1D) program

1. Define structure parameters, number of elements N, and wavelength λ.
2. Construct vectors with sampling points x_i and structure dielectric constant ε_i.
3. Construct a vector with incident field at sampling points: $E_{0,i} = \exp(ik_0 n_{\text{ref}} x_i)$
4. Construct the matrix $\bar{\bar{A}}$ in (3.17).
5. Solve the matrix equation (3.18).
6. Obtain the field $E(x)$ at a number of discrete positions before and after the barrier by using the result from Step 5 in Eq. (3.20).
7. Use the results from Steps 5 and 6 to plot the field.
8. Calculate reflectance and transmittance by using Eqs. (3.21) and (3.22).

Transmittance and reflectance spectra may be obtained by wrapping the whole procedure in a for-loop where the wavelength is varied.

3.6 EXERCISES

1. Prove that the Green's function in Eq. (3.4) is a solution of Eq. (3.2).
One approach is to consider the integral

$$\int_x \left(\frac{\partial^2}{\partial x^2} + k_0^2 \varepsilon_{\text{ref}} \right) g(x,x')dx. \tag{3.34}$$

Show that when $x \neq x'$ the integrand vanishes, and that consequently the integral is reduced to

$$\lim_{\delta \to 0} \int_{x=x'-\delta}^{x=x'+\delta} \frac{\partial^2}{\partial x^2} g(x,x')dx. \tag{3.35}$$

Carrying out the integration leads to

$$\left[\frac{\partial}{\partial x} g(x,x') \right]_{x=x'-\delta}^{x=x'+\delta}. \tag{3.36}$$

Evaluate this expression and show that (3.4) is a solution of Eq. (3.2).

2. Reproduce the results in Fig. 3.2 by following the steps outlined in Section 3.5.

3. Modify the procedure from Problem 2 to use the conjugate gradient method combined with the FFT and IFFT following Sec. 3.4 and Appendix B. In this case it is not necessary to construct the matrix $\overline{\overline{A}}$. Instead, construct the vector $\overline{g}$.

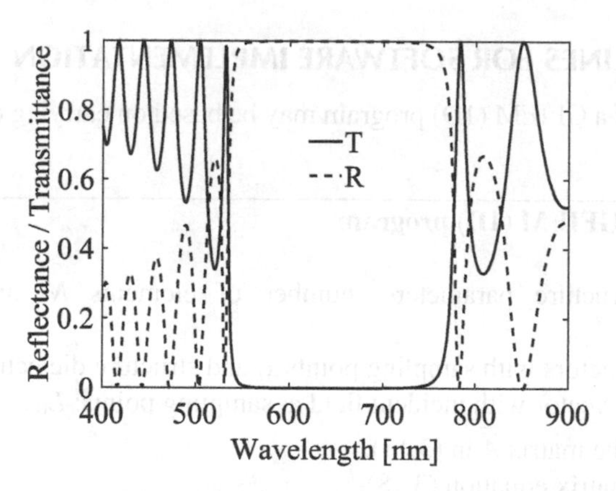

Figure 3.3 Reflectance and transmittance spectra for a multilayer structure with alternating 10 layers of refractive index 2.32 and thickness 633 nm / 4 / 2.32, and 10 layers of refractive index 1.38 and thickness 633 nm / 4 / 1.38. The calculation was made using 20 elements per layer ($N = 400$) and a wavelength resolution of 1 nm.

4. Consider a multilayer structure with 10 periods each consisting of a layer of refractive index $n_{\text{H}} = 2.32$ and thickness $a_{\text{H}} = \lambda_{\text{d}}/4n_{\text{H}}$, and a layer of refractive index

$n_L = 1.38$ and thickness $a_L = \lambda_d/4n_L$. Use the design wavelength $\lambda_d = 633$ nm. Calculate the reflectance and transmittance spectra for wavelengths from 400 nm to 900 nm. The solution is given in Fig. 3.3.

4 Surface integral equation method for 2D scattering problems

In this chapter, two-dimensional scattering problems are considered. Light will be restricted to propagation in the xy-plane, and the electric and magnetic fields, and the dielectric constant of the structure of interest, are invariant along the z-axis. For the position vector we will use

$$\mathbf{r} \equiv \hat{\mathbf{x}}x + \hat{\mathbf{y}}y, \tag{4.1}$$

and the position vector thus does not have a z-component. With these assumptions the solutions to Maxwell's equations separate into two cases, namely s and p polarization. For s polarization, the electric field only has a component along the z-axis, $\mathbf{E}(\mathbf{r}) = \hat{\mathbf{z}}E(\mathbf{r})$, which must satisfy the wave equation

$$\left(\nabla^2 + k_0^2\varepsilon(\mathbf{r})\right)E(\mathbf{r}) = 0, \quad (s-\text{polarization}). \tag{4.2}$$

Note that $\nabla \cdot \mathbf{E}(\mathbf{r}) = 0$ holds for all positions $\mathbf{r}$ for s polarization.

For p polarization, it is instead the magnetic field that only has a z-component, $\mathbf{H}(\mathbf{r}) = \hat{\mathbf{z}}H(\mathbf{r})$, which must satisfy the wave equation

$$\left(\nabla^2 + k_0^2\varepsilon(\mathbf{r})\right)H(\mathbf{r}) = \frac{\nabla\varepsilon(\mathbf{r}) \cdot \nabla H(\mathbf{r})}{\varepsilon(\mathbf{r})}, \quad (p-\text{polarization}). \tag{4.3}$$

We will only consider dielectric constants $\varepsilon(\mathbf{r})$ that are piecewise constant. Thus,

$$\varepsilon(\mathbf{r}) = \varepsilon_i \ \text{ for } \ \mathbf{r} \in \Omega_i, \ \ i = 1,2,3,\ldots, \tag{4.4}$$

where Ω_i is a region with dielectric constant ε_i. The right-hand side of (4.3) is thus only non-zero across interfaces between different media. The electric field for s polarization, and the magnetic field for p polarization, must thus satisfy the same wave equation within each region Ω_i. However, boundary conditions across interfaces are different.

In this chapter the general approach will be that the electric or magnetic fields at a positon $\mathbf{r}$ in a region Ω_i are expressed in terms of the field and its normal derivative at the surface of that region $\partial\Omega_i$ by using an integral equation. This is combined with boundary conditions across interfaces between different media, and by also applying periodic and / or radiating boundary conditions.

4.1 SCATTERER IN A HOMOGENEOUS MEDIUM

In this section we will consider the scattering from an object with dielectric constant ε_2 placed in a homogeneous reference medium with dielectric constant ε_1, and to begin with we will consider *p*-polarized light, and the object is illuminated by an incident field given by $\mathbf{H}_0(\mathbf{r}) = \hat{\mathbf{z}}H_0(\mathbf{r})$. The situation is illustrated in Fig. 4.1. The scattering object is bounded by the closed solid curve. In addition a number of surfaces and normal vectors are included in the figure for the following discussion.

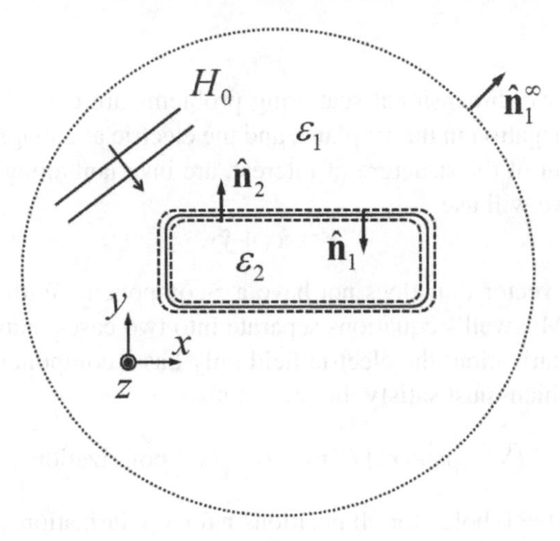

Figure 4.1 Schematic of two-dimensional scattering problem including surfaces and surface normal vectors used in the discussion of the two-dimensional GFSIEM.

4.1.1 GREEN'S FUNCTION INTEGRAL EQUATIONS

The magnetic field inside the scatterer (region Ω_2 with $\varepsilon = \varepsilon_2$) must satisfy the wave equation

$$\left(\nabla^2 + k_0^2\varepsilon_2\right) H(\mathbf{r}) = 0, \quad \mathbf{r} \in \Omega_2, \tag{4.5}$$

where $\nabla^2 = \frac{\partial^2}{\partial x^2} + \frac{\partial^2}{\partial y^2}$. In addition we shall use a Green's function g_2 being a solution of the equation

$$\left(\nabla^2 + k_0^2\varepsilon_2\right) g_2(\mathbf{r},\mathbf{r}') = -\delta(\mathbf{r}-\mathbf{r}'). \tag{4.6}$$

When H and g_2 satisfy these relations, it can be shown that the magnetic field at any position $\mathbf{r} \in \Omega_2$ is given by the following integral over the inner dashed curve C_2:

$$H(\mathbf{r}) = \oint_{C_2} \left\{ g_2(\mathbf{r},\mathbf{r}')\hat{\mathbf{n}}' \cdot \nabla' H(\mathbf{r}') - H(\mathbf{r}')\hat{\mathbf{n}}' \cdot \nabla' g_2(\mathbf{r},\mathbf{r}') \right\} dl', \quad \mathbf{r} \in \Omega_2, \tag{4.7}$$

where the normal vector $\hat{\mathbf{n}} = \hat{\mathbf{n}}_2$. This expression can be obtained by starting out with the integral on the right-hand side. Gauss's law is then used to convert the surface integral to an area integral over Ω_2, and afterward Eqs. (4.5) and (4.6) are used to rewrite the integrand into a product of the delta function and H, which then reduces to the left-hand side of Eq. (4.7).

The magnetic field outside the scatterer (region Ω_1 with $\varepsilon = \varepsilon_1$) must similarly satisfy the wave equation

$$\left(\nabla^2 + k_0^2 \varepsilon_1\right) H(\mathbf{r}) = 0, \quad \mathbf{r} \in \Omega_1. \tag{4.8}$$

In addition we shall use a Green's function g_1 that is a solution of the equation

$$\left(\nabla^2 + k_0^2 \varepsilon_1\right) g_1(\mathbf{r}, \mathbf{r}') = -\delta(\mathbf{r} - \mathbf{r}'). \tag{4.9}$$

The magnetic field at any position $\mathbf{r} \in \Omega_1$ is also similarly given by the following integral over the outer dashed curve C_1 and the dotted curve C_1^∞:

$$H(\mathbf{r}) = \oint_{C_1 + C_1^\infty} \left\{ g_1(\mathbf{r}, \mathbf{r}') \hat{\mathbf{n}}_1' \cdot \nabla' H(\mathbf{r}') - H(\mathbf{r}') \hat{\mathbf{n}}_1' \cdot \nabla' g_1(\mathbf{r}, \mathbf{r}') \right\} dl', \tag{4.10}$$

where $\hat{\mathbf{n}}_1$ is pointing out of Ω_1, which means that $\hat{\mathbf{n}}_1 = -\hat{\mathbf{n}} = -\hat{\mathbf{n}}_2$ when integrating over C_1, and $\hat{\mathbf{n}}_1 = \hat{\mathbf{n}}_1^\infty$ when integrating over C_1^∞. Here, it is assumed that the position $\mathbf{r}$ is located between the two curves C_1 and C_1^∞, which is no restriction since the curve C_1^∞ can be placed as far away from the scatterer as needed.

Outside the scatterer, the total field must be the sum of the incident field H_0 and a scattered field which is propagating away from the scatterer H_{scat}. This is the radiating boundary condition. At the curve C_1^∞ in the far field, the scattered field will locally propagate like a plane wave in the direction $\hat{\mathbf{r}} = \mathbf{r}/r$, where $r = \sqrt{x^2 + y^2}$. This means that the scattered field will be of the form

$$H_{\text{scat}}(r, \theta) \approx \frac{e^{ik_0 n_1 r}}{\sqrt{r}} f_H(\theta), \tag{4.11}$$

where we have used cylindrical coordinates ($\mathbf{r} = \hat{\mathbf{x}} r \cos\theta + \hat{\mathbf{y}} r \sin\theta$).

It is now advantageous to choose the following Green's function

$$g_1(\mathbf{r}, \mathbf{r}') = \frac{i}{4} H_0^{(1)}(k_0 n_1 |\mathbf{r} - \mathbf{r}'|), \tag{4.12}$$

where $H_0^{(1)}$ is the Hankel function of zero order and type 1, and the refractive index of medium u is given by $n_u = \sqrt{\varepsilon_u}$. This Green's function has the property that it describes a cylindrical wave propagating away from the position $\mathbf{r}'$ (see Appendix C). For positions $\mathbf{r}'$ in the far field being much farther away from the scatterer than $\mathbf{r}$, this Green's function will also locally resemble a plane wave and be of the form

$$g_1(\mathbf{r}, r', \theta') \approx \frac{e^{ik_0 n_1 r'}}{\sqrt{r'}} f_g(\theta', \mathbf{r}). \tag{4.13}$$

By inserting these far-field expressions for H_{scat} and g_1 into the integral

$$\oint_{C_1^\infty} \left\{ g_1(\mathbf{r}, \mathbf{r}') \hat{\mathbf{n}}_1' \cdot \nabla' H_{\text{scat}}(\mathbf{r}') - H_{\text{scat}}(\mathbf{r}') \hat{\mathbf{n}}_1' \cdot \nabla' g_1(\mathbf{r}, \mathbf{r}') \right\} dl' \tag{4.14}$$

it is found that the integrand, and therefore this integral, vanishes. In addition,

$$\oint_{C_1^\infty} \{g_1(\mathbf{r},\mathbf{r}')\hat{\mathbf{n}}_1' \cdot \nabla' H_0(\mathbf{r}') - H_0(\mathbf{r}')\hat{\mathbf{n}}_1' \cdot \nabla' g_1(\mathbf{r},\mathbf{r}')\}\, dl' = H_0(\mathbf{r}). \tag{4.15}$$

This leads finally to the following integral equation for the field outside the scatterer:

$$H(\mathbf{r}) = H_0(\mathbf{r}) - \oint_{C_1} \{g_1(\mathbf{r},\mathbf{r}')\hat{\mathbf{n}}' \cdot \nabla' H(\mathbf{r}') - H(\mathbf{r}')\hat{\mathbf{n}}' \cdot \nabla' g_1(\mathbf{r},\mathbf{r}')\}\, dl', \quad \mathbf{r} \in \Omega_1. \tag{4.16}$$

Note that g_2 can be any solution of Eq. (4.6) but for convenience we may choose

$$g_2(\mathbf{r},\mathbf{r}') = \frac{i}{4} H_0^{(1)}(k_0 n_2 |\mathbf{r}-\mathbf{r}'|). \tag{4.17}$$

If we now let the position $\mathbf{r}$ approach the surfaces C_1 and C_2 in Eqs. (4.7) and (4.16) from the side of the media Ω_1 and Ω_2, respectively, we will only be concerned with H and $\hat{\mathbf{n}} \cdot \nabla H$ on each of these surfaces. This amounts to having two equations with four unknowns. By supplementing with the boundary conditions (Sec. 2.1.1) that connect H and $\hat{\mathbf{n}} \cdot \nabla H$ on C_1 with H and $\hat{\mathbf{n}} \cdot \nabla H$ on C_2 we end up with two equations and two unknowns, which can be solved. For s polarization the electric field must satisfy the same wave equations in each region Ω_1 and Ω_2. The only difference therefore is that H should be replaced with E, and the electromagnetics boundary conditions connecting fields at C_1 and C_2 will be different.

For p polarization it is often the electric field which is of interest, and it can be obtained directly from the magnetic field by using

$$\mathbf{E}(\mathbf{r}) = \frac{i}{\omega\varepsilon_0\varepsilon(\mathbf{r})} \nabla \times \hat{\mathbf{z}} H(\mathbf{r}) = \frac{-i}{\omega\varepsilon_0\varepsilon(\mathbf{r})} \hat{\mathbf{z}} \times \nabla H(\mathbf{r}), \tag{4.18}$$

and when applying this to the expressions (4.16) and (4.7) it is found that

$$\mathbf{E}(\mathbf{r}) = -i \oint_{C_2} \left\{ \frac{\hat{\mathbf{z}} \times \nabla g_2(\mathbf{r},\mathbf{r}')}{\omega\varepsilon_0\varepsilon_2} \hat{\mathbf{n}}' \cdot \nabla' H(\mathbf{r}') \right.$$

$$\left. - H(\mathbf{r}') \frac{\hat{\mathbf{z}} \times \nabla \hat{\mathbf{n}}' \cdot \nabla' g_2(\mathbf{r},\mathbf{r}')}{\omega\varepsilon_0\varepsilon_2} \right\} dl', \quad \mathbf{r} \in \Omega_2, \tag{4.19}$$

$$\mathbf{E}(\mathbf{r}) = \mathbf{E}_0(\mathbf{r}) + i \oint_{C_1} \left\{ \frac{\hat{\mathbf{z}} \times \nabla g_1(\mathbf{r},\mathbf{r}')}{\omega\varepsilon_0\varepsilon_1} \hat{\mathbf{n}}' \cdot \nabla' H(\mathbf{r}') \right.$$

$$\left. - H(\mathbf{r}') \frac{\hat{\mathbf{z}} \times \nabla \hat{\mathbf{n}}' \cdot \nabla' g_1(\mathbf{r},\mathbf{r}')}{\omega\varepsilon_0\varepsilon_1} \right\} dl', \quad \mathbf{r} \in \Omega_1. \tag{4.20}$$

Alternatively, the electric field can also be obtained by numerical differentiation of H.

For the calculation of scattering and extinction cross sections (Sec. 2.3.1) the scattered far field is needed. Here it is convenient to use that for positions $\mathbf{r}$ in the far-field, and thus with the structure and $\mathbf{r}'$ much closer to the origin, the Green's function

g_1 can be written

$$g_1(\mathbf{r},\mathbf{r}') \approx g_1^{(\mathrm{ff})}(\mathbf{r},\mathbf{r}') = \frac{1}{4}\sqrt{\frac{2}{\pi k_0 n_1 r}} e^{ik_0 n_1 r} e^{i\pi/4} e^{-ik_0 n_1 \hat{\mathbf{r}}\cdot\mathbf{r}'}, \tag{4.21}$$

where $\hat{\mathbf{r}} = \mathbf{r}/r$. This follows from (C.11) and $|\mathbf{r}-\mathbf{r}'| \approx r - \hat{\mathbf{r}}\cdot\mathbf{r}'$.

The scattered far field is then obtained by inserting (4.21) into Eq. (4.16) leading to

$$H_{\mathrm{scat}}^{(\mathrm{ff})}(r,\theta) = -\frac{1}{4}\sqrt{\frac{2}{\pi k_0 n_1 r}} e^{i\pi/4} e^{ik_0 n_1 r} \times$$

$$\oint_{C_1} e^{-ik_0 n_1 \hat{\mathbf{r}}\cdot\mathbf{r}'} \left\{ \hat{\mathbf{n}}'\cdot\nabla' H(\mathbf{r}') - H(\mathbf{r}')\hat{\mathbf{n}}'\cdot(-ik_0 n_1 \hat{\mathbf{r}}) \right\} dl', \tag{4.22}$$

where the angle θ enters through $\hat{\mathbf{r}}$.

The integral equations and boundary conditions are summarized in the textbox:

Green's function surface integral equations (2D):
A scatterer with dielectric constant ε_2 placed in a homogeneous medium with dielectric constant ε_1 is illuminated with an incident *p*-polarized magnetic field $\mathbf{H}_0(\mathbf{r}) = \hat{\mathbf{z}} H_0(\mathbf{r})$ propagating in the *xy*-plane. The magnetic field at any position inside the scatterer is given by

$$H(\mathbf{r}) = \oint_{C_2} \left\{ g_2(\mathbf{r},\mathbf{r}')\hat{\mathbf{n}}'\cdot\nabla' H(\mathbf{r}') - H(\mathbf{r}')\hat{\mathbf{n}}'\cdot\nabla' g_2(\mathbf{r},\mathbf{r}') \right\} dl', \quad \mathbf{r}\in\Omega_2$$

where C_2 is a curve just inside the scatterer surface, Ω_2 is the region inside the scatterer, $\hat{\mathbf{n}}$ is the surface normal vector pointing out of the scatterer, and the Green's function is given by

$$g_u(\mathbf{r},\mathbf{r}') = \frac{i}{4} H_0^{(1)}(k_0 n_u |\mathbf{r}-\mathbf{r}'|).$$

The magnetic field at any position outside the scatterer is given by

$$H(\mathbf{r}) = H_0(\mathbf{r}) - \oint_{C_1} \left\{ g_1(\mathbf{r},\mathbf{r}')\hat{\mathbf{n}}'\cdot\nabla' H(\mathbf{r}') - H(\mathbf{r}')\hat{\mathbf{n}}'\cdot\nabla' g_1(\mathbf{r},\mathbf{r}') \right\} dl', \quad \mathbf{r}\in\Omega_1$$

where C_1 is a curve just outside the scatterer surface, and Ω_1 is the region outside the scatterer. The fields on C_1 and C_2 are related by the boundary conditions:

$$H_1 = H_2, \quad \left(\frac{1}{\varepsilon_1}\hat{\mathbf{n}}\cdot\nabla H_1\right) = \left(\frac{1}{\varepsilon_2}\hat{\mathbf{n}}\cdot\nabla H_2\right),$$

where the subscripts 1 and 2 refer to positions on C_1 and C_2, respectively, being just inside and just outside the scatterer surface. For *s* polarization the same equations apply except that H must be replaced by E and the boundary conditions across the scatterer surface are instead

$$E_1 = E_2, (\hat{\mathbf{n}}\cdot\nabla E_1) = (\hat{\mathbf{n}}\cdot\nabla E_2).$$

4.1.2 FINITE-ELEMENT-BASED DISCRETIZATION APPROACHES

We are now going to consider finite-element-based discretization approaches for representing the field and its normal derivative at the surfaces. For notational convenience we shall introduce

$$\phi_u(\mathbf{r}) = \hat{\mathbf{n}} \cdot \nabla H_u(\mathbf{r}) \tag{4.23}$$

to represent the normal derivative of the field at the surface C_u. The surface C_1 is illustrated in Fig. 4.2, where it has been divided into N sections of finite length. The

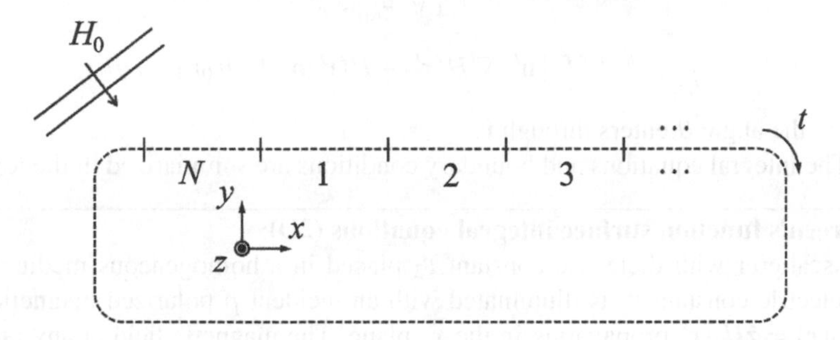

Figure 4.2 Schematic of discretization of surface into N sections of finite length.

position on the scatterer surface, $\mathbf{s}$, can be parameterized by a single parameter t. In this chapter it is chosen that $\mathbf{s}(t)$ will represent the position on the surface obtained when moving clockwise around the surface a distance t from the start of section 1. The start and end of section i will be identified by $t = t^{(s,i)}$ and $t = t^{(e,i)}$, respectively.

There are now different finite-element schemes for representing $H(\mathbf{s}(t))$ or $\phi(\mathbf{s}(t))$. The simplest is to assume that H or ϕ is a constant within each line section. This can expressed by using pulse-type finite elements given by

$$f_i^{(\mathrm{p})}(t) = \begin{cases} 1, & t^{(s,i)} < t < t^{(e,i)} \\ 0, & \text{otherwise} \end{cases}, \tag{4.24}$$

and being illustrated in Fig. 4.3(a). By using these elements, the fields along surface C_u may be approximated by

$$H_u(\mathbf{s}(t)) \approx \sum_{i=1}^{N} H_{u,i} f_i^{(\mathrm{p})}(t) \quad \text{and} \quad \phi_u(\mathbf{s}(t)) \approx \sum_{i=1}^{N} \phi_{u,i} f_i^{(\mathrm{p})}(t). \tag{4.25}$$

The approximation to the fields are thus governed by a finite number of coefficients $H_{u,i}$ and $\phi_{u,i}$. The pulse-type finite-elements lead to a stair-cased representation of fields, which is illustrated in Fig. 4.3(b).

The next-simplest finite-element scheme is to assume that H and ϕ vary linearly within each section. This can be expressed by using the following two linearly varying finite elements that are equal to 1 at one end of the line section and equal to 0 at

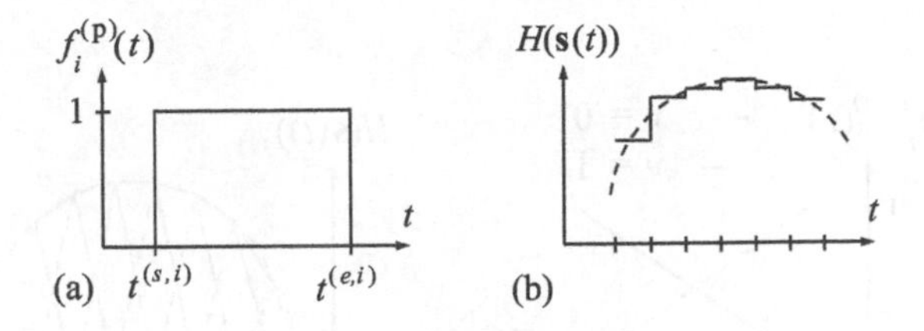

Figure 4.3 Illustration of 0-order finite elements or pulse elements.

the other end:

$$f_i^{(L,v)}(t) = \begin{cases} 1-\left(t-t^{(s,i)}\right)/\left(t^{(e,i)}-t^{(s,i)}\right), & t^{(s,i)}<t<t^{(e,i)},\ v=0 \\ \left(t-t^{(s,i)}\right)/\left(t^{(e,i)}-t^{(s,i)}\right), & t^{(s,i)}<t<t^{(e,i)},\ v=1 \\ 0, & \text{otherwise} \end{cases}. \tag{4.26}$$

These elements are illustrated in Fig. 4.4(a).

The fields along surface C_u may now be expressed as

$$H_u(\mathbf{s}(t)) \approx \sum_{i=1}^{N}\sum_{v=0}^{1} H_{u,i,v} f_i^{(L,v)}(t) \quad \text{and} \quad \phi_u(\mathbf{s}(t)) \approx \sum_{i=1}^{N}\sum_{v=0}^{1} \phi_{u,i,v} f_i^{(L,v)}(t), \tag{4.27}$$

which is a piece-wise linear approximation as illustrated in Fig. 4.4(b).

Continuity of the magnetic field along the boundary can be enforced by the condition

$$H_{u,i,0} = H_{u,j,1} \quad \text{when} \quad \mathbf{s}(t^{(e,j)}) = \mathbf{s}(t^{(s,i)}). \tag{4.28}$$

This condition will be automatically enforced if point-matching is applied at the ends of line sections, and then the number of unknown coefficients in the field representation (4.27) will be no greater than the number of unknown coefficients in (4.25). Now the fields can be continuous along the boundary but the tangential derivative will, however, jump from one section to the next. It can be noted that the normal component of the electric field $\hat{\mathbf{n}}\cdot\mathbf{E}$ is proportional to the tangential derivative of H, and for a smooth boundary the tangential derivative of H should thus in principle not jump. If another method than point-matching is used, we may still ensure continuity of fields along the boundary by instead using the so-called hat functions defined by

$$f_i^{(LH)}(t) = \begin{cases} \left(t-t^{(s,j)}\right)/\left(t^{(e,j)}-t^{(s,j)}\right), & t^{(s,j)}<t<t^{(e,j)},\ \mathbf{s}(t^{(e,j)})=\mathbf{s}(t^{(s,i)}) \\ 1-\left(t-t^{(s,i)}\right)/\left(t^{(e,i)}-t^{(s,i)}\right), & t^{(s,i)}<t<t^{(e,i)} \\ 0, & \text{otherwise} \end{cases}, \tag{4.29}$$

and illustrated in Fig. 4.4(c). In terms of these finite elements the magnetic field can be expressed as

$$H_u(\mathbf{s}(t)) \approx \sum_{i=1}^{N} H_{u,i} f_i^{(LH)}(t). \tag{4.30}$$

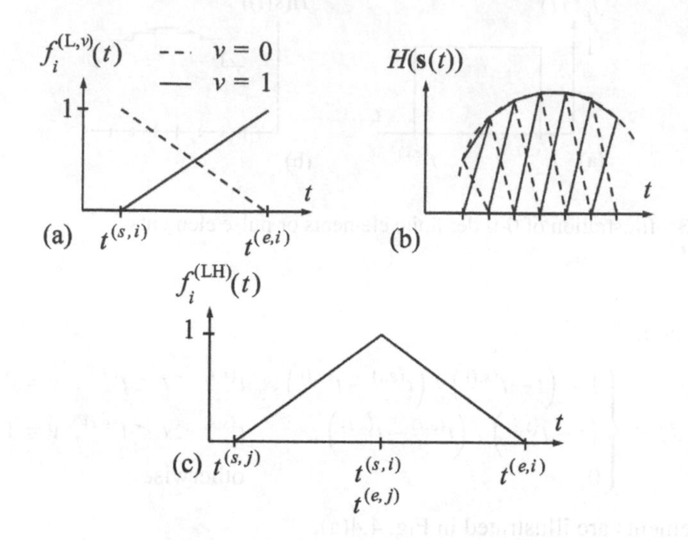

Figure 4.4 Illustration of 1st-order finite elements.

It is also possible to use the hat functions for ϕ if the boundary is smooth. However, if the boundary, for example, has a 360-degree corner between two neighbor sections of the boundary, then ϕ should change sign across this corner, while H should still be continuous. This is a consequence of the surface normal vector at the two sides of the corner having opposite sign. This situation is illustrated in Fig. 4.5, and in that case the hat functions are not applicable. With Eq. (4.27) we may, in such a situation, use

$$\phi_{u,i,0} = -\phi_{u,j,1} \quad \text{when} \quad \mathbf{s}(t^{(e,j)}) = \mathbf{s}(t^{(s,i)}) \quad \text{and} \quad \alpha = 360^\circ. \tag{4.31}$$

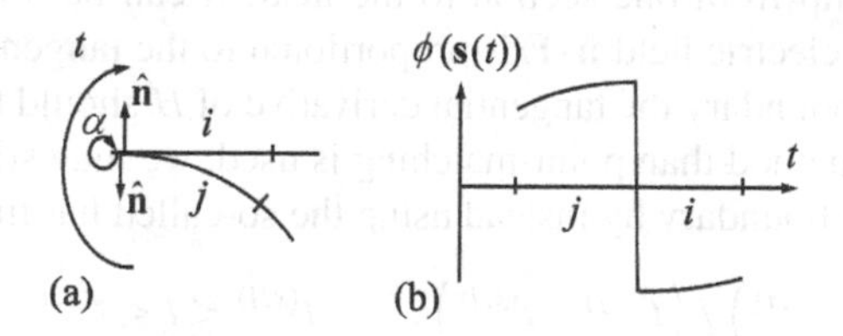

Figure 4.5 Illustration of boundary condition for ϕ at a $\alpha = 360^\circ$ corner between neighbor line sections.

The finite elements considered so far are lower-order polynomial functions. Higher-order polynomial finite elements of order m can be obtained by having $m+1$ sampling points within each element, where two of the points are at the ends of the element. If we denote the sampling points x_ν, $\nu = 0, 1, \ldots, m$, and if we order them such that $x_0 < x_1 < x_2 \ldots < x_m$, and if we furthermore require that the polynomial functions must equal 1 at one of the sampling points, and must equal zero at all other sampling points, then the finite elements can be constructed as follows:

$$f^{(m,v)}(x) = \frac{\prod_{k=0,k\neq v}^{m}(x_k - x)}{\prod_{k=0,k\neq v}^{m}(x_k - x_v)}, \quad v = 0,1,2,\ldots,m, \quad x_0 \leq x \leq x_m. \tag{4.32}$$

Examples for quadratic ($m = 2$) and cubic ($m = 3$) elements are shown in Figs. 4.6(a) and (b), respectively. The elements shown in Fig. 4.6 are defined for an interval from

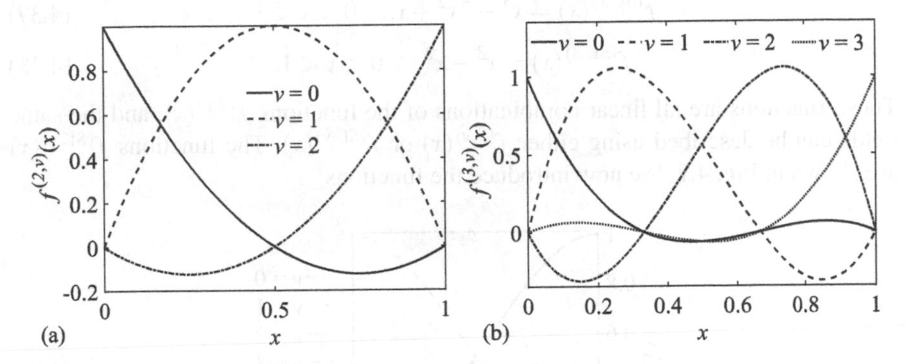

Figure 4.6 (a) Finite elements of order $m = 2$ with $x_0 = 0$, $x_1 = 0.5$, $x_2 = 1$, and (b) of order $m = 3$ with $x_0 = 0$, $x_1 = 1/3$, $x_2 = 2/3$, and $x_3 = 1$.

$x_0 = 0$ to $x_m = 1$ but can straightforwardly be adapted to another interval corresponding to section i on the boundary, by using

$$f_i^{(m,v)}(t) = \begin{cases} f^{(m,v)}\left(\frac{t-t^{(s,i)}}{t^{(e,i)}-t^{(s,i)}}\right), & t^{(s,i)} < t < t^{(e,i)} \\ 0, & \text{otherwise} \end{cases}. \tag{4.33}$$

When using the higher-order elements the fields along surface C_u may now be expressed as

$$H_u(\mathbf{s}(t)) \approx \sum_{i=1}^{N}\sum_{v=0}^{m} H_{u,i,v} f_i^{(m,v)}(t) \quad \text{and} \quad \phi_u(\mathbf{s}(t)) \approx \sum_{i=1}^{N}\sum_{v=0}^{m} \phi_{u,i,v} f_i^{(m,v)}(t). \tag{4.34}$$

The functions (4.34) are highly suitable for a point-matching scheme since the coefficients $H_{u,i,v}$ and $\phi_{u,i,v}$ directly correspond to the value of H or ϕ at the sampling points.

For a smooth surface with no sharp corners, both the field and its tangential derivative must be continuous along the boundary. Continuity can be ensured when using elements of order $m = 1$ or higher. However, when matching the value of fields at specific points it cannot be ensured that the tangential derivative of the field along the surface will be continuous from one section to the next. Instead of matching the fields at specific points it is also possible to match the fields and the slopes of the fields at end points of elements. However, this will require at least elements of order $m = 3$, since four parameters must be controlled for each section i. It may then for $m = 3$ be preferable to use the following 4 finite elements suitable for controlling the value of fields at end points, and the value of the tangential derivative at end points:

$$f^{(\mathrm{SL},0)}(x) = 1 - 3x^2 + 2x^3, \quad 0 < x < 1, \tag{4.35}$$

$$f^{(\mathrm{SL},1)}(x) = 3x^2 - 2x^3, \quad 0 < x < 1, \tag{4.36}$$

$$f^{(\mathrm{SL},2)}(x) = x^3 - 2x^2 + x, \quad 0 < x < 1, \tag{4.37}$$

$$f^{(\mathrm{SL},3)}(x) = x^3 - x^2, \quad 0 < x < 1. \tag{4.38}$$

These functions are all linear combinations of the functions $f^{(3,\nu)}(x)$, and the same fields can be described using either $f^{(3,\nu)}(x)$ or $f^{(\mathrm{SL},\nu)}(x)$. The functions $f^{(\mathrm{SL},\nu)}(x)$ are shown in Fig. 4.7. We now introduce the functions

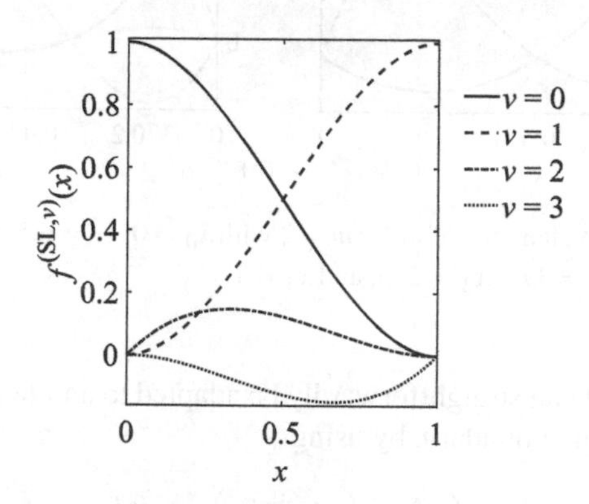

Figure 4.7 Finite elements suitable for point-matching of either the value of the field, or the slope of the field, at end points of a line section.

$$f_i^{(\mathrm{SL},\nu)}(t) = \begin{cases} f^{(\mathrm{SL},\nu)}\left(\frac{t - t^{(s,i)}}{|t^{(e,i)} - t^{(s,i)}|}\right), & t^{(s,i)} < t < t^{(e,i)}, \\ 0, & \text{otherwise} \end{cases}, \nu = 0, 1, \tag{4.39}$$

$$f_i^{(\mathrm{SL},\nu)}(t) = \begin{cases} |t^{(e,i)} - t^{(s,i)}| f^{(\mathrm{SL},\nu)}\left(\frac{t - t^{(s,i)}}{|t^{(e,i)} - t^{(s,i)}|}\right), & t^{(s,i)} < t < t^{(e,i)}, \\ 0, & \text{otherwise} \end{cases}, \nu = 2, 3. \tag{4.40}$$

When the fields are described by

$$H_u(\mathbf{s}(t)) \approx \sum_{i=1}^{N} \sum_{\nu=0}^{3} H_{u,i,\nu} f_i^{(\mathrm{SL},\nu)}(t) \quad \text{and} \quad \phi_u(\mathbf{s}(t)) \approx \sum_{i=1}^{N} \sum_{\nu=0}^{3} \phi_{u,i,\nu} f_i^{(\mathrm{SL},\nu)}(t), \tag{4.41}$$

then the coefficients $H_{u,i,v}$ and $\phi_{u,i,v}$ can be directly interpreted as either the value of the field at one of the ends of section i, or the value of the tangential slope at one of those ends.

If we combine the functions $f_i^{(\mathrm{SL},0)}$ and $f_j^{(\mathrm{SL},1)}$ for two neighbor sections i and j of equal length we can construct the function $f_i^{(\mathrm{CH},0)}$ shown in Fig. 4.8(a). If we combine the two functions $f_i^{(\mathrm{SL},2)}$ and $f_j^{(\mathrm{SL},3)}$ we can obtain the function $f_i^{(\mathrm{CH},1)}$ shown in Fig. 4.8(b). These are the higher-order equivalents of the hat function from the linear case, which are applicable to a smooth scatterer surface, i.e., a surface having no sharp corners. By multiplying the function in Fig. 4.8(a) with the value of the field at position $\mathbf{s}(t^{(s,i)})$, and further adding the function in Fig. 4.8(b) multiplied with the tangential derivative of the field, we have constructed a function that has the correct field value and tangential derivative at the sampling point $\mathbf{s}(t^{(s,i)})$. By further adding similar functions corresponding to all other sampling points we obtain a description of the field with a match of field values and tangential derivatives at N sampling points. An example of the construction of the higher-order hat functions for the case of neighbor sections of different lengths is shown in Figs. 4.8(c) and (d).

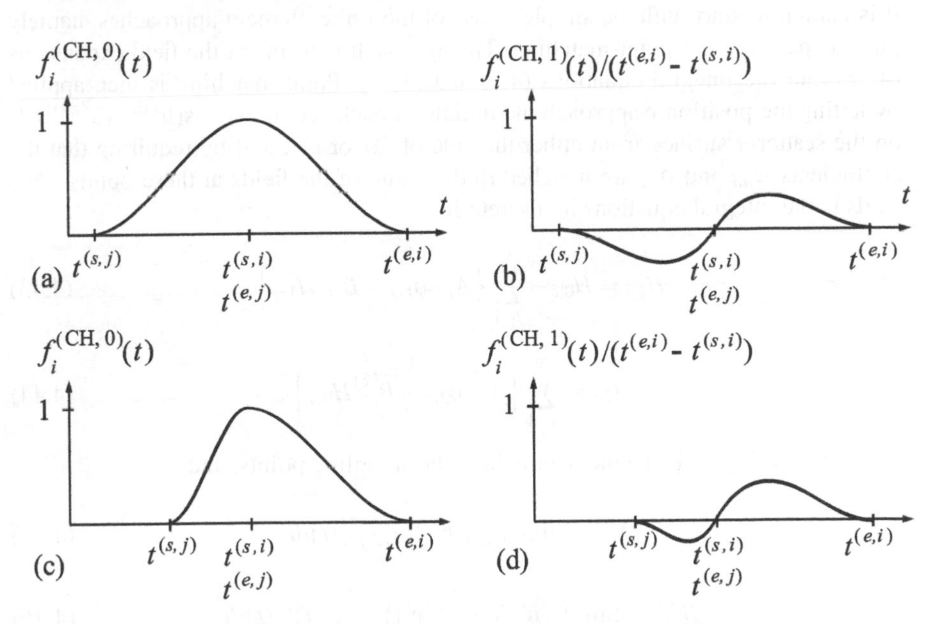

Figure 4.8 Alternative 3rd-order finite elements for (a) matching of the value of the field at the point of intersection between sections i and j, and (b) matching of the slope of the field at the same position. (c) and (d) are similar to (a) and (b) except that the neighbor sections have different lengths.

With the new functions $f_i^{(\mathrm{CH},v)}(t)$ the fields can be expressed by

$$H_u(\mathbf{s}(t)) \approx \sum_{i=1}^{N}\sum_{v=0}^{1} H_{u,i,v} f_i^{(\mathrm{CH},v)}(t) \quad \text{and} \quad \phi_u(\mathbf{s}(t)) \approx \sum_{i=1}^{N}\sum_{v=0}^{1} \phi_{u,i,v} f_i^{(\mathrm{CH},v)}(t), \quad (4.42)$$

where it should be noted that $f_i^{(\mathrm{CH},v)}$ is non-zero both over section i and the section that is a neighbor to the start of section i. $H_{u,i,v}$ and $\phi_{u,i,v}$ may be interpreted as the v-th-order tangential derivative of H and ϕ, respectively, at position $\mathbf{s}_i^{(s)}$. It turns out that the expressions (4.42) are more practical than those in (4.41) when discretizing the equations. The problems with the expansion (4.41) are considered in the exercises.

We can now insert one of the field approximations (4.25), (4.27), (4.34) or (4.42) into the integral expressions (4.7) and (4.16), and apply point-matching or other approaches for obtaining equations for the field coefficients. This will be investigated in the following subsections.

4.1.3 PULSE EXPANSION AND POINT-MATCHING

It is natural to start with the simplest case of the finite-element approaches, namely pulse expansion and point-matching. The approach is to insert the field expansions (4.25) into the integral equations (4.7) and (4.16). Point-matching is then applied by letting the position $\mathbf{r}$ approach the middle of each section $\mathbf{s}_i \equiv \mathbf{s}([t^{(s,i)} + t^{(e,i)}]/2)$ on the scatterer surface from either the side of Ω_1 or Ω_2, and by requiring that the coefficients $H_{u,i}$ and $\phi_{u,i}$ are matched to the value of the fields at these points. This leads to the integral equations in discrete form:

$$H_{1,i} = H_{0,i} - \sum_{j=1}^{N} \left\{ A_{ij}^{(1)} \phi_{1,j} - B_{ij}^{(1)} H_{1,j} \right\}, \quad (4.43)$$

$$H_{2,i} = \sum_{j=1}^{N} \left\{ A_{ij}^{(2)} \phi_{2,j} - B_{ij}^{(2)} H_{2,j} \right\}, \quad (4.44)$$

where $H_{0,i} = H_0(\mathbf{s}_i)$ is the incident field at the sampling points, and

$$A_{ij}^{(u)} = \lim_{\mathbf{r}\to\mathbf{s}_i} \int g_u(\mathbf{r},\mathbf{s}(t)) f_j^{(\mathrm{p})}(t)dt, \quad (4.45)$$

$$B_{ij}^{(u)} = \lim_{\mathbf{r}\to\mathbf{s}_i} \int \left(\hat{\mathbf{n}}' \cdot \nabla' g_u(\mathbf{r},\mathbf{r}')\right)_{\mathbf{r}'=\mathbf{s}(t)} f_j^{(\mathrm{p})}(t)dt, \quad (4.46)$$

are matrix elements of matrices that we will refer to as $\overline{\overline{A}}^{(u)}$ and $\overline{\overline{B}}^{(u)}$. It is understood that the limiting value $\mathbf{r} \to \mathbf{s}_i$ is to be taken from the side of Ω_u.

It is convenient to introduce the vectors

$$\overline{H}_u = \begin{bmatrix} H_{u,1} & H_{u,2} & \dots & H_{u,N} \end{bmatrix}^T, \quad (4.47)$$

$$\overline{\phi}_u = \begin{bmatrix} \phi_{u,1} & \phi_{u,2} & \dots & \phi_{u,N} \end{bmatrix}^T, \quad (4.48)$$

representing the field and its normal derivative on the surface C_u. The discretized integral equations can then be written in matrix form:

$$\left(\overline{\overline{I}}-\overline{\overline{B}}^{(1)}\right)\overline{H}_1+\overline{\overline{A}}^{(1)}\overline{\phi}_1=\overline{H}_0, \tag{4.49}$$

$$\left(\overline{\overline{I}}+\overline{\overline{B}}^{(2)}\right)\overline{H}_2-\overline{\overline{A}}^{(2)}\overline{\phi}_2=\overline{0}. \tag{4.50}$$

This represents two matrix equations with four vector unknowns.

By further using the electromagnetics boundary conditions connecting the fields on C_1 and C_2

$$\overline{H}_1=\overline{H}_2\equiv\overline{H} \quad \text{and} \quad \frac{1}{\varepsilon_1}\overline{\phi}_1=\frac{1}{\varepsilon_2}\overline{\phi}_2, \tag{4.51}$$

the matrix equation that can be solved by straightforward matrix inversion is obtained

$$\begin{bmatrix}\left(\overline{\overline{I}}-\overline{\overline{B}}^{(1)}\right) & \overline{\overline{A}}^{(1)}\\ \left(\overline{\overline{I}}+\overline{\overline{B}}^{(2)}\right) & -\frac{\varepsilon_2}{\varepsilon_1}\overline{\overline{A}}^{(2)}\end{bmatrix}\begin{bmatrix}\overline{H}\\ \overline{\phi}_1\end{bmatrix}=\begin{bmatrix}\overline{H}_0\\ \overline{0}\end{bmatrix}. \tag{4.52}$$

By solving this equation we find the coefficients of the field expansions (4.25) at the surface C_1. The electromagnetics boundary conditions are then used to obtain the corresponding fields at C_2, and by inserting the fields at the scatterer surface into the integral equations (4.7) and (4.16), the fields can be calculated at all other positions.

It is necessary to discuss a bit further the calculation of the matrix elements $A_{ij}^{(u)}$ and $B_{ij}^{(u)}$ since the involved Green's function has a singularity. When the limit $\mathbf{r}\to\mathbf{s}_i$ is taken in the expressions (4.45) and (4.46) then for the case of $i\neq j$ this simply amounts to setting $\mathbf{r}=\mathbf{s}_i$. However, in the case of $i=j$ it is necessary to take this limit. The Green's function itself only has a logarithmic singularity, and due to the weakness of such a singularity it is allowed, when calculating $A_{ii}^{(u)}$, to use $\mathbf{r}=\mathbf{s}_i$ and simply exclude the part of the integration right at the singularity. Thus, we find

$$A_{ij}^{(u)}=P\int g_u(\mathbf{s}_i,\mathbf{s}(t))f_j^{(\mathrm{p})}(t)dt, \tag{4.53}$$

where P indicates that this is the principal value integral, where, in the event that $\mathbf{s}(t)=\mathbf{s}_i$ for some t, this value of t is excluded from the integration. The contribution from the excluded part of the surface when taking the limit $\mathbf{r}\to\mathbf{s}_i$ will itself have a limiting value of 0. For $i=j$ we are still left with having to integrate near the logarithmic singularity. For an element described as a single straight line, or two straight lines of equal length connected at a corner, and length $\Delta_i=t^{(e,i)}-t^{(s,i)}$ with sampling in the center of the element, we find

$$A_{ii}^{(u)}=P\int_{x'=x-\Delta_i/2}^{x'=x+\Delta_i/2}\frac{i}{4}H_0^{(1)}(k_0n_u|x-x'|)dx'. \tag{4.54}$$

If we now use an approximation to the Hankel function in the case of small arguments (see Appendix C) this reduces to

$$A_{ii}^{(u)} \approx 2 \lim_{\varepsilon \to 0} \int_{x'=x+\varepsilon}^{x+\Delta_i/2} \frac{i}{4} \left(1 + i\left(\frac{2}{\pi}\left(\log(k_0 n_u |x - x'|/2) + \gamma\right)\right)\right) dx', \tag{4.55}$$

where $\gamma = 0.5772156...$ is Euler's constant. This integral can be calculated analytically resulting in

$$A_{ii}^{(u)} \approx \Delta_i \left(\frac{i}{4}\left(1 + i\frac{2}{\pi}\gamma\right) - \frac{1}{2\pi}\left(\log(k_0 n_u \Delta_i/4) - 1\right) \right). \tag{4.56}$$

A precise calculation of $A_{ii}^{(u)}$ that can also be applied for curved elements can be obtained by using

$$\log(k_0 n_u |\mathbf{s}(t) - \mathbf{s}_i|)dt = d\left(\log(k_0 n_u |\mathbf{s}(t) - \mathbf{s}_i|)(t - t^{(i)})\right) - \frac{t - t^{(i)}}{|\mathbf{s}(t) - \mathbf{s}_i|} \frac{\mathbf{s}(t) - \mathbf{s}_i}{|\mathbf{s}(t) - \mathbf{s}_i|} \cdot \hat{\mathbf{t}}dt, \tag{4.57}$$

where $t^{(i)} \equiv [t^{(s,i)} + t^{(e,i)}]/2$, and

$$\hat{\mathbf{t}}dt \equiv d\mathbf{s}(t). \tag{4.58}$$

Thus, $\hat{\mathbf{t}}$ is the surface tangential unit vector in the direction of increasing t. These considerations suggest the following expression for the $A_{ii}^{(u)}$ coefficients:

$$\begin{aligned} A_{ii}^{(u)} = & \int \left(\frac{i}{4} H_0^{(1)}(k_0 n_u |\mathbf{s}(t) - \mathbf{s}_i|) + \frac{1}{2\pi}\log(k_0 n_u |\mathbf{s}(t) - \mathbf{s}_i|) \right) f_i^{(\mathrm{p})}(t)dt \\ & - \frac{1}{2\pi}\left(\log(k_0 n_u |\mathbf{s}(t^{(e,i)}) - \mathbf{s}_i|) + \log(k_0 n_u |\mathbf{s}(t^{s,i}) - \mathbf{s}_i|) \right) \Delta_i/2 \\ & + \frac{1}{2\pi} \int \frac{t - t^{(j)}}{|\mathbf{s}(t) - \mathbf{s}_i|} \frac{\mathbf{s}(t) - \mathbf{s}_i}{|\mathbf{s}(t) - \mathbf{s}_i|} \cdot \hat{\mathbf{t}} f_i^{(\mathrm{p})}(t)dt. \end{aligned} \tag{4.59}$$

Note that the integrands in Eq. (4.59) will not vanish in the limit when $\mathbf{s}(t) = \mathbf{s}_i$. In that case

$$\lim_{\mathbf{s}(t) \to \mathbf{s}_i} \left(\frac{i}{4} H_0^{(1)}(k_0 n_u |\mathbf{s}(t) - \mathbf{s}_i|) + \frac{1}{2\pi}\log(k_0 n_u |\mathbf{s}(t) - \mathbf{s}_i|) \right)$$

$$= \frac{i}{4}\left(1 + i\frac{2}{\pi}\gamma\right) + \frac{1}{2\pi}\log(2), \tag{4.60}$$

and

$$\lim_{t \to t^{(i)}} \frac{t - t^{(i)}}{|\mathbf{s}(t) - \mathbf{s}_i|} \frac{\mathbf{s}(t) - \mathbf{s}_i}{|\mathbf{s}(t) - \mathbf{s}_i|} \cdot \hat{\mathbf{t}} = 1. \tag{4.61}$$

In the case of calculating the $B_{ii}^{(u)}$ coefficients using (4.46) the singularity of $(\hat{\mathbf{n}}' \cdot \nabla' g_u(\mathbf{r},\mathbf{r}'))_{\mathbf{r}'=\mathbf{s}(t)}$ is much stronger. For small distances between $\mathbf{r}$ and $\mathbf{r}'$ the normal derivative of the Green's function can be approximated by

$$(\hat{\mathbf{n}}' \cdot \nabla' g_u(\mathbf{r},\mathbf{r}'))_{\mathbf{r}'=\mathbf{s}(t)} \approx \frac{-1}{2\pi} \frac{\mathbf{s}(t)-\mathbf{r}}{|\mathbf{s}(t)-\mathbf{r}|^2} \cdot \hat{\mathbf{n}}(t), \tag{4.62}$$

where $\hat{\mathbf{n}}(t)$ is the surface normal vector at the position $\mathbf{s}(t)$. The situation is illustrated in the left part of Fig. 4.9. The infinitesimal angle $d\theta$ in the figure is given by

$$d\theta = -\frac{\mathbf{s}(t)-\mathbf{r}}{|\mathbf{s}(t)-\mathbf{r}|^2} \cdot \hat{\mathbf{n}}(t)dt. \tag{4.63}$$

Thus, in the limit of $\mathbf{r} \to \mathbf{s}_i$, where $\mathbf{s}_i$ is approached from the side of Ω_1 with $\varepsilon = \varepsilon_1$, the infinitesimal part of the surface around $\mathbf{s}_i$ will contribute $\theta_i/2\pi$ to the integral, where θ_i is the corner angle shown in Fig. 4.9 on the inside of the corner. If the position $\mathbf{s}_i$ is approached instead from the side with $\varepsilon = \varepsilon_2$ the contribution will instead be $-(2\pi - \theta_i)/2\pi$. This means

$$B_{ij}^{(u)} = P\int (\hat{\mathbf{n}}' \cdot \nabla' g_u(\mathbf{s}_i,\mathbf{r}'))_{\mathbf{r}'=\mathbf{s}(t)} f_j^{(\mathrm{p})}(t)dt + \delta_{ij}\left(\delta_{u,1}\frac{\theta_i}{2\pi} - \delta_{u,2}\frac{2\pi-\theta_i}{2\pi}\right). \tag{4.64}$$

Again, P refers to the principal value integral, meaning that the singularity point, where $\mathbf{s}(t) = \mathbf{s}_i$, must be excluded from the integration. The contribution from this part of the integral is given explicitly by the last terms. In the case of structures with no sharp corners we have $\theta_i = \pi$ for all i. For the case of an element described as a single straight line, the above expression even reduces to

$$B_{ii}^{(u)} = \frac{1}{2}(\delta_{u,1} - \delta_{u,2}). \tag{4.65}$$

The integral in (4.64) does not contribute to $B_{ii}^{(u)}$ for a straight line section i, or in the case of sampling at a corner between two straight lines, because, except at the excluded part of the integral, the normal vector $\hat{\mathbf{n}}'$ will be perpendicular to $(\nabla' g_u(\mathbf{s}_i,\mathbf{r}'))_{\mathbf{r}'=\mathbf{s}(t)}$. Therefore, even though $(\nabla' g_u(\mathbf{s}_i,\mathbf{r}'))_{\mathbf{r}'=\mathbf{s}(t)}$ is singular, and becomes very large in magnitude as $\mathbf{s}(t) \to \mathbf{s}_i$, the principal value integral in (4.64) is quite unproblematic from a numerical point of view.

A crude approximation to the matrix elements can be acceptable in many cases. However, an accurate calculation may very significantly improve the accuracy of a calculation. This will be illustrated with an example in the rest of this subsection. We will consider scattering of a p-polarized plane wave by a circular gold nanocylinder of radius 50 nm in water ($n_2 = 1.33$). The refractive index of gold is obtained from [40]. The advantage of this structure is that the exact analytic results are available for comparison (see Appendix D). The structure has a smooth surface, and a simple and crude analytic approximation to the matrix elements is then given by using (4.56) and (4.65) when $i = j$, and for $i \neq j$ the matrix elements can be approximated by just replacing the integrand in their expressions with a constant determined

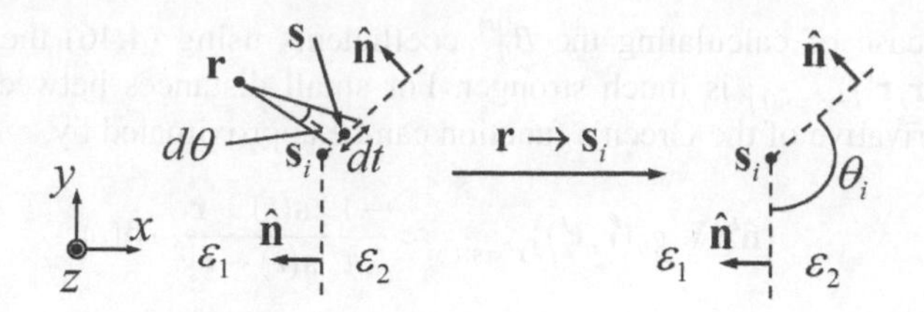

Figure 4.9 Schematic illustrating the contribution from the singularity of the Green's function when calculating $B_{ii}^{(u)}$ with sampling at a corner.

from the centers of sections i and j. Thus,

$$A_{ij}^{(u)} \approx \begin{cases} \frac{i}{4} H_0^{(1)}(k_0 n_u |\mathbf{s}_i - \mathbf{s}_j|)\Delta_j & i \neq j \\ \Delta_i \left(\frac{i}{4}(1 + i\frac{2}{\pi}\gamma) - \frac{1}{2\pi}\left(\log(k_0 n_u \Delta_i/4) - 1\right)\right) & i = j \end{cases}, \tag{4.66}$$

$$B_{ij}^{(u)} \approx \begin{cases} -\frac{i}{4} H_1^{(1)}(k_0 n_u |\mathbf{s}_i - \mathbf{s}_j|) k_0 n_u \frac{\mathbf{s}_j - \mathbf{s}_i}{|\mathbf{s}_j - \mathbf{s}_i|} \cdot \hat{\mathbf{n}}_j \Delta_j & i \neq j \\ \frac{1}{2}\left(\delta_{u,1} - \delta_{u,2}\right) & i = j \end{cases}, \tag{4.67}$$

where $\hat{\mathbf{n}}_j$ is the surface normal vector at position $\mathbf{s}_j$. After solving the matrix equations the absorption cross section (2.92) can be calculated within the crude approximation by using

$$\sigma_{\text{abs}} \approx \frac{-1}{k_0 n_1} \frac{1}{|H_0|^2} \sum_{i=1}^{N} \text{Imag}\left\{\phi_{1,i} H_i^*\right\} \Delta_i, \tag{4.68}$$

and the scattered far field can be calculated using

$$H_{\text{scat}}^{(\text{ff})}(r,\theta) \approx -\frac{1}{4}\sqrt{\frac{2}{\pi k_0 n_1 r}} e^{ik_0 n_1 r} e^{i\pi/4} \sum_{j=1}^{N} e^{-ik_0 n_1 \hat{\mathbf{r}} \cdot \mathbf{s}_j} \left\{\phi_{1,j} - H_j \hat{\mathbf{n}}_j \cdot (-ik_0 n_1 \hat{\mathbf{r}})\right\} \Delta_j, \tag{4.69}$$

which may subsequently be inserted into the expressions (2.89) and (2.90) to obtain numerically the extinction, absorption, and scattering cross sections. The calculation of the optical cross sections versus wavelength is shown in Fig. 4.10(a). The result, referred to as Method 1, is based on using the simple and crude expressions (4.66) and (4.67). The cylinder surface is divided into $N = 50$ sections. The result referred to as Method 2 is, on the other hand, based on using the expressions (4.53) and (4.64), where the integrals have been carried out numerically to high precision using Gauss-Kronrod quadrature. The solid lines are the exact analytic results obtained using Appendix D. The wavelength range covers the plasmon resonance seen as a peak in cross sections near $\lambda = 520$ nm. It is not a particularly strong resonance due to the properties of gold at this wavelength. For the wavelength 550 nm the relative error in the optical cross sections obtained with the GFSIEM is studied in more detail as a

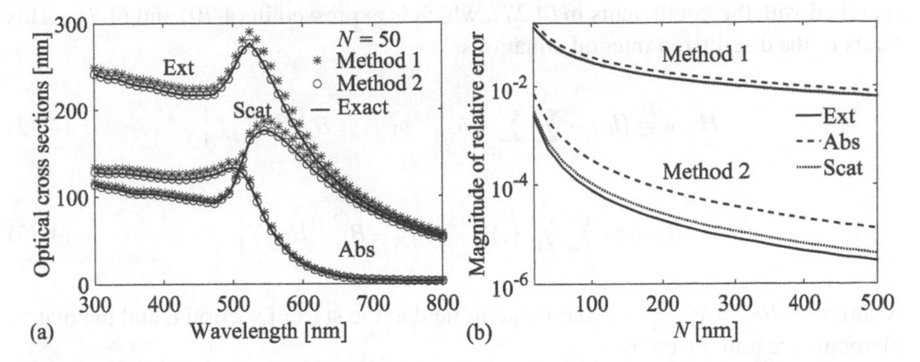

Figure 4.10 Example of performance of GFSIEM (2D) for a gold cylinder in water of radius 50 nm. (a) Exact optical cross-section spectra (solid lines), and calculations using the GFSIEM with pulse expansion of fields using two methods: Method 1, where matrix elements are calculated using Eqs. (4.66) and (4.67), and Method 2 based on Eqs. (4.53) and (4.64) calculated to high precision. (b) Error in cross sections vs. N for the two methods for a fixed wavelength $\lambda = 550$ nm.

function of N in Fig. 4.10(b). Relative error here means that the deviation or error is normalized with the exact result.

It is clear that the simple and crude approach to calculate matrix elements (Method 1), and using only $N = 50$, is already reasonably accurate with an error well below 10%. However, rigorous calculation of the matrix elements (Method 2) gives much higher precision, and again for $N = 50$ the error is now below 0.15%. This comes at a cost of an increased calculation time to obtain the matrix elements.

4.1.4 LINEAR-FIELD EXPANSION AND POINT-MATCHING

In this section we will again consider dividing the scatterer surface into N sections but now we will consider the 1st-order finite-element scheme (4.27). The linearly varying field in section i is determined by the value of fields at the beginning and end of the section. Using the same notation as in Section 4.1.3 we have

$$H_{u,i,0} = H_u(\mathbf{s}(t^{(s,i)})), \qquad H_{u,i,1} = H_u(\mathbf{s}(t^{(e,i)})), \tag{4.70}$$

$$\phi_{u,i,0} = \phi_u(\mathbf{s}(t^{(s,i)})), \qquad \phi_{u,i,1} = \phi_u(\mathbf{s}(t^{(e,i)})). \tag{4.71}$$

When constructing the integral equations we should recall that every end point of section i is equivalent to the starting point of another section j. It is therefore sufficient to use point-matching at, for example, the beginning of all elements, namely at the positions $\mathbf{s}_i^{(s)} \equiv \mathbf{s}(t^{(s,i)})$. Discretized integral equations are now obtained by inserting the field approximations (4.27) in the integral equations, and taking the limit of $\mathbf{r} \to \mathbf{s}_i^{(s)}$ from the side of Ω_1 and Ω_2. The fields at the positions $\mathbf{s}_i^{(s)}$ must then be matched with the coefficients in (4.27), which is expressed in (4.70) and (4.71). This leads to the discretized integral equations

$$H_{1,i,0} = H_{0,i} - \sum_{j=1}^{N}\sum_{v=0}^{1}\left\{A_{i,j}^{(1,v)}\phi_{1,j,v} - B_{ij}^{(1,v)}H_{1,j,v}\right\}, \tag{4.72}$$

$$H_{2,i,0} = \sum_{j=1}^{N}\sum_{v=0}^{1}\left\{A_{ij}^{(2,v)}\phi_{2,j,v} - B_{ij}^{(2,v)}H_{2,j,v}\right\}, \tag{4.73}$$

where now $H_{0,i} = H_0(\mathbf{s}_i^{(s)})$ is the incident field at the start of section i, and the matrix elements are now given by

$$A_{ij}^{(u,v)} = P\int_t g_u(\mathbf{s}_i^{(s)},\mathbf{s}(t))f_j^{(\mathrm{L},v)}(t)dt, \tag{4.74}$$

$$B_{ij}^{(u,v)} = P\int_t\left(\hat{\mathbf{n}}'\cdot\nabla' g_u(\mathbf{s}_i^{(s)},\mathbf{r}')\right)_{\mathbf{r}'=\mathbf{s}(t)} f_j^{(\mathrm{L},v)}(t)dt + \delta_{i,j}\delta_{v,0}\left(\delta_{u,1}\frac{\theta_i}{2\pi} - \delta_{u,2}\frac{2\pi-\theta_i}{2\pi}\right). \tag{4.75}$$

The corner angle θ_i is the same as the one shown in Fig. 4.9 in the previous subsection, but with the difference that there, the sampling point was at the center of an element, and here, the sampling point is at the start of an element. It is important that the contribution from the corner is included only once, which is ensured by the Kronecker delta functions $\delta_{i,j}$ and $\delta_{v,0}$.

Continuity of the tangential magnetic and electric field across the scatterer surface can again be applied to obtain

$$H_{1,i,v} = H_{2,i,v} \equiv H_{i,v} \quad \text{and} \quad \frac{1}{\varepsilon_1}\phi_{1,i,v} = \frac{1}{\varepsilon_2}\phi_{2,i,v}. \tag{4.76}$$

In order to rewrite the discretized integral equations in matrix form, the following vectors are introduced:

$$\overline{H}_0 = \left[H_{0,1} \quad H_{0,2} \quad \ldots \quad H_{0,N}\right]^T, \tag{4.77}$$

$$\overline{H}^{(v)} = \left[H_{1,v} \quad H_{2,v} \quad \ldots \quad H_{N,v}\right]^T, \tag{4.78}$$

$$\overline{\phi}_u^{(v)} = \left[\phi_{u,1,v} \quad \phi_{u,2,v} \quad \ldots \quad \phi_{u,N,v}\right]^T. \tag{4.79}$$

With the ordering of elements in Fig. 4.2 the fields at the beginning and end of the elements are furthermore related by the continuity boundary condition as follows:

$$H_{i,1} = H_{i+1,0}, \quad i = 1,2,...,N-1, \quad \text{and} \quad H_{N,1} = H_{1,0}. \tag{4.80}$$

This can also be expressed by the matrix relation

$$\overline{H}^{(1)} = \overline{\overline{D}}_H\overline{H}^{(0)}. \tag{4.81}$$

For the ordering of elements shown in Fig. 4.2 the matrix $\overline{\overline{D}}_H$ is given by

$$\overline{\overline{D}}_H = \begin{bmatrix} 0 & 1 & 0 & 0 & \dots & 0 \\ 0 & 0 & 1 & 0 & \dots & 0 \\ \vdots & \vdots & \vdots & \vdots & \vdots & \vdots \\ 0 & 0 & 0 & 0 & \dots & 1 \\ 1 & 0 & 0 & 0 & \dots & 0 \end{bmatrix}, \tag{4.82}$$

but any ordering of elements is in principle possible, and then $\overline{\overline{D}}_H$ must be constructed accordingly. For a structure with no sharp corners (all $\theta_i = \pi$) we also find

$$\overline{\phi}_u^{(1)} = \overline{\overline{D}}_\phi \overline{\phi}_u^{(0)} \tag{4.83}$$

with $\overline{\overline{D}}_\phi = \overline{\overline{D}}_H$. If some corners in the structure have corner angle $\theta_i = 0$, while the rest have angles $\theta_i = \pi$, then some 1's in $\overline{\overline{D}}_\phi$ must be replaced by -1, as illustrated in Fig. 4.5.

For a general choice of corner angles the calculation of $\overline{\phi}_u^{(1)}$ cannot be straightforwardly related to $\overline{\phi}_u^{(0)}$. In fact, for p polarization the electromagnetics boundary conditions can only be satisfied across the scatterer surface from C_1 to C_2, and along the surfaces C_1 and C_2 at the same time, for special angles $\theta_i = 0, \pi$. What will happen for other corner angles is that there will be a field singularity at the corner [43]. However, that cannot be described with linearly varying fields in elements. A reasonable way of dealing with the boundary conditions is to always use corner angles that are either $\theta_i = 0$ or π, and otherwise, instead of using a sharp corner, use a smooth corner with a very small corner radius.

The equations (4.72) and (4.73) can now be formulated in matrix form

$$\begin{bmatrix} \left(\overline{\overline{I}} - \overline{\overline{B}}^{(1,0)} - \overline{\overline{B}}^{(1,1)}\overline{\overline{D}}_H\right) & \overline{\overline{A}}^{(1,0)} + \overline{\overline{A}}^{(1,1)}\overline{\overline{D}}_\phi \\ \left(\overline{\overline{I}} + \overline{\overline{B}}^{(2,0)} + \overline{\overline{B}}^{(2,1)}\overline{\overline{D}}_H\right) & -\frac{\varepsilon_2}{\varepsilon_1}\left(\overline{\overline{A}}^{(2,0)} + \overline{\overline{A}}^{(2,1)}\overline{\overline{D}}_\phi\right) \end{bmatrix} \begin{bmatrix} \overline{H}^{(0)} \\ \overline{\phi}_1^{(0)} \end{bmatrix} = \begin{bmatrix} \overline{H}_0 \\ \overline{0} \end{bmatrix}. \tag{4.84}$$

This equation can be straightforwardly solved for the field coefficients $H_{1,i,0}$ and $\phi_{1,i,0}$. The corresponding fields at end points of elements are then given straightforwardly using (4.81) and (4.83), and fields on C_2 can then be obtained using (4.76).

4.1.5 HIGHER-ORDER POLYNOMIAL FIELD EXPANSION AND POINT MATCHING

In this section we are going to use the polynomial functions (4.33) and the field expressions (4.34) which also define the meaning of the field coefficients $H_{u,i,v}$ and $\phi_{u,i,v}$. The scatterer surface is still divided into N sections, and now each section has $m+1$ surface sampling points. The sampling points of section i shall be denoted $\mathbf{s}_i^{(v)}$, where $v = 0, 1, \dots, m$. The position $\mathbf{s}_i^{(0)}$ will be at the start of section i and $\mathbf{s}_i^{(m)}$ will be at the end of section i, while $\mathbf{s}_i^{(v)}$ for $v = 1,2,\ldots,m-1$ will be sampling points placed between the end points of section i. The field coefficients can be expressed in a point-matching approach as

$$H_{u,i,v} \equiv H_u(\mathbf{s}_i^{(v)}) \quad \text{and} \quad \phi_{u,i,v} \equiv \phi_u(\mathbf{s}_i^{(v)}), \tag{4.85}$$

and thus represent the values of fields at the sampling points. We note that the case of linear elements considered in the previous section is also included as a special case, namely for $m = 1$.

We now organize the field coefficients in the following vectors

$$\overline{H}_u^{(v)} \equiv \begin{bmatrix} H_{u,1,v} & H_{u,2,v} & \cdots & H_{u,N,v} \end{bmatrix}^T, \tag{4.86}$$

$$\overline{\phi}_u^{(v)} = \begin{bmatrix} \phi_{u,1,v} & \phi_{u,2,v} & \cdots & \phi_{u,N,v} \end{bmatrix}^T, \tag{4.87}$$

and the incident field at positions $\mathbf{s}_i^{(v)}$ in the vector

$$\overline{H}_0^{(v)} \equiv \begin{bmatrix} H_0(\mathbf{s}_1^{(v)}) & H_0(\mathbf{s}_2^{(v)}) & \ldots & H_0(\mathbf{s}_N^{(v)}) \end{bmatrix}^T. \tag{4.88}$$

The field approximations (4.34) may now be inserted into the integral equations, and after taking the limit of $\mathbf{r} \to \mathbf{s}_i^{(v)}$ from the side of Ω_1 and Ω_2, and matching the fields at the positions $\mathbf{s}_i^{(v)}$ with the coefficients in (4.34), we arrive at the following discretized integral equations

$$\overline{H}^{(v)} + \sum_{v'=0}^{m} \left\{ \overline{\overline{A}}^{(1,v,v')} \overline{\phi}_1^{(v')} - \overline{\overline{B}}^{(1,v,v')} \overline{H}^{(v')} \right\} = \overline{H}_0^{(v)}, \quad v = 0,1,\ldots,m-1, \tag{4.89}$$

$$\overline{H}^{(v)} - \sum_{v'=0}^{m} \left\{ \overline{\overline{A}}^{(2,v,v')} \frac{\varepsilon_2}{\varepsilon_1} \overline{\phi}_1^{(v')} - \overline{\overline{B}}^{(2,v,v')} \overline{H}^{(v')} \right\} = \overline{0}, \quad v = 0,1,\ldots,m-1, \tag{4.90}$$

where the matrix elements are now given by the expressions

$$A_{ij}^{(u,v,v')} = P \int_t g_u(\mathbf{s}_i^{(v)}, \mathbf{s}(t)) f_j^{(m,v')}(t) dt, \tag{4.91}$$

$$B_{ij}^{(u,v,v')} = P \int_t \left(\hat{\mathbf{n}}' \cdot \nabla' g_u(\mathbf{s}_i^{(v)}, \mathbf{r}') \right)_{\mathbf{r}'=\mathbf{s}(t)} f_j^{(m,v')}(t) dt$$
$$+ \delta_{i,j} \delta_{v,v'} \left(\delta_{u,1} \frac{\theta_{i,v}}{2\pi} - \delta_{u,2} \frac{2\pi - \theta_{i,v}}{2\pi} \right). \tag{4.92}$$

The angle $\theta_{i,v}$ is equivalent to the angles θ_i considered previously, but here the angle should be obtained at the position $\mathbf{s}_i^{(v)}$. If we assume that a sharp corner will only ever occur between neighbor surface sections, while individual surface sections themselves are smooth and have no sharp corners, then $\theta_{i,v} = \pi$ if $v = 1,2,\ldots,m-1$, and $\theta_{i,0} = \theta_i$, where θ_i is the same as in the case of the linear-field approximation.

Similar to the previous section we may use the boundary condition between neighbor sections on the surface

$$\overline{H}^{(m)} = \overline{\overline{D}}_H \overline{H}^{(0)} \quad \text{and} \quad \overline{\phi}_1^{(m)} = \overline{\overline{D}}_\phi \overline{\phi}_1^{(0)}. \tag{4.93}$$

With these boundary conditions, and the following grand A- and B-matrices

$$\overline{\overline{A}}^{(u)} = \begin{bmatrix} \left(\overline{\overline{A}}^{(u,0,0)} + \overline{\overline{A}}^{(u,0,m)}\overline{\overline{D}}_\phi\right) & \overline{\overline{A}}^{(u,0,1)} & \dots & \overline{\overline{A}}^{(u,0,m-1)} \\ \left(\overline{\overline{A}}^{(u,1,0)} + \overline{\overline{A}}^{(u,1,m)}\overline{\overline{D}}_\phi\right) & \overline{\overline{A}}^{(u,1,1)} & \dots & \overline{\overline{A}}^{(u,1,m-1)} \\ \vdots & \vdots & \dots & \vdots \\ \left(\overline{\overline{A}}^{(u,m-1,0)} + \overline{\overline{A}}^{(u,m-1,m)}\overline{\overline{D}}_\phi\right) & \overline{\overline{A}}^{(u,m-1,1)} & \dots & \overline{\overline{A}}^{(u,m-1,m-1)} \end{bmatrix}, \tag{4.94}$$

$$\overline{\overline{B}}^{(u)} = \begin{bmatrix} \left(\overline{\overline{B}}^{(u,0,0)} + \overline{\overline{B}}^{(u,0,m)}\overline{\overline{D}}_H\right) & \overline{\overline{B}}^{(u,0,1)} & \dots & \overline{\overline{B}}^{(u,0,m-1)} \\ \left(\overline{\overline{B}}^{(u,1,0)} + \overline{\overline{B}}^{(u,1,m)}\overline{\overline{D}}_H\right) & \overline{\overline{B}}^{(u,1,1)} & \dots & \overline{\overline{B}}^{(u,1,m-1)} \\ \vdots & \vdots & \dots & \vdots \\ \left(\overline{\overline{B}}^{(u,m-1,0)} + \overline{\overline{B}}^{(u,m-1,m)}\overline{\overline{D}}_H\right) & \overline{\overline{B}}^{(u,m-1,1)} & \dots & \overline{\overline{B}}^{(u,m-1,m-1)} \end{bmatrix}, \tag{4.95}$$

the discretized integral equations can finally be formulated as the following grand matrix equation:

$$\begin{bmatrix} \left(\overline{\overline{I}} - \overline{\overline{B}}^{(1)}\right) & \overline{\overline{A}}^{(1)} \\ \left(\overline{\overline{I}} + \overline{\overline{B}}^{(2)}\right) & -\frac{\varepsilon_2}{\varepsilon_1}\overline{\overline{A}}^{(2)} \end{bmatrix} \begin{bmatrix} \overline{H}^{(0)} \\ \vdots \\ \overline{H}^{(m-1)} \\ \overline{\phi}_1^{(0)} \\ \vdots \\ \overline{\phi}_1^{(m-1)} \end{bmatrix} = \begin{bmatrix} \overline{H}_0^{(0)} \\ \vdots \\ \overline{H}_0^{(m-1)} \\ \overline{0} \\ \vdots \\ \overline{0} \end{bmatrix}. \tag{4.96}$$

By solving this equation numerically, and by using the electromagnetics boundary conditions, we finally obtain the fields at the scatterer surfaces C_1 and C_2 in the mth-order polynomial approximation (4.34). These fields can then be used in the integral equations to calculate the fields everywhere else and to calculate the different optical cross sections.

The absorption cross section can be obtained by inserting the field approximations (4.34) into (2.92). This leads to

$$\sigma_{\text{abs}} = \frac{-1}{k_0 n_1} \sum_{i=1}^{N} \sum_{v=0}^{m} \sum_{v'=0}^{m} M_{i,v,v'}^{(m)} \text{Imag}\left\{\phi_{1,i,v}\left(H_{1,i,v'}\right)^*\right\}, \tag{4.97}$$

where

$$M_{i,v,v'}^{(m)} \equiv \int_t f_i^{(m,v)}(t) f_i^{(m,v')}(t) dt = \left(t^{(e,i)} - t^{(s,i)}\right) \int_0^1 f^{(m,v)}(x) f^{(m,v')}(x) dx. \quad (4.98)$$

The latter coefficients can be calculated analytically since $f^{(m,v)(x)}$ are polynomial functions. Assuming that all sections i use equidistant subsampling when constructing the higher-order elements, then $M_{i,v,v'}^{(m)} / \left(t^{(e,i)} - t^{(s,i)}\right)$ is given in the following tables for $m = 1, 2, 3$ and 4:

$m=1$	$v=0$	$v=1$
$v'=0$	1/3	1/6
$v'=1$	1/6	1/3

$m=2$	$v=0$	$v=1$	$v=2$
$v'=0$	2/15	1/15	-1/30
$v'=1$	1/15	8/15	1/15
$v'=2$	-1/30	1/15	2/15

$m=3$	$v=0$	$v=1$	$v=2$	$v=3$
$v'=0$	8/105	33/560	-3/140	19/1680
$v'=1$	33/560	27/70	-27/560	-3/140
$v'=2$	-3/140	-27/560	27/70	33/560
$v'=3$	19/1680	-3/140	33/560	8/105

$m=4$	$v=0$	$v=1$	$v=2$	$v=3$	$v=4$
$v'=0$	146/2835	148/2835	-29/945	4/405	-29/5670
$v'=1$	148/2835	128/405	-64/945	128/2835	4/405
$v'=2$	-29/945	-64/945	104/315	-64/945	-29/945
$v'=3$	4/405	128/2835	-64/945	128/405	148/2835
$v'=4$	-29/5670	4/405	-29/945	148/2835	146/2835

In order to calculate the scattering and extinction cross sections we need the scattered far field, which here can be obtained by inserting the field approximations (4.34) into (4.22). This leads to

$$H_{\text{sc}}^{(\text{ff})}(r,\theta) = -\frac{1}{4}\sqrt{\frac{2}{\pi k_0 n_1 r}} e^{i\pi/4} e^{ik_0 n_1 r} \times$$

$$\sum_{i=1}^{N} \sum_{v=0}^{m} \int_t e^{-ik_0 n_1 \hat{\mathbf{r}} \cdot \mathbf{s}(t)} \left\{ \phi_{1,i,v} + H_{1,i,v} ik_0 n_1 \hat{\mathbf{n}}(t) \cdot \hat{\mathbf{r}} \right\} f_i^{(m,v)}(t) dt. \quad (4.99)$$

The scattered far field can subsequently be applied in (2.89) and (2.90) to obtain the scattering and extinction cross sections.

It will now be interesting to investigate the performance when using different orders m. As an example we will again consider the example of a gold cylinder of radius 50 nm in water being illuminated by a plane wave of wavelength 550 nm. The relative error in the extinction cross section, defined as the difference between

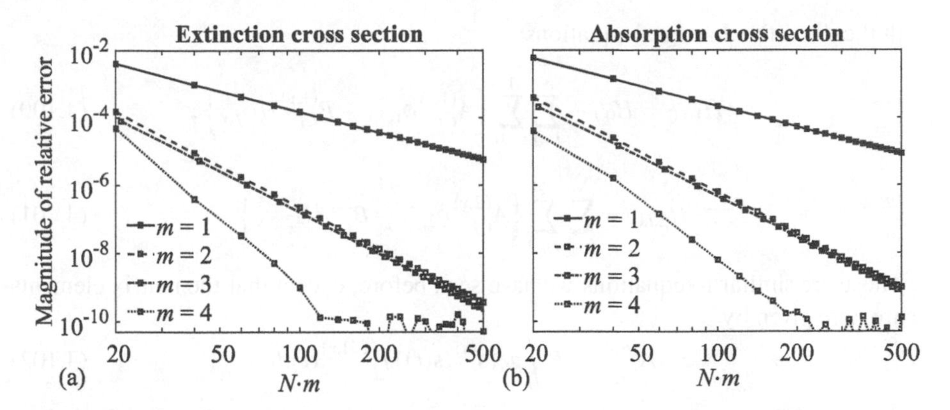

Figure 4.11 Relative error in extinction cross section for a plane wave of wavelength 550 nm being incident on a gold cylinder of radius 50 nm in water when using the GFSIEM (2D) with N surface sections and m-th-order polynomial expansion of fields in each section.

the calculated and exact cross section divided by the exact cross section, is shown in Fig. 4.11(a) versus $N \cdot m$ for different orders m. By showing the error versus $N \cdot m$ the results for different m are directly comparable, since $N \cdot m$ represents the total number of sampling points. By using double-logarithmic axes, the error versus $N \cdot m$ becomes approximately straight lines. The error for $m = 1$ is significantly higher compared with $m = 2, 3, 4$. For $m = 1$ and $N = 20$, the error is 0.004 but for $m = 2$ and $N = 10$, it is 0.00015, which is more than 10 times better using the same number of sampling points. As $N \cdot m$ increases, the advantage of using a higher m is seen from the slope of the curves for different m. A larger m means faster convergence. For the specific example there is not much difference in the slope for $m = 2$ and $m = 3$ implying that third-order terms are not so important for the specific problem, but then for $m = 4$ the slope is again steeper. For $m = 4$ and $N = 20$, the relative error is already below 10^{-8}. As the relative error reaches approximately 10^{-10} numerical issues leads to a noisy curve, and improving the relative error further in a consistent way requires higher numerical precision (more digits). Note that in order to see the very small errors, such as 10^{-10}, the matrix elements in the GFSIEM must be calculated to very high accuracy. The relative error for the absorption cross section versus N and m is shown in Fig. 4.11(b), and it clearly follows the same trend. This is also the case for the scattering cross section.

For $m = 3$ we may use the alternative expressions for H and ϕ given in (4.42), when a scatterer surface with no sharp corners is considered. In this expression $H_{u,i,0}$ and $\phi_{u,i,0}$ represent the value of H and ϕ_u at position $\mathbf{s}_i^{(s)}$, and $H_{u,i,1}$ and $\phi_{u,i,1}$ represent the tangential derivative of H and ϕ at the same position. The field is thus described directly in terms of its field value and its tangential derivative at the start of each section i. Point-matching in the integral equations at positions $\mathbf{s}_i^{(s)}$, representing the start of section i, leads to expressions that can be point-matched with $H_{u,i,0}$, resulting in the discretized integral equations:

$$H_{1,i,0} = H_{0,i} - \sum_{j=1}^{N}\sum_{v=0}^{1}\left\{A_{i,j}^{(1,v)}\phi_{1,j,v} - B_{i,j}^{(1,v)}H_{1,j,v}\right\}, \tag{4.100}$$

$$H_{2,i,0} = \sum_{j=1}^{N}\sum_{v=0}^{1}\left\{A_{i,j}^{(2,v)}\phi_{2,j,v} - B_{i,j}^{(2,v)}H_{2,j,v}\right\}. \tag{4.101}$$

These are similar to equations we have seen before, except that the matrix elements are here given by

$$A_{i,j}^{(u,v)} = P\int_t g_u(\mathbf{s}_i^{(s)}, \mathbf{s}(t)) f_j^{(\mathrm{CH},v)}(t)dt, \tag{4.102}$$

$$B_{i,j}^{(u,v)} = P\int_t \left(\hat{\mathbf{n}}'\cdot\nabla' g_u(\mathbf{s}_i^{(s)}, \mathbf{r}')\right)_{\mathbf{r}'=\mathbf{s}(t)} f_j^{(\mathrm{CH},v)}(t)dt + \delta_{i,j}\delta_{v,0}\left(\delta_{u,1}\frac{\theta_i}{2\pi} - \delta_{u,2}\frac{2\pi-\theta_i}{2\pi}\right). \tag{4.103}$$

Further expressions that can be point-matched with $H_{u,i,1}$, representing the tangential derivative of H, can now be obtained by letting $\mathbf{r}$ approach $\mathbf{s}_i^{(s)}$ from either side of the scatterer surface in the integral equations, and then taking the tangential derivative. This approach leads to the following further equations

$$H_{1,i,1} = H'_{0,i} - \sum_{j=1}^{N}\sum_{v=0}^{1}\left\{C_{i,j}^{(1,v)}\phi_{1,j,v} - D_{i,j}^{(1,v)}H_{1,j,v}\right\}, \tag{4.104}$$

$$H_{2,i,1} = \sum_{j=1}^{N}\sum_{v=0}^{1}\left\{C_{i,j}^{(2,v)}\phi_{2,j,v} - D_{i,j}^{(2,v)}H_{2,j,v}\right\}. \tag{4.105}$$

Here,

$$H'_{0,i} \equiv \left(\frac{dH_0(\mathbf{s}(t))}{dt}\right)_{t=t^{(s,i)}}, \tag{4.106}$$

with the additional matrix elements given by

$$C_{i,j}^{(u,v)} = \left(\frac{d}{dt}P\int_{t'} g_u(\mathbf{s}(t), \mathbf{s}(t')) f_j^{(\mathrm{CH},v)}(t')dt'\right)_{t=t^{(s,i)}}, \tag{4.107}$$

$$D_{i,j}^{(u,v)} = \left(\frac{d}{dt}P\int_{t'} (\hat{\mathbf{n}}'\cdot\nabla' g_u(\mathbf{s}(t), \mathbf{r}'))_{\mathbf{r}'=\mathbf{s}(t')} f_j^{(\mathrm{CH},v)}(t')dt'\right)_{t=t^{(s,i)}}. \tag{4.108}$$

By inserting the homogeneous-medium Green's function for $u = 1,2$, the matrix elements can be written

$$C_{i,j}^{(u,v)} = P\int_{t'} \frac{-i}{4} H_1^{(1)}(k_0 n_u|\mathbf{s}_i^{(s)} - \mathbf{s}(t')|) k_0 n_u \frac{\mathbf{s}_i^{(s)} - \mathbf{s}(t')}{|\mathbf{s}_i^{(s)} - \mathbf{s}(t')|}\cdot\hat{\mathbf{t}}_i^{(s)} f_j^{(\mathrm{CH},v)}(t')dt', \tag{4.109}$$

where

$$\hat{\mathbf{t}}_i^{(s)} \equiv \left(\frac{d\mathbf{s}(t)}{dt}\right)_{t=t^{(s,i)}} \tag{4.110}$$

is the tangent unit vector of the surface at position $\mathbf{s}_i^{(s)}$. Convergence of the integral (4.109) is unproblematic for $i \neq j$, since, although $H_1^{(1)}(k_0 n_u |\mathbf{s}_i^{(s)} - \mathbf{s}(t')|)$ is singular for the value of $t' = t^{(s,i)}$ where $\mathbf{s}(t') = \mathbf{s}_i^{(s)}$, $f_j^{(\mathrm{CH},v)}(t')$ is zero for the same value of t'. For $i = j$ and in the limit of small values of $|\mathbf{s}_i^{(s)} - \mathbf{s}(t')|$ we may note that $H_1^{(1)}(k_0 n_u |\mathbf{s}_i^{(s)} - \mathbf{s}(t')|) \propto 1/|\mathbf{s}_i^{(s)} - \mathbf{s}(t')|$ is an even function of t' across $t' = t^{(s,i)}$, while

$$\frac{\mathbf{s}_i^{(s)} - \mathbf{s}(t')}{|\mathbf{s}_i^{(s)} - \mathbf{s}(t')|} \cdot \hat{\mathbf{t}}_i^{(s)} \tag{4.111}$$

is an odd function. Then in the case of $i = j$ and $v = 0$ the function $f_j^{(\mathrm{CH},0)}(t')$ is also even across this value of t'. The integral is thus convergent. In the case of $i = j$ and $v = 1$ the function $f_j^{(\mathrm{CH},1)}(t') \propto |\mathbf{s}_i^{(s)} - \mathbf{s}(t')|$ is odd but also cancels the singularity of $H_1^{(1)}(k_0 n_u |\mathbf{s}_i^{(s)} - \mathbf{s}(t')|)$, and the integral thus also converges in this case.

By inserting the same Green's function in (4.108) it can be shown that

$$\begin{aligned} D_{i,j}^{(u,v)} = P \int_{t'} \Bigg\{ & \frac{-i}{8} \left(H_0^{(1)}(q_u(t')) - H_2^{(1)}(q_u(t')) \right) (k_0 n_u)^2 \left(\frac{\mathbf{s}_i^{(s)} - \mathbf{s}(t')}{|\mathbf{s}_i^{(s)} - \mathbf{s}(t')|} \cdot \hat{\mathbf{t}}_i^{(s)} \right) \times \\ & \left(\frac{\mathbf{s}(t') - \mathbf{s}_i^{(s)}}{|\mathbf{s}_i^{(s)} - \mathbf{s}(t')|} \cdot \hat{\mathbf{n}}(t') \right) + \frac{-i}{4} H_1^{(1)}(q_u(t')) k_0 n_u \left(\frac{-\hat{\mathbf{n}}(t') \cdot \hat{\mathbf{t}}_i^{(s)}}{|\mathbf{s}_i^{(s)} - \mathbf{s}(t')|} \right. \\ & \left. - \left[\frac{\hat{\mathbf{n}}(t') \cdot (\mathbf{s}(t') - \mathbf{s}_i^{(s)})}{|\mathbf{s}_i^{(s)} - \mathbf{s}(t')|^2} \right] \left[\frac{\hat{\mathbf{t}}_i^{(s)} \cdot (\mathbf{s}_i^{(s)} - \mathbf{s}(t'))}{|\mathbf{s}_i^{(s)} - \mathbf{s}(t')|} \right] \right) \Bigg\} f_j^{(CH,v)}(t') dt', \end{aligned} \tag{4.112}$$

where

$$q_u(t') \equiv k_0 n_u |\mathbf{s}_i^{(s)} - \mathbf{s}(t')|. \tag{4.113}$$

A detailed analysis of the integrand in (4.112), which has some terms that are singular at a certain t', and some terms that go to zero at the same t', shows that the integrand is non-singular, and numerical calculation of the integral (4.112) is thus straightforward. In order to solve these equations it is convenient to introduce the vectors

$$\overline{H}_u = \left[H_{u,1,0} \quad H_{u,2,0} \quad \cdots \quad H_{u,N,0} \right]^T, \tag{4.114}$$

$$\overline{H}'_u = \left[H_{u,1,1} \quad H_{u,2,1} \quad \cdots \quad H_{u,N,1} \right]^T, \tag{4.115}$$

$$\overline{\phi}_u = \left[\phi_{u,1,0} \quad \phi_{u,2,0} \quad \cdots \quad \phi_{u,N,0} \right]^T, \tag{4.116}$$

$$\overline{\phi}'_u = \left[\phi_{u,1,1} \quad \phi_{u,2,1} \quad \cdots \quad \phi_{u,N,1} \right]^T. \tag{4.117}$$

The electromagnetics boundary conditions can be expressed as

$$\overline{H}_1 = \overline{H}_2, \quad \overline{H}'_1 = \overline{H}'_2, \quad \frac{1}{\varepsilon_1}\overline{\phi}_1 = \frac{1}{\varepsilon_2}\overline{\phi}_2, \quad \text{and} \quad \frac{1}{\varepsilon_1}\overline{\phi}'_1 = \frac{1}{\varepsilon_2}\overline{\phi}'_2, \tag{4.118}$$

and by combining the boundary conditions and Eqs. (4.100), (4.101), (4.104) and (4.105), the following matrix equation is obtained

$$\begin{bmatrix} \left(\overline{\overline{I}}-\overline{\overline{B}}^{(1,0)}\right) & \overline{\overline{A}}^{(1,0)} & -\overline{\overline{B}}^{(1,1)} & \overline{\overline{A}}^{(1,1)} \\ \left(\overline{\overline{I}}+\overline{\overline{B}}^{(2,0)}\right) & -\frac{\varepsilon_2}{\varepsilon_1}\overline{\overline{A}}^{(2,0)} & \overline{\overline{B}}^{(2,1)} & -\frac{\varepsilon_2}{\varepsilon_1}\overline{\overline{A}}^{(2,1)} \\ -\overline{\overline{D}}^{(1,0)} & \overline{\overline{C}}^{(1,0)} & \left(\overline{\overline{I}}-\overline{\overline{D}}^{(1,1)}\right) & \overline{\overline{C}}^{(1,1)} \\ \overline{\overline{D}}^{(2,0)} & -\frac{\varepsilon_2}{\varepsilon_1}\overline{\overline{C}}^{(2,0)} & \left(\overline{\overline{I}}+\overline{\overline{D}}^{(2,1)}\right) & -\frac{\varepsilon_2}{\varepsilon_1}\overline{\overline{C}}^{(2,1)} \end{bmatrix} \begin{bmatrix} \overline{H} \\ \overline{\phi}_1 \\ \overline{H}' \\ \overline{\phi}'_1 \end{bmatrix} = \begin{bmatrix} \overline{H}_0 \\ \overline{0} \\ \overline{H}'_0 \\ \overline{0} \end{bmatrix}. \tag{4.119}$$

After solving this matrix equation the absorption cross section can be calculated using

$$\sigma_{\text{abs}} = \frac{-1}{k_0 n_1}\text{Imag}\left\{\sum_{i=1}^{N}\sum_{i'=1}^{N}\sum_{v=0}^{1}\sum_{v'=0}^{1}\int_t \phi_{1,i,v} f_i^{(\text{CH},v)}(t)\left(H_{1,i',v'} f_{i'}^{(\text{CH},v')}(t)\right)^* dt\right\}. \tag{4.120}$$

The performance of the GFSIEM with the field expansion based on point-matching of fields and tangential derivatives is examined in Fig. 4.12 considering again the example of the gold cylinder in water being illuminated by a plane wave of wavelength 550 nm. In this case the relative error for extinction, absorption, and scattering cross sections, is shown versus $2N$ to facilitate direct comparison with the other previously considered methods. It can be seen that the relative error in ab-

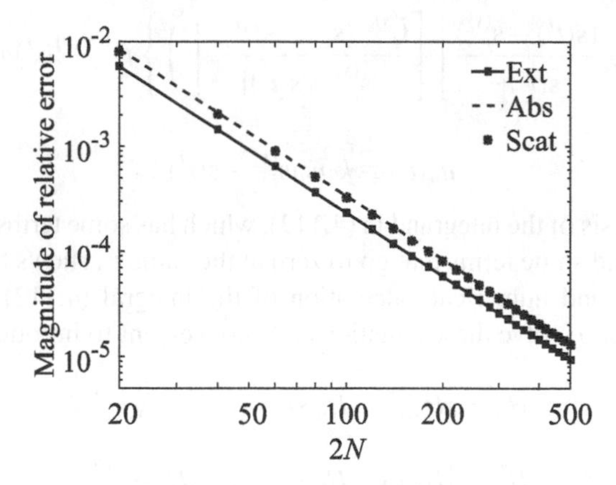

Figure 4.12 Relative error in extinction, absorption, and scattering cross section for a plane wave of wavelength 550 nm that is incident on a gold cylinder of radius 50 nm in water when using the GFSIEM (2D) with N surface sections and point-matching of the field and m-th order polynomial expansion of fields in each section.

sorption cross sections decreases with $2N$ similar to the performance versus N when using linear elements in each section ($m = 1$). We may conclude that the method works fine. However, based on the considered example, a better performance can be obtained with a higher-order polynomial expansion method.

4.1.6 FOURIER EXPANSION METHODS

In the case of a single scatterer with no sharp corners, another possibility is to use a Fourier expansion of the fields on the scatterer surface. If the total length of the scatterer surface is L then we can, for example, use the following expansions of the surface fields

$$H(\mathbf{s}(t)) = \sum_{n=-N}^{N} \tilde{h}_n e^{i\frac{2\pi}{L}nt}, \tag{4.121}$$

$$\phi_u(\mathbf{s}(t)) = \sum_{n=-N}^{N} \tilde{\phi}_{u,n} e^{i\frac{2\pi}{L}nt}. \tag{4.122}$$

This will give smooth continuous distributions of H and ϕ with no jumps along the surface, and with no jumps in the tangential derivative as well. We may in addition use

$$H_0(\mathbf{s}(t)) = \sum_{j=-N}^{N} \tilde{h}_{0,n} e^{i\frac{2\pi}{L}jt}. \tag{4.123}$$

Note that the Fourier coefficients $\tilde{h}_n$ are given by

$$\tilde{h}_n = \frac{1}{L}\int_{t=0}^{L} H(\mathbf{s}(t)) e^{-i\frac{2\pi}{L}nt} dt, \tag{4.124}$$

and similar expressions apply for $\tilde{\phi}_{u,n}$ and $\tilde{h}_{0,n}$.

In principle any function can be described with the expansions (4.121)-(4.123) if enough terms are included. However, for a surface with sharp corners, or a surface where the normal vector $\hat{\mathbf{n}}(t)$ is varying fast with t, as in the case of rounded corners with a small corner radius, then $\phi(\mathbf{s}(t))$ will vary fast in some regions. This can only be described when including high spatial frequencies in the expansion (4.122), and thus N must be large. For structures without such corners, the expansions (4.121) and (4.122) may lead to convergence using only a few terms, as will be shown in the following.

We now insert the expressions (4.121)-(4.123) in the integral equations and subsequently apply (4.124). This leads to the integral equations in discrete form

$$\tilde{h}_n = \tilde{h}_{0,n} - \sum_{n'} \left\{ A^{(1)}_{n,n'} \tilde{\phi}_{1,n'} - B^{(1)}_{n,n'} \tilde{h}_{n'} \right\}, \tag{4.125}$$

$$\tilde{h}_n = \sum_{n'} \left\{ A^{(2)}_{n,n'} \tilde{\phi}_{2,n'} - B^{(2)}_{n,n'} \tilde{h}_{n'} \right\}. \tag{4.126}$$

Here, the matrix elements are given by

$$A^{(u)}_{n,n'} = \frac{1}{L}\int_{t=0}^{L} e^{-i\frac{2\pi}{L}nt} \left[P\int_{t'=0}^{L} g_u(\mathbf{s}(t), \mathbf{s}(t')) e^{i\frac{2\pi}{L}n't'} dt' \right] dt, \tag{4.127}$$

$$B_{n,n'}^{(u)} = \frac{1}{L}\int_{t=0}^{L} e^{-i\frac{2\pi}{L}nt}\left[P\int_{t'=0}^{L}\left(\hat{\mathbf{n}}'\cdot\nabla' g_u(\mathbf{r},\mathbf{r}')\right)_{\mathbf{r}=\mathbf{s}(t),\,\mathbf{r}'=\mathbf{s}(t')} e^{i\frac{2\pi}{L}n't'}dt'\right]dt$$

$$+\frac{1}{2}\delta_{n,n'}\left(\delta_{u,1}-\delta_{u,2}\right), \tag{4.128}$$

where the last term is obtained when assuming that the surface is smooth.

The field coefficients may be conveniently collected in vectors

$$\overline{H} = \left[\tilde{h}_{-N} \quad \tilde{h}_{-N+1} \quad \dots \quad \tilde{h}_{N}\right]^T, \tag{4.129}$$

$$\overline{H}_0 = \left[\tilde{h}_{0,-N} \quad \tilde{h}_{0,-N+1} \quad \dots \quad \tilde{h}_{0,N}\right]^T, \tag{4.130}$$

$$\overline{\phi}_u = \left[\tilde{\phi}_{u,-N} \quad \tilde{\phi}_{u,-N+1} \quad \dots \quad \tilde{\phi}_{u,N}\right]^T. \tag{4.131}$$

We have only defined one surface magnetic field (4.121) due to continuity of H across the scatterer surface, and for the normal derivative the usual boundary condition applies, i.e., $\overline{\phi}_1/\varepsilon_1 = \overline{\phi}_2/\varepsilon_2$. If we now define matrices $\overline{\overline{A}}^{(u)}$ and $\overline{\overline{B}}^{(u)}$ such that the matrix element at position i,j is given by $A_{i-N-1,j-N-1}^{(u)}$ and $B_{i-N-1,j-N-1}^{(u)}$, respectively, the discrete integral equations can be put in matrix form:

$$\begin{bmatrix} \overline{\overline{I}} - \overline{\overline{B}}^{(1)} & \overline{\overline{A}}^{(1)} \\ \overline{\overline{I}} + \overline{\overline{B}}^{(2)} & -\frac{\varepsilon_2}{\varepsilon_1}\overline{\overline{A}}^{(2)} \end{bmatrix} \begin{bmatrix} \overline{H} \\ \overline{\phi}_1 \end{bmatrix} = \begin{bmatrix} \overline{H}_0 \\ \overline{0} \end{bmatrix}. \tag{4.132}$$

Note that the calculation of matrix elements here requires the evaluation of double integrals, and both integrals are in principle over the entire scatterer surface. When using a polynomial field expansion and point-matching, only single integrals are needed, and each integral is over only a small part of the scatterer surface. The evaluation of matrix elements is here thus very time consuming, but apart from that, the method may be rather efficient in the number of elements that are required to obtain good convergence.

For circular symmetric structures, such as the example of a gold nanocylinder in water, the matrix elements greatly simplify. First of all, for a homogeneous medium we have that $g_u(\mathbf{s}(t),\mathbf{s}(t'))$ only depends on $|\mathbf{s}(t)-\mathbf{s}(t')|$, and for cylindrically symmetric surfaces it then follows that, e.g., the integral

$$P\int_{t'=0}^{L} g_u(\mathbf{s}(t),\mathbf{s}(t'))e^{i\frac{2\pi}{L}n'(t'-t)}dt'$$

does not depend on t. From (4.127) it can then be seen that

$$A_{n,n'}^{(u)} = \delta_{n,n'}P\int_{t'=0}^{L} g_u(\mathbf{s}(0),\mathbf{s}(t'))e^{iGnt'}dt'. \tag{4.133}$$

Similarly, it can be shown that

$$B_{n,n'}^{(u)} = \delta_{n,n'}\left(P\int_{t'=0}^{L}\left(\hat{\mathbf{n}}'\cdot\nabla' g_u(\mathbf{r},\mathbf{r}')\right)_{\mathbf{r}=\mathbf{s}(0),\,\mathbf{r}'=\mathbf{s}(t')} e^{i\frac{2\pi}{L}n't'}dt' + \frac{\delta_{u,1}-\delta_{u,2}}{2}\right). \tag{4.134}$$

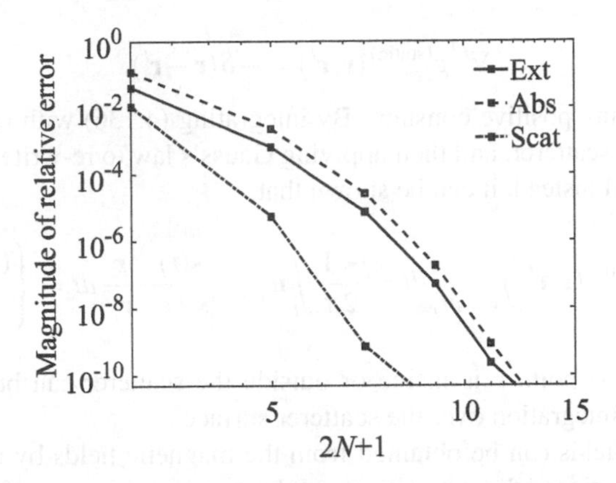

Figure 4.13 Relative error in extinction, absorption, and scattering cross section for a plane wave of wavelength 550 nm that is incident on a gold cylinder of radius 50 nm in water when using the GFSIEM (2D) with a Fourier expansion of the field having $2N+1$ terms.

The convergence for the gold cylinder in water is shown in Fig. 4.13. In this case a formidable relative error below 10^{-10} is possible using $2N+1=15$ terms. For structures with cylindrical symmetry the approach is in fact very similar to the analytic approach considered in Appendix D. For other geometries the matrices $\overline{\overline{A}}^{(u)}$ and $\overline{\overline{B}}^{(u)}$ will in general not be diagonal matrices, and the calculation of matrix elements will require double integrals.

4.1.7 CALCULATING ELECTRIC AND MAGNETIC FIELD DISTRIBUTIONS

In this section we consider the calculation of maps of the electric and magnetic fields for the scattering problem. It will be assumed that the magnetic field and its normal derivative have already been calculated at the surface of the scatterer using one of the methods from the previous sections, such as, for example, simple or higher-order polynomial field expansion, and then Eqs. (4.7) and (4.16) will be used to obtain the magnetic field at all positions inside or outside the scatterer. Usually, it will be convenient to calculate the field on a grid of points in the xy-plane, store the field values in a matrix, and then plot the field versus x and y to obtain a 2D map of the field.

The first step for each position $\mathbf{r}$ in the xy-plane is to determine which of the equations (4.7) or (4.16) should be used to obtain the field. This is equivalent to determining if the position is inside ($\mathbf{r} \in \Omega_2$) or outside ($\mathbf{r} \in \Omega_1$) the scatterer. For a simple geometry this may be straightforward. For a general structure, one possibility is to use the simple static-limit Green's function

$$g^{(\text{static})}(\mathbf{r},\mathbf{r}') = \frac{-1}{2\pi}\log(k|\mathbf{r}-\mathbf{r}'|), \tag{4.135}$$

which satisfies

$$\nabla'^2 g^{(\text{static})}(\mathbf{r},\mathbf{r}') = -\delta(\mathbf{r}-\mathbf{r}'). \tag{4.136}$$

Here, k can be any positive constant. By integrating (4.136) with respect to $\mathbf{r}'$ over the inside of the scatterer, and then applying Gauss's law to re-write the integral into a surface integral instead, it can be shown that

$$\int_t \left(\hat{\mathbf{n}}'\cdot\nabla' g^{(\text{static})}(\mathbf{r},\mathbf{r}')\right)_{\mathbf{r}'=\mathbf{s}(t)} dt = \frac{-1}{2\pi}\int_t \hat{\mathbf{n}}(t)\cdot\frac{\mathbf{s}(t)-\mathbf{r}}{|\mathbf{s}(t)-\mathbf{r}|^2}dt = \begin{cases} 0, & \mathbf{r}\in\Omega_1 \\ -1, & \mathbf{r}\in\Omega_2 \end{cases}. \tag{4.137}$$

Thus, whether the point $\mathbf{r}$ is inside or outside the scatterer can be determined by straightforward integration over the scatterer surface.

The electric fields can be obtained from the magnetic fields by using (4.19) and (4.20). For convenience these equations are here written in terms of $\phi_u(\mathbf{s}(t))$:

$$\mathbf{E}(\mathbf{r}) = \frac{-i}{\omega\varepsilon_0\varepsilon_2}\oint_{C_2}\Big\{\hat{\mathbf{z}}\times\nabla g_2(\mathbf{r},\mathbf{s}(t))\phi(\mathbf{s}(t))$$

$$-H(\mathbf{s}(t))\hat{\mathbf{z}}\times\nabla\left(\hat{\mathbf{n}}'\cdot\nabla' g_2(\mathbf{r},\mathbf{r}')\right)_{\mathbf{r}'=\mathbf{s}(t)}\Big\}dt, \quad \mathbf{r}\in\Omega_2, \tag{4.138}$$

$$\mathbf{E}(\mathbf{r}) = \mathbf{E}_0(\mathbf{r}) + \frac{i}{\omega\varepsilon_0\varepsilon_1}\oint_{C_1}\Big\{\hat{\mathbf{z}}\times\nabla g_1(\mathbf{r},\mathbf{s}(t))\phi(\mathbf{s}(t))$$

$$-H(\mathbf{s}(t))\hat{\mathbf{z}}\times\nabla\left(\hat{\mathbf{n}}'\cdot\nabla' g_1(\mathbf{r},\mathbf{r}')\right)_{\mathbf{r}'=\mathbf{s}(t)}\Big\}dt, \quad \mathbf{r}\in\Omega_1. \tag{4.139}$$

Here, it can be used that

$$\hat{\mathbf{z}}\times\nabla g_u(\mathbf{r},\mathbf{r}') = \left(\hat{\mathbf{x}}\frac{-(y-y')}{|\mathbf{r}-\mathbf{r}'|} + \hat{\mathbf{y}}\frac{x-x'}{|\mathbf{r}-\mathbf{r}'|}\right)B, \tag{4.140}$$

$$\hat{\mathbf{z}}\times\nabla\left(\hat{\mathbf{n}}'\cdot\nabla' g_u(\mathbf{r},\mathbf{r}')\right) =$$

$$\hat{\mathbf{x}}\left(A\frac{-(y-y')}{|\mathbf{r}-\mathbf{r}'|} + B\frac{\hat{\mathbf{y}}\cdot\hat{\mathbf{n}}'}{|\mathbf{r}-\mathbf{r}'|}\right) + \hat{\mathbf{y}}\left(A\frac{x-x'}{|\mathbf{r}-\mathbf{r}'|} - B\frac{\hat{\mathbf{x}}\cdot\hat{\mathbf{n}}'}{|\mathbf{r}-\mathbf{r}'|}\right), \tag{4.141}$$

where

$$A = \left(\frac{-i}{8}\left(H_0^{(1)}(k_0 n_u|\mathbf{r}-\mathbf{r}'|) - H_2^{(1)}(k_0 n_u|\mathbf{r}-\mathbf{r}'|)\right)(k_0 n_u)^2\right.$$

$$\left.-\frac{B}{|\mathbf{r}-\mathbf{r}'|}\right)\frac{\mathbf{r}'-\mathbf{r}}{|\mathbf{r}'-\mathbf{r}|}\cdot\hat{\mathbf{n}}', \tag{4.142}$$

and

$$B = -\frac{i}{4}H_1^{(1)}(k_0 n_u|\mathbf{r}-\mathbf{r}'|)k_0 n_u. \tag{4.143}$$

If the incident magnetic field is a plane wave given by

$$H_0(\mathbf{r}) = Ae^{ik_0 n_1(\cos(\alpha)x+\sin(\alpha)y)} \tag{4.144}$$

then the incident electric field will be given by

$$\mathbf{E}_0 = \sqrt{\frac{\mu_0}{\varepsilon_0}}\frac{1}{n_1}(-\hat{\mathbf{x}}\sin(\alpha)+\hat{\mathbf{y}}\cos(\alpha))H_0(\mathbf{r}). \tag{4.145}$$

The term $\hat{\mathbf{z}}\times\nabla(\hat{\mathbf{n}}'\cdot\nabla' g_1(\mathbf{r},\mathbf{r}'))_{\mathbf{r}'=\mathbf{s}(t)}$ is highly singular as $\mathbf{r}\rightarrow\mathbf{s}(t)$, and thus an extra careful integration is required if $\mathbf{r}$ is close to the scatterer surface. With pulse expansion of the surface magnetic field, the jumps in the field along the surface from one section to the next will cause problems when calculating the electric field very close to the scatterer surface. There is no problem though for positions at a distance from the surface on the order of the length of the nearby sections. When using at least a linear field approximation within sections, then right at the scatterer surface near, e.g., element j, both the normal derivative $\phi = \hat{\mathbf{n}}\cdot\nabla H$ and the tangential derivative $dH(\mathbf{s}(t))/dt$, are known, and thus one might as well use

$$\left(\frac{\partial H}{\partial x}\right)_u \approx \hat{\mathbf{x}}\cdot\left(\hat{\mathbf{n}}\phi_{u,j}+\frac{d\mathbf{s}(t)}{dt}\frac{dH(\mathbf{s})(t)}{dt}\right), \tag{4.146}$$

$$\left(\frac{\partial H}{\partial y}\right)_u \approx \hat{\mathbf{y}}\cdot\left(\hat{\mathbf{n}}\phi_{u,j}+\frac{d\mathbf{s}(t)}{dt}\frac{dH(\mathbf{s})(t)}{dt}\right), \tag{4.147}$$

instead of the integration. The electric field can then be obtained by using (4.146) and (4.147) in (2.17).

4.1.8 EXAMPLES OF METAL NANOSTRIP RESONATORS

In this section examples of the optics of metal nanostrip resonators will be presented. All calculations of optical cross sections and fields of nanostrip resonators presented in this section have been made using the GFSIEM. This will be supplemented with calculations of the optical modes of metal films made using the methods in Appendix F. Metal nanostrips are interesting nanostructures because they can function as an optical resonator (or an optical antenna), and the electric field can be significantly enhanced near the metal nanostructure. The field enhancement is interesting for obtaining stronger light-matter interaction and has applications in sensors and for enhanced Raman scattering, etc.

Two types of metal nanostrip resonators are illustrated in Figs. 4.14(a) and (b) based on either a single nanostrip or two nanostrips. The nanostrips are characterized by a width w, thickness d, and dielectric constant ε_{m}, and the surrounding dielectric has dielectric constant ε_d. In the case of two nanostrips, they are separated vertically by a distance δ.

The single nanostrip can be thought of as a nanometer-thin metal film being terminated at two ends. A thin metal film supports p-polarized guided modes with magnetic field of the form

$$H(\mathbf{r}) = Ae^{\pm ik_0 n_{\mathrm{m}} x} f(y), \tag{4.148}$$

where $f(y)$ is a function that will decrease exponentially as y moves away from the metal film and into the surrounding dielectric. The phase thus changes upon propagation along the x-axis in the same way as a plane wave would change its phase

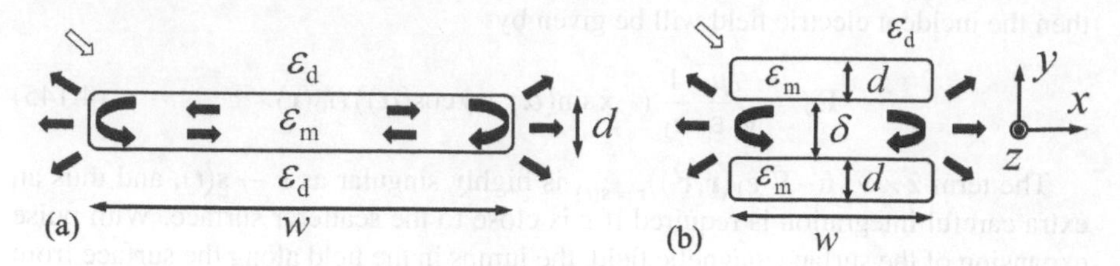

Figure 4.14 Illustration of nanostrip resonators based on (a) a single metal nanostrip of width w, thickness d, and dielectric constant ε_m surrounded by a dielectric with dielectric constant ε_d, and (b) two nanostrips separated vertically by a gap δ.

if propagating in a medium with refractive index n_m. The thin-metal film will only support two types of guided modes (for appropriate values of the metal dielectric constant), which are known as long-range and short-range surface plasmon polaritons (SR-SPPs and LR-SPPs) [44, 45]. In the case of the nanostrip an SPP propagating along the strip and reaching a termination may be reflected into a backward propagating SPP and will incur a reflection phase of ϕ_1 and ϕ_2 at the two ends, respectively. The situation is thus quite similar to a Fabry-Perot resonator [38], and strong resonant fields can be built up when the resonance condition

$$k_0 n_m(2w) + \phi_1 + \phi_2 = m2\pi, \qquad m = 0, \pm 1 \pm 2, \ldots \tag{4.149}$$

is satisfied. The field should thus be in phase with itself after propagating one round-trip back and forth. This is the same condition that applies for the resonance of, e.g., a guitar string, where a mechanical wave must be in phase with itself after one round-trip between the two ends of the string. It is also the condition that applies for longitudinal laser modes in a laser cavity.

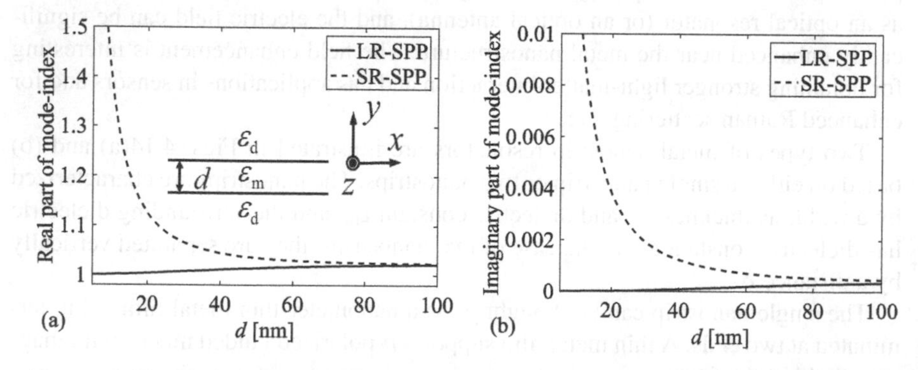

Figure 4.15 Mode index of SPPs supported by a silver film of thickness d surrounded by vacuum at the wavelength 700 nm. (a) Real part, and (b) imaginary part.

The mode index of SR-SPPs and LR-SPPs for a silver film surrounded by air

($\varepsilon_d = 1$) is shown in Fig. 4.15 versus the film thickness d. This figure was calculated using the method of Appendix F with refractive index of silver at wavelength 700 nm given by $0.0412 + 4.7951i$ [40]. The real part of the mode index is shown in Fig. 4.15(a), and the imaginary part is shown in Fig. 4.15(b). The mode index of the LR-SPP approaches the refractive index of vacuum ($n = 1$) as $d \to 0$. This implies that the LR-SPP becomes delocalized. On the other hand the SR-SPP mode-index increases drastically as $d \to 0$ implying that the SR-SPP becomes strongly bound to the metal film. For SR-SPPs the imaginary part of the mode index also increases for thinner films, which means higher propagation losses. The round-trip distance in a nanostrip can be quite short though, being in the few 100-nm range, and the round-trip propagation loss does not hinder the existence of a resonance. For increasing d the mode-index difference between the SR-SPP and the LR-SPP diminishes, and for both modes the mode-index approaches the mode index of an SPP on a semi-infinite metal given by [16]

$$n_m = \sqrt{\frac{\varepsilon_d \varepsilon_m}{\varepsilon_d + \varepsilon_m}}. \tag{4.150}$$

The magnetic-field profiles ($f(y)$) for the LR-SPP and SR-SPP are shown in Figs. 4.16(a) and (b), respectively. The magnetic field of the LR-SPP is symmetric, and for the SR-SPP it is antisymmetric. Both fields decrease exponentially away from the metal film. The SR-SPP is clearly much more strongly localized to the metal film as compared with the LR-SPP. The LR-SPP has most of its energy placed outside the film, and thus ohmic losses due to interaction with the metal will be small, which explains the small imaginary part of the mode index. Although the LR-SPP propagation losses are small, which is advantageous, on the other hand, as the LR-SPP reaches a termination of the strip, the reflection into a backward propagating LR-SPP will also be small due to its weakly bound nature. This mode thus does not cause any significant resonances due to a large round-trip reflection loss. Note that the SR-SPP is really restricted to propagating along the x-axis. The resonance wavelengths of the nanostrip resonator should thus not be very sensitive to the direction of the incident light.

It is usually the electric field which is of most interest. The y-component of the SR-SPP electric field is shown in Fig. 4.16(c). Due to the boundary condition (2.23) the y-component of the electric field increases in magnitude by a factor of $|\varepsilon_m/\varepsilon_d|$ across the metal-dielectric interface, which amounts to a factor of 23 for the present case. On the other hand, the x-component of the SR-SPP electric field, shown in Fig. 4.16(d), does not jump across the metal-dielectric interfaces, which follows from the boundary condition (2.21). The x-component of the field thus dominates inside the metal film and is also not small compared with the y-component outside the film. The SR-SPP thus has a significant electric field component along its axis of propagation, which is a consequence of its strongly bound nature. Note that the x- and y-components are mutually out of phase.

The first nanostrip that will be considered is a silver nanostrip of width $w = 168$ nm, thickness $d = 10$ nm, and with corners that are quarter circles with radius 2 nm. The optical cross sections for the case of illuminating the nanostrip with a plane wave

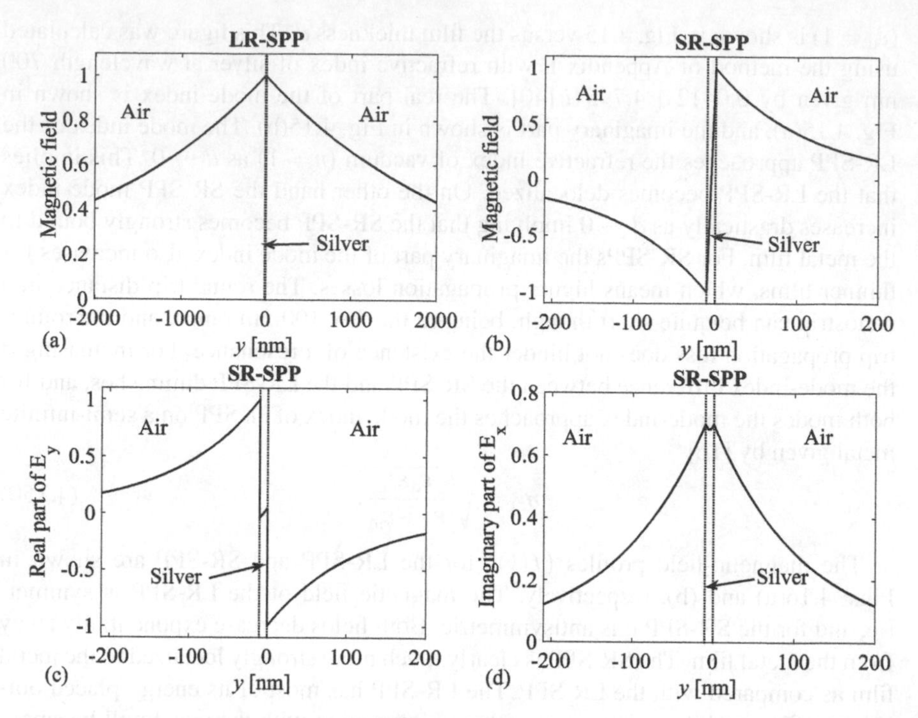

Figure 4.16 Real part of magnetic field along y for (a) LR-SPP, and (b) SR-SPP, supported by a silver film of thickness $d = 10$ nm surrounded by air. The wavelength is 700 nm. (c) Real part of the y-component of the electric field of the SR-SPP. (d) Imaginary part of the x-component of the electric field of the SR-SPP.

propagating at an angle of $-45°$ to the x-axis are shown in Fig. 4.17(a) versus the wavelength of the incident light. The extinction and scattering cross sections peak at the wavelength 690 nm. The absorption cross section is significantly smaller than the scattering cross section, and the extinction cross section is thus mainly due to scattering. The scattering cross section has previously been shown for this geometry in Ref. [50]. A similar calculation for incident light at an angle of $-90°$ to the x-axis is shown in Fig. 4.17(b). It can be seen that changing the angle of light incidence does not lead to much of a change in the wavelength where the scattering and extinction cross sections peak. This is in accordance with the interpretation of the nanostrip as a resonator with resonances due to SR-SPPs propagating back and forth between strip terminations along the x-axis only. Since no other waves can be excited that will give a resonance, the angle of light incidence only has an effect on how efficiently the resonance is excited.

The magnitude of the electric field for the case of incident light with wavelength 690 nm propagating at an angle of $-90°$ is shown in Fig. 4.17(c) for a region in the xy-plane that includes the nanostrip, and in Fig. 4.17(d) along the line $y = 0$ going

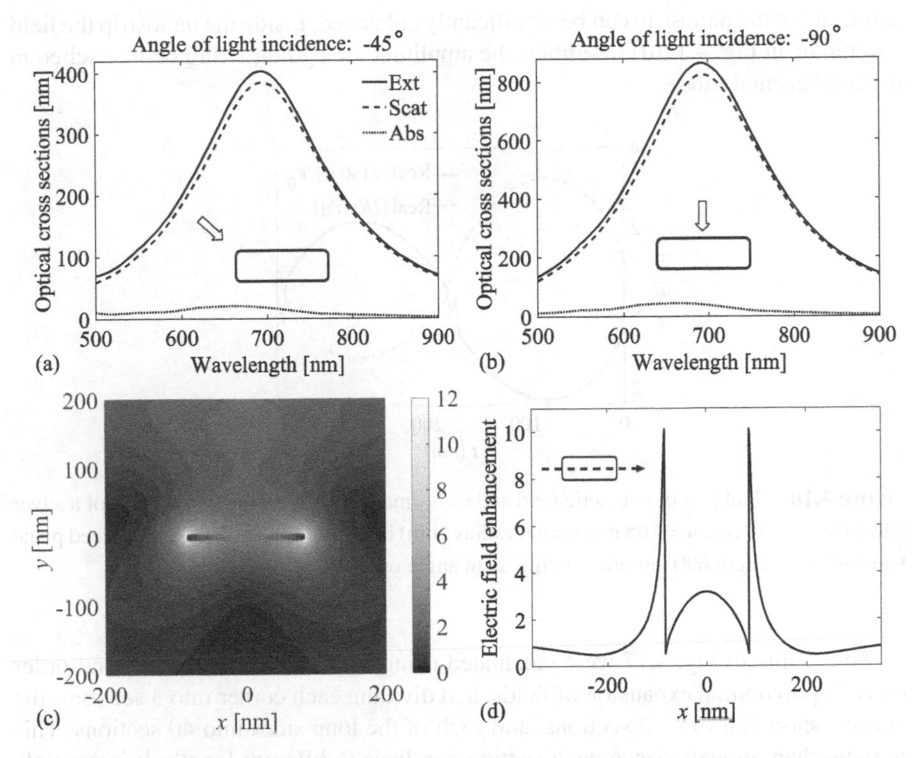

Figure 4.17 (a,b) Optical cross sections for a silver nanostrip of thickness d = 10 nm surrounded by air being illuminated by a p-polarized plane wave propagating at an angle of $-45°$ and $-90°$ to the x-axis, respectively. (c,d) Enhancement of the electric field magnitude relative to the incident electric field in the xy-plane and along the line $y = 0$ going through the nanostrip, respectively, in both cases at the wavelength 690 nm.

through the middle of the nanostrip. The magnitude of the field is normalized with the magnitude of the incident field, and the figure thus shows the enhancement of the field magnitude. The total field in Fig. 4.17(d) is dominated by the x-component of the electric field, and since this field component is normal to the nanostrip surface at the nanostrip terminations, the field magnitude thus jumps at $x = \pm$ 84 nm such that the electric field magnitude just outside the nanostrip is increased by a factor $|\varepsilon_m/\varepsilon_d|$ compared with the electric field just inside the nanostrip. This effect easily gives a factor of 10 in field enhancement. The energy density built up in the electric field is given by

$$u_E = \frac{1}{2}\varepsilon_0\varepsilon|\mathbf{E}|^2, \tag{4.151}$$

which is part of the total energy density being integrated in Eq. (2.36). The factor-of-10 field enhancement thus translates to a factor 10^2 enhancement in energy density. This already promises that the coupling between incident light and a molecule placed at the end of the nanostrip can be significantly enhanced. Inside the nanostrip the field magnitude in Fig. 4.17(d) resembles the amplitude of a guitar string being excited in the fundamental mode.

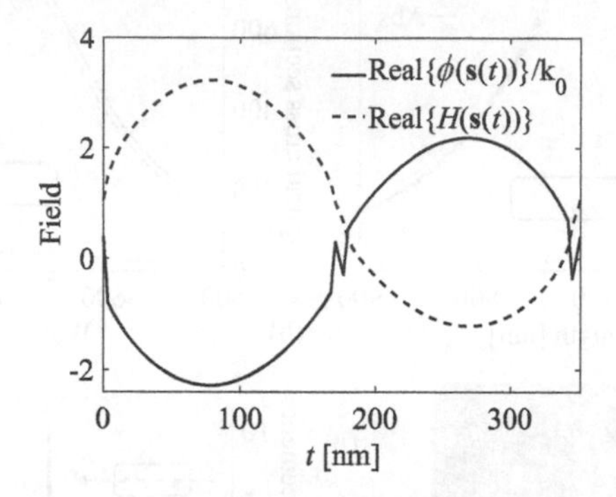

Figure 4.18 Real part of magnetic field and its normal derivative along the surface of a silver nanostrip ($d = 10$ nm, $w = 168$ nm, corner radius 2nm) being illuminated by a *p*-polarized plane wave of wavelength 690 nm propagating at an angle of $-45°$.

The results in Fig. 4.17 were calculated using the GFSIEM with a second-order ($m = 2$) polynomial expansion of fields, and dividing each corner into 3 sections, the vertical short sides into 5 sections, and each of the long sides into 40 sections. This is more than enough. Since each section can have a different length, it is possible to give special attention to regions where fields change fast with distance along the surface. The surface magnetic field and the normal derivative versus distance t along the nanostrip surface at resonance (wavelength 690 nm) and angle of incidence $-45°$ are shown in Fig. 4.18.

The point with this figure is that at corners and short sides, the field $\phi(\mathbf{s}(t))$ may change fast. A description of $\phi(\mathbf{s}(t))$ using the simple Fourier-expansion approach Eq. (4.122) thus requires a large number of terms N. With the polynomial expansion approach it is easy to give particular attention to those regions. It is of course also possible to develop more advanced Fourier-expansion approaches giving extra attention to these regions as well.

If the width of the nanostrip is increased slightly, the resonance wavelength will increase as well. The exact peak wavelength of cross sections for another strip width w is difficult to predict though from Eq. (4.149) because the mode index n_{m} and the reflection phases ϕ_1 and ϕ_2 depend on the wavelength. However, according to Eq. (4.149) the peak wavelength for the next excitation mode ($m \rightarrow m+1$) will appear at the same wavelength when increasing the strip width by half an SR-SPP wavelength. Thus, by increasing the strip width by $\Delta w = \lambda_{\mathrm{p}}/\mathrm{Real}\{n_{\mathrm{m}}(\lambda_{\mathrm{p}})\}$, where $\lambda_{\mathrm{p}} = 690$ nm is the resonance wavelength, a new resonance must be found at approximately the same wavelength. This leads to increasing the width of the nanostrip from 168 nm to 413 nm. The optical cross-section spectra for this nanostrip are shown in Figs. 4.19(a)-(d) for the angles of light incidence $-30°$, $-45°$, $-70°$, and

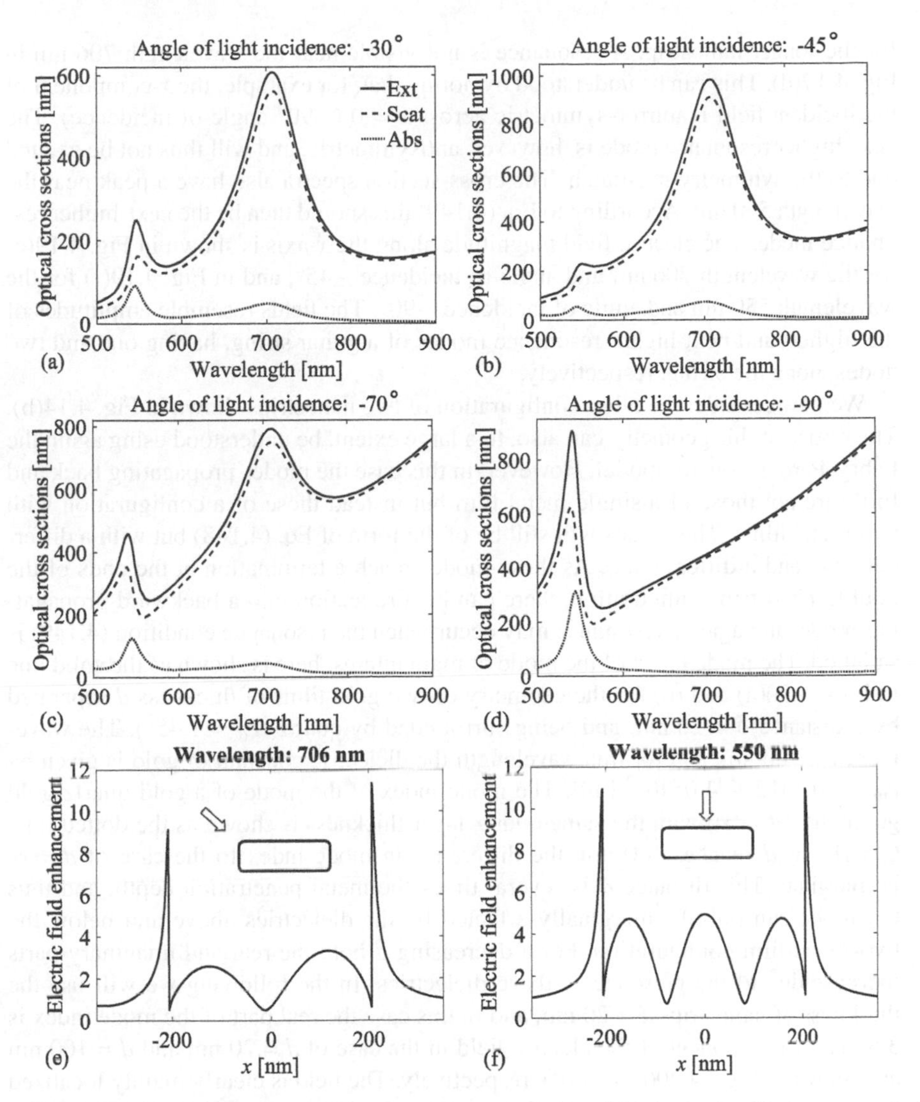

Figure 4.19 Optical cross sections for a silver nanostrip of thickness $d = 10$ nm surrounded by air being illuminated by a p-polarized plane wave propagating at an angle of (a) $-30°$, (b) $-45°$, (c) $-70°$, and (d) $-90°$ to the x-axis. Enhancement of the electric field magnitude along the line $y = 0$ going through the middle of the nanostrip for (e) wavelength 706 nm and angle of incidence $-45°$, and (f) wavelength 550 nm and angle of light incidence $-90°$.

$-90°$, respectively. For the angles of incidence of $-30°$, $-45°$, and $-70°$, a resonance peak is found near the wavelength 706 nm, which is quite close to the peak wavelength 690 nm found for the shorter nanostrip. The interpretation of resonances given in Eq. (4.149) thus predicts reasonably well the resonance peak wavelength for the longer nanostrip. A resonance is not visible near the wavelength 706 nm in Fig. 4.17(d). This can be understood by noting that, for example, the x-component of the incident field is mirror-symmetric across $x = 0$ (-90° angle of incidence). The next-higher resonance mode is, however, antisymmetric, and will thus not be excited due to the symmetry mismatch. The cross-section spectra also have a peak near the wavelength 550 nm. According to Eq. (4.149) this should then be the next-higher resonance mode. The electric field magnitude along the x-axis is shown in Fig. 4.19(e) for the wavelength 706 nm and angle of incidence -45°, and in Fig. 4.19(f) for the wavelength 550 nm and angle of incidence -90°. The fields resemble amplitudes of the higher- and next-higher resonance modes of a guitar string, having one and two nodes along the string, respectively.

We now consider the other configuration of two nanostrips shown in Fig. 4.14(b). The optics of this geometry can also, to a large extent, be understood using again the Fabry-Perot resonator model. However, in this case the modes propagating back and forth are not those of a single metal film but instead those of a configuration with two metal films. The modes will still be of the form of Eq. (4.148) but with a different $f(y)$ and a different n_{m}. As these modes reach a termination at the ends of the double-nanostrip configuration, there can be a reflection into a backward propagating mode, and again a resonance may occur when the resonance condition (4.149) is satisfied. The mode index of the mode of main interest here is shown as the solid line in Figs. 4.20(a) and (b) for the geometry of two gold films of thickness d separated by a distance $\delta = 20$ nm, and being surrounded by quartz ($\varepsilon_{\mathrm{d}} = 1.45^2$). The wavelength is 700 nm, and for this wavelength the dielectric constant of gold is given by $\varepsilon_{\mathrm{m}} = (0.1312 + i4.0540)^2$ [40]. The mode index of the mode of a gold-quartz-gold geometry ($d = \infty$) with the same quartz-layer thickness is shown as the dotted line. Clearly, as d reaches 100 nm, the difference in mode index to the case of $d = \infty$ is marginal. This distance d is several times the metal penetration depth, and thus the mode can only be marginally affected by the dielectrics above and below the two-metal-film configuration. For a decreasing d both the real and imaginary parts increase due to the presence of these dielectrics. In the following we will use the thickness of nanostrips $d = 20$ nm, and in this case the real part of the mode index is 3.8. The y-component of the electric field in the case of $d = 20$ nm and $d = 100$ nm are shown in Figs. 4.20(c) and (d), respectively. The field is clearly mainly localized to the gap between the two metal films, and for $d = 100$ nm the field in the dielectrics on both sides of the two-metal-film configuration is very small. The modes are thus often referred to as gap SPPs.

The scattering cross-section spectra are shown in Fig. 4.21(a) for the nanostrip widths $w = 66$, 158, 250, and 342 nm, and for different angles of light incidence relative to the x-axis indicated in the legends. These geometries have previously been considered in [46]. For $w = 66$ nm a small resonance peak can be seen at the wavelength 700 nm. The other strip widths have been chosen by noting that the resonance condition (4.149) predicts that the next-higher resonances can be found at the same wavelength for a strip that is longer by an integer times the amount 700 nm / 2 / 3.8 = 92 nm. A small resonance is also seen at the wavelength 700 nm for $w = 158$ nm. For

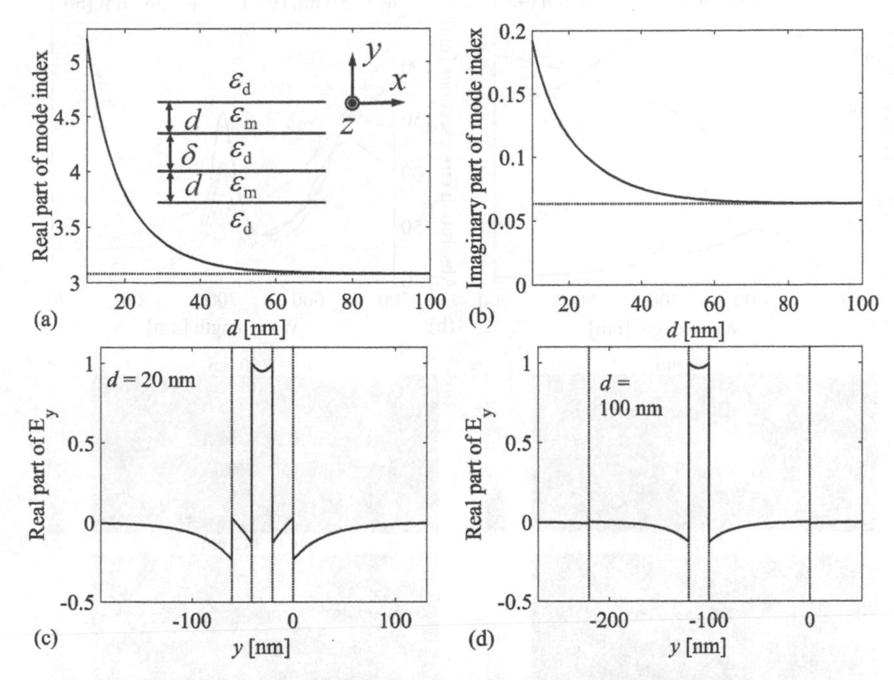

Figure 4.20 (a,b) Real and imaginary part of mode index of the gap-SPP mode in a waveguide consisting of two gold films of thickness d separated by a distance of $\delta = 20$ nm. Between and outside the gold layers, the material is quartz ($n = 1.5$). The wavelength is 700 nm. The inset illustrates the waveguide geometry. (c,d) The y-component of the electric field of the gap-SPP mode for the case of $d = 20$ nm, and $d = 100$ nm, respectively.

the longer strips of $w = 250$ and 342, nm the resonance at this wavelength is not easily identifiable from the scattering cross section. The absorption cross section shown in Fig. 4.21(b) on the other hand shows a clear resonance peak at the wavelength 700 nm for all considered strip widths. This difference is because of high reflection of the gap SPP at strip terminations, and the round-trip loss of gap SPPs propagating back and forth along the x-axis in the two-strip resonator is thus dominated by absorption losses, and especially so for increasing w. Fabry-Perot resonances due to gap-plasmons have also been considered in, for example, Refs. [47, 48, 49].

Maps of the magnitude of the electric in the xy-plane at wavelength 700 nm is shown for the four considered strip widths in Fig. 4.21(c). The angles of light incidence are the same as were considered in Fig. 4.21(a) and (b). It can be seen that there is a relatively large field in the gap between the two metal nanostrips. In the case of $w = 66$ nm the field magnitude inside the gap has one node. For $w = 158$, 250, and 342 nm, the field magnitude has two, three, and four nodes, respectively. This is also the type of behavior that can be expected for resonance modes that follow

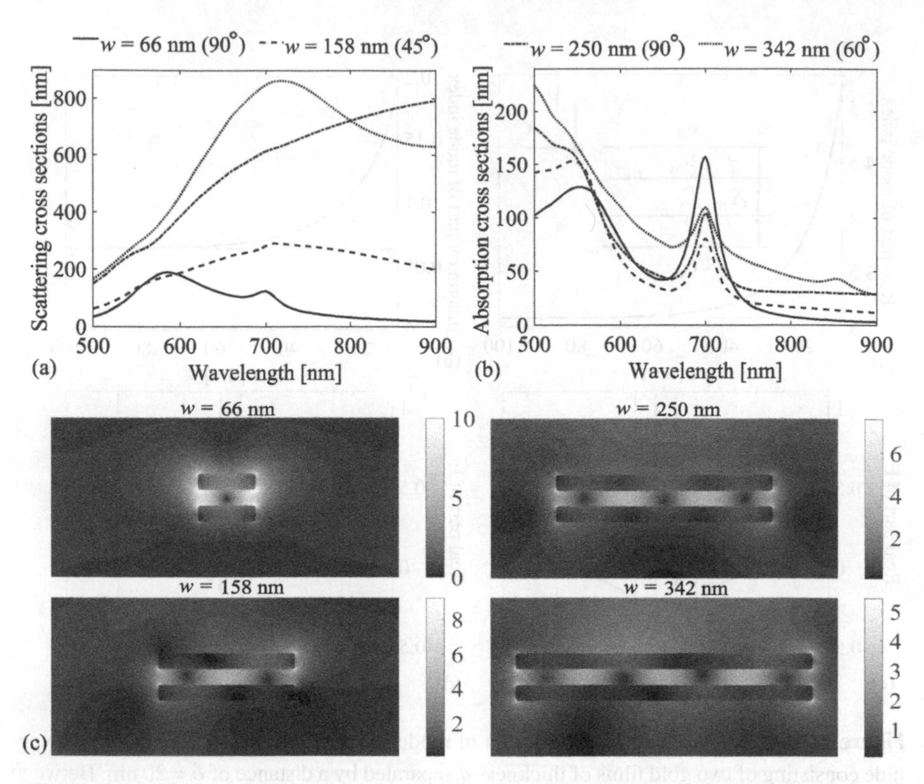

Figure 4.21 (a) Scattering cross-section spectra, and (b) absorption cross-section spectra, for two gold nanostrips of thickness 20 nm separated vertically with a gap of 20 nm, surrounded by quartz. The width of the strips w and the angle of light incidence for each spectrum are indicated in the legend. (c) Corresponding magnitude of electric field in the xy-plane at the wavelength 700 nm for each considered strip width.

Eq. (4.149). Here, the fundamental mode has a node in the middle of the resonator, which was not the case for the single silver nanostrip. The difference lies in the sign of the reflection coefficient of the relevant field component of modes propagating back and forth when they are reflected at the terminations.

We now return to the case of the single silver nanostrip and consider an approach for obtaining an even higher field enhancement by introducing a small gap in the nanostrip at positions with large resonant fields [50, 51, 52, 53]. As an example of this approach, the nanostrip of width w = 168 nm and thickness d = 10 nm is considered once more, but it is divided into two nanostrips of width w = 84 nm with a small gap of δ = 5 nm between the strips. This geometry is illustrated in Fig. 4.22(a). The extinction cross-section spectrum is shown as the solid line in Fig. 4.22(b). Due to the gap the resonance wavelength is blue-shifted by approximately 100 nm. It is also shown that the resonance wavelength can be shifted back again to the original resonance wavelength of 690 nm by increasing the width of the strips to w = 110

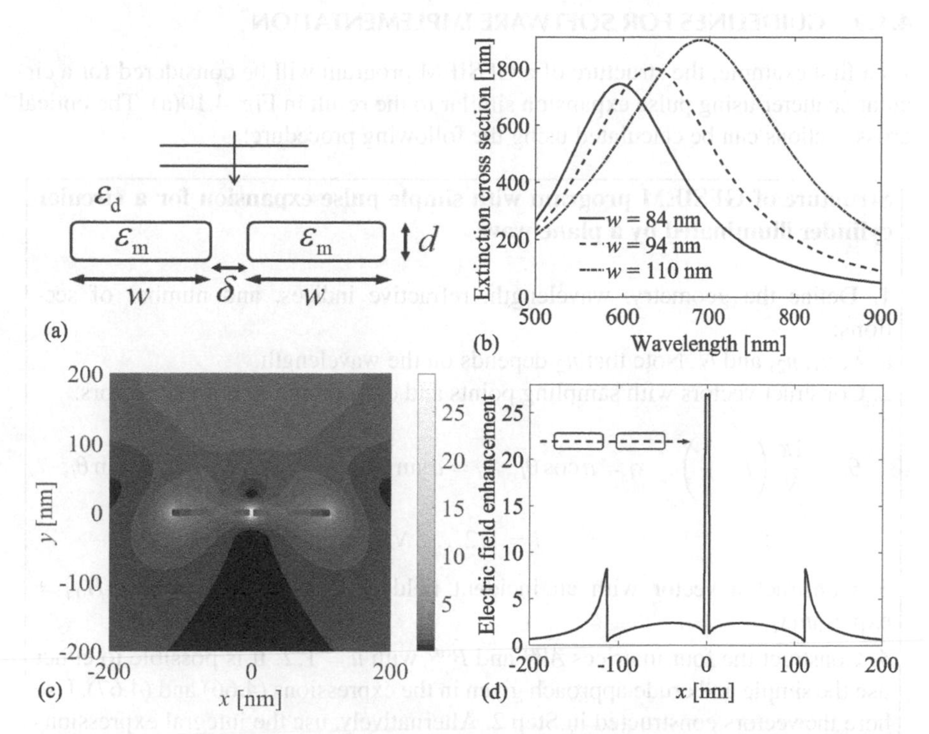

Figure 4.22 (a) Schematic of two silver nanostrips of thickness d and width w separated by a gap δ along the x-axis being illuminated by a normally incident p-polarized plane wave. (b) Corresponding optical cross-section spectra for d = 10 nm, and w = 84 nm, 94 nm and 110 nm. (c) Map of the electric field magnitude enhancement in the xy-plane for w = 110 nm, d = 10 nm, and wavelength 690 nm. (d) Corresponding electric field magnitude along the x-axis going through the center of both nanostrips.

nm. The spectra were calculated for light propagating at -90° to the x-axis, which was also optimum for exciting the fundamental resonance of the strip of width 168 nm. The electric-field-magnitude enhancement at wavelength 690 nm is shown in Figs. 4.22(c) and (d) showing a 2D map of the field or the field along the line $y = 0$, respectively. The field enhancement in the small gap between strips reaches a factor of approximately 27, which is a significant improvement over the field enhancement at the ends of the single strip. The improvement is obtained because now as the field jumps across the metal-dielectric interface at the gap, it jumps from a higher value inside the metal, which is a consequence of having placed the gap at a field anti-node.

While focus here has been on metal nanostructures, the GFSIEM has also been applied for modeling of, for example, diffractive optical elements [54].

4.1.9 GUIDELINES FOR SOFTWARE IMPLEMENTATION

As a first example, the structure of a GFSIEM program will be considered for a circular scatterer using pulse expansion similar to the result in Fig. 4.10(a). The optical cross sections can be calculated using the following procedure:

Structure of GFSIEM program with simple pulse expansion for a circular cylinder illuminated by a plane wave

1. Define the geometry, wavelength, refractive indices, and number of sections:
a, λ, n_1, n_2, and N. Note that n_2 depends on the wavelength.
2. Construct vectors with sampling points and corresponding normal vectors:

$$\theta_i \equiv \frac{2\pi}{N}\left(i-\frac{1}{2}\right), \quad x_i = a\cos\theta_i, \quad y_i = a\sin\theta_i, \quad n_{x,i} = \cos\theta_i, \quad n_{y,i} = \sin\theta_i,$$

$$i = 1,2,\ldots,N.$$

3. Construct a vector with an incident field at the sampling points: $H_{0,i} = \exp(ik_0 n_1 x_i)$.
4. Construct the four matrices $\overline{\overline{A}}^{(u)}$ and $\overline{\overline{B}}^{(u)}$ with $u = 1,2$. It is possible to either use the simple and crude approach given in the expressions (4.66) and (4.67). Use here the vectors constructed in Step 2. Alternatively, use the integral expressions (4.53) and (4.64) and carry out the integrals to high precision by defining analytic expressions for the integrands, seting up the integration limits, and then using these in, for example, Gauss-Kronrod integration routines.
5. Solve the matrix equation (4.52) and find H_i and $\phi_{1,i}$ which are the magnetic field and its normal derivative at each sampling point.
6. Use H_i and $\phi_{1,i}$ to calculate the absorption cross section (4.68).
7. Calculate the scattered far field in the direction $\theta = 0°$ by using Eq. (4.69) and subsequently use this result in Eq. (2.89) to obtain the extinction cross section. Alternatively, in the case of setting up matrix elements by rigorous integration, the scattered far field may also be constructed using a more precise evaluation based on integration using (4.22).
8. Calculate the scattered far field for, e.g., 500 directions θ, and use this to numerically calculate the scattering cross section using Eq. (2.90).
9. Wrap the whole program in a for-loop where a new wavelength is considered in each step. Store the cross sections for each wavelength in vectors.
10. Plot the optical cross sections versus wavelength.

As a second example, the structure of a GFSIEM program will be considered for one or two nanostrips using higher-order polynomial expansion of fields. Note that this program can be applied for the circular cylinder as well. The optical cross sections and field plots can be calculated using the following procedure:

Structure of GFSIEM program with higher-order polynomial expansion of fields for nanostrips illuminated by a plane wave

1. Define the nanostrip dimensions and gap: w, d, r_c, δ, number of sections for each corner and each straight side: N_c, N_w, N_d, refractive indices: n_1, n_2, wavelength and angle of light incidence: λ, α, polynomial expansion order: m, the region in the xy-plane for field calculation and number of sampling points along x and y: $x_{\text{p, min}}$, $x_{\text{p, max}}$, $y_{\text{p, min}}$, $y_{\text{p, max}}$, $N_{\text{p}, x}$, and $N_{\text{p}, y}$.
2. Construct a matrix where each row represents a section. The first column identifies if the section is part of a circle or is a straight line. For a circle, the next 6 columns describe the circle center x_c, y_c, radius r_c, start and end corner angles θ_1 and θ_2, and direction of the normal vector. For a line, the 6 columns contain the start and end of the line and the normal vector, x_1, y_1, x_2, y_2, n_x and n_y.
3. Construct matrices with sampling points $\mathbf{s}_i^{(v)}$. Let each row refer to an i, and columns to v. Use data from Step 2.
4. Construct vectors describing a sub-sampling of the scatterer surface. Each section is subdivided into $N_s \cdot m$ subsections, and vectors are constructed containing for each subsection the positions, normal vector, and length x, y, n_x, n_y, Δt.
5. Plot the structure for a visual check.
6. Construct $\overline{H}_0^{(v)}$, $v = 0, 1, \ldots, m-1$, as in (4.88). Use sampling points from Step 3.
7. Construct the matrices $\overline{\overline{D}}_H$ and $\overline{\overline{D}}_\phi$.
8. Construct the matrices $\overline{\overline{A}}^{(u,v,v')}$ and $\overline{\overline{B}}^{(u,v,v')}$. Use (4.91) and (4.92) for the matrix elements $A_{i,j}^{(u,v,v')}$ and $B_{i,j}^{(u,v,v')}$. Loop over $v = 0 : m-1$, $v' = 0 : m$, $i = 1 : N$, $j = 1 : N$. If $|\mathbf{s}_i^{(v)} - \mathbf{s}(t)|$ is larger than a threshold for all t on j, then calculate matrix elements as a simple summation over $N_s \cdot m$ terms using data from Step 4. If not, apply an analytic expression for the integrand and integration limits in an integration routine (e.g. Gauss-Kronrod). Give special attention when $\mathbf{s}(t)$ crosses the point $\mathbf{s}_i^{(v)}$.
9. Construct and solve the matrix equation (4.96) using data from Steps 8 and 6.
10. Use the result from Step 9 in (4.97) to calculate the absorption cross section.
11. Calculate the extinction and absorption cross sections. Construct vectors with H and ϕ_1 at positions from Step 4. Use this to calculate the scattered far field (4.99) as a summation over $N \cdot N_s \cdot m$ terms. Alternatively, use rigorous integration.
12. Calculate the fields. Loop over all positions $\mathbf{r}$ where the field should be calculated. Determine if $\mathbf{r}$ is inside or outside the scatterer using (4.137). Calculate the field at $\mathbf{r}$ using (4.7), (4.138), or (4.16), (4.139). If the distance to the scatterer is beyond a threshold, use simple summation based on data from Steps 4 and 11. Otherwise use rigorous integration.
13. Wrap the whole program in a for-loop scanning a wavelength interval.
14. Plot the optical cross sections versus wavelength. Plot the calculated fields.

4.1.10 EXERCISES

1. Prove that the Green's function (4.12) is a solution of Eq. (4.9). Hints: For simplicity choose $\mathbf{r}' = \mathbf{0}$. Insert (4.12) into the left-hand side of (4.9) and integrate over the xy-plane. Notice that only an infinitesimal part of the xy-plane will contribute. Evaluate $\nabla H_0^{(1)}(k_0 n_1 r)$ in the limit of $r \to 0$. Furthermore, note that $\nabla^2 f(\mathbf{r}) = \nabla \cdot \nabla f(\mathbf{r})$ and apply Gauss's law to rewrite the area integral into a surface integral. Show that the integral has the value -1.

2. Show that the integral (4.14) vanishes.

3. Give expressions for the finite elements in Eq. (4.32) for $m = 4$, and plot the functions similar to the plots for $m = 2$ and 3 in Fig. 4.6.

4. Verify Eqs. (4.63) and (4.64).

5. Verify Eqs. (4.97) and (4.99).

6. Insert (4.41) in the integral equations. Let $\mathbf{r} \to \mathbf{s}_i^{(s)}$ from either side of the scatterer surface, and calculate the tangential derivative. This leads to, for example,

$$H_{1,i,3} = H'_{0,i} - \sum_{j=1}^{N} \sum_{v=0}^{3} \left\{ C_{i,j}^{(1,v)} \phi_{1,j,v} - D_{i,j}^{(1,v)} H_{1,j,v} \right\}. \tag{4.152}$$

Here, $H'_{0,i} \equiv \left(\frac{dH_0(\mathbf{s}(t))}{dt} \right)_{t=t^{(s,i)}}$, with the matrix elements given by

$$C_{i,j}^{(u,v)} = \left(\frac{d}{dt} P \int_{t'} g_u(\mathbf{s}(t), \mathbf{s}(t')) f_j^{(\mathrm{SL},v)}(t') dt' \right)_{t=t^{(s,i)}}, \tag{4.153}$$

$$D_{i,j}^{(u,v)} = \left(\frac{d}{dt} P \int_{t'} \left(\hat{\mathbf{n}}' \cdot \nabla' g_u(\mathbf{s}(t), \mathbf{r}') \right)_{\mathbf{r}'=\mathbf{s}(t')} f_j^{(\mathrm{SL},v)}(t') dt' \right)_{t=t^{(s,i)}}. \tag{4.154}$$

Verify these expressions and show that for, e.g., $i = j$ and $v = 0$ some of these matrix elements will be singular. The approach of using the field expansion (4.41) and point-matching at ends of sections is thus not suitable for numerical computation.

7. Follow the guidelines in Sec. 4.1.9 and develop a GFSIEM code capable of calculating optical cross sections for a circular cylinder. Use the simplest possible approach. Reproduce Fig. 4.10(a).

8. Follow the guidelines in Sec. 4.1.9 and develop a GFSIEM code capable of calculating optical cross sections and fields for one or two metal nanostrips. Experiment with making calculations for nanostrips of different widths and thicknesses and compare with Eq. (4.149).

4.2 SCATTERER ON OR NEAR PLANAR SURFACES

In this section the GFSIEM (2D) method is considered for the type of geometry illustrated in Fig. 4.23, where a scattering object is placed above or on a layered reference structure. The scatterer with dielectric constant ε_2 is placed in the upper layer with dielectric constant ε_1. The other layers have dielectric constants ε_{L2}, ε_{L3}, ..., and thicknesses d_2, d_3, ...

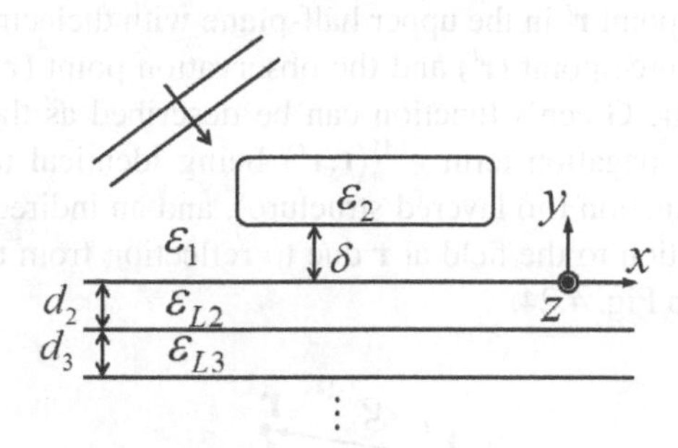

Figure 4.23 Schematic of scattering by an object placed above or near a layered structure.

The field inside the scatterer can be obtained using the same integral equation (4.7) and the same Green's function $g_2(\mathbf{r},\mathbf{r}')$ as in the case of the homogeneous-medium reference structure. The field outside the scatterer can also still be obtained using the same integral equation (4.16) except that the Green's function $g_1(\mathbf{r},\mathbf{r}')$ must now govern a layered reference structure, and the reference field (e.g. H_0 for p polarization) must also be a solution for the layered reference geometry.

This section is organized as follows. In subsection 4.2.1 the Green's function $g_1(\mathbf{r},\mathbf{r}')$ will be derived and expressed in terms of integrals that take into account Fresnel reflection and transmission from the layers. The numerical calculation of the integrals will be considered including the handling of poles in involved Fresnel reflection and transmission coefficients. In addition, far-field approximations to the Green's function will be obtained that are needed for calculating optical cross sections including cross sections for the excitation of guided modes. In subsection 4.2.2 the GFSIEM will be considered for a layered reference structure. The limit ($\delta \rightarrow 0$ in Fig. 4.23) will be of special interest, since the presence of the layers leads to an additional singularity in the Green's function for p polarization in the limit $x - x' \rightarrow 0$ and $y + y' \rightarrow 0^+$. The calculation of fields using the angular spectrum representation is considered in Sec. 4.2.3. Finally, the subsections 4.2.4-4.2.8 consider a number of examples of applying the GFSIEM for layered reference structures. This includes a metal nanostrip on a single dielectric substrate supporting no guided modes (Sec. 4.2.4), a silver nanostrip placed above a silver surface supporting SPP modes (Sec. 4.2.5), a single groove in a metal surface that can be a resonator or broadband absorber or scatterer (Sec. 4.2.6), a silver nanostrip on a thin-film silicon-on-silver waveguide supporting from one to many guided modes (Sec. 4.2.7), and finally a microstructured gradient-index lens for THz photoconductive antennas for out-coupling and collimating light emitted by a dipole (Sec. 4.2.8).

4.2.1 GREEN'S FUNCTION FOR A LAYERED REFERENCE STRUCTURE WITH PLANAR SURFACES

In this subsection the Green's function for layered reference structures will be obtained for a source point $\mathbf{r}'$ in the upper half-plane with dielectric constant ε_1. In the case of both the source point ($\mathbf{r}'$) and the observation point ($\mathbf{r}$) being placed in the upper half-plane, the Green's function can be described as the sum of two terms, namely a direct propagation term $g^{(d)}(\mathbf{r},\mathbf{r}')$ being identical to the homogeneous-medium Green's function (no layered structure), and an indirect part $g^{(i)}(\mathbf{r},\mathbf{r}')$ governing the contribution to the field at $\mathbf{r}$ due to reflection from the layered structure. This is illustrated in Fig. 4.24.

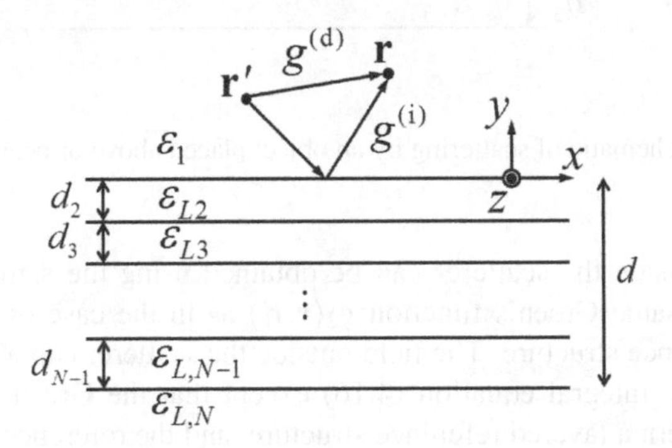

Figure 4.24 Schematic of Green's function for a layered reference structure illustrating that the radiation generated at **r** by a source at **r**′ can be divided into a direct and an indirect propagation term.

For s polarization, the Green's function must be a solution of

$$\left(\nabla^2 + k_0^2\varepsilon_{\text{ref}}(\mathbf{r})\right) g(\mathbf{r},\mathbf{r}') = -\delta(\mathbf{r}-\mathbf{r}'), \tag{4.155}$$

where the dielectric constant of the layered reference structure $\varepsilon_{\text{ref}}(\mathbf{r})$ now depends on the position. When both $\mathbf{r}$ and $\mathbf{r}'$ are in the upper layer with dielectric constant ε_1 a particular solution of (4.155) is given by the homogeneous-medium Green's function that we shall now refer to as the direct Green's function:

$$g^{(d)}(\mathbf{r},\mathbf{r}') = \frac{i}{4} H_0^{(1)}(k_0 n_1 |\mathbf{r}-\mathbf{r}'|). \tag{4.156}$$

This term will not satisfy the boundary conditions at the interfaces between layers, and thus in the upper layer, homogeneous solutions of the wave equation must be added, while in other layers a solution must be constructed, such that boundary conditions across interfaces are satisfied. In addition, the radiating boundary condition must be satisfied, which here means that all terms in the bottom layer (N) must propagate downward, while those terms added in the upper layer must propagate upward. For s polarization, $g(\mathbf{r},\mathbf{r}')$ must be continuous as $\mathbf{r}$ crosses the boundary between two layers, since the electric field must behave that way. In addition, since g here describes the z-component of the electric field, we must also require that the tangential component of the corresponding magnetic field, $\nabla \times \hat{\mathbf{z}} g(\mathbf{r},\mathbf{r}')/i\omega\mu_0$, is continuous across interfaces.

The first step in constructing the indirect part of the Green's function is to expand the direct Green's function into plane waves:

$$g^{(\mathrm{d})}(\mathbf{r},\mathbf{r}') = \frac{i}{2\pi}\int_0^{+\infty} \frac{\cos(k_x[x-x'])e^{ik_{y,1}|y-y'|}}{k_{y,1}} dk_x, \tag{4.157}$$

where $k_{y,i} = \sqrt{k_0^2 n_i^2 - k_x^2}$ with $\mathrm{Im}(k_{y,i}) \geq 0$. This expression shows that for $y > y'$ there will be plane wave components propagating upward, while for $y < y'$ there will be plane wave components propagating downward. By thinking of $g^{(d)}(\mathbf{r},\mathbf{r}')$ as the field generated by a point source at $\mathbf{r}'$ the source emits radiation going both upward and downward.

The plane-wave expansion (4.157) can be obtained by using the mode-expansion method. In this method, the Green's function is constructed using a complete set of eigenmodes to the left-hand-side operator in Eq. (4.155). The eigenmodes $E_\lambda(\mathbf{r})$ with eigenvalue λ are thus a solution of

$$\left(\nabla^2 + k_0^2\varepsilon_1\right) E_\lambda(\mathbf{r}) = \lambda E_\lambda(\mathbf{r}). \tag{4.158}$$

This leads to the equivalent expression

$$\left(\nabla^2 + k^2\right) E_{\mathbf{k}}(\mathbf{r}) = 0, \tag{4.159}$$

where $k^2 = k_0^2\varepsilon_1 - \lambda_{\mathbf{k}}$. This expression has a complete set of solutions of the form

$$E_{\mathbf{k}}(\mathbf{r}) = e^{i\mathbf{k}\cdot\mathbf{r}}. \tag{4.160}$$

Here $\mathbf{k} = \hat{\mathbf{x}}k_x + \hat{\mathbf{y}}k_y$ is a wave vector, and $k^2 = \mathbf{k}\cdot\mathbf{k} = k_x^2 + k_y^2$. These modes satisfy the following normalization and orthogonality condition:

$$\int E_{\mathbf{k}}(\mathbf{r})\left(E_{\mathbf{k}'}(\mathbf{r})\right)^* dA = N_{\mathbf{k}}\delta(\mathbf{k}-\mathbf{k}'), \tag{4.161}$$

where $N_{\mathbf{k}} = (2\pi)^2$, and $\delta(\mathbf{k}-\mathbf{k}') \equiv \delta(k_x - k_x')\delta(k_y - k_y')$. The result (4.161) follows directly from the following expression for the Dirac delta function

$$\delta(x) = \frac{1}{2\pi}\int_k e^{ikx} dk. \tag{4.162}$$

The Green's function satisfying Eq. (4.155) with $\varepsilon_{\mathrm{ref}}(\mathbf{r}) = \varepsilon_1$ is now given by

$$g(\mathbf{r},\mathbf{r}') = -\int_{\mathbf{k}} \frac{E_{\mathbf{k}}(\mathbf{r})\left(E_{\mathbf{k}}(\mathbf{r}')\right)^*}{N_{\mathbf{k}}\lambda_{\mathbf{k}}} d^2k. \tag{4.163}$$

This can be seen by first applying the operator $(\nabla^2 + k_0^2\varepsilon_1)$ and then using Eq. (4.158), which leads to

$$(\nabla^2 + k_0^2\varepsilon_1)g(\mathbf{r},\mathbf{r}') = -\int_{\mathbf{k}} \frac{E_{\mathbf{k}}(\mathbf{r})\left(E_{\mathbf{k}}(\mathbf{r}')\right)^*}{N_{\mathbf{k}}} d^2k. \tag{4.164}$$

Now consider a function $f(\mathbf{r}')$ being expanded in the complete set of eigenmodes

$$f(\mathbf{r}') = \int_{\mathbf{k}} A(\mathbf{k}) E_{\mathbf{k}}(\mathbf{r}') d^2k. \tag{4.165}$$

An overlap integral between the right-hand side of (4.164) and $f(\mathbf{r}')$ with respect to $\mathbf{r}'$ now leads to $-f(\mathbf{r})$ due to (4.161), and thus

$$(\nabla^2 + k_0^2\varepsilon_1)g(\mathbf{r},\mathbf{r}') = -\delta(\mathbf{r}-\mathbf{r}'). \tag{4.166}$$

Alternatively, this can also be shown by inserting expressions for $E_{\mathbf{k}}$ and $N_{\mathbf{k}}$ in (4.164), which leads to

$$(\nabla^2 + k_0^2\varepsilon_1)g(\mathbf{r},\mathbf{r}') = -\int_{\mathbf{k}} \frac{e^{i\mathbf{k}\cdot(\mathbf{r}-\mathbf{r}')}}{(2\pi)^2} d^2k, \tag{4.167}$$

and then using (4.162).

The particular Green's function that satisfies the radiating boundary condition can be obtained by adding a small imaginary part η in the denominator of (4.163) in the following way:

$$g(\mathbf{r},\mathbf{r}') = -\lim_{\eta'\to 0^+} \int \frac{e^{i\mathbf{k}\cdot(\mathbf{r}-\mathbf{r}')}}{(2\pi)^2(k_0^2\varepsilon_1 - k_x^2 - k_y^2 + i\eta')} d^2k. \tag{4.168}$$

This is equivalent to adding a homogeneous solution to Eq. (4.155), and Eq. (4.155) is thus still satisfied.

The integration over k_y can be carried out by using the residue theorem (see Appendix A). First note that the integrand has poles at

$$k_{y,\text{pole}} = \pm\left(\sqrt{k_0^2\varepsilon_1 - k_x^2} + i\eta\right). \tag{4.169}$$

For k_y in the complex plane, one pole is above the real axis, and the other is below. This is illustrated in Fig. 4.25, where $k_{y,r}$ is the real part of k_y, and $k_{y,i}$ is the imaginary part of k_y. The integral over k_y along the whole of the real axis can be replaced by an integral over the real axis from $-R$ to R and a semi-circle of radius R going either into the upper or lower complex half-plane, depending on in which half-plane the magnitude of $\exp(ik_y(y-y'))$ will decrease as k_y moves into the half-plane. The total integration path will be a closed path, and in the limit $R \to \infty$ the integral over the semi-circle will vanish due to the properties of the integrand. The choice of semi-circle depends on the sign of $y-y'$. If $y-y' > 0$ the upper half-plane must be chosen

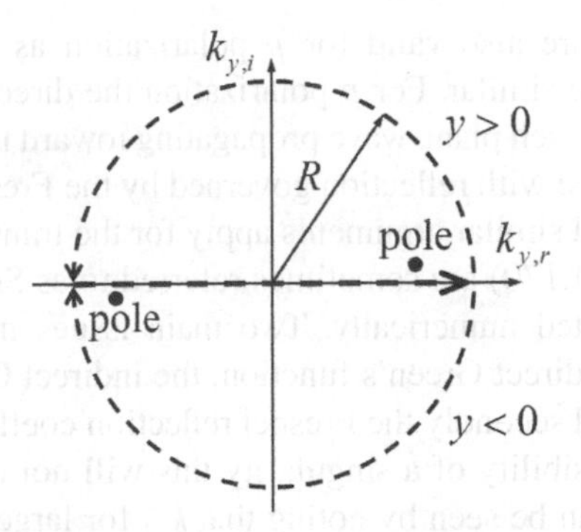

Figure 4.25 Schematic of poles in the integrand of Eq. (4.168) and replacing the integral over k_y along the real axis with a closed path going into the complex plane, in which case the residue theorem can be applied.

since the magnitude of $\exp(ik_y(y-y'))$ will decrease as k_y moves into the upper complex half-plane. The other half-plane cannot be chosen since then the magnitude of $\exp(ik_y(y-y'))$ will blow up, and the integral over the semi-circle will not vanish. For $y-y'<0$ the lower half-plane must be chosen instead. The replacement of the integral along the real axis with the integral over one or the other closed path means that the integral can be evaluated straightforwardly using the residue theorem, and finally the plane-wave expansion of the direct Green's function (4.157) is obtained.

The indirect part of the Green's function can now be constructed by noting that each of the downward propagating plane waves in the integrand of (4.157) will be partly reflected with a reflection coefficient given by ordinary Fresnel reflection as in Sec. 2.2.1, and this reflection must be added to the Green's function for $y>0$. The reflection coefficient will depend on k_x. In other layers a solution must be constructed such that the boundary conditions are satisfied. The indirect part of the Green's function thus becomes

$$g^{(\mathrm{i})}(\mathbf{r},\mathbf{r}') = \frac{i}{2\pi}\int_0^{+\infty} \frac{\cos(k_x[x-x'])r^{(\mathrm{u})}(k_x)e^{ik_{y,1}(y+y')}}{k_{y,1}}dk_x, \quad y,\, y'>0 \qquad (4.170)$$

where $u=s$ or p, and $r^{(\mathrm{u})}(k_x)$ is the Fresnel reflection coefficient for u polarized light being incident on the layered structure.

For $\mathbf{r}$ below all interfaces on the transmission side in layer N, i.e., for $y<-d$, where $d=(d_2+d_3+\ldots+d_{N-1})$ is the total thickness of all layers, the Green's function can instead be expressed as

$$g(\mathbf{r},\mathbf{r}') = \frac{i}{2\pi}\int_0^{+\infty} \frac{\cos(k_x[x-x'])t^{(u)}(k_x)e^{ik_{y,1}y'-ik_{y,LN}(y+d)}}{k_{y,1}}dk_x, \quad y'>0,\ y<-d, \qquad (4.171)$$

where $t^{(u)}(k_x)$ is the Fresnel transmission coefficient for the layered structure and u polarization. Although we have so far considered s polarization the expressions

(4.170) and (4.171) are also valid for p polarization as implied by the index u. The derivation is quite similar. For p polarization the direct Green's function is unchanged, and now for each plane wave propagating toward the layered structure there will be a reflected wave with reflection governed by the Fresnel reflection coefficient for p polarization, and similar arguments apply for the transmission.

Integrals such as (4.170) are sometimes referred to as Sommerfeld integrals, and they must be calculated numerically. Two main issues must then be considered. Firstly, similar to the direct Green's function, the indirect Green's function can also have a singularity, and secondly the Fresnel reflection coefficient can have poles.

Regarding the possibility of a singularity this will not occur for $y+y'>0$ by a finite amount. This can be seen by noting that $k_{y,i}$ for large k_x can be approximated as

$$k_{y,i} = \sqrt{k_0^2\varepsilon_i - k_x^2} \approx ik_x\left(1 - \frac{1}{2}\frac{k_0^2\varepsilon_i}{k_x^2}\right). \tag{4.172}$$

Thus, $\exp(ik_{y,1}(y+y')) \approx \exp(-k_x(y+y'))$ will decrease sufficiently fast with increasing k_x to ensure uniform convergence of the integral, and no singularity can occur in that case. However, in the limit $y+y'\to 0$ this term will not decay exponentially with k_x, and this term alone can then not ensure convergence. By using (4.172) the single-interface Fresnel reflection coefficients for s and p polarization can be approximated by

$$r_{1,L2}^{(s)}(k_x) = \frac{k_{y,1}-k_{y,L2}}{k_{y,1}+k_{y,L2}} \approx -\frac{1}{4}\frac{k_0^2(\varepsilon_1-\varepsilon_{L2})}{k_x^2}, \tag{4.173}$$

$$r_{1,L2}^{(p)}(k_x) = \frac{k_{y,1}\varepsilon_{L2}-k_{y,L2}\varepsilon_1}{k_{y,1}\varepsilon_{L2}+k_{y,L2}\varepsilon_1} \approx \frac{\varepsilon_{L2}-\varepsilon_1}{\varepsilon_{L2}+\varepsilon_1} + \frac{k_0^2\varepsilon_1\varepsilon_{L2}}{k_x^2}\frac{\varepsilon_{L2}-\varepsilon_1}{(\varepsilon_{L2}+\varepsilon_1)^2}. \tag{4.174}$$

Clearly, the Fresnel reflection coefficient for s polarization vanishes fast with increasing k_x, and the Sommerfeld integral (4.170) converges uniformly for any $x-x'$ and $y+y'$. The same is true for layered structures with several interfaces. The Fresnel reflection coefficient for p polarization on the other hand approaches a constant for large k_x, and thus as $x-x'\to 0$ and $y+y'\to 0$ the indirect Green's function can be expressed as

$$g^{(\mathrm{i})}(\mathbf{r},\mathbf{r}') \approx \frac{\varepsilon_{L2}-\varepsilon_1}{\varepsilon_{L2}+\varepsilon_1}\frac{i}{2\pi}\int_0^{+\infty} \frac{\cos(k_x[x-x'])e^{ik_{y,1}(y+y')}}{k_{y,1}}dk_x. \tag{4.175}$$

By comparison with (4.157) it is clear that

$$g^{(\mathrm{i})}(\mathbf{r},\mathbf{r}') \approx \frac{\varepsilon_{L2}-\varepsilon_1}{\varepsilon_{L2}+\varepsilon_1}\frac{i}{4}H_0^{(1)}(k_0 n_1|\tilde{\mathbf{r}}-\mathbf{r}'|),\ x-x'\to 0,\ y+y'\to 0, \tag{4.176}$$

where $\tilde{\mathbf{r}} = \hat{\mathbf{x}}x - \hat{\mathbf{y}}y$ is the mirror position of $\mathbf{r}$ in the plane $y=0$. In the case of a scatterer placed directly on the first interface, this singularity must be dealt with in the GFSIEM similar to the way the singularity of $g^{(d)}(\mathbf{r},\mathbf{r}')$ is dealt with. Note that the expressions (4.173) and (4.174) also hold for layered structures with several interfaces if k_x is large enough.

The indirect part of the Green's function for p polarization can now be calculated as follows:

$$g^{(\mathrm{i})}(\mathbf{r},\mathbf{r}') = \frac{\varepsilon_{L2}-\varepsilon_1}{\varepsilon_{L2}+\varepsilon_1}\frac{i}{4}H_0^{(1)}(k_0 n_1|\tilde{\mathbf{r}}-\mathbf{r}'|)$$

$$+\frac{i}{2\pi}\int_0^{+\infty}\frac{\cos(k_x[x-x'])\left(r^{(p)}(k_x)-\frac{\varepsilon_{L2}-\varepsilon_1}{\varepsilon_{L2}+\varepsilon_1}\right)e^{ik_{y,1}(y+y')}}{k_{y,1}}dk_x,\quad y,\, y'>0. \quad (4.177)$$

When $y+y'=0$ the magnitude of the integrands in (4.170) for s polarization, and in (4.177) for p polarization, are asymptotically proportional to $\cos(k_x[x-x'])/k_x^3$, and the integrals are thus clearly uniformly convergent and thus suitable for numerical calculation. In the GFSIEM we will also need the derivatives of g with respect to x' and y' in order to construct the surface normal derivative. Now, in the case of $y+y'=0$ the integrands for those derivatives of the integrals vanish asymptotically as $\cos(k_x[x-x'])/k_x^2$ or $\sin(k_x[x-x'])/k_x^2$, which is still fast enough to ensure uniform convergence.

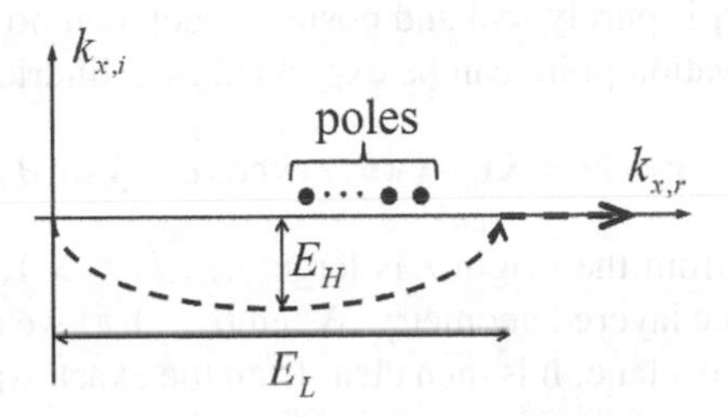

Figure 4.26 Schematic of integration path for Sommerfeld integrals in the case of poles in reflection and transmission coefficients.

The second issue to consider when calculating the Sommerfeld integrals is that the reflection coefficients $r^{(s)}(k_x)$ or $r^{(p)}(k_x)$ may have poles at specific values of $k_x = k_{x,\text{pole}}$ corresponding to guided modes in the layered reference structure. The integral with respect to k_x in (4.170) and (4.177) is taken along the real axis. If the layered structure is lossy, the poles, if any exist, may be shifted away from the real axis by a finite amount, and direct numerical integration is unproblematic. However, in the limit of small or vanishing propagation losses, as will be the case when the imaginary part of all dielectric constants approach zero, the poles will be displaced only infinitesimally from the real axis. In that case direct numerical integration is difficult. However, residue calculus can be applied to carry out the integral along a path which is at a distance from the poles. This is illustrated in Fig. 4.26. The idea is that the integral over part of the real axis can be replaced with the integral over half of the surface of, for example, an ellipse going into the lower complex half-plane. This follows from residue calculus because there are no poles in the closed domain bounded by the part of the real axis being replaced in the integral and the half surface of the ellipse. Numerically, the poles can be handled in the integration by making the following replacement, assuming that $f(k_x)$ is the integrand:

$$\int_{k_x=0}^{\infty} f(k_x)dk_x = \int_{\alpha=-\pi}^{0} f(k_x(\alpha))\frac{dk_x}{d\alpha}d\alpha + \int_{k_x=E_L}^{\infty} f(k_x)dk_x, \tag{4.178}$$

where

$$k_x(\alpha) = (1+\cos(\alpha))E_L/2 + iE_H\sin(\alpha). \tag{4.179}$$

When cast in this form the integrals can be carried out using, for example, Gauss-Kronrod quadrature. The width and height of the ellipse, E_L and E_H, must be chosen such that the ellipse is wide enough to cover all poles, and E_H cannot be chosen too large since then the term $\cos(k_x[x-x'])$ will blow up, and more so for large $|x-x'|$.

In the last part of this subsection, far-field approximations to the Green's function will be obtained that are useful for evaluating scattering and extinction cross sections. The first case considered is an observation point $\mathbf{r}$ at a very large distance from the source position $\mathbf{r}'$, and a source position which is much closer to the surface of the layered geometry than the observation point. We will furthermore assume that the dielectric constant ε_1 is purely real and positive, such that no light is absorbed in this medium. The observation point can be expressed in cylindrical coordinates as

$$\mathbf{r} = \hat{\mathbf{r}}r = \hat{\mathbf{x}}x + \hat{\mathbf{y}}y = r(\hat{\mathbf{x}}\cos\theta + \hat{\mathbf{y}}\sin\theta), \tag{4.180}$$

where the distance from the origin r is large, i.e., $kr >> 1$, with the origin placed near the surface of the layered geometry. When $\theta \in]0;\pi[$ we consider an observation point in the upper half-plane. It is then clear from the exact expression for the Green's function that for values of θ and r, such that $|y-y'| = (y-y')$ and $y+y'$ are also large distances, the part of the integral Eq. (4.170) where $k_{y,1}$ is complex can be neglected, i.e., the total Green's function $g(\mathbf{r},\mathbf{r}') = g^{(d)}(\mathbf{r},\mathbf{r}') + g^{(i)}(\mathbf{r},\mathbf{r}')$ can be approximated by

$$g^{(\text{ff})}(\mathbf{r},\mathbf{r}') \approx \frac{i}{2\pi}\int_0^{k_0 n_1} \frac{\cos(k_x[x-x'])\left(e^{ik_{y1}|y-y'|} + r^{(u)}(k_x)e^{ik_{y1}(y+y')}\right)}{k_{y1}} dk_x. \tag{4.181}$$

By making the variable change

$$k_x = k_0 n_1 \cos\alpha, \quad k_{y1} = k_0 n_1 \sin\alpha, \tag{4.182}$$

resulting in $dk_x = -k_{y1}d\alpha$, and using $\cos(k_x[x-x']) = (e^{ik_x(x-x')} + e^{-ik_x(x-x')})/2$, the expression (4.181) becomes

$$g^{(\text{ff})}(\mathbf{r},\mathbf{r}') \approx \frac{i}{4\pi}e^{ik_0 n_1 r}\int_{\alpha=0}^{\pi} e^{-ik_0 n_1 r(1-\cos(\alpha-\theta))} f(\alpha,x',y')d\alpha, \tag{4.183}$$

where $f(\alpha,x',y') = e^{-ik_0 n_1 \cos\alpha x'}\left(e^{-ik_0 n_1 \sin\alpha y'} + r^{(u)}(\alpha)e^{ik_0 n_1 \sin\alpha y'}\right)$. Here we should notice that when r is very large, the term $e^{-ik_0 n_1 r(1-\cos(\alpha-\theta))}$ will oscillate very fast with α except when $\alpha \approx \theta$, whereas the rest of the integrand, f, will be comparably slowly varying with α. This means that in the limit of very large r, the only part of the integral that will contribute is a small range of values for α around θ, in which case

$$1-\cos(\alpha-\theta)\approx\frac{1}{2}(\alpha-\theta)^2. \tag{4.184}$$

We can thus replace $e^{-ik_0n_1r(1-\cos(\alpha-\theta))}$ with $e^{-i\frac{1}{2}k_0n_1r(\alpha-\theta)^2}$, and further extend the integration range from $-\infty$ to $+\infty$, since the added integration does not contribute anything due to the fast oscillatory behavior of $e^{-i\frac{1}{2}k_0n_1r(\alpha-\theta)^2}$ with α when $\alpha\neq\theta$. Furthermore, since there will only be a contribution for $\alpha\approx\theta$ we can insert $\alpha=\theta$ into the slowly varying term and place this outside the integral. This leads to

$$g^{(\mathrm{ff})}(\mathbf{r},\mathbf{r}')\approx\frac{e^{ik_0n_1r}}{\sqrt{r}}\frac{e^{i\frac{\pi}{4}}}{4}\sqrt{\frac{2}{\pi k_0n_1}}e^{-ik_0n_1\cos\theta x'}\left(e^{-ik_0n_1\sin\theta y'}+r^{(u)}(\theta)e^{ik_0n_1\sin\theta y'}\right), \tag{4.185}$$

where

$$r^{(u)}(\theta)\equiv r^{(u)}(k_x)|_{k_x=k_0n_1\cos\theta}. \tag{4.186}$$

In the case when light can propagate into the substrate without loss, implying that ε_{LN} is real and positive, we can in a similar way obtain a far-field expression, where here we will again use $\mathbf{r}=\hat{\mathbf{r}}r=\hat{\mathbf{x}}x+\hat{\mathbf{y}}y=r(\hat{\mathbf{x}}\cos\theta+\hat{\mathbf{y}}\sin\theta)$ but with $\theta\in\,]-\pi;0[$ such that we consider a position deep into the semi-infinite medium below the layered geometry. Also, in this case, when $-(y+d)$ is a very large distance into the substrate along the y-axis, we can neglect the part of the integral of Eq. (4.171) where $k_{y,LN}$ is imaginary, i.e.,

$$g(\mathbf{r},\mathbf{r}')\approx\frac{i}{2\pi}\int_0^{k_0n_{LN}}\frac{\cos(k_x[x-x'])\left(e^{i(k_{y,1}y'-k_{y,LN}(y+d))}t^{(u)}(k_x)\right)}{k_{y1}}dk_x, \tag{4.187}$$

where $n_{LN}^2=\varepsilon_{LN}$. We will make a similar change of variables as before

$$k_x=k_0n_{LN}\cos\alpha,\quad k_{y,LN}=k_0n_{LN}\sin\alpha, \tag{4.188}$$

where now $dk_x=-k_{y,LN}d\alpha$, and a similar derivation leads to

$$g^{(\mathrm{ff})}(\mathbf{r},\mathbf{r}')\approx\frac{e^{ik_0n_{LN}r}}{\sqrt{r}}\frac{e^{i\pi/4}}{4}\sqrt{\frac{2}{\pi k_0n_{LN}}}\frac{k_{y,LN}}{k_{y,1}}e^{-ik_0n_{LN}\cos\theta x'}e^{ik_{y,1}y'}e^{-ik_{y,LN}d}t^{(u)}(\theta), \tag{4.189}$$

where

$$t^{(u)}(\theta)\equiv t^{(u)}(k_x)|_{k_x=k_0n_{LN}\cos\theta}. \tag{4.190}$$

Note that here $k_{y,i}=\sqrt{k_0^2\varepsilon_i-k_x^2}$ with $\mathrm{Imag}\{k_{y,i}\}\geq0$ still applies.

If we instead consider the far field but in a situation where $y+y'>0$ is small we can no longer neglect the part of the integral in, e.g., Eq. (4.170) for $k_x>k_0n_1$. It

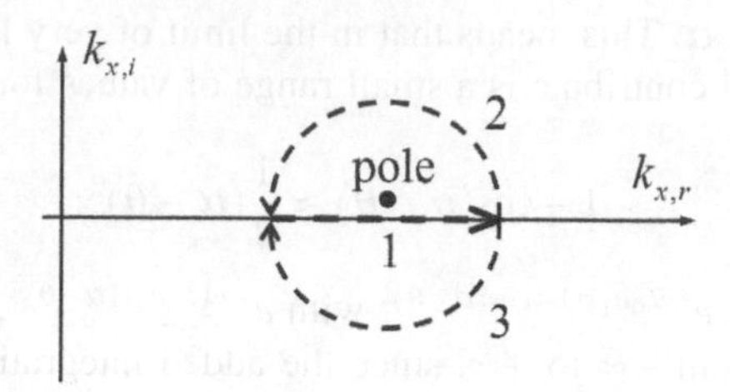

Figure 4.27 Schematic of integration path for Sommerfeld integrals for extracting the guided-mode far-field contribution to the Green's function being related to poles in reflection and transmission coefficients.

can then be useful to divide the integral into two integrals by using $\cos(k_x[x-x']) = (\exp(ik_x[x-x'])+\exp(-ik_x[x-x']))/2$. Now consider one of those integrals

$$\frac{i}{2\pi}\int_{k_0 n_1}^{+\infty} \frac{e^{ik_x(x-x')} r^{(\mathrm{u})}(k_x) e^{ik_{y,1}(y+y')}}{k_{y,1}} dk_x \tag{4.191}$$

for $x-x'>0$. In the far field with large values of $|x-x'|$ the factor $\exp(ik_x(x-x'))$ will oscillate very fast with increasing k_x, and for integration intervals where the rest of the integrand is slowly varying the integral will average out to approximately zero. The integrand will not be slowly varying though for k_x at poles of $r^{(u)}(k_x)$. Now consider replacing part of the integral over path 1 in Fig. 4.27 with an integral over a closed surface including both path 1 and 2. Since $\exp(ik_x|x-x'|)$ will decrease exponentially very fast as k_x moves into the upper half of the complex plane (due to the large $|x-x'|$), the integration over path 2 will not contribute, and the integral over path 1 can be evaluated as being the same as the integral over the closed curve, where the latter can be evaluated straightforwardly using residue calculus. Regarding the other integral

$$\frac{i}{2\pi}\int_{k_0 n_1}^{+\infty} \frac{e^{-ik_x(x-x')} r^{(\mathrm{u})}(k_x) e^{ik_{y,1}(y+y')}}{k_{y,1}} dk_x \tag{4.192}$$

and still $x-x'>0$, the integral over path 1 can be replaced by an integral over a closed curve including path 1 and 3, since $\exp(-ik_x|x-x'|)$ will decrease very fast as k_x moves into the lower half-plane. From residue calculus this integral vanishes. Similar considerations apply for the case of $x-x'<0$. These considerations lead to the following pole contribution to the integral

$$g^{(\mathrm{ff,pole})}(\mathbf{r},\mathbf{r}') = A(k_{x,\mathrm{p}}) e^{ik_{x,\mathrm{p}}|x-x'|} e^{ik_{y,1,\mathrm{p}}(y+y')}, \quad y,\ y'>0, \tag{4.193}$$

where

$$A(k_{x,\mathrm{p}}) = -\frac{1}{2}\frac{1}{k_{y,1,\mathrm{p}}} \lim_{k_x\to k_{x,\mathrm{p}}} r^{(u)}(k_x)(k_x-k_{x,\mathrm{p}}). \tag{4.194}$$

Here $k_{x,\mathrm{p}}$ is the pole of the reflection coefficient $r^{(u)}(k_x)$ with $u = s$ or p, and $k_{y,i,\mathrm{p}} = \sqrt{k_0^2\varepsilon_i - k_{x,\mathrm{p}}^2}$. A term of the form (4.193) will be obtained for each pole. The expression (4.194) can be evaluated numerically by taking the limit. Analytical expressions can also be constructed. In both cases the poles must be known, and they can be obtained numerically using the methods of Appendix F.

The only case with an analytic expression for the pole is for p polarization and a reference geometry with a metal-dielectric interface. Such a structure supports surface plasmon polaritons with the pole given by [16]

$$k_{x,\mathrm{p}} = k_0\sqrt{\frac{\varepsilon_1\varepsilon_{L2}}{\varepsilon_1+\varepsilon_{L2}}}. \tag{4.195}$$

In the case of ε_{L2} being real and negative the pole exists, and will be on the real axis when $\varepsilon_1 > 0$ and $\varepsilon_2 < -\varepsilon_1$. Usually, the metal dielectric constant is complex and the pole will thus be displaced from the real axis. By inserting $r_{1,L2}^{(p)}$ into (4.194) and using (4.195) the following simple analytic expression is obtained

$$A(k_{x,\mathrm{p}}) = \frac{1}{2}\frac{k_{y,L2,\mathrm{p}}}{k_{x,\mathrm{p}}}\frac{\varepsilon_{L2}k_{y,1,\mathrm{p}} - \varepsilon_1 k_{y,L2,\mathrm{p}}}{\varepsilon_{L2}k_{y,L2,\mathrm{p}} + \varepsilon_1 k_{y,1,\mathrm{p}}}. \tag{4.196}$$

In this case the Green's function represents a magnetic field, and the boundary conditions at the interface at $y = 0$ and the wave equation can be used to construct the rest of the guided-mode contribution to the Green's function leading to

$$g^{(\text{ff, pole})}(\mathbf{r},\mathbf{r}') = A(k_{x,\mathrm{p}})e^{ik_{x,\mathrm{p}}|x-x'|}e^{ik_{y,1,\mathrm{p}}y'}\begin{cases} e^{ik_{y,1,\mathrm{p}}y}, & y>0 \\ e^{-ik_{y,2,\mathrm{p}}y}, & y<0 \end{cases}. \tag{4.197}$$

For a three-layer geometry things do not simplify quite as much. Here

$$r^{(p)}(k_x) = \frac{r_{1,L2}^{(p)}(k_x) + r_{L2,L3}^{(p)}e^{i2k_{y,L2}d}}{f(k_x)}, \tag{4.198}$$

with

$$f(k_x) = 1 + r_{1,L2}^{(p)}r_{L2,L3}^{(p)}e^{i2k_{y,L2}d}, \tag{4.199}$$

and we find that near the pole

$$f(k_x) \approx (k_x - k_{x,\mathrm{p}})f'(k_{x,\mathrm{p}}), \tag{4.200}$$

where

$$\begin{aligned} f'(k_x) = \{& r_{1,L2}^{(p)\,\prime}(k_x)r_{L2,L3}^{(p)}(k_x) + r_{1,L2}^{(p)}(k_x)r_{L2,L3}^{(p)\,\prime}(k_x) + \\ & r_{1,L2}^{(p)}(k_x)r_{L2,L3}^{(p)}(k_x)(-2i)d\frac{k_x}{k_{y,L2}}\}e^{2ik_{y,L2}d} \end{aligned} \tag{4.201}$$

with

$$r_{i,j}^{(p)\prime}(k_x) = \frac{d}{dk_x}\left(\frac{\varepsilon_j k_{y,i} - \varepsilon_i k_{y,j}}{\varepsilon_j k_{y,i} + \varepsilon_i k_{y,j}}\right) = \frac{2k_x\varepsilon_i\varepsilon_j k_0^2(\varepsilon_i - \varepsilon_j)}{k_{y,i}k_{y,j}(\varepsilon_j k_{y,i} + \varepsilon_i k_{y,j})^2}. \tag{4.202}$$

Thus,

$$A(k_{xp}) = -\frac{1}{2k_{y,1,\mathrm{p}}}\frac{r^{(p)}_{1,L2}(k_{x,\mathrm{p}}) + r^{(p)}_{L2,L3}e^{i2\kappa_{y,L2,\mathrm{p}}d}}{f'(k_{xp})}. \tag{4.203}$$

While inserting the expression for $f'(k_x)$ results in a long expression it is nevertheless analytical. We thus find that

$$g^{(\mathrm{pole})}(\mathbf{r},\mathbf{r}') = A(k_{x,\mathrm{p}})e^{ik_{x,\mathrm{p}}|x-x'|}e^{ik_{y,1,\mathrm{p}}y'}\begin{cases} e^{ik_{y,1,\mathrm{p}}y}, & y>0 \\ Be^{ik_{y,L2,\mathrm{p}}y} + Ce^{-ik_{y,L2,\mathrm{p}}y}, & -d<y<0\,. \\ De^{-ik_{y,L3,\mathrm{p}}(y+d)}, & y<-d \end{cases} \tag{4.204}$$

where

$$B = \frac{1}{2}\left(1 + \frac{\varepsilon_{L2}k_{y,1,\mathrm{p}}}{\varepsilon_1 k_{y,L2,\mathrm{p}}}\right), \quad C = \frac{1}{2}\left(1 - \frac{\varepsilon_{L2}k_{y,1,\mathrm{p}}}{\varepsilon_1 k_{y,L2,\mathrm{p}}}\right), \tag{4.205}$$

$$D = Be^{-ik_{y,L2,\mathrm{p}}d} + Ce^{+ik_{y,2,\mathrm{p}}d}, \tag{4.206}$$

which is the same as (2.71) and (2.72) except for a slightly different notation.

For s polarization and $k_x \approx k_{x,\mathrm{p}}$ the three-layer reflection coefficient can be written

$$r^{(s)}(k_x) \approx \frac{1}{k_x - k_{x,\mathrm{p}}}\frac{r^{(s)}_{1,L2}(k_x) + r^{(s)}_{L2,L3}(k_x)e^{i2\kappa_{y,L2}d}}{f'(k_{x,\mathrm{p}})}, \tag{4.207}$$

where

$$f(k_x) = 1 + r^{(s)}_{1,L2}(k_x)r^{(s)}_{L2,L3}(k_x)e^{i2\kappa_{y,L2}d}, \tag{4.208}$$

and $f'(k_x) = df(k_x)/dk_x$. It can be shown after some algebra that

$$\frac{d}{dk_x}r^{(s)}_{ij}(k_x) = \frac{2k_x}{k_{y,i}k_{y,j}}r^{(s)}_{i,j}(k_x), \tag{4.209}$$

which together with $f(k_{x,\mathrm{p}}) = 0$ can be used to derive the fairly simple expression

$$f'(k_{x,\mathrm{p}}) = -\frac{2k_{x,\mathrm{p}}}{k_{y,L2,\mathrm{p}}}\left(\frac{1}{k_{y,1,\mathrm{p}}} + \frac{1}{k_{y,L3,\mathrm{p}}} - id\right). \tag{4.210}$$

Here,

$$g^{(\mathrm{ff,pole})}(\mathbf{r},\mathbf{r}') = A(k_{x,\mathrm{p}})e^{ik_{x,\mathrm{p}}|x-x'|}e^{ik_{y,1,\mathrm{p}}(y+y')}, \; y,\, y'>0, \tag{4.211}$$

where $A(k_{x,\mathrm{p}})$ is given by (4.194), and can be obtained analytically as

$$A(k_{x,\mathrm{p}}) = \frac{1}{4k_{y,1,\mathrm{p}}}\frac{r^{(s)}_{1,L2}(k_x) + r^{(s)}_{L2,L3}(k_x)e^{i2\kappa_{y,L2}d}}{\frac{k_{x,\mathrm{p}}}{k_{y,L2,\mathrm{p}}}\left(\frac{1}{k_{y,1}} + \frac{1}{k_{y,L3}} - id\right)}. \tag{4.212}$$

By using the boundary conditions at surfaces and the wave equation it is also here straightforward to extend the expression (4.211) to values of $y<0$ ($y'>0$). Thus, here for the three-layer geometry and s polarization

$$g^{(\text{ff,pole})}(\mathbf{r},\mathbf{r}') = A(k_{x,\text{p}})e^{ik_{x,\text{p}}|x-x'|}e^{ik_{y,1,\text{p}}y'}\begin{cases} e^{ik_{y,1,\text{p}}y}, & y>0 \\ Be^{ik_{y,L2,\text{p}}y}+Ce^{-ik_{y,L2,\text{p}}y}, & -d<y<0 \\ De^{-ik_{y,L3,\text{p}}(y+d)} & y<-d \end{cases}. \tag{4.213}$$

where

$$B = \frac{1}{2}\left(1+\frac{k_{y,1,\text{p}}}{k_{y,L2,\text{p}}}\right), \quad C = \frac{1}{2}\left(1-\frac{k_{y,1,\text{p}}}{k_{y,L2,\text{p}}}\right), \tag{4.214}$$

$$D = Be^{-ik_{y,L2,\text{p}}d}+Ce^{+ik_{y,L2,\text{p}}d}. \tag{4.215}$$

4.2.2 GFSIEM FOR A LAYERED REFERENCE STRUCTURE

In this section the GFSIEM will be considered for a scatterer on or near a layered reference geometry as in Fig. 4.23. As already mentioned the main difference compared with the case of a homogeneous reference medium is that the Green's function $g_1(\mathbf{r},\mathbf{r}')$ must be the one that governs the new reference structure, and the reference field (e.g. $H_0(\mathbf{r})$ for p polarization) must be a solution for the reference geometry as well.

In the first step when calculating the fields just inside and just outside the scatterer surface we only need to consider the Green's function for the case of having both $\mathbf{r}$ and $\mathbf{r}'$ in the upper layer (y, $y'>0$). In that case, as explained in the previous section, the Green's function g_1 is given by the sum of two terms

$$g_1(\mathbf{r},\mathbf{r}') = g^{(d)}(\mathbf{r},\mathbf{r}')+g^{(i)}(\mathbf{r},\mathbf{r}'), \tag{4.216}$$

where $g^{(d)}$ is the direct Green's function given in Eq. (4.156) and $g^{(i)}$ is the indirect Green's function given in (4.170). The indirect Green's function must be calculated numerically, and one strategy is to tabulate the Green's function and then interpolate values for the Green's function from this table. Here, it is clearly an advantage to divide the Green's function into $g^{(d)}$ and $g^{(i)}$ because these terms have the property

$$g^{(d)}(\mathbf{r},\mathbf{r}') = g^{(d)}(|x-x'|,|y-y'|), \tag{4.217}$$

$$g^{(i)}(\mathbf{r},\mathbf{r}') = g^{(i)}(|x-x'|,y+y'), \tag{4.218}$$

while the total Green's function would have the form

$$g_1(\mathbf{r},\mathbf{r}') = g_1(|x-x'|,y,y'). \tag{4.219}$$

It is thus sufficient to tabulate the indirect Green's function for a two-dimensional array of values for $|x-x'|$ and $y+y'$. The total Green's function would instead require tabulation for a three-dimensional array of values for $|x-x'|$, y and y', which is clearly a much bigger task. The derivatives of $g^{(i)}$ have similar properties

$$\frac{\partial}{\partial x'}g^{(i)}(\mathbf{r},\mathbf{r}') = \left(\frac{\partial}{\partial|x-x'|}g^{(i)}(|x-x'|,y+y')\right)\text{sign}\{x'-x\}, \tag{4.220}$$

$$\frac{\partial}{\partial y'} g^{(i)}(\mathbf{r},\mathbf{r}') = \frac{\partial}{\partial(y'+y)} g^{(i)}(|x-x'|, y+y'), \tag{4.221}$$

and it is thus also sufficient to tabulate these for a two-dimensional array of values for $x-x'>0$ and $y+y'$.

For s polarization we found in Section 4.2.1 that even in the limit of $y+y'\to 0$, which can be of relevance when the scatterer is placed directly on the layered reference geometry ($\delta\to 0$ in Fig. 4.23), there will be no singularity in the indirect Green's function $g^{(i)}$. For s polarization it is thus reasonable to directly tabulate the indirect Green's function.

For p polarization it was shown in the previous subsection in Eq. (4.177) that there is an additional singularity in the indirect Green's function, and that the Green's function has the form

$$g^{(i)}(\mathbf{r},\mathbf{r}') = \frac{\varepsilon_{L2}-\varepsilon_1}{\varepsilon_{L2}+\varepsilon_1}\frac{i}{4}H_0^{(1)}(k_0 n_1|\tilde{\mathbf{r}}-\mathbf{r}'|) + g^{(\mathrm{ns1})}(|x-x'|, y+y'), \quad y>0 \quad y'>0, \tag{4.222}$$

where $g^{(\mathrm{ns1})}$ is a non-singular term, and $\tilde{\mathbf{r}} = \hat{\mathbf{x}}x - \hat{\mathbf{y}}y$ is the mirror position of $\mathbf{r}$ in the plane $y=0$. It is preferable to tabulate only the slowly varying or non-singular part of a function and handle the singularity separately. Since the first term is analytic this can be kept as it is, and only the second term which is well behaved needs to be tabulated.

Now consider the method of Sec. 4.1.5 with higher-order polynomial field expansion. The expression (4.91) for the matrix elements $A_{ij}^{(u,v,v')}$ can be used directly as it is except that here the Green's function g_1 to be used in this expression is different. The additional singularity for p polarization poses no problem since it is a logarithmic singularity similar to the singularity of the direct Green's function. It is also possible to use the expression (4.92) for the matrix elements $B_{ij}^{(u,v,v')}$ as long as the considered $y+y'>0$. However, in the limit $y+y'\to 0$ this expression needs modification. In the limit of very small $|x-x'|$ and $y+y'$ the indirect Green's function can be expressed as

$$g^{(i)}(\mathbf{r},\mathbf{r}') \approx \frac{\varepsilon_{L2}-\varepsilon_1}{\varepsilon_{L2}+\varepsilon_1}\frac{-1}{2\pi}\log(k_0 n_1|\tilde{\mathbf{r}}-\mathbf{r}'|). \tag{4.223}$$

Now consider the integral equation (4.16) in the limit of $\mathbf{r}\to\mathbf{s}_i^{(v)}$, where $\mathbf{s}_i^{(v)}$ is a position at the bottom of the surface of the scatterer, and then consider the limit $\delta\to 0$ in Fig. 4.23. In this limit the position $\mathbf{s}_i^{(v)}$ will also be placed directly on the surface of the reference structure ($\hat{\mathbf{y}}\cdot\mathbf{s}_i^{(v)}\to 0^+$). We may further assume that the scatterer surface has a corner at this position. Now consider the surface integral for the infinitesimal part of the integral where $\mathbf{r}'$ or $\mathbf{s}(t)$ moves across the corner. In that case the contribution due to the indirect part of the Green's function can be expressed as

$$-\lim_{\mathbf{r}\to\mathbf{s}_i^{(v)}, y_i^{(v)}\to 0}\int_t \frac{1}{2\pi}\frac{\varepsilon_{L2}-\varepsilon_1}{\varepsilon_{L2}+\varepsilon_1}\frac{1}{|\mathbf{s}(t)-\tilde{\mathbf{r}}|}\left(\frac{\mathbf{s}(t)-\tilde{\mathbf{r}}}{|\mathbf{s}(t)-\tilde{\mathbf{r}}|}\cdot\hat{\mathbf{n}}(t)\right)dt. \tag{4.224}$$

Here $y_i^{(v)} = \hat{\mathbf{y}} \cdot \mathbf{s}_i^{(v)}$. The integral near a corner of the structure is illustrated in Fig. 4.28. Notice that $\mathbf{r}$ and $\tilde{\mathbf{r}}$ are mirror positions in the interface of the reference structure at $y = 0$. Similar to the homogeneous-reference-medium case, here

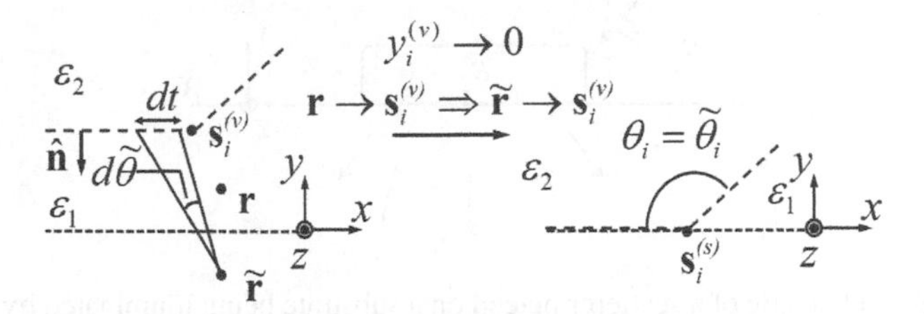

Figure 4.28 Schematic of integration across a bottom corner of the scatterer. The limit of the scatterer approaching the layered structure is considered, and the limit of the observation point approaching the corner. In this limit $\tilde{\mathbf{r}}$ and $\mathbf{r}$ will coincide.

$$-\frac{1}{|\mathbf{s}(t)-\tilde{\mathbf{r}}|}\left(\frac{\mathbf{s}(t)-\tilde{\mathbf{r}}}{|\mathbf{s}(t)-\tilde{\mathbf{r}}|}\cdot\hat{\mathbf{n}}(t)\right)dt = d\tilde{\theta}. \tag{4.225}$$

In the limit considered, the integral (4.224) over the infinitesimal corner will result in

$$\frac{\theta_i}{2\pi}\frac{\varepsilon_{L2}-\varepsilon_1}{\varepsilon_{L2}+\varepsilon_1}, \tag{4.226}$$

where θ_i is here the same angle considered previously for corners in the case of the direct part of the Green's function or homogeneous-medium Green's function. $\tilde{\theta}_i = \theta_i$ occurs because in the considered limit $\tilde{\mathbf{r}} = \mathbf{r}$. These considerations now lead to the following expression for the $B_{ij}^{(u,v,v')}$ matrix elements

$$B_{ij}^{(u,v,v')} = P\int_t \left(\hat{\mathbf{n}}'\cdot\nabla' g_u(\mathbf{s}_i^{(v)},\mathbf{r}')\right)_{\mathbf{r}'=\mathbf{s}(t)} f_j^{(m,v')}(t)dt$$
$$+\delta_{i,j}\delta_{v,v'}\left(\delta_{u,1}\frac{\theta_{i,v}}{2\pi}-\delta_{u,2}\frac{2\pi-\theta_{i,v}}{2\pi}\right)+\delta_{i,j}\delta_{v,v'}\delta_{\hat{\mathbf{y}}\cdot\mathbf{s}_i^{(v)}=0}\delta_{u,1}\frac{\theta_{i,v}}{2\pi}\frac{\varepsilon_{L2}-\varepsilon_1}{\varepsilon_{L2}+\varepsilon_1}. \tag{4.227}$$

Here, the following shorthand notation has been used

$$\delta_{\hat{\mathbf{y}}\cdot\mathbf{s}_i^{(v)}=0} = \begin{cases} 1, & \hat{\mathbf{y}}\cdot\mathbf{s}_i^{(v)} = 0 \\ 0, & \text{otherwise} \end{cases}. \tag{4.228}$$

Note that in the case where all corners are rounded the angle $\theta_{i,v} = \pi$.

4.2.3 CALCULATION OF FIELDS USING THE ANGULAR SPECTRUM REPRESENTATION

In this section the angular spectrum representation will be introduced, and a method for calculating fields based on the angular spectrum representation will be described in detail.

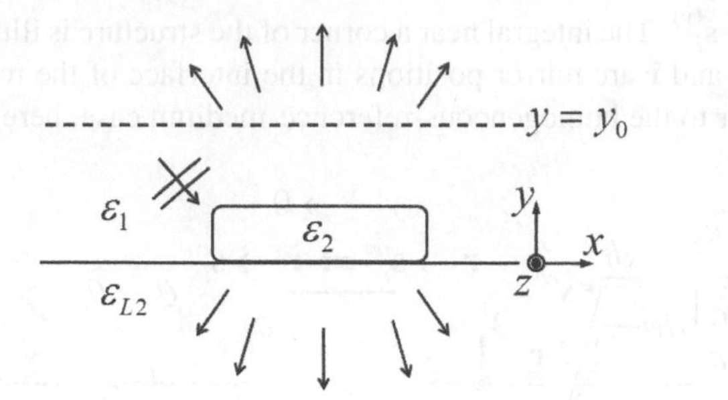

Figure 4.29 Schematic of a scatterer placed on a substrate being illuminated by a downward propagating beam. As a result the scattered part of the field for $y > y_0$ being above the scatterer will propagate upward, and for $y < 0$ the scattered field will propagate downward.

Consider the scattering situation illustrated in Fig. 4.29, where a scatterer placed on a substrate is illuminated from above with a p-polarized beam. Above the scatterer, the scattered part of the field can only propagate upward. This applies to, for example, positions above the dashed line at $y = y_0$. Likewise, for $y < 0$ the scattered part of the field can only propagate downward.

The magnetic field along any line given by a constant value of y can be expressed as

$$H^{(s)}(x,y) = \int_{k_x} \tilde{H}^{(s)}(k_x;y)e^{ik_x x}dk_x, \tag{4.229}$$

where the angular spectrum, or the Fourier transform of the field, is given by

$$\tilde{H}^{(s)}(k_x;y) = \frac{1}{2\pi}\int_x H^{(s)}(x,y)e^{-ik_x x}dx. \tag{4.230}$$

The field (4.229) must satisfy the wave equation

$$\left(\nabla^2 + k_0^2\varepsilon_1\right)H^{(s)}(x,y) = 0, \quad y \geq y_0, \tag{4.231}$$

which leads to

$$\frac{\partial^2 \tilde{H}^{(s)}(k_x;y)}{\partial y^2} = -(k_0^2\varepsilon_1 - k_x^2)\tilde{H}^{(s)}(k_x;y), \tag{4.232}$$

and thus, since the scattered field must propagate upward for $y \geq y_0$, the field can be expressed as

$$H^{(s)}(x,y) = \int_{k_x} \tilde{H}^{(s)}(k_x;y=y_0)e^{ik_x x}e^{ik_{y1}(y-y_0)}dk_x, \quad y \geq y_0, \tag{4.233}$$

where $k_{y1} = \sqrt{k_0^2\varepsilon_1 - k_x^2}$. A similar technique has already been used in Section 2.3 for deriving expressions for optical cross sections, although there the beams considered were wide Gaussian beams, where only a small range of values for k_x was needed.

Here, right above the scatterer, a wide range of k_x may be needed in order to correctly calculate sharp features in the field profile. As y increases, and a position farther away from the scatterer is considered, then it is sufficient to use a much smaller range of values for k_x, since, if $|k_x| > k_0 n_1$ then the corresponding k_{y1} will be purely imaginary, and the contribution from spectral field components with $|k_x| > k_0 n_1$ will vanish exponentially fast with y.

In order to calculate the fields at a given position $\mathbf{r}$ the first step is to determine if $\mathbf{r}$ is inside or outside the scatterer. This can be done in exactly the same way as was suggested for a scatterer in a homogeneous reference medium. If the position is inside the scatterer the approach for calculating the field is also no different from that with a scatterer in a homogeneous reference structure. For positions outside the scatterer the main difference compared with the case of a homogeneous reference structure is that the Green's function will be different. In general the Green's function must be calculated numerically. Here, instead of calculating the Green's function directly, the angular spectrum of part of the field is calculated instead, which can subsequently be applied for fast field calculation.

For $y > 0$ and p polarization it is relevant to recall the expression for the Green's function of the layered structure

$$g(\mathbf{r},\mathbf{r}') = \frac{i}{4}H_0^{(1)}(k_0 n_1|\mathbf{r}-\mathbf{r}'|) + \frac{\varepsilon_{L2}-\varepsilon_1}{\varepsilon_{L2}+\varepsilon_1}\frac{i}{4}H_0^{(1)}(k_0 n_1|\tilde{\mathbf{r}}-\mathbf{r}'|)$$

$$+\frac{i}{2\pi}\int_0^{+\infty}\frac{\cos(k_x[x-x'])\left(r^{(p)}(k_x)-\frac{\varepsilon_{L2}-\varepsilon_1}{\varepsilon_{L2}+\varepsilon_1}\right)e^{ik_{y,1}(y+y')}}{k_{y,1}}dk_x, \quad y, y' > 0, \quad (4.234)$$

where the first two terms have a singularity, while the last term is non-singular.

The total magnetic field can be divided into the sum of terms

$$H(\mathbf{r}) = H_0(\mathbf{r}) + H^{(A)}(\mathbf{r}) + H^{(i,\mathrm{ns})}(\mathbf{r}), \quad \mathbf{r}\in\Omega_1, \quad (4.235)$$

where as usual $H_0(\mathbf{r})$ is the reference field, and

$$H^{(A)}(\mathbf{r}) = -\int\left\{\phi(\mathbf{r}')g^{(A)}(\mathbf{r},\mathbf{r}') - H(\mathbf{r}')\hat{\mathbf{n}}'\cdot\nabla' g^{(A)}(\mathbf{r},\mathbf{r}')\right\}dl', \quad \mathbf{r}\in\Omega_1, \quad (4.236)$$

is the part of the scattered field that relates to the analytic part of the Green's function

$$g^{(A)}(\mathbf{r},\mathbf{r}') = \frac{i}{4}H_0^{(1)}(k_0 n_1|\mathbf{r}-\mathbf{r}'|) + \frac{\varepsilon_{L2}-\varepsilon_1}{\varepsilon_{L2}+\varepsilon_1}\frac{i}{4}H_0^{(1)}(k_0 n_1|\tilde{\mathbf{r}}-\mathbf{r}'|). \quad (4.237)$$

The remaining term $H^{(i,\mathrm{ns})}(\mathbf{r})$ will be a term that propagates upward, and this term can thus be expressed using the angular spectrum representation as follows

$$H^{(i,\mathrm{ns})}(\mathbf{r}) = \int_{k_x}\tilde{H}^{(i,\mathrm{ns})}(k_x; y=0^+)\left(r^{(p)}(k_x) - \frac{\varepsilon_{L2}-\varepsilon_1}{\varepsilon_{L2}+\varepsilon_1}\right)e^{ik_x x}e^{ik_{y1}y}dk_x, \quad (4.238)$$

where the angular spectrum is given by

$$\tilde{H}^{(i,\mathrm{ns})}(k_x; y=0^+) = \frac{-i}{4\pi}\int_{C_1}\left\{\frac{e^{-ik_x x'}e^{ik_{y1}y'}}{k_{y1}}\left(\phi(\mathbf{r}') - H(\mathbf{r}')\hat{\mathbf{n}}'\cdot(-ik_x\hat{\mathbf{x}} + ik_{y1}\hat{\mathbf{y}})\right)\right\}dl'. \quad (4.239)$$

Notice that since $(r^{(p)}(k_x) - (\varepsilon_{L2} - \varepsilon_1)/(\varepsilon_{L2} + \varepsilon_1))$ vanishes relatively fast with $|k_x|$ the spectrum does not need to be calculated for very large ranges of k_x. This is even true for $y = 0$. This is because the singular part of the field is already taken care of in the term (4.236).

The calculation of the electric field is now going to be considered. It may be convenient to normalize the electric field using the normalization factor $n_1\sqrt{\varepsilon_0/\mu_0}$ and thus introduce the normalized electric field

$$\tilde{\mathbf{E}}(\mathbf{r}) = \mathbf{E}(\mathbf{r})\sqrt{\frac{\varepsilon_0}{\mu_0}} n_1. \tag{4.240}$$

The normalized incident electric field and the magnetic field will have the same units. The normalized reference electric field is denoted $\tilde{\mathbf{E}}_0(\mathbf{r}) = \hat{\mathbf{x}}\tilde{E}_{0,x} + \hat{\mathbf{y}}\tilde{E}_{0,y}$.

The components of the electric field can now, similar to the magnetic field, be split into three terms:

$$\tilde{E}_x(\mathbf{r}) = \tilde{E}_{0,x}(\mathbf{r}) + \tilde{E}_x^{(A)}(\mathbf{r}) + \tilde{E}_x^{(i,\mathrm{ns})}(\mathbf{r}), \tag{4.241}$$

$$\tilde{E}_y(\mathbf{r}) = \tilde{E}_{0,y}(\mathbf{r}) + \tilde{E}_y^{(A)}(\mathbf{r}) + \tilde{E}_y^{(i,\mathrm{ns})}(\mathbf{r}). \tag{4.242}$$

The part of the electric field components that can be obtained using the analytic part of the Green's function is given by

$$\tilde{E}_x^{(A)}(\mathbf{r}) = -\frac{in_1}{k_0\varepsilon_1}\int\left\{\phi(\mathbf{r}')\frac{\partial g^{(A)}(\mathbf{r},\mathbf{r}')}{\partial y} - H(\mathbf{r}')\left(\frac{\partial}{\partial y}\hat{\mathbf{n}}'\cdot\nabla' g^{(A)}(\mathbf{r},\mathbf{r}')\right)\right\}dl', \ \mathbf{r}\in\Omega_1 \tag{4.243}$$

and

$$\tilde{E}_y(\mathbf{r}) = \frac{in_1}{k_0\varepsilon_1}\int\left\{\phi(\mathbf{r}')\frac{\partial g^{(A)}(\mathbf{r},\mathbf{r}')}{\partial x} - H(\mathbf{r}')\left(\frac{\partial}{\partial x}\hat{\mathbf{n}}'\cdot\nabla' g^{(A)}(\mathbf{r},\mathbf{r}')\right)\right\}dl', \ \mathbf{r}\in\Omega_1. \tag{4.244}$$

The remaining term can be derived from (4.238) using the usual relations between the magnetic and the electric fields. Thus,

$$\tilde{E}_x^{(i,\mathrm{ns})}(\mathbf{r}) = \frac{in_1}{k_0\varepsilon_1}\int_{k_x}\tilde{H}^{(i,\mathrm{ns})}(k_x; y=0^+)\left(r^{(p)}(k_x) - \frac{\varepsilon_{L2}-\varepsilon_1}{\varepsilon_{L2}+\varepsilon_1}\right)e^{ik_x x}e^{ik_{y1}y}(ik_{y1})dk_x. \tag{4.245}$$

$$\tilde{E}_y^{(i,\mathrm{ns})}(\mathbf{r}) = \frac{-in_1}{k_0\varepsilon_1}\int_{k_x}\tilde{H}^{(i,\mathrm{ns})}(k_x; y=0^+)\left(r^{(p)}(k_x) - \frac{\varepsilon_{L2}-\varepsilon_1}{\varepsilon_{L2}+\varepsilon_1}\right)e^{ik_x x}e^{ik_{y1}y}(ik_x)dk_x. \tag{4.246}$$

It can be noted that if the integrals in, for example, equation (4.238) are taken along the real axis, then a problem can occur if $r^{(p)}(k_x)$ has poles near the real axis, which will be the case if the layered reference structure supports guided modes with small propagation losses. In addition, the factor $1/k_{y1}$ in (4.239) means that $\tilde{H}^{(i,\mathrm{ns})}(k_x; y=0^+)$ will have spikes near $k_x = \pm k_0 n_1$, since here $k_{y1} = 0$. These spikes can be readily dealt with, e.g., by using $k_x = k_0 n_1 \sin(\alpha)$ with $\alpha \in [-\pi/2 : \pi/2]$ for the integration interval from $-k_0 n_1$ to $+k_0 n_1$, in which case $dk_x = k_{y1} d\alpha$. This eliminates the k_{y1} in the denominator of (4.239). An approach that can be used to manage both spikes near $k_x = \pm k_0 n_1$ and the poles of $r^{(p)}(k_x)$ is to move the integral over k_x into the complex plane. The integration along part of the real axis is replaced by an integration path going into the complex plane as illustrated in Fig. 4.30. This will give the same result as the integration along the real axis, which follows from the residue theorem, since no poles are located between the original path along the real axis and the new path. The poles of the reflection (or transmission) coefficient are thus dealt with by keeping the integration path at a distance from the poles.

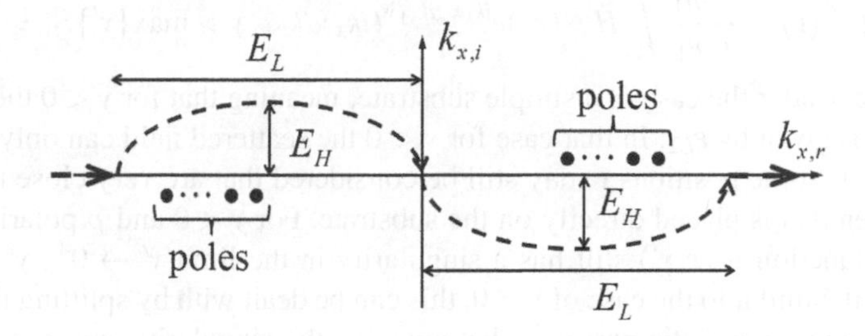

Figure 4.30 Illustration of integration path going into the complex plane when calculating fields using the angular spectrum representation.

One possible approach when replacing the integration path along the part of the real axis from $k_x = -E_L$ to $k_x = +E_L$ with an integral going into the complex plane is to use

$$k_x = \text{sign}(\alpha)\frac{E_L}{2}(1-\cos(\alpha)) - iE_H \sin(\alpha), \quad \alpha \in [-\pi : \pi], \tag{4.247}$$

in which case

$$dk_x = \left(\frac{E_L}{2}|\sin(\alpha)| - iE_H \sin(\alpha)\right) d\alpha, \tag{4.248}$$

and k_{y1}, $k_{y,L2}$ and $r^{(p)}$ also become functions of α.

For positions high above the nanostrip, e.g. $y > y_T$, the angular spectrum for the total scattered field will be quite limited due to the exponential decrease with y of high-spatial-frequency terms. In addition all scattered field components will propagate upward. In that case it may be faster to apply the angular spectrum representation to the total scattered field:

$$H^{(s)}(\mathbf{r}) = H(\mathbf{r}) - H_0(\mathbf{r}) = \int_{k_x} \tilde{H}^{(s)}(k_x) e^{ik_x x} e^{ik_{y1}(y-y_T)} dk_x, \quad y > y_T > \max\{y'\}, \tag{4.249}$$

where now

$$\tilde{H}^{(s)}(k_x) = \frac{-i}{4\pi} \int_{C_1} \left\{ \frac{e^{-ik_x x'} \left(e^{ik_{y1}(y_T - y')} + r^{(p)}(k_x) e^{ik_{y1}(y_T + y')} \right)}{k_{y1}} \right.$$

$$\times\left(\phi(\mathbf{r}') - H(\mathbf{r}')\hat{\mathbf{n}}'\cdot(-ik_x\hat{\mathbf{x}} + ik_{y1}\hat{\mathbf{y}})\right)\Big\} dl'. \quad (4.250)$$

Note that this cannot be applied at $y = 0$ since here the direct part of the scattered field violates the assumption of an upward propagating scattered field.

For $y > y_T$ the total scattered part of the electric field can be expressed as

$$\tilde{E}_x^{(s)}(\mathbf{r}) = \frac{in_1}{k_0\varepsilon_1}\int_{k_x} \tilde{H}^{(s)}(k_x)e^{ik_x x}e^{ik_{y1}y}(ik_{y1})dk_x, \ y > \max\{y'\}, \quad (4.251)$$

$$\tilde{E}_y^{(s)}(\mathbf{r}) = \frac{-in_1}{k_0\varepsilon_1}\int_{k_x} \tilde{H}^{(s)}(k_x)e^{ik_x x}e^{ik_{y1}y}(ik_x)dk_x, \ y > \max\{y'\}. \quad (4.252)$$

Now consider the case of a simple substrate, meaning that for $y < 0$ the dielectric constant is given by ε_{L2}. In that case for $y < 0$ the scattered field can only propagate downward. Some positions $\mathbf{r}$ may still be considered that are very close to the scatterer when this is placed directly on the substrate. For $y < 0$ and p polarization, the Green's function $g_1(\mathbf{r},\mathbf{r}')$ still has a singularity in the limit $y' \to 0^+$, $y \to 0^-$, and $x - x' \to 0$. Similar to the case of $y > 0$, this can be dealt with by splitting the Green's function into an analytic part $g^{(A)}$ that contains the singularity, and a non-singular part $g^{(t,\text{ns})}$, i.e.,

$$g_1(\mathbf{r},\mathbf{r}') = g^{(A)}(\mathbf{r},\mathbf{r}') + g^{(t,\text{ns})}(\mathbf{r},\mathbf{r}'), \quad (4.253)$$

where the analytic part is given by

$$g^{(A)}(\mathbf{r},\mathbf{r}') = \frac{2\varepsilon_{L2}}{\varepsilon_{L2}+\varepsilon_1}\frac{i}{4}H_0^{(1)}(k_0 n_1|\mathbf{r}-\mathbf{r}'|). \quad (4.254)$$

The magnetic field can thus be expressed as

$$H(\mathbf{r}) = H_0(\mathbf{r}) + H^{(A)}(\mathbf{r}) + H^{(t,\text{ns})}(\mathbf{r}), \ y < 0, \quad (4.255)$$

where the part of the scattered field related to the analytic part of the Green's function is given by

$$H^{(A)}(\mathbf{r}) = -\int \left\{\phi(\mathbf{r}')g^{(A)}(\mathbf{r},\mathbf{r}') - H(\mathbf{r}')\hat{\mathbf{n}}'\cdot\nabla' g^{(A)}(\mathbf{r},\mathbf{r}')\right\} dl', \ y < 0. \quad (4.256)$$

The remaining part of the magnetic field can be obtained by using the angular spectrum representation as follows:

$$H^{(t,\text{ns})}(\mathbf{r}) = \int_{k_x} \tilde{H}^{(t,\text{ns})}(k_x)\tilde{t}^{(p)}(k_x)e^{ik_x x}dk_x, \ y < 0, \quad (4.257)$$

where the angular spectrum is given by

$$\tilde{H}^{(t,\text{ns})}(k_x) = \tilde{H}^{(i,\text{ns})}(k_x), \quad (4.258)$$

and

$$\tilde{t}^{(p)}(k_x) = t^{(p)}(k_x)e^{-ik_{y,L2}y} - \frac{2\varepsilon_{L2}}{\varepsilon_{L2}+\varepsilon_1}e^{-ik_{y1}y}. \quad (4.259)$$

The angular spectrum used for calculating the reflected field can thus be reused for calculating the transmitted field.

Similarly, the normalized transmitted electric field can be expressed as

$$\tilde{\mathbf{E}}(\mathbf{r}) = \tilde{\mathbf{E}}_0(\mathbf{r}) + \tilde{\mathbf{E}}^{(A)}(\mathbf{r}) + \tilde{\mathbf{E}}^{(t,\text{ns})}(\mathbf{r}),\ \ y < 0, \tag{4.260}$$

where now

$$\tilde{E}_x^{(A)}(\mathbf{r}) = \tilde{E}_{0,x}(\mathbf{r}) - \frac{in_1}{k_0\varepsilon_{L2}} \int \left\{ \phi(\mathbf{r}') \frac{\partial g^{(A)}(\mathbf{r},\mathbf{r}')}{\partial y} - H(\mathbf{r}') \left(\frac{\partial}{\partial y} \hat{\mathbf{n}}' \cdot \nabla' g^{(A)}(\mathbf{r},\mathbf{r}') \right) \right\} dl',$$
$$y < 0. \tag{4.261}$$

and

$$\tilde{E}_y^{(A)}(\mathbf{r}) = \tilde{E}_{0,y}(\mathbf{r}) + \frac{in_1}{k_0\varepsilon_{L2}} \int \left\{ \phi(\mathbf{r}') \frac{\partial g^{(A)}(\mathbf{r},\mathbf{r}')}{\partial x} - H(\mathbf{r}') \left(\frac{\partial}{\partial x} \hat{\mathbf{n}}' \cdot \nabla' g^{(A)}(\mathbf{r},\mathbf{r}') \right) \right\} dl',$$
$$y < 0. \tag{4.262}$$

Again, the remaining part of the electric field can be obtained from the magnetic field as follows:

$$\tilde{E}_x^{(t,\text{ns})}(\mathbf{r}) = \frac{in_1}{k_0\varepsilon_{L2}} \int_{k_x} \tilde{H}^{(t,\text{ns})}(k_x) \tilde{t}^{(p)}(k_x) \times e^{ik_x x} dk_x,\ \ y < 0, \tag{4.263}$$

$$\tilde{E}_y^{(t,\text{ns})}(\mathbf{r}) = \frac{-in_1}{k_0\varepsilon_{L2}} \int_{k_x} \tilde{H}^{(t,\text{ns})}(k_x) \tilde{t}^{(p)}(k_x) e^{ik_x x} (ik_x) dk_x,\ \ y < 0. \tag{4.264}$$

For smaller values of y (farther below the interface at y = 0) the factor $\exp(-ik_{y,L2}y)$ will lead to exponential damping of the high-spatial frequency components, and it is possible to use

$$H(\mathbf{r}) = H_0(\mathbf{r}) + H^{(s)}(\mathbf{r}), \tag{4.265}$$

where

$$H^{(s)}(\mathbf{r}) = \int_{k_x} \tilde{H}^{(t,\text{ns})}(k_x) t^{(p)}(k_x) e^{ik_x x} e^{-ik_{y,L2}y} dk_x,\ \ y < 0, \tag{4.266}$$

and similarly

$$\tilde{\mathbf{E}}(\mathbf{r}) = \tilde{\mathbf{E}}_0(\mathbf{r}) + \tilde{\mathbf{E}}^{(s)}(\mathbf{r}), \tag{4.267}$$

where

$$\tilde{E}_x^{(s)}(\mathbf{r}) = \frac{in_1}{k_0\varepsilon_{L2}} \int_{k_x} \tilde{H}^{(t,\text{ns})}(k_x) t^{(p)}(k_x) e^{ik_x x} e^{-ik_{y,L2}y} (-ik_{y,L2}) dk_x,\ \ y < 0, \tag{4.268}$$

$$\tilde{E}_y^{(s)}(\mathbf{r}) = \frac{-in_1}{k_0\varepsilon_{L2}} \int_{k_x} \tilde{H}^{(t,\text{ns})}(k_x) t^{(p)}(k_x) e^{ik_x x} e^{-ik_{y,L2}y} (ik_x) dk_x,\ \ y < 0. \tag{4.269}$$

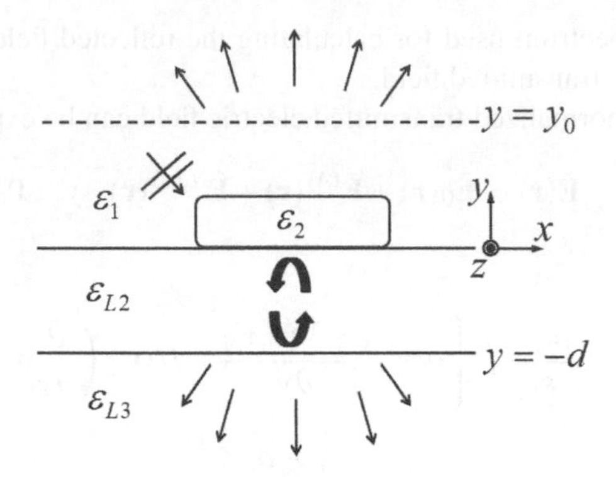

Figure 4.31 Schematic of a scatterer placed on a film of thickness d on a substrate being illuminated by a downward propagating beam. In the region $0 > y > -d$, the scattered part of the field consists of both upward and downward propagating components. For $y > y_0$ the scattered field will propagate upward, and for $y < -d$ the scattered field will propagate downward.

We shall now consider the calculation of the fields for a three-layer geometry as illustrated in Fig. 4.31. For $y > 0$ the only difference compared with the two-layer reference geometry is that the reflection coefficient should be the one for the three-layer reference geometry. For $y < -d$ the situation is also the same except that the transmission coefficient for the three-layer geometry should be used. In addition, since the positions considered are at least at a distance d from the scatterer, there is usually no need for special handling of any singular part of the Green's function. We can thus for $y < -d$ use the equivalent of Eqs. (4.265)-(4.269) with the transmission coefficient replaced with the one for the three-layer geometry, and ε_{L2}, $k_{y,L2}$, and y replaced by ε_{L3}, $k_{y,L3}$, and $(y+d)$.

For positions $0 > y > -d$ and $y' > 0$, the Green's function can be expressed as

$$g_1(\mathbf{r},\mathbf{r}') = \frac{i}{2\pi}\int_{k_x=0}^{\infty} \frac{\cos(k_x[x-x'])e^{ik_{y1}y'}}{k_{y1}} f(k_x)dk_x, \;\; y' > 0, \;\; 0 > y > -d, \quad (4.270)$$

where

$$f(k_x) = \frac{\left(t^{(p)}_{1,L2}(k_x)e^{-ik_{y,L2}y} + t^{(p)}_{1,L2}(k_x)r^{(p)}_{L2,L3}(k_x)e^{ik_{y,L2}(y+2d)}\right)}{1+r^{(p)}_{1,L2}(k_x)r^{(p)}_{L2,L3}(k_x)e^{ik_{y,L2}(2d)}}. \quad (4.271)$$

Here $r^{(p)}_{1,L2}$, $r^{(p)}_{L2,L3}$, and $t^{(p)}_{1,L2}$, are the single-interface Fresnel reflection and transmission coefficients.

This expression can be understood by starting out with the direct part of the Green's function in medium 1 expanded in plane-wave components as described in Eq. (4.157). For each plane-wave component propagating downward of the form

$$e^{ik_{y1}y'}e^{-ik_{y1}y} \quad (4.272)$$

there will be a downward propagating component in layer $L2$ that can be described as the sum of several terms. First, there will be a single transmission of the incident wave through the interface between layer 1 and layer $L2$. In addition, this transmitted field will propagate down to the next interface between $L2$ and $L3$; here it will be reflected at the interface, propagate back to the upper interface, and be reflected once more. This adds another term to the downward propagating field. The field can go back and forth several times in layer $L2$ in this way, and there will thus be multiple contributions to the downward propagating field. The downward propagating field in $L2$ can thus be described as the series

$$e^{ik_{y1}y'} e^{-ik_{y,L2}y} t^{(p)}_{1,L2} \sum_{n=0}^{\infty} \left(r^{(p)}_{L2,L3}(k_x) r^{(p)}_{L2,1}(k_x) e^{2ik_{y,L2}d} \right)^n , \tag{4.273}$$

which reduces to

$$e^{ik_{y1}y'} e^{-ik_{y,L2}y} \frac{t^{(p)}_{1,L2}}{1 + r^{(p)}_{1,L2}(k_x) r^{(p)}_{L2,L3}(k_x) e^{2ik_{y,L2}d}} . \tag{4.274}$$

The upward propagating part of the field in layer $L2$ can be constructed in a similar way through an infinite series. By combining both upward and downward propagating field components we arrive at the expression (4.271).

In layer $L2$ it can also be convenient to handle separately the singular part of the Green's function. A singularity may still occur when $x \to x'$, $y' \to 0$ and $y \to 0$. This can be dealt with by rewriting the Green's function as the sum of two terms:

$$g_1(\mathbf{r},\mathbf{r}') = g_1^{(A)}(\mathbf{r},\mathbf{r}') + g_1^{(ns)}(\mathbf{r},\mathbf{r}'), \tag{4.275}$$

where

$$g_1^{(A)}(\mathbf{r},\mathbf{r}') = \frac{2\varepsilon_{L2}}{\varepsilon_1 + \varepsilon_{L2}} \frac{i}{4} H_0^{(1)}(k_0 n_1 |\mathbf{r} - \mathbf{r}'|), \tag{4.276}$$

and

$$g_1^{(ns)}(\mathbf{r},\mathbf{r}') = \frac{i}{2\pi} \int_{k_x=0}^{\infty} \frac{\cos(k_x[x-x'])e^{ik_{y1}y'}}{k_{y1}} \left(f(k_x) - \frac{2\varepsilon_{L2}}{\varepsilon_1 + \varepsilon_{L2}} \right) dk_x,$$

$$y' > 0,\ 0 > y > -d. \tag{4.277}$$

The magnetic field can thus be expressed as

$$H(\mathbf{r}) = H_0(\mathbf{r}) + H^{(A)}(\mathbf{r}) + H^{(ns,-)}(\mathbf{r}) + H^{(ns,+)}(\mathbf{r}),\ 0 > y > -d, \tag{4.278}$$

where

$$H^{(A)}(\mathbf{r}) = -\int \left\{ \phi(\mathbf{r}') g_1^{(A)}(\mathbf{r},\mathbf{r}') - H(\mathbf{r}') \hat{\mathbf{n}}' \cdot \nabla' g_1^{(A)}(\mathbf{r},\mathbf{r}') \right\} dl',\ 0 > y > -d. \tag{4.279}$$

The remaining terms $H^{(\mathrm{ns},-)}$ and $H^{(\mathrm{ns},+)}$ are downward and upward propagating waves, respectively, inside the layer, which can be handled using the angular spectrum representation:

$$H^{(\mathrm{ns},-)}(\mathbf{r}) = \int_{k_x} \tilde{H}^{(i,\mathrm{ns})}(k_x)e^{ik_x x}\left(B(k_x)e^{-ik_{y,L2}y} - \frac{2\varepsilon_{L2}}{\varepsilon_1+\varepsilon_{L2}}e^{-ik_{y1}y}\right)dk_x, \quad (4.280)$$

$$H^{(\mathrm{ns},+)}(\mathbf{r}) = \int_{k_x} \tilde{H}^{(i,\mathrm{ns})}(k_x)e^{ik_x x}C(k_x)e^{ik_{y,L2}(y+2d)}dk_x, \quad (4.281)$$

where the angular spectrum is still given by Eq. (4.239), and

$$B(k_x) = \frac{t^{(p)}_{1,L2}(k_x)}{1+r^{(p)}_{1,L2}(k_x)r^{(p)}_{L2,L3}(k_x)e^{2ik_{y,L2}d}}, \quad (4.282)$$

$$C(k_x) = \frac{t^{(p)}_{1,L2}(k_x)r^{(p)}_{L2,L3}(k_x)}{1+r^{(p)}_{1,L2}(k_x)r^{(p)}_{L2,L3}(k_x)e^{2ik_{y,L2}d}}. \quad (4.283)$$

For p polarization, expressions for $\tilde{E}_x$ and $\tilde{E}_y$ follow by using

$$\mathbf{E}(\mathbf{r}) = \frac{i}{\omega\varepsilon_0\varepsilon_{L2}}\nabla\times\hat{\mathbf{z}}H(\mathbf{r}). \quad (4.284)$$

Only minor adjustments are required for the case of s polarization instead. For s polarization the indirect part of the Green's function does not have a singularity, and there is thus no need for special handling of singularities as was the case for p polarization.

4.2.4 EXAMPLE: GOLD NANOSTRIP ON A DIELECTRIC SUBSTRATE

In this section we will consider the case of a gold nanostrip placed directly on a dielectric substrate as illustrated in Fig. 4.32. The nanostrip will be illuminated by a p-polarized plane wave from above, and the optical properties of the geometry will be studied by calculating a range of optical cross-section spectra, differential scattering, and fields. The substrate is considered infinitely thick and does not support any waveguiding of modes along the x-axis. The substrates quartz and silicon will be considered.

The first step in a calculation of optical cross sections and fields is to determine the fields (and normal derivatives) on the surface of the nanostrip using the Green's function surface integral equation method. This has been described in Section 4.2.2. In the present example the incident field is a p-polarized plane wave being incident from above and propagating along $-y$. The incident field thus has the form

$$H_i(\mathbf{r}) = Ae^{-ik_1 y}. \quad (4.285)$$

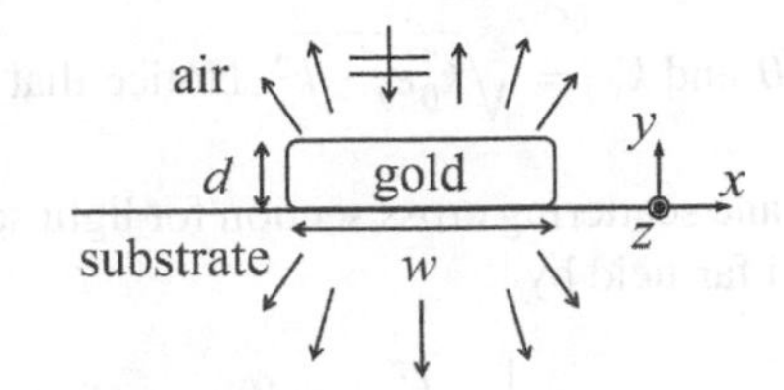

Figure 4.32 Schematic of a gold strip of width w and thickness d on a dielectric substrate. The nanostrip is illuminated from above. It is illustrated that there will be scattering into both the upper and lower half-plane.

The reference field H_0 to be used in the Green's function integral equation must be a solution for the layered reference geometry. Thus, for positions above the interface the reflection from the interface between air and the dielectric substrate must be added, and for positions below the interface the transmitted field must be used. The reference field thus has the form

$$H_0(\mathbf{r}) = \begin{cases} A\left(e^{-ik_1y} + r^{(p)}(0)e^{+ik_1y}\right), & y > 0 \\ At^{(p)}(0)e^{-ik_2y}, & y < 0 \end{cases}, \tag{4.286}$$

where

$$r^{(p)}(k_x) = \frac{\varepsilon_{L2}k_{y1} - \varepsilon_1 k_{y,L2}}{\varepsilon_{L2}k_{y1} + \varepsilon_1 k_{y,L2}}, \tag{4.287}$$

$$t^{(p)}(k_x) = r^{(p)}(k_x) + 1, \tag{4.288}$$

and $k_{y,i} = \sqrt{k_0^2\varepsilon_i - k_x^2}$.

In the following it will be assumed that the total magnetic field H and its normal derivative ϕ have been calculated on the outer surface of the nanostrip using the GFSIEM, and we shall now consider how to use the Green's function integral equations to obtain optical cross sections. As usual the scatterer surface is given by $\mathbf{s}(t)$. By inserting H and ϕ at the scatterer surface in the Green's function surface integral equation, and using the far-field Green's function (4.185), the scattered far field going into the upper half-plane is obtained:

$$H_{\text{sc}}^{(\text{ff})}(r,\theta) = -\frac{e^{ik_0n_1r}}{\sqrt{r}}\frac{e^{i\pi/4}}{4}\sqrt{\frac{2}{\pi k_0 n_1}}\int_t \Big(\phi(\mathbf{s}(t))f_1(\mathbf{s}(t)) -$$
$$H(\mathbf{s}(t))\hat{\mathbf{n}}(t)\cdot(\hat{\mathbf{x}}(-ik_x)f_1(\mathbf{s}(t)) + \hat{\mathbf{y}}(-ik_{y1})f_2(\mathbf{s}(t)))\Big)dt, \quad 0 < \theta < \pi \tag{4.289}$$

where

$$f_1(\mathbf{r}) = e^{-ik_xx}\left(e^{-ik_{y1}y} + r^{(p)}(k_x)e^{ik_{y1}y}\right), \tag{4.290}$$

$$f_2(\mathbf{r}) = e^{-ik_xx}\left(e^{-ik_{y1}y} - r^{(p)}(k_x)e^{ik_{y1}y}\right), \tag{4.291}$$

and here $k_x = k_0 n_1 \cos\theta$ and $k_{y,j} = \sqrt{k_0^2 \varepsilon_j - k_x^2}$. Notice that here k_x can be negative for certain values of θ.

The partial out-of-plane scattering cross section for light scattered upward is given in terms of the scattered far field by

$$\sigma_{\text{OUP, up}} = \frac{1}{|A|^2} \int_{\theta=0}^{\pi} |H_{\text{sc}}^{(\text{ff})}(r,\theta)|^2 r d\theta. \tag{4.292}$$

Similarly, the scattered far field going into the substrate side can be obtained by inserting (4.189) into (4.16), which leads to

$$H_{\text{sc}}^{(\text{ff})}(r,\theta) = -\frac{e^{ik_0 n_{L2} r}}{\sqrt{r}} \frac{e^{i\pi/4}}{4} \sqrt{\frac{2}{\pi k_0 n_{L2}}} \frac{k_{y,L2}}{k_{y1}} t^{(p)}(k_x)$$
$$\times \int_t e^{-ik_x x(t)} e^{ik_{y1} y(t)} \left(\phi(\mathbf{s}(t)) - H(\mathbf{s}(t)) \hat{\mathbf{n}}(t) \cdot [\hat{\mathbf{x}}(-ik_x) + \hat{\mathbf{y}} ik_{y1}]\right) dt, \quad -\pi < \theta < 0, \tag{4.293}$$

where here, $k_x = k_0 n_{L2} \cos\theta$, and $k_{y,j} = \sqrt{k_0^2 \varepsilon_j - k_x^2}$. Here $x(t)$ and $y(t)$ are the components of the vector $\mathbf{s}(t)$. The partial out-of-plane scattering cross section for light scattered downward is given by

$$\sigma_{\text{OUP, down}} = \frac{1}{|A|^2} \frac{n_1}{n_{L2}} \int_{\theta=-\pi}^{0} |H_{\text{sc}}^{(\text{ff})}(r,\theta)|^2 r d\theta. \tag{4.294}$$

The differential scattering cross section defined as

$$\frac{\partial \sigma_{\text{OUP}}}{\partial \theta} = \begin{cases} |H_{\text{scat}}^{(\text{ff})}(r,\theta)|^2 r / |A|^2, & 0 < \theta < \pi \\ |H_{\text{scat}}^{(\text{ff})}(r,\theta)|^2 r (n_1/n_{L2}) / |A|^2, & 0 > \theta > -\pi \end{cases} \tag{4.295}$$

represents the scattered power per unit angle, which is also the radiation pattern.

There will also be two contributions to the extinction cross section, namely a cross section for the power removed from the specularly reflected beam

$$\sigma_{\text{EXT, up}} = -2 \frac{\sqrt{\frac{2\pi}{k_0 n_1}}}{|A|^2} \text{Re}\left\{ r^{(p)}(0) \left(H_{\text{sc}}^{(\text{ff})}(r,\pi/2) \right)^* \sqrt{r} e^{ik_0 n_1 r} e^{-i\pi/4} A \right\}, \tag{4.296}$$

and a cross section for the power removed from the transmitted beam

$$\sigma_{\text{EXT, down}} = -2 \frac{n_1}{n_{L2}} \frac{\sqrt{\frac{2\pi}{k_0 n_{L2}}}}{|A|^2} \text{Re}\left\{ t^{(p)}(0) \left(H_{\text{sc}}^{(\text{ff})}(r,-\pi/2) \right)^* \sqrt{r} e^{ik_0 n_{L2} r} e^{-i\pi/4} A \right\}. \tag{4.297}$$

As the first example we shall consider the case of a gold nanostrip of width 150 nm, thickness 17 nm, and with corners of radius 2 nm being placed on a quartz substrate. The geometry is illuminated with normally incident p-polarized light. The optical cross-section spectra obtained for this geometry using the GFSIEM are shown

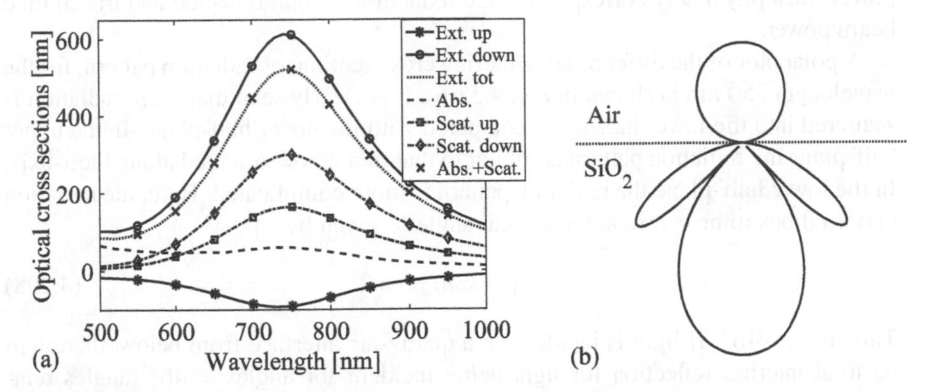

Figure 4.33 (a) Optical cross-section spectra for a gold nanostrip of width w = 150 nm and thickness d = 17 nm being placed directly on a quartz substrate ($n_{L2} = 1.452$) and being illuminated by a p-polarized plane wave propagating along the negative y-axis. (b) Polar plot of differential scattering cross section for the wavelength 750 nm.

in Fig. 4.33(a). The differential scattering cross section at the wavelength 750 nm is shown in Fig. 4.33(b). The refractive index of the quartz substrate is set to n_{L2} = 1.452 for all considered wavelengths. Only the out-of-plane scattering cross section for scattering into the upper half-plane has been previously considered for this geometry [55]. In Fig. 4.33(a) all the cross sections are shown. It can be seen that there is a resonance peak in, e.g., the Scat. up ($\sigma_{\text{OUP, up}}$) cross-section spectrum near the wavelength 740 nm. The out-of-plane scattering cross section for light scattered into the lower half-plane Scat. down ($\sigma_{\text{OUP, down}}$) is seen to be larger than the corresponding cross section for light scattered up. It is a general trend that more light is scattered into the higher-refractive-index substrate compared with the lower-refractive-index upper reference medium.

The extinction cross section is also divided into terms for the power removed from the reflected beam duc to the presence of the scatterer (Ext. up), and the power removed from the transmitted beam due to the scatterer (Ext. down). The reflected beam extinction cross section, and thus also the power removed from the reflected beam, is negative, which means that the power in the reflected beam is increased by the presence of the nanostrip. This is quite reasonable since a quartz-air interface is reflecting less than 4% of the light, while metal is highly reflecting, and thus the reflected beam power can increase. The transmitted beam power is on the other hand reduced due to the presence of the scatterer since the corresponding extinction cross section is positive. The absorption cross section (Abs.) shown as the dashed curve is not very large but also not negligible. The total extinction (Ext. up + Ext. down) is shown as the dotted line, and the sum of absorption and scattering (Abs + Scat = Abs. + Scat. up + Scat. down) shown with crosses, is seen to coincide with the total extinction. This serves as a check of the calculation since the absorbed and scattered power must physically correspond to the reduction in total reflected and transmitted beam power.

A polar plot of the differential scattering cross section, or radiation pattern, for the wavelength 750 nm is shown in Fig. 4.33(b). It is clearly seen that more radiation is scattered into the lower half-plane compared with the upper half-plane. In the upper half-plane the radiation pattern is similar to that of a dipole oriented along the x-axis. In the lower half-plane the radiation pattern is more complicated. Here, the radiation pattern drops to near zero at the critical angle θ_c given by

$$k_0 n_1 = k_0 n_{L2} \cos\theta_c. \tag{4.298}$$

Thus $\theta_c \approx -46^\circ$. If light is incident on a quartz-air interface from below, there will be total internal reflection for light being incident for angles $< 46^\circ$ (angles relative to the x-axis). Likewise a plane wave incident from the air-side propagating in any direction θ_i cannot be transmitted into the substrate for this angular range, as a consequence of the relation between the angle of incident and transmitted light $k_0 n_1 \cos\theta_i = k_0 n_{L2} \cos\theta_t$. It appears though that the scatterer placed in the upper medium can in fact scatter light into the substrate into directions with total internal reflection. This is a near-field effect, and it will not occur if the nanostrip is placed a distance above the air-quartz interface which is comparable to a few wavelengths. The near-field effect can be explained by considering the field generated initially by a point source at a distance y' above the surface. Such a field is given by a Green's function of the form (4.157), and the field contains downward propagating waves with any value of k_x including those with $|k_x| > k_0 n_1$. If the point source is placed high above the surface no plane waves with $|k_x| > k_0 n_1$ will in fact reach the surface due to the y-dependence of such waves being of the form $\exp(ik_{y,1}(y' - y))$. These waves are exponentially damped with distance toward the interface, and more so for larger k_x. If, on the other hand, the point source is placed right on the substrate surface ($y' = 0$), these waves will reach the interface and can be transmitted. Conservation of k_x, as must be the case according to the electromagnetics boundary condition at the interface, means that waves are also transmitted into the usually forbidden range of directions in the substrate.

The same situation but considering a range of different widths w from 150 nm to 450 nm is considered in Fig. 4.34. The absorption cross-section spectra are shown in Fig. 4.34(a). Some sharp peaks at wavelengths near 600 nm correspond to higher-order resonances. Overall the absorption cross sections are small compared with other cross sections. The extinction cross-section spectra are shown in Fig. 4.34(b). A dominant resonance peak is seen to move to longer wavelengths as the strip width is increased, which is in accordance with the Fabry-Perot resonator interpretation of a nanostrip that was introduced in Sec. 4.1.8. Again, the extinction up cross section is negative because the reflected beam power actually increases due to the reflection from the nanostrip, while the extinction down cross section is positive since overall power is removed from the transmitted beam due to the nanostrip. The scattering up and down cross-section spectra are shown in Figs 4.34(c) and (d), respectively. The trend of a resonance peak shifting to longer wavelengths with increasing strip width

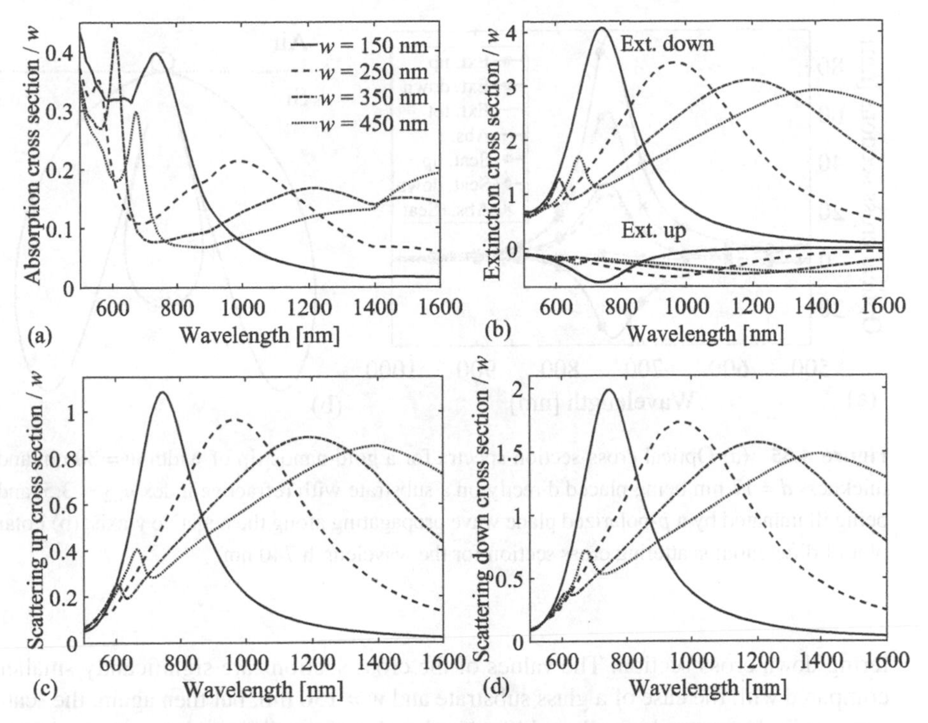

Figure 4.34 Optical cross-section spectra for a gold nanostrip of thickness d = 17 nm and different widths w = 150, 250, 350 and 450 nm placed directly on a quartz substrate (n_{L2} = 1.452), and being illuminated by a p-polarized plane wave propagating along the negative y-axis. All optical cross sections are normalized with the width of the nanostrip w. (a) Absorption, (b) extinction up and extinction down, (c) scattering up, and (d) scattering down cross sections.

is also seen here, and generally more light is scattered into the substrate than into the air region.

The effect of using a substrate with an even larger refractive index of $n_{L2} = 3.5$ is considered in Fig. 4.35. This refractive index can be representative of a range of semiconductors such as, for example, silicon. Here, the absorption in the substrate material is ignored though. Since the material surrounding the nanostrip now has a much larger refractive index, it is necessary to reduce the strip width w in order to have the fundamental resonance at a similar wavelength to the case considered above with a glass substrate. Here, a width of 30 nm leads to a resonance peak wavelength at 740 nm. If we would keep the original strip width of 150 nm, the resonance peak wavelength would instead be 1400 nm. Now, due to the higher refractive index of the substrate, the scattering up cross section is very small compared with other cross sections. The absorption is now not small and is only slightly smaller than the scat-

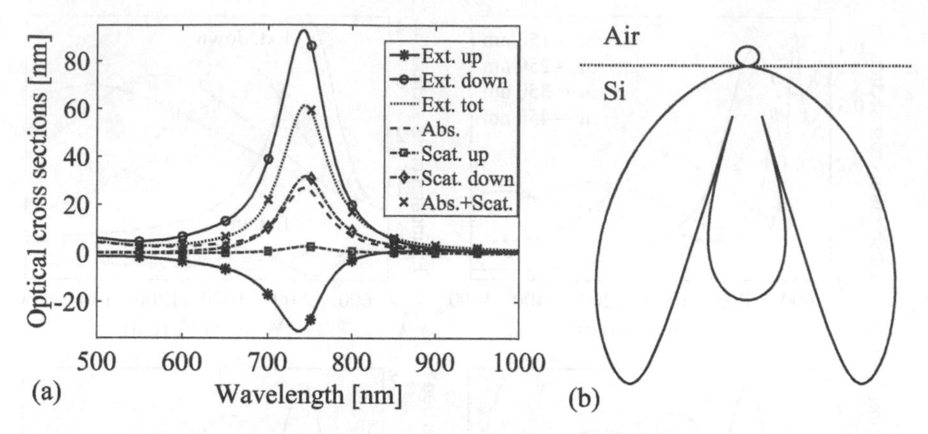

Figure 4.35 (a) Optical cross-section spectra for a gold nanostrip of width w = 30 nm and thickness d = 17 nm being placed directly on a substrate with refractive index $n_{L2} = 3.5$, and being illuminated by a p-polarized plane wave propagating along the negative y-axis. (b) Polar plot of differential scattering cross section for the wavelength 740 nm.

tering down cross section. The values of the cross sections are significantly smaller compared with the case of a glass substrate and w = 150 nm, but then again, the scatterer is also has a much smaller width. A polar plot of the differential scattering cross section at the wavelength 740 nm is shown in Fig. 4.35(b). Clearly, very little light is scattered up into the air region. Comparing with Fig. 4.33 the effect of asymmetric scattering with more going into the substrate clearly has increased significantly with the increase of the substrate refractive index.

The fields are considered in Fig. 4.36 for a gold nanostrip of width 150 nm and thickness 17 nm on a quartz substrate, which is being illuminated by normally incident light with wavelength 740 nm. This corresponds to resonant excitation of the gold nanostrip. A map of the resonant electric field is shown in Fig. 4.36(a), while a cross section of the electric field magnitude through the center of the nanostrip is shown in Fig. 4.36(b). Both follow the same trends as resonant fields for a nano-strip in a homogeneous reference medium. Strong values of the field are found just outside the two ends of the nanostrip at $x = \pm$ 75 nm, and inside the strip the field tends to a maximum in the center of the strip as expected for the field of the fundamental resonance. The magnitude of the y-component of the electric field is shown in Fig. 4.36(c). The reference electric field being considered does not have a y-component, and the y-component of the field seen here is entirely due to scattering by the nanostrip. For completeness the magnitude of the magnetic field is shown in Fig. 4.36(d). Note that regarding the maps of the electric field, an upper limit has been imposed on the values of the field magnitude being shown in order to make features at smaller field magnitude values away from corners more visible.

Field plots for the same situation except that the strip width is 450 nm and the

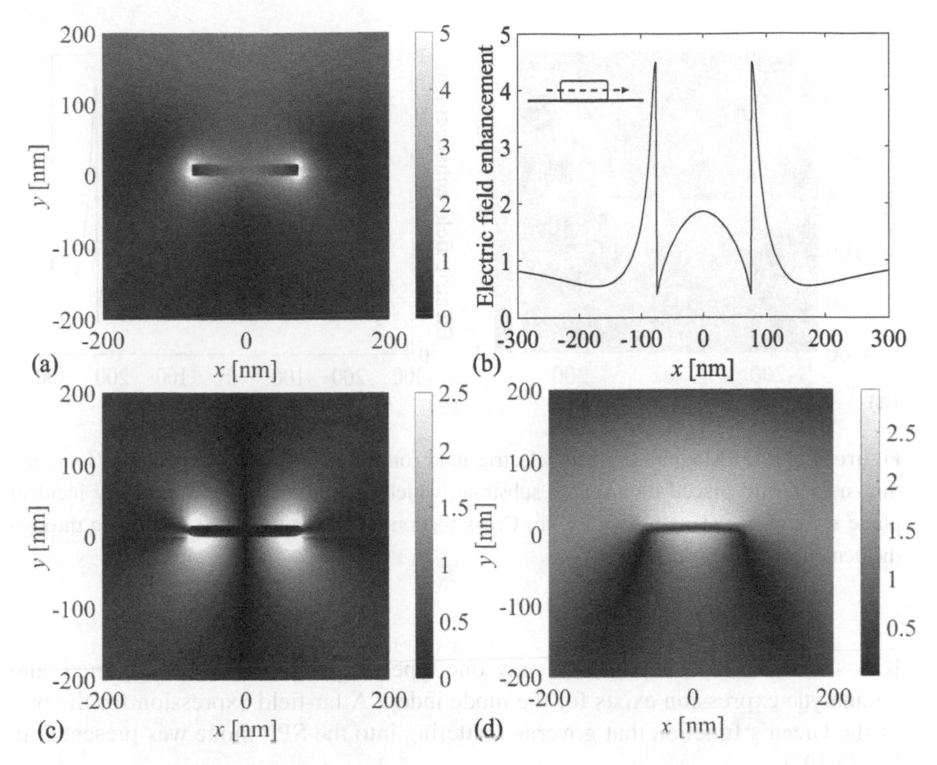

Figure 4.36 Calculation of fields for a gold nanostrip of width 150 nm and thickness 17 nm placed on quartz substrate and being illuminated by a normally incident plane wave with wavelength 740 nm. (a) Magnitude of electric field, (b) cross section of the electric field magnitude through the center of the nanostrip at $y = 8.5$ nm, (c) magnitude of the y-component of electric field, and (d) magnitude of the magnetic field.

wavelength is 673 nm are shown in Fig. 4.37. The wavelength has been selected by observing that there is a higher-order resonance peak in cross-section spectra in Fig. 4.34 at this wavelength. The map of the electric field magnitude shown in Fig. 4.37(a), and the cross section through the center of the nanostrip shown in Fig. 4.37(b), have two minima in the electric field magnitude inside the nanostrip corresponding to a third-order resonance.

4.2.5 EXAMPLE: SILVER NANOSTRIP ABOVE A SILVER SURFACE

In this section an example will be considered where the reference structure is again a two-layer structure as in the previous section, but now the substrate is made of silver, and the material above the silver surface is quartz. The main new feature of this reference structure is that it supports a guided mode, namely the surface plasmon polariton (SPP), which is bound to and propagating along the silver-quartz interface.

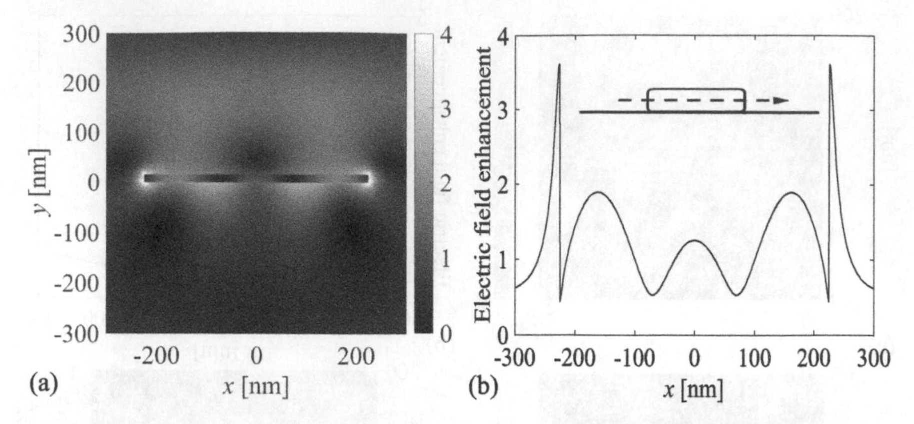

Figure 4.37 (a) Magnitude of the electric field for a gold nanostrip of width 450 nm and thickness 17 nm placed on a quartz substrate, which is illuminated by a normally incident plane wave of wavelength 673 nm. (b) Cross section of the electric field magnitude through the center of the nanostrip.

It is still relatively simple since only one type of guided mode is supported, and an analytic expression exists for the mode index. A far-field expression for the part of the Green's function that governs scattering into the SPP mode was presented in Eq. (4.197).

The scatterer being considered is a silver nanostrip of width $w = 80$ nm and thickness $d = 10$ nm placed at a distance δ above the silver surface inside the quartz region. The nanostrip is illuminated from above at an angle of $-45°$ with respect to the x-axis. The scattering situation illustrated in Fig. 4.38 has been previously studied in Ref. [56]. For small gaps δ the resonances are mainly governed by gap SPPs propagating back and forth in the gap between the metal nanostrip and the metal surface.

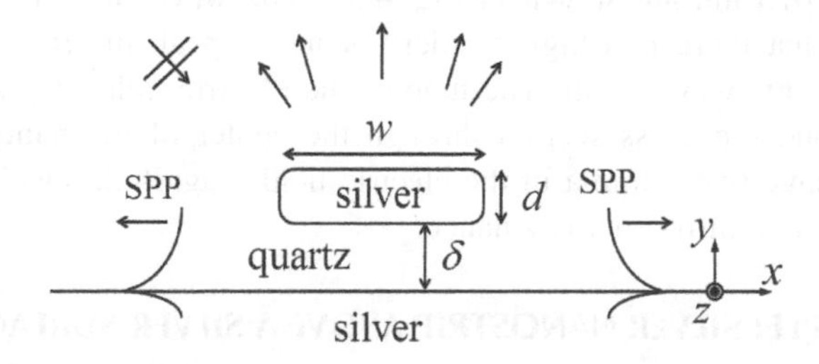

Figure 4.38 Schematic of scattering by a silver nanostrip of width w and thickness d placed a distance δ above a silver surface. The nanostrip is surrounded by quartz, and is illuminated by a plane wave propagating at an angle of $-45°$ with respect to the x-axis.

There will now be no scattering into the substrate, but there will still be out-of-plane scattering into the upper half-plane, and there will be scattering into SPPs propagating away from the nanostrip along the quartz-silver interface. In addition, power will be absorbed in the nanostrip, and there will be an extinction of power from the specularly reflected beam due to scattering and absorption. Except for the excitation of SPP waves, the other cross sections are defined in the same way as in the previous subsection. Since there is no extinction and scattering cross section related to propagation into the lower half-plane, the extinction and scattering going into the upper half-plane will be referred to as simply extinction and scattering. Note that the metal substrate is also lossy, and the absorption cross section only accounts for absorption in the nanostrip, while any excess absorption in the substrate is not included in the absorption cross section.

The SPP wave excitation may be quantified by a scattering cross section for scattering into the SPP mode. This can be further divided into cross sections for scattering going into a right- and a left-propagating SPP. Since other scattering cross sections are already described in the previous subsection, the first thing that will be focused on here is to derive expressions for the SPP scattering cross section.

The first step is to note that the far field will now contain a right-propagating and a left-propagating SPP component propagating along the air-metal interface given by

$$H^{(\text{SPP, ff})}(\mathbf{r}) = -\int_{C_1} \left\{ \phi(\mathbf{r}')g^{(\text{SPP})}(\mathbf{r},\mathbf{r}') - H(\mathbf{r}')\hat{\mathbf{n}}' \cdot \nabla' g^{(\text{SPP})}(\mathbf{r},\mathbf{r}') \right\} dl', \quad (4.299)$$

where the far-field SPP Green's function component $g^{(\text{SPP})}$ was obtained in Eq. (4.197).

This leads to the following right-propagating far-field SPP magnetic field

$$H_r^{(\text{SPP, ff})}(\mathbf{r}) = H_{r,0}^{(\text{SPP, ff})} e^{ik_{x,p}x} \begin{cases} e^{ik_{y1,p}y} & y > 0 \\ e^{-ik_{y,L2,p}y} & y < 0 \end{cases}, \quad x > \max\{x'\} \quad (4.300)$$

where the front factor is given by

$$H_{r,0}^{(\text{SPP, ff})} = -A^{(\text{SPP})} \int_{C_1} e^{-ik_{x,p}x'} e^{ik_{y1,p}y'} \left(\phi(\mathbf{r}') - H(\mathbf{r}')\hat{\mathbf{n}}' \cdot \left[-ik_{x,p}\hat{\mathbf{x}} + ik_{y1,p}\hat{\mathbf{y}} \right] \right) dl'. \quad (4.301)$$

The front-factor contains the coefficient

$$A^{(\text{SPP})} = \frac{1}{2} \frac{k_{y,L2,p}}{k_{x,p}} \frac{\varepsilon_{L2}k_{y,1,p} - \varepsilon_1 k_{y,L2,p}}{\varepsilon_{L2}k_{y,L2,p} + \varepsilon_1 k_{y,1,p}} \quad (4.302)$$

that can be recognized as the coefficient that also enters in the expression for the far-field SPP part of the Green's function (4.196). Here $k_{x,p} = k_0\sqrt{\varepsilon_1\varepsilon_{L2}/(\varepsilon_1 + \varepsilon_{L2})}$, and $k_{y,j,p} = \sqrt{k_0^2\varepsilon_j - k_{x,p}^2}$.

The power in the right-propagating SPP wave at position x can be obtained from the flux of the Poynting vector as

$$P_r^{(\text{SPP})}(x) = \frac{1}{2}\text{Real}\left\{ \int_y \mathbf{E}^{(\text{SPP})}(\mathbf{r}) \times \left(\mathbf{H}^{(\text{SPP})}(\mathbf{r}) \right)^* \cdot \hat{\mathbf{x}} dy \right\}$$

$$= \frac{1}{2}\text{Real}\left\{\int_y \frac{k_{x,\mathrm{p}}}{\omega\varepsilon_0\varepsilon_{\mathrm{ref}}(y)}|H_r^{\mathrm{SPP,\,ff}}(\mathbf{r})|^2 dy\right\}. \tag{4.303}$$

This leads to

$$P_r^{(\mathrm{SPP})}(x) = |H_{r,0}^{(\mathrm{SPP,\,ff})}e^{ik_{x,\mathrm{p}}x}|^2\left(\frac{\text{Real}\{k_{x,\mathrm{p}}/\omega\varepsilon_0\varepsilon_1\}}{4\text{Imag}\{k_{y1,\mathrm{p}}\}}+\frac{\text{Real}\{k_{x,\mathrm{p}}/\omega\varepsilon_0\varepsilon_{L2}\}}{4\text{Imag}\{k_{y,L2,\mathrm{p}}\}}\right). \tag{4.304}$$

The power per unit area of the incident plane wave is given by

$$I_i = \frac{1}{2}\frac{k_0 n_1}{\omega\varepsilon_0\varepsilon_1}|H_i|^2, \tag{4.305}$$

where H_i is the amplitude of the incident magnetic field. The ratio of (4.304) and (4.305) leads finally to the right-propagating SPP scattering cross section

$$\sigma_{\mathrm{SPP,\,r}}(x) = \frac{|H_{r,0}^{(\mathrm{SPP,\,ff})}e^{ik_{x,\mathrm{p}}x}|^2}{|H_i|^2}\left(\frac{\text{Real}\{k_{x,\mathrm{p}}/k_0n_1\}}{2\text{Imag}\{k_{y1,\mathrm{p}}\}}+\frac{\text{Real}\{k_{x,\mathrm{p}}\varepsilon_1/k_0n_1\varepsilon_{L2}\}}{2\text{Imag}\{k_{y,L2,\mathrm{p}}\}}\right). \tag{4.306}$$

The scattering cross section for the left-propagating SPP is similarly given by

$$\sigma_{\mathrm{SPP,\,l}}(x) = \frac{|H_{l,0}^{(\mathrm{SPP,\,ff})}e^{-ik_{x,\mathrm{p}}x}|^2}{|H_i|^2}\left(\frac{\text{Real}\{k_{x,\mathrm{p}}/k_0n_1\}}{2\text{Imag}\{k_{y1,\mathrm{p}}\}}+\frac{\text{Real}\{k_{x,\mathrm{p}}\varepsilon_1/k_0n_1\varepsilon_{L2}\}}{2\text{Imag}\{k_{y,L2,\mathrm{p}}\}}\right), \tag{4.307}$$

where

$$H_{l,0}^{(\mathrm{SPP,\,ff})} = -A^{(\mathrm{SPP})}\int_{C_1} e^{+ik_{x,\mathrm{p}}x'}e^{ik_{y1,\mathrm{p}}y'}\left(\phi(\mathbf{r}') - H(\mathbf{r}')\hat{\mathbf{n}}'\cdot\left[ik_{x,\mathrm{p}}\hat{\mathbf{x}}+ik_{y1,\mathrm{p}}\hat{\mathbf{y}}\right]\right)dl'. \tag{4.308}$$

Note that it has been explicitly pointed out in the expressions (4.306) and (4.307) that these cross sections depend on x. For a metal substrate where absorption losses are ignored there will be no x dependence. However, for a lossy metal substrate the power in the SPP mode will decrease with propagation distance.

The optical cross-section spectra for the considered geometry are presented in Fig. 4.39 for four different gap sizes δ = 5 nm, 10 nm, 20 nm, and 50 nm. Note that the extinction cross section has not previously been shown [56]. The wavelength of the first-order resonance peak is seen to be highly sensitive to the gap size δ, and especially so for small gaps. The resonance peak red-shifts with decreasing δ, which can be understood by considering a geometry with a silver film of thickness d placed above a silver surface. Such a structure supports gap-SPP modes, where the field to a large extent is localized to the gap between the metal film and metal surface. This is very similar to the gap SPPs studied previously in Fig. 4.20 for a gap between two metal films. As the gap decreases the mode index increases. According to the Fabry-Perot resonator model, where resonances are due to counter-propagating gap SPPs being reflected at structure terminations forming standing-wave resonances, the resonance wavelength must increase with decreasing δ as observed. For very

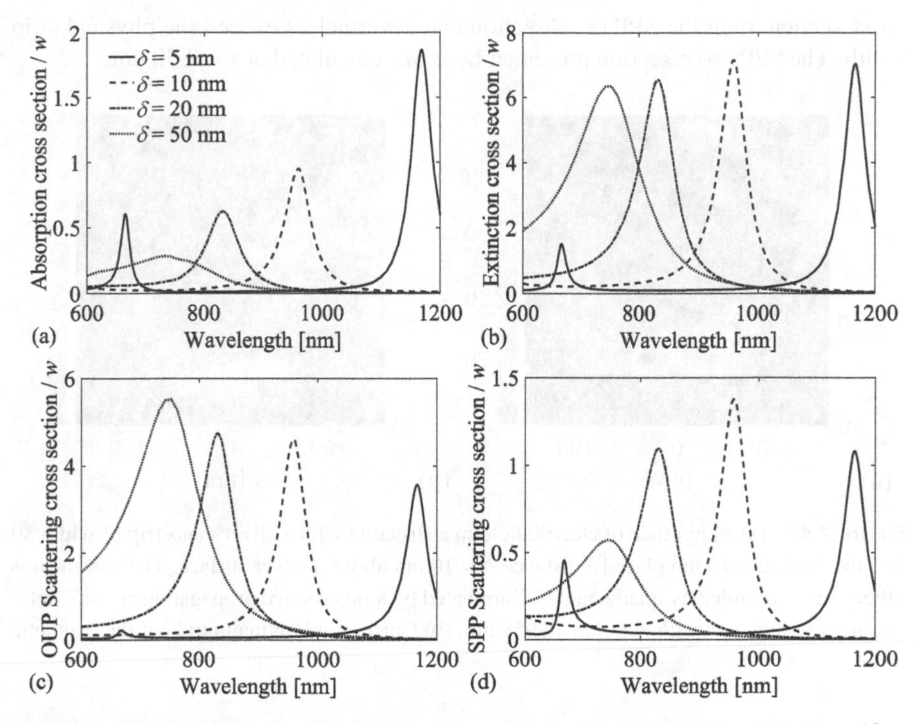

Figure 4.39 Optical cross sections for a silver nanostrip of width 80 nm and thickness 10 nm placed a distance δ = 5 nm, 10 nm, 20 nm, or 50 nm, above a silver surface. The nanostrip is surrounded by quartz and is illuminated by a plane wave propagating at an angle of −45° with respect to the x-axis. (a) Absorption cross section, (b) extinction cross section, (c) out-of-plane (OUP) scattering cross section, and (d) total SPP scattering cross section.

large gaps the resonances will ultimately be the same as for the silver strip in a homogeneous quartz medium. We do not go to that limit here though.

For the gap size $\delta = 5$ nm there is a second-order resonance at the wavelength of approximately 665 nm. For symmetry reasons this resonance will not be excited for normally incident light but can be excited by light incident at an oblique angle. The resonance is barely visible in the OUP scattering cross-section spectrum but is easily noticed in the other cross-section spectra. Since the absorption cross section only includes absorption in the nanostrip, and does not include any extra absorption in the substrate, the sum of absorption and total scattering is slightly smaller than the extinction cross section. It was checked, however, that in the case of setting the imaginary part of ε_{L2} to zero, such that no absorption occurs in the substrate, while absorption still occurs in the nanostrip, then the extinction cross section, and the sum of scattering and absorption cross sections, coincide, meaning that $\sigma_{\text{Ext}} = \sigma_{\text{Abs}} + \sigma_{\text{OUP}} + \sigma_{\text{SPP, r}} + \sigma_{\text{SPP, l}}$. Also notice that while the SPP scattering cross section at most wavelengths is several times smaller than, for example, the OUP scattering cross section, then the SPP cross section can nevertheless exceed the physical strip width. The SPP cross section presented here was calculated at $x = \pm 50$ nm.

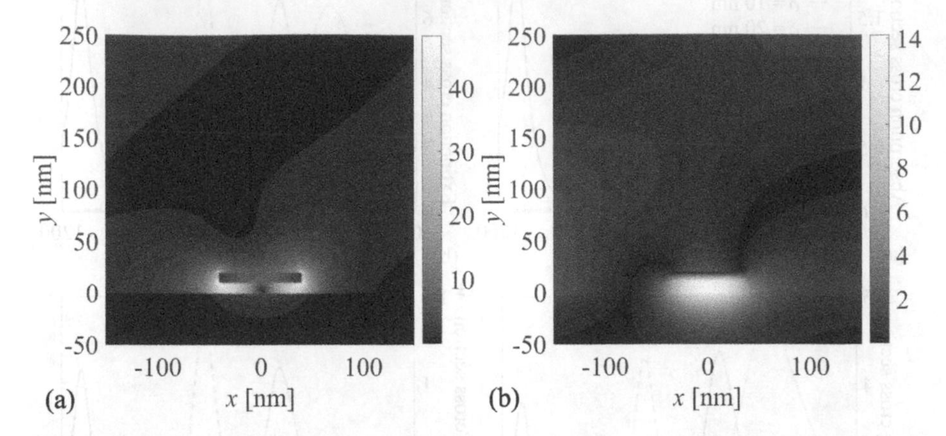

Figure 4.40 (a) Magnitude of electric field in a structure with a silver nanostrip of width 80 nm and thickness 10 nm placed a distance $\delta = 10$ nm above a silver surface. The nanostrip is otherwise surrounded by quartz, and is illuminated by a plane wave propagating at $-45°$ relative to the x-axis and with wavelength 958 nm. (b) Corresponding magnitude of the magnetic field.

The magnitude of the electric field at the first-order resonance is shown in Fig. 4.40 for the gap size 10 nm and wavelength 958 nm. The structure is resonant at this wavelength. The electric field magnitude is strong near the ends of the gap between strip and substrate, while it has a node in the middle. A similar field was seen for the first-order resonance for a geometry consisting of two nanostrips in Fig. 4.21, where the resonant field was also interpreted as being due to counter-propagating gap SPPs forming a fundamental standing wave in the gap. The corresponding magnitude of the magnetic field is shown in Fig. 4.40(b). A strong magnetic field is also present in the gap. Gap-plasmon resonances for a geometry of two metal strips, and a geometry with a metal strip and a metal surface are similar, although as discussed in Ref. [56], the terminations at the strip ends can be considered to be softer for a strip above a metal substrate, which leads to a slight red-shift of resonance wavelengths and a broadening of the resonance peaks.

A final word of caution on the limiting case of placing the metal nanostrip directly on the metal surface is in order. Consider the two geometries in Fig. 4.41. While the rounding of corners for a nanostrip placed above the surface leads to a geometry with no sharp corners, then in the limit of placing the nanostrip directly on the metal substrate as illustrated in Fig. 4.41(a) an ultra-sharp groove in metal is formed at the strip ends. Gap SPPs can propagate into this ultra-sharp groove, and the response depends strongly on the bottom of the corner. The bottom of the corner is, however, difficult to resolve properly in a numerical method due to the ultra-sharp nature of the groove. In this limit it is better to use the bottom corner geometry illustrated in

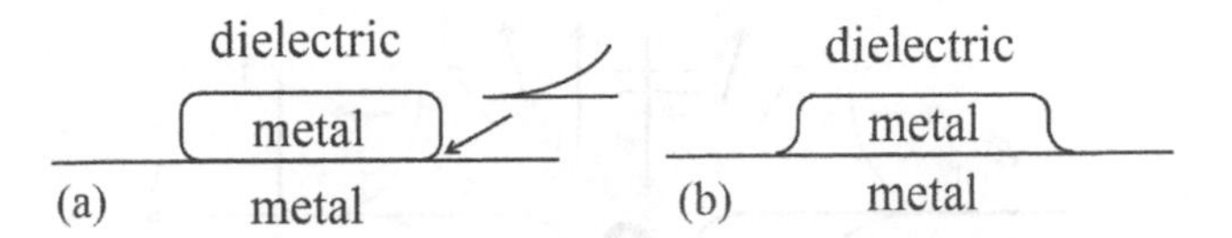

Figure 4.41 Schematic of metal nanostrip placed directly on a metal surface. (a) The usual rounding of corners leading to a sharp groove between strip and metal surface. (b) Choice of bottom corners that lead to a total structure without sharp edges or grooves.

Fig. 4.41(b), since the resulting total geometry of strip and substrate will now not have any sharp corners, and convergence will be unproblematic.

Finally, it can be mentioned that the Green's function for a thin metal film surrounded by a dielectric was derived in Ref. [57], and later used to study scattering by a metal nanostrip placed above a nanometer-thin metal film in Ref. [58]. In this case the incident light can couple to not just one but two types of SPP waves, namely a short-range and a long-range SPP, bound to and propagating along the metal film. In this case there will still be resonances related to gap-SPP waves in the gap between the metal nanostrip and the metal film. In addition, the case of a gold nanostrip placed above a gold surface was considered in Ref. [59].

4.2.6 EXAMPLE: SINGLE GROOVE IN METAL

In this section we will consider a single wavelength-scale groove in a metal surface (Fig. 4.42) being illuminated by a normally incident plane wave. Similar to a scatterer placed above a metal surface, there will be out-of-plane scattering, and scattering into surface plasmon polaritons. The reference structure is again a metal substrate for $y < 0$ and a dielectric, in this case air, for $y > 0$. The modification to the reference structure is the groove itself, and the scatterer is thus an air region. There is no point in calculating the usual absorption cross section here for a scatterer since no absorption will take place inside the groove. However, the groove will nevertheless be the cause of additional absorption in the surrounding metal beyond that of the case with no groove. The additional absorption, scattering out of the plane, and scattering into SPPs lead to a reduction in the reflected beam power, which is quantified by the extinction cross section. The absorption cross section may here be estimated as $\sigma_{\mathrm{ABS}} = \sigma_{\mathrm{EXT}} - \sigma_{\mathrm{OUP}} - \sigma_{\mathrm{SPP}}$.

The groove may play the role as a resonator. The air layer between groove walls supports gap-SPP waves that can propagate both upward and downward in the groove. These waves can be reflected at the top and bottom groove terminations, and standing waves may form in the groove at certain resonance wavelengths. This is entirely similar to other resonators considered so far. At resonance, the absorption in the metal can be significant.

Resonances in periodic arrays of grooves in metal surfaces have been considered for making selective narrow-band thermal emitters [60, 61, 62, 63], which may have applications as emitter surfaces in thermophotovoltaics [64]. In thermophotovoltaics

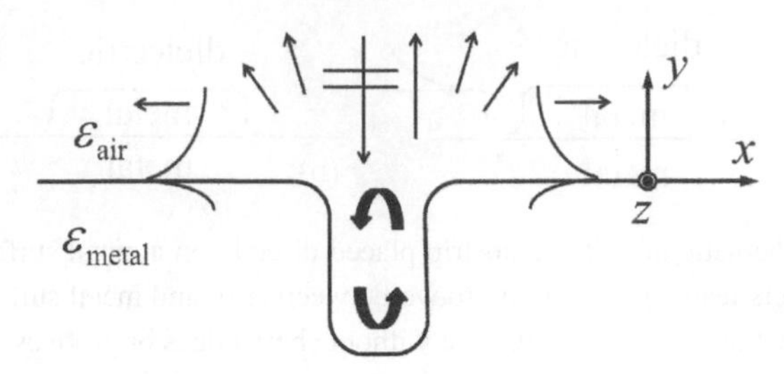

Figure 4.42 Schematic of a groove in a flat metal surface being illuminated by a plane wave. The presence of the groove leads to out-of-plane scattering, scattering into SPP waves, and additional absorption.

a special solar cell is not necessarily converting sunlight into electricity. Instead it may convert thermal radiation coming from a thermal emitter surface. This also implies a solar cell operating at longer wavelengths than those aimed at conversion of sunlight. Special thermal emitter surfaces that predominantly emit thermal emission at a wavelength matched for optimum conversion efficiency in the solar cell are of interest here. Any selective emitter is also a selective absorber, and vice versa, due to Kirchoff's law stating equality between spectral absorptivity and emissivity of a surface. Thus, strong resonant absorption in the groove at a certain wavelength and small absorption at other wavelengths implies good thermal emission at the same wavelength, and small thermal emission at other wavelengths. Grooves have also been considered as selective absorbers [65, 66]. Many other types of structured surfaces involving metals, e.g. [67, 68], can also be used for making selective absorbers.

Long and possibly bent individual tapered grooves in metal surfaces have been investigated as optical waveguides [69, 70], including waveguide components such as Mach-Zehnder interferometers. Periodic arrays of subwavelength ultra-sharp tapered grooves in metal surfaces have been considered for making black metals [71, 72, 73], that are broadband non-resonant absorbers, where the groove arrays change a high-reflectivity shiny metal surface into a black surface. A GFSIEM was developed for the modeling of such groove arrays [74]. Black metals can also be made with crossed grooves or pillar arrays that are not tapered much if absorption is high enough [75].

Apart from waveguiding, most recent work on grooves in metal surfaces concerning reflectivity and scattering is for periodic arrays of grooves. There is not much work on the optics of individual grooves, which is also more difficult. Scattering measurements on a single groove is difficult because the measured signal will be small compared with an array of grooves. In addition, regarding theoretical work, a calculation assuming a single groove in an otherwise flat surface of infinite extent is with most theoretical methods not straightforward, since one needs to set up, e.g., perfectly matched layers or absorbing boundary conditions to truncate the computational problem, and this must be done across a metal interface. GFIEMs do not require truncation of the computational domain, since the radiating boundary condition is automatically satisfied via the choice of Green's function. The GFSIEM has been used by the author and co-workers for studying the optics of a single rectangular or ultra-sharp groove in Refs. [76, 77].

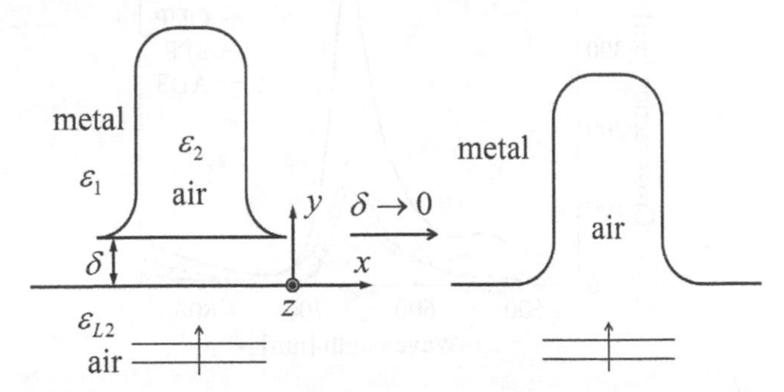

Figure 4.43 Schematic of a groove in a metal introduced as an air void in the metal of a reference geometry with a planar interface between air and metal. The air void is at a distance δ from the air-metal reference structure interface. By taking the limit $\delta \to 0$, a groove opens up in the metal surface.

In order to treat the geometry (Fig. 4.42) in the same way as in some previous sections, that is as a scatterer introduced in the upper layer of a two-layer geometry, we shall turn the problem upside-down as shown in Fig. 4.43. Here, we use metal in the upper layer ($\varepsilon_1 = \varepsilon_{\text{metal}}$), and air in the lower layer ($\varepsilon_{L2} = 1$). The scatterer is an air region being introduced in the upper layer ($\varepsilon_2 = 1$). The air scatterer can be thought of as being at a distance δ from the interface of the reference geometry, and by taking the limit $\delta \to 0$ a groove opens up in the metal surface. This procedure is the same as when, e.g., considering a nanostrip placed directly on a quartz surface.

Notice the two ultra-sharp corners of the air scatterer at the end of a quarter circle. This choice of corners leads to a total resulting structure with no sharp corners, which is appropriate for numerical calculations. When using such corners, the matrix $\overline{\overline{D}}_\phi$ [see Eq. (4.83)] will have two elements that are equal to -1, since clearly the normal vector on one and the other side of the corner point in exact opposite directions, and the normal derivative of the field will thus change sign across the corner.

Examples will now be given for a resonant rectangular groove, and later for an ultra-sharp tapered groove. The optical cross sections are shown in Fig. 4.44 for a rectangular groove in gold of width 50 nm and depth 500 nm. Groove corners are rounded with a 2-nm radius. The dielectric constant of gold is obtained from [40]. The main point of this figure is that a relatively sharp resonance peak is seen in cross-section spectra near the wavelength 650 nm. The resonance can be explained as a result of gap SPPs being excited at the top of the groove and propagating back and forth in the groove being reflected at the bottom and top groove terminations. When the total round-trip phase equals an integer number of 2π, a Fabry-Perot type resonance occurs [76], seen here as the sharp peak. It has been checked that when the imaginary part of the metal dielectric constant is set to zero, in which case there are no absorption losses, the extinction is equal to the sum of out-of-plane scattering

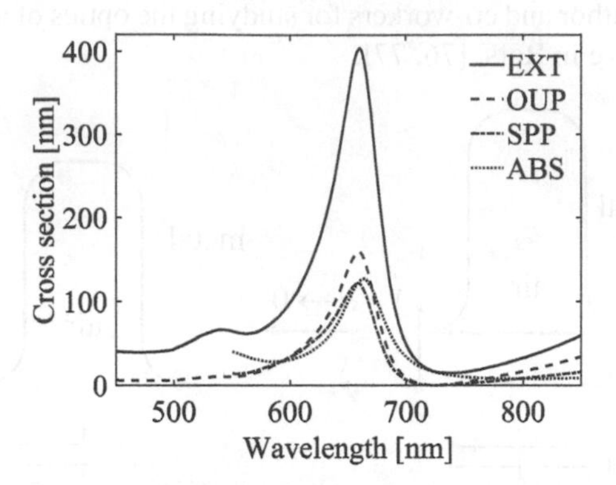

Figure 4.44 Optical cross-section spectra for a groove of width 50 nm and depth 500 nm in a gold surface. Groove corners are rounded with a 2-nm corner radius.

and scattering into SPP waves. In this case the resonance peak in extinction and scattering cross sections will still be there. On the other hand, if one considers a periodic array of rectangular grooves as in e.g. [66], and considers a case with no absorption, the reflectivity will always be unity, and no resonance shows up. However, when absorption is present the resonance will be clearly seen also for periodic structures as a minimum in the reflectivity at the resonance wavelength. For the single groove the resonances can be seen no matter if absorption occurs or not. If the groove depth is increased or decreased, the resonance wavelength will also increase or decrease. Notice that the cross sections at resonance are significantly larger than the groove width.

The optical cross sections for the case of an ultra-sharp groove [77] in gold are considered in Fig. 4.45. The groove is shown as an inset in Fig. 4.45(b) and has an opening in the top of width 250 nm, a depth of 500 nm, and a bottom groove width of 0.3 nm. The groove is illuminated by normally incident light. Near the bottom the groove walls are parallel making the groove ultra-sharp.

The optical cross-section spectra are radically different compared with those of the rectangular groove. There is no single sharp resonance peak but there is still some oscillatory behavior with wavelength. The optics of ultra-sharp grooves can also be understood in terms of gap-SPP waves propagating back and forth in the groove and forming standing wave resonances. However, due to the higher absorption in the narrow groove, especially near the groove bottom, those are not good resonances. Notice that apart from some minor oscillations, the absorption cross section (= EXT - OUP - SPP) is here of a broadband nature, such that a relatively high absorption is found over a broad range of wavelengths, rather than at a narrow band of wavelengths as for the rectangular groove.

A few examples of calculations of fields for the ultra-sharp groove are shown in Fig. 4.46. For the case of wavelength 705 nm the magnitude of the electric field near the bottom part of the groove is shown in Fig. 4.46(a) and (b). The field magnitude

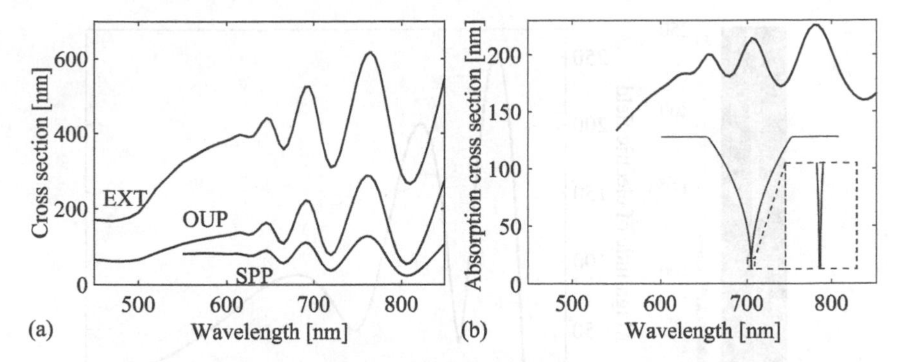

Figure 4.45 Optical cross-section spectra for an ultra-sharp groove of top width 250 nm, groove depth 500 nm, and bottom groove width of 0.3 nm. (a) Extinction (EXT), out-of-plane scattering (OUP), and SPP cross section (SPP) spectra. (b) Absorption cross-section spectrum (= EXT - OUP - SPP). The inset shows the groove along with a zoom-in on the ultra-sharp bottom part of the groove.

is seen to reach values as high as 250 near the groove bottom, where the light has been strongly focused. The adiabatic tapering of the groove means that the incident light is concentrated in an extremely narrow gap near the groove bottom with a resulting high field magnitude. In order to see the scattered field above the groove, the y-component of the electric field is considered in Figs. 4.46(c) and (d). Since the reference electric field is normally incident, and does not have a y-component, the y-component of the field is entirely due to scattering by the groove. The magnitude $|E_y|$ is shown in Fig. 4.46(d) for the wavelength 695 nm. Notice that part of the scattered light is confined near the groove surface. This is the SPP waves being excited at the groove and propagating away from the groove. The field seen high above the groove represents out-of-plane scattered light. Fig. 4.46(c) shows the real part of the y-component of the field instead, which is equivalent to a snapshot in time of the real field.

4.2.7 EXAMPLE: SILVER NANOSTRIP ON A THIN-FILM-SILICON-ON-SILVER WAVEGUIDE

In this section we will consider a silver nanostrip placed on a thin-film-silicon-on-silver waveguide geometry as illustrated in Fig. 4.47. The nanostrip is illuminated by normally incident light. The main new feature compared with the previous two sections is that the reference structure is now a three-layer structure. This reference structure may support guided modes for both s and p polarization. SPPs are still supported for p polarization but the structure may also support additional guided modes. The number of guided modes supported by the waveguide geometry depends on the thickness of the waveguide d relative to the wavelength, and on the refractive indices at the considered wavelength. Due to scattering by the nanostrip there will be out-

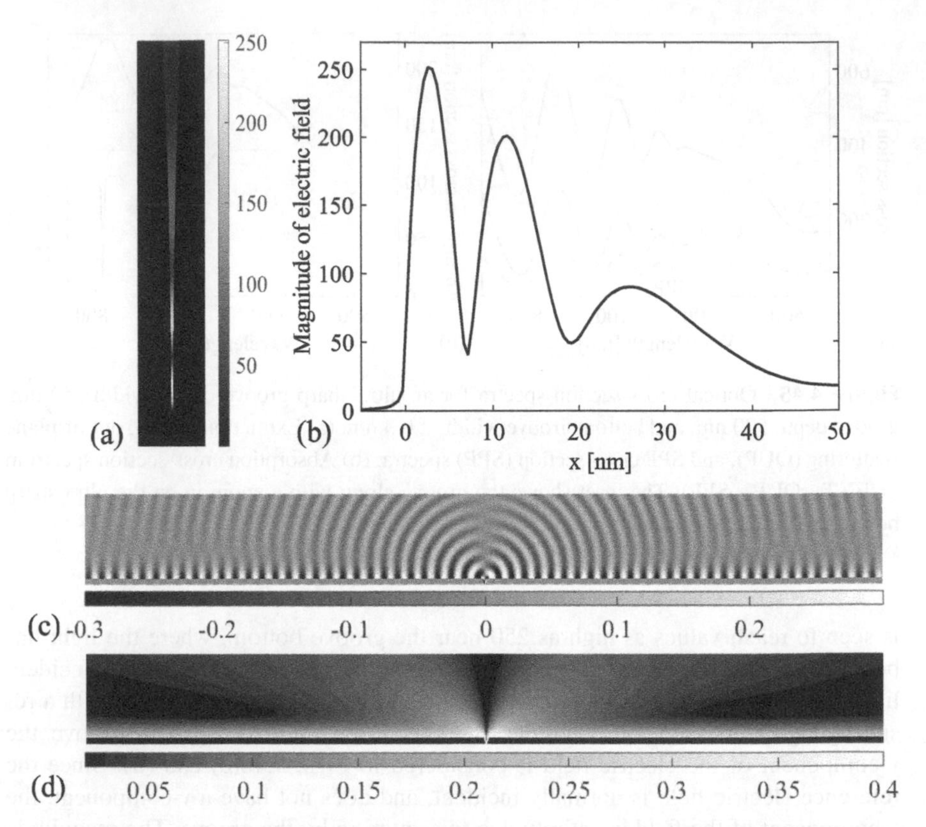

Figure 4.46 Field plots for an ultra-sharp groove of top width 250 nm, groove depth 500 nm, and bottom groove width of 0.3 nm being illuminated by a normally incident plane wave. (a) Magnitude of electric field at the wavelength 705 nm in the bottom part of the groove. (b) Magnitude of electric field along the center of the groove ($x = 0$) for the wavelength 705 nm. (c) and (d) show the magnitude of the y-component of the electric field, and the real part of the y-component of the electric field, for a large region above the groove and including part of the groove. Here the wavelength is 695 nm instead.

of-plane scattering, and scattering into the guided modes of the waveguide reference structure. In addition, there can be absorption in the silver nanostrip. These processes are quantified by their own scattering cross sections. In addition, the extinction cross section is of interest, which describes the power removed from the reflected beam due to scattering and absorption, where absorption both includes absorption in the nanostrip and any additional absorption in the silicon film and silver substrate caused by the presence of the nanostrip. Except for the guided mode cross sections for the three-layer geometry, which will be derived in this section, all other cross sections have already been described in previous sections.

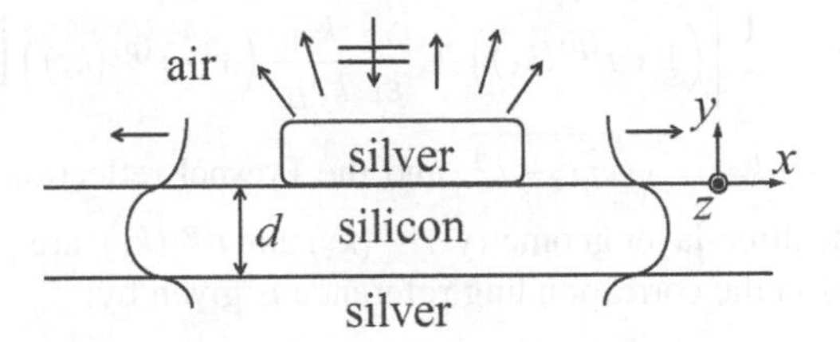

Figure 4.47 Schematic of a silver nanostrip placed on a thin-film-silicon-on-silver waveguide geometry. The nanostrip is illuminated by normally incident light, and it is illustrated that this leads to out-of-plane scattering, and scattering into guided modes of the waveguide geometry.

The geometry in Fig. 4.47 has been considered in Ref. [78] as a part of thin-film solar cell research, where the objective was to obtain light-trapping of incident light by scattering of light into guided modes. This is desirable, since, when no scatterer is present, and the wavelength of light is in a range with small absorption of light in the silicon, then most of the incident light will not be absorbed but will instead just be reflected back out of the solar cell. Even if the light propagates back and forth in the silicon layer a couple of times, the absorption will still be small for some wavelengths near the bandgap of the silicon. However, the presence of the scatterer can lead to scattering of light into guided modes of the waveguide geometry, in which case the light will propagate long distances in the silicon before it can be scattered out again, and thus absorption can be much higher. The main result of investigating the effect of a single scatterer in Ref. [78] is that the scattering cross sections have a resonant peak at the wavelengths where the waveguide geometry has a cut-off wavelength for one of the guided modes. At such a wavelength the guided mode is weakly bound, which can explain why scattering into such a mode is efficient. Investigations of light-trapping and absorption using a periodic array of metal particles instead of a single scatterer have been considered in, e.g., Refs. [79, 80, 81, 82, 83, 84]. The scattering by a single scatterer on a silicon surface with no back reflector, in which case the reference geometry is a two-layer geometry which supports no guided modes, has been studied in Refs. [85, 86].

In the case of an incident plane wave being p polarized and propagating in a direction given by the angle θ relative to the x-axis, the (z-component) of the magnetic field of the reference geometry can be expressed as

$$H_0(\mathbf{r}) = Ae^{ik_x x}\begin{cases} e^{-ik_{y1}y} + r^{(p)}(k_x)e^{ik_{y1}y}, & y \geq 0 \\ Be^{ik_{y,L2}y} + Ce^{-ik_{y,L2}y}, & 0 > y > -d \\ t^{(p)}(k_x)e^{-ik_{y,L3}(y+d)}, & y < -d \end{cases}, \tag{4.309}$$

where here

$$B = \frac{1}{2}\left[\left(1 + r^{(p)}(k_x)\right) - \frac{\varepsilon_{L2}}{\varepsilon_1}\frac{k_{y1}}{k_{y,L2}}\left(1 - r^{(p)}(k_x)\right)\right], \tag{4.310}$$

$$C = \frac{1}{2}\left[\left(1 + r^{(p)}(k_x)\right) + \frac{\varepsilon_{L2}}{\varepsilon_1}\frac{k_{y1}}{k_{y,L2}}\left(1 - r^{(p)}(k_x)\right)\right], \tag{4.311}$$

and $k_x = k_0 n_1 \cos\theta$. $k_{y,i} = \sqrt{k_0^2\varepsilon_i - k_x^2}$, and the Fresnel reflection and transmission coefficients for the three-layer geometry, $r^{(p)}(k_x)$ and $t^{(p)}(k_x)$, are given in Eq. (2.54).

For s polarization the corresponding reference is given by:

$$E_0(\mathbf{r}) = Ae^{ik_x x}\begin{cases} e^{-ik_{y1}y} + r^{(s)}(k_x)e^{ik_{y1}y}, & y \geq 0 \\ Be^{ik_{y,L2}y} + Ce^{-ik_{y,L2}y}, & 0 > y > -d \\ t^{(s)}(k_x)e^{-ik_{y,L3}(y+d)}, & y < -d \end{cases}, \tag{4.312}$$

where here

$$B = \frac{1}{2}\left[\left(1 + r^{(s)}(k_x)\right) - \frac{k_{y1}}{k_{y,L2}}\left(1 - r^{(s)}(k_x)\right)\right], \tag{4.313}$$

$$C = \frac{1}{2}\left[\left(1 + r^{(s)}(k_x)\right) + \frac{k_{y1}}{k_{y,L2}}\left(1 - r^{(s)}(k_x)\right)\right]. \tag{4.314}$$

The reflection and transmission coefficients for s polarization are also given in Eq. (2.54).

We shall now consider the calculation of scattering into guided modes of the reference geometry. The guided modes are characterized by a mode index n_{m}, or in-plane wave number $k_{x,\mathrm{p}} = k_0 n_{\mathrm{m}}$, where $k_{x,\mathrm{p}}$ is a pole of the Fresnel reflection coefficient of the reference geometry. Scattering into a guided mode is governed by the far-field guided-mode component of the Green's function given in Eqs. (4.204) and (4.213). For $y' > 0$ this component is given by

$$g^{(\mathrm{p})}(\mathbf{r},\mathbf{r}') = A(k_{x,\mathrm{p}})e^{ik_{x,\mathrm{p}}|x-x'|}e^{ik_{y,1,\mathrm{p}}y'}\begin{cases} e^{ik_{y,1,\mathrm{p}}y}, & y > 0 \\ Be^{ik_{y,L2,\mathrm{p}}y} + Ce^{-ik_{y,L2,\mathrm{p}}y}, & -d < y < 0 \\ De^{-ik_{y,L3,\mathrm{p}}(y+d)}, & y < -d \end{cases}, \tag{4.315}$$

where $k_{y,j,\mathrm{p}} = \sqrt{k_0^2\varepsilon_j - k_{x,\mathrm{p}}^2}$, and the coefficients A, B, C, and D depend on the polarization being considered and on the pole $k_{x,\mathrm{p}}$. The coefficients are given in Eqs. (4.203), (4.205) and (4.206) for p polarization, and in Eqs. (4.212), (4.214) and (4.215) for s polarization.

For p polarization this leads to the following scattered guided-mode contribution to the far-field field, where the position x must be far to the right or to the left of the scatterer:

$$H^{(\mathrm{p,\,ff})}(\mathbf{r}) = -\int_{C_1}\left\{\phi(\mathbf{r}')g^{(\mathrm{p})}(\mathbf{r},\mathbf{r}') - H(\mathbf{r}')\hat{\mathbf{n}}'\cdot\nabla' g^{(\mathrm{p})}(\mathbf{r},\mathbf{r}')\right\}dl'. \tag{4.316}$$

The right-propagating far-field component related to the specific guided mode can thus be expressed as

$$H_r^{(\mathrm{p,\,ff})} = H_{r,0}^{(\mathrm{p,\,ff})}e^{ik_{x,\mathrm{p}}x}\begin{cases} e^{ik_{y1,\mathrm{p}}y}, & y > 0 \\ Be^{ik_{y,L2,\mathrm{p}}y} + Ce^{-ik_{y,L2,\mathrm{p}}y}, & 0 > y > -d \\ De^{-ik_{y,L3,\mathrm{p}}(y+d)}, & y < -d \end{cases}, \tag{4.317}$$

where

$$H_{r,0}^{(\text{p, ff})} = -A(k_{x,\text{p}})\int_{C_1} e^{-ik_{x,\text{p}}x'} e^{ik_{y1,\text{p}}y'} \left(\phi(\mathbf{r}') - H(\mathbf{r}')\hat{\mathbf{n}}' \cdot \left[-ik_{x,\text{p}}\hat{\mathbf{x}} + ik_{y1,\text{p}}\hat{\mathbf{y}}\right]\right) dl'. \tag{4.318}$$

This is very similar to Eqs. (4.300) and (4.301) except that here a three-layer structure is considered, and the mode index is different.

The power in the right-propagating guided mode at position x and for p polarization is given by

$$P_r^{(\text{p})} = \frac{1}{2}\text{Real}\left\{\int_y \frac{k_{x,\text{p}}}{\omega\varepsilon_0\varepsilon_{\text{ref}}(y)} |H_r^{\text{p, ff}}(\mathbf{r})|^2 dy\right\}. \tag{4.319}$$

By inserting (4.317) into Eq. (4.319) we find

$$P_r^{(\text{p})} = |H_{r,0}^{(\text{p, ff})} e^{ik_{x,\text{p}}x}|^2 \left(\frac{\text{Real}\{k_{x,\text{p}}/\omega\varepsilon_0\varepsilon_1\}}{4\text{Imag}\{k_{y1,\text{p}}\}} + |D|^2 \frac{\text{Real}\{k_{x,\text{p}}/\omega\varepsilon_0\varepsilon_{L3}\}}{4\text{Imag}\{k_{y,L3,\text{p}}\}} + \right.$$

$$\left. \frac{1}{2}\text{Real}\left\{\frac{k_{x,\text{p}}}{\omega\varepsilon_0\varepsilon_{L2}}\right\} \int_{y=-d}^{0} \left| B e^{ik_{y,L2,\text{p}}y} + C e^{-ik_{y,L2,\text{p}}y}\right|^2 dy \right). \tag{4.320}$$

In principle the integral in (4.320) can be evaluated analytically. However, then we have to distinguish between three cases, namely the cases where both the real and imaginary parts of $k_{y,L2,\text{p}}$ are non-zero, the case where $k_{y,L2,\text{p}}$ is purely real, and the case where $k_{y,L2,\text{p}}$ is purely imaginary. The latter case occurs for an SPP wave bound to the silver-silicon interface when absorption in silver and silicon is neglected.

The power in the left-propagating mode will be given by the same expression except that $H_{r,0}^{(\text{p, ff})} e^{ik_{x,\text{p}}x}$ must be replaced by $H_{l,0}^{(\text{p, ff})} e^{-ik_{x,\text{p}}x}$, where

$$H_{l,0}^{(\text{p, ff})} = -A(k_{x,\text{p}})\int_{C_1} e^{+ik_{x,\text{p}}x'} e^{ik_{y1,\text{p}}y'} \left(\phi(\mathbf{r}') - H(\mathbf{r}')\hat{\mathbf{n}}' \cdot \left[+ik_{x,\text{p}}\hat{\mathbf{x}} + ik_{y1,\text{p}}\hat{\mathbf{y}}\right]\right) dl'. \tag{4.321}$$

By normalizing these powers with the power per unit area of the incident beam

$$I_i = \frac{1}{2}\frac{k_0 n_1}{\omega\varepsilon_0\varepsilon_1}|H_i|^2, \tag{4.322}$$

the cross sections for scattering into the right- and left-propagating guided modes are obtained:

$$\sigma_{r,\text{p}} = \frac{P_r^{(\text{p})}}{I_i}, \text{ and } \sigma_{l,\text{p}} = \frac{P_l^{(\text{p})}}{I_i}. \tag{4.323}$$

Thus

$$\sigma_r^{(\text{p})} = \frac{|H_{r,0}^{(\text{p, ff})} e^{ik_{x,\text{p}}x}|^2}{|H_i|^2} \left(\frac{\text{Real}\{k_{x,\text{p}}/k_0 n_1\}}{2\text{Imag}\{k_{y1,\text{p}}\}} + |D|^2 \frac{\text{Real}\{k_{x,\text{p}}\varepsilon_1/k_0 n_1\varepsilon_{L3}\}}{2\text{Imag}\{k_{y,L3,\text{p}}\}} + \right.$$

$$\text{Real}\left\{\frac{k_{x,\text{p}}\varepsilon_1}{k_0 n_1 \varepsilon_{L2}}\right\}\int_{y=-d}^{0}\left|Be^{ik_{y,L2,\text{p}}y}+Ce^{-ik_{y,L2,\text{p}}y}\right|^2 dy\Bigg). \tag{4.324}$$

For s polarization, a similar procedure can be followed. In that case, though, the mode index or guided-mode wave number will be for the s-polarized guided modes. The right-propagating guided mode will have the electric field (z-component)

$$E_r^{(\text{p, ff})} = E_{r,0}^{(\text{p, ff})} e^{ik_{x,\text{p}}x}\begin{cases} e^{ik_{y1,\text{p}}y}, & y>0 \\ Be^{ik_{y,L2,\text{p}}y}+Ce^{-ik_{y,L2,\text{p}}y}, & 0>y>-d\,, \\ De^{-ik_{y,L3,\text{p}}(y+d)}, & y<-d \end{cases} \tag{4.325}$$

where

$$E_{r,0}^{(\text{p, ff})} = -A(k_{x,\text{p}})\int_{C_1} e^{-ik_{x,\text{p}}x'} e^{ik_{y1,\text{p}}y'}\left(\phi(\mathbf{r}')-H(\mathbf{r}')\hat{\mathbf{n}}'\cdot\left[-ik_{x,\text{p}}\hat{\mathbf{x}}+ik_{y1,\text{p}}\hat{\mathbf{y}}\right]\right)dl'. \tag{4.326}$$

Here, the coefficient $A(k_{x,\text{p}})$ is the coefficient for s polarization.

The power in the right-propagating guided mode is for this polarization given by

$$P_r^{(\text{p})} = \frac{1}{2}\text{Real}\left\{\int_y \frac{i}{\omega\mu_0}E\left(\frac{\partial E}{\partial x}\right)^* dy\right\}, \tag{4.327}$$

and the power per unit area of the incident plane wave is given by

$$I_i = \frac{k_0 n_1}{\omega\mu_0}|E_i|^2. \tag{4.328}$$

Inserting (4.326) into (4.327) leads to

$$P_r^{(\text{p})} = |E_{r,0}^{(\text{p, ff})}e^{ik_{x,\text{p}}x}|^2\left(\frac{\text{Real}\left\{k_{x,\text{p}}/\omega\mu_0\right\}}{4\text{Imag}\left\{k_{y1,\text{p}}\right\}}+|D|^2\frac{\text{Real}\left\{k_{x,\text{p}}/\omega\mu_0\right\}}{4\text{Imag}\left\{k_{y,L3,\text{p}}\right\}}+\right.$$

$$\left.\frac{1}{2}\text{Real}\left\{\frac{k_{x,\text{p}}}{\omega\mu_0}\right\}\int_{y=-d}^{0}\left|Be^{ik_{y,L2,\text{p}}y}+Ce^{-ik_{y,L2,\text{p}}y}\right|^2 dy\right). \tag{4.329}$$

Also in this case we can use (4.323) for obtaining the cross sections for scattering into the guided modes. Thus

$$\sigma_r^{(\text{p})} = \frac{|E_{r,0}^{(\text{p, ff})}e^{ik_{x,\text{p}}x}|^2}{|E_i|^2}\text{Real}\left\{\frac{k_{x,\text{p}}}{k_0 n_1}\right\}\left(\frac{1}{2\text{Imag}\left\{k_{y1,\text{p}}\right\}}+\frac{|D|^2}{2\text{Imag}\left\{k_{y,L3,\text{p}}\right\}}+\right.$$

$$\left.\int_{y=-d}^{0}\left|Be^{ik_{y,L2,\text{p}}y}+Ce^{-ik_{y,L2,\text{p}}y}\right|^2 dy\right), \tag{4.330}$$

and the cross section $\sigma_l^{(\text{p})}$ can be obtained by replacing $|E_{r,0}^{(\text{p, ff})}e^{ik_{x,\text{p}}x}|^2$ in (4.330) with $|E_{l,0}^{(\text{p, ff})}e^{-ik_{x,\text{p}}x}|^2$.

The expressions for the guided-mode scattering cross sections require the mode-indices of the guided modes. These can be obtained using the techniques in Appendix F.

The optical cross-section spectra for s polarization and silicon-film thicknesses of 50 and 250 nm are shown in Figs. 4.48(c) and (d), respectively. The corresponding

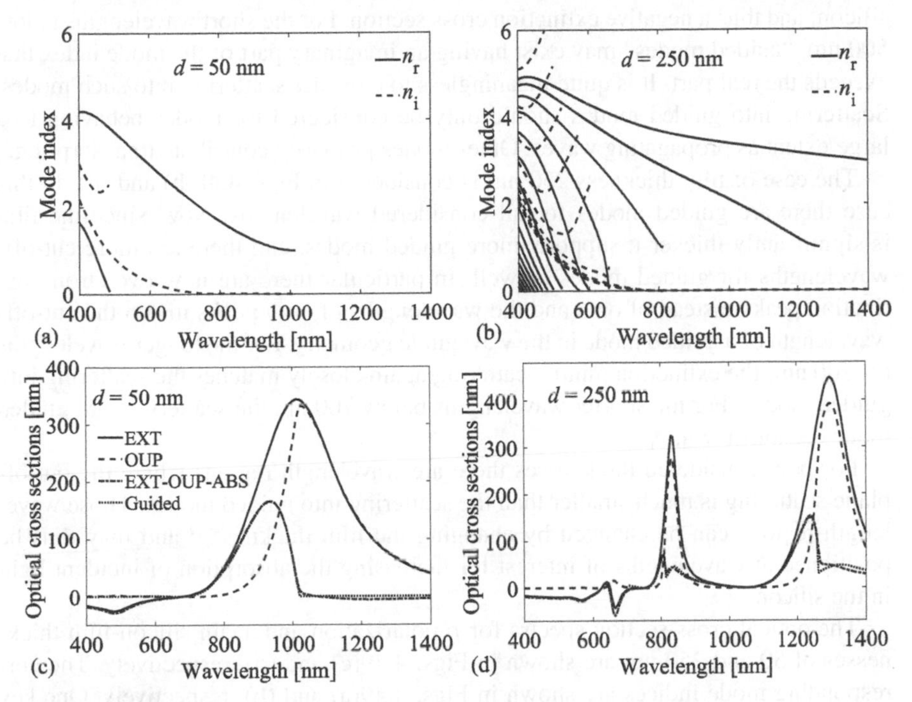

Figure 4.48 (a) Mode index for s-polarized guided modes of a reference geometry with a silicon slab of thickness d = 50 nm surrounded by air above and silver below. n_r and n_i represent the real and imaginary parts of the mode index, respectively. (c) Optical cross-section spectra for a silver nanostrip of width 60 nm and thickness 25 nm placed on top of the reference geometry with d = 50 nm, and being illuminated by a normally incident s-polarized plane wave. (b) and (d) Corresponding mode-index and optical cross sections for the same geometry with a silicon slab of thickness d = 250 nm.

mode indices for guided modes versus wavelength are shown in Figs. 4.48(a) and (b), respectively. Considering first the case of film thickness 50 nm, it can be seen that there are no guided modes for wavelengths longer than 1015 nm. For these wavelengths the extinction and out-of-plane scattering cross sections practically coincide. For wavelengths between 700 nm and 1015 nm the extinction approximately equals the sum of scattering into the guided mode and out-of-plane scattering. The extinction minus out-of-plane scattering (minus absorption) matches quite well the scattering into guided modes. The scattering into guided modes is not directly measurable, while extinction and out-of-plane scattering are directly measurable. The difference between the latter thus give a convenient indirect method of obtaining the scattering into guided modes. For the shorter wavelengths, where especially the absorption in the silicon is significant, the extinction can be negative. Here, this means that the silver nanostrip reflects some light and thus prevents absorption of part of the light in the silicon. This leads to a stronger reflected beam due to the reduced absorption in silicon, and thus a negative extinction cross section. For the short wavelengths below 500 nm, "guided modes" may exist having an imaginary part of the mode index that exceeds the real part. It is quite meaningless to consider scattering into such modes. Scattering into guided modes should only be considered for modes behaving to a large extent as propagating waves. Other modes primarily contribute to absorption.

The case of film thickness 250 nm is considered in Figs. 4.48(b) and (d). In this case there are guided modes for all considered wavelengths. Now, since the film is significantly thicker it supports more guided modes, and there are more cut-off-wavelengths for guided modes as well. In particular there are now two strong extinction peaks instead of one, and the wavelengths of both peaks match the cut-off-wavelength of a guided mode in the waveguide geometry. For the longer wavelengths ($>$ 700 nm) the extinction minus scattering again closely matches the scattering into guided modes. For the shorter wavelengths below 700 nm the scattering into guided modes is not shown.

For both considered thicknesses there are wavelength regions where the out-of-plane scattering is much smaller than the scattering into guided modes. Those wavelength regions can be changed by changing the film thickness d and may thus be positioned at wavelengths of interest for increasing the absorption of incident light in the silicon.

The optical cross-section spectra for p polarization and again silicon-film thicknesses of 50 and 250 nm are shown in Figs. 4.49(c) and (d), respectively. The corresponding mode indices are shown in Figs. 4.49(a) and (b), respectively. One key difference now is that the waveguide not only supports ordinary waveguide modes but may also support surface plasmon polaritons being bound to the silver-silicon interface of the reference geometry. These SPP waves may also extend farther into the air above the silicon. For the case of film thickness 50 nm [Figs. 4.49(a) and (c)] and wavelengths $>$ 700 nm the reference geometry only supports an SPP wave. Near the wavelength 1400 nm the mode index approaches unity, which implies that the SPP is weakly bound and extends far into the air region with refractive index of unity. The effect of the silicon layer is thus small here, and we could almost just as well consider an air-silver interface. As the wavelength decreases, the SPP becomes more strongly confined to the silicon layer, and eventually for wavelengths approaching 700 nm the SPP wave can be buried below the air-silicon interface, in which case scattering into the SPP wave becomes weak. Similar to the case of s polarization it can also be seen here that the scattering into the SPP guided mode matches quite well the extinction minus scattering.

For the case of film thickness 250 nm [Figs. 4.49(b) and (d)] the SPP mode has a mode index approaching 4 even for the wavelength 1400 nm. In this case the SPP

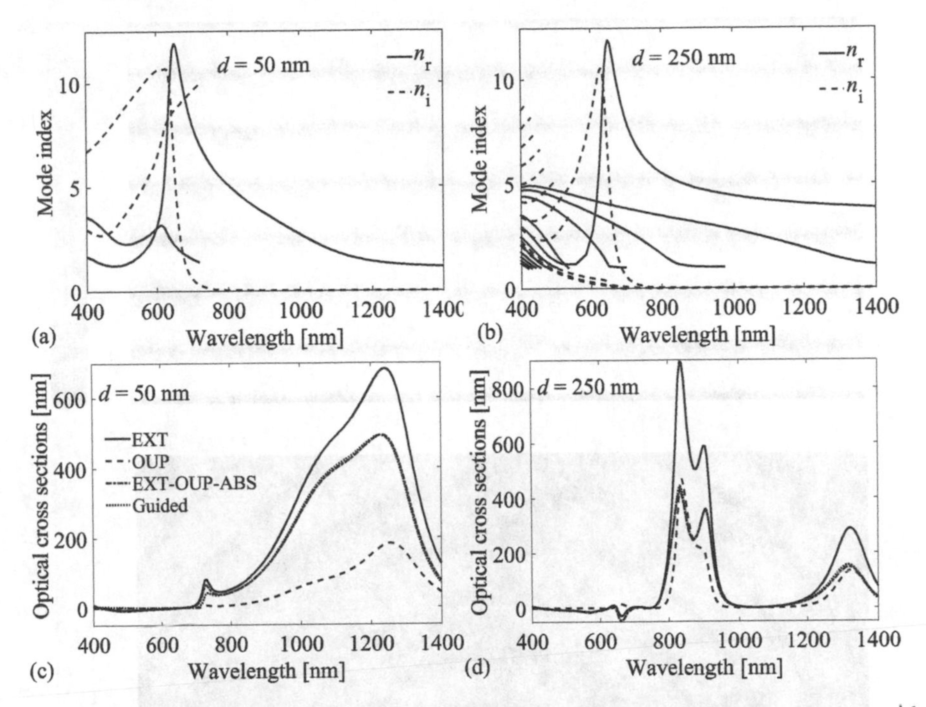

Figure 4.49 (a, b) Mode index for p-polarized guided modes of a reference geometry with a silicon slab of thickness d surrounded by air above and silver below. n_r and n_i represent the real and imaginary parts of the mode index, respectively. (c, d) Optical cross sections for a silver nanostrip of width 60 nm and thickness 25 nm placed on top of the reference geometry, and being illuminated by a normally incident p-polarized plane wave. (a,c) d = 50 nm. (b, d) d = 250 nm.

is buried below the air-silicon interface for all wavelengths above 700 nm, at least. No strong scattering into this SPP mode can be observed in Fig. 4.49(d). Instead an effective cut-off-wavelength can be seen in Fig. 4.49(c) for two ordinary waveguide modes at wavelengths near 1300 nm and 850 nm. An extinction peak is found at these wavelengths similar to the case of s polarization. In Ref. [78] it was found that measured extinction and scattering considering silver strips with 10 μm between them on an ultra-flat 290-nm-silicon-film on silver matched very well with the theoretically predicted extinction and scattering obtained using the method considered in this chapter.

Examples of field plots for the case of normally incident p-polarized light of wavelength 1200 nm and a silicon film thickness of 50 nm are presented in Fig. 4.50. Fig 4.50(a) shows the absolute value of the electric field $|\mathbf{E}|$. This field plot is dominated by a standing-wave interference pattern in the reference field component. In addition the excitation of right- and left-propagating SPPs near the metal surface in-

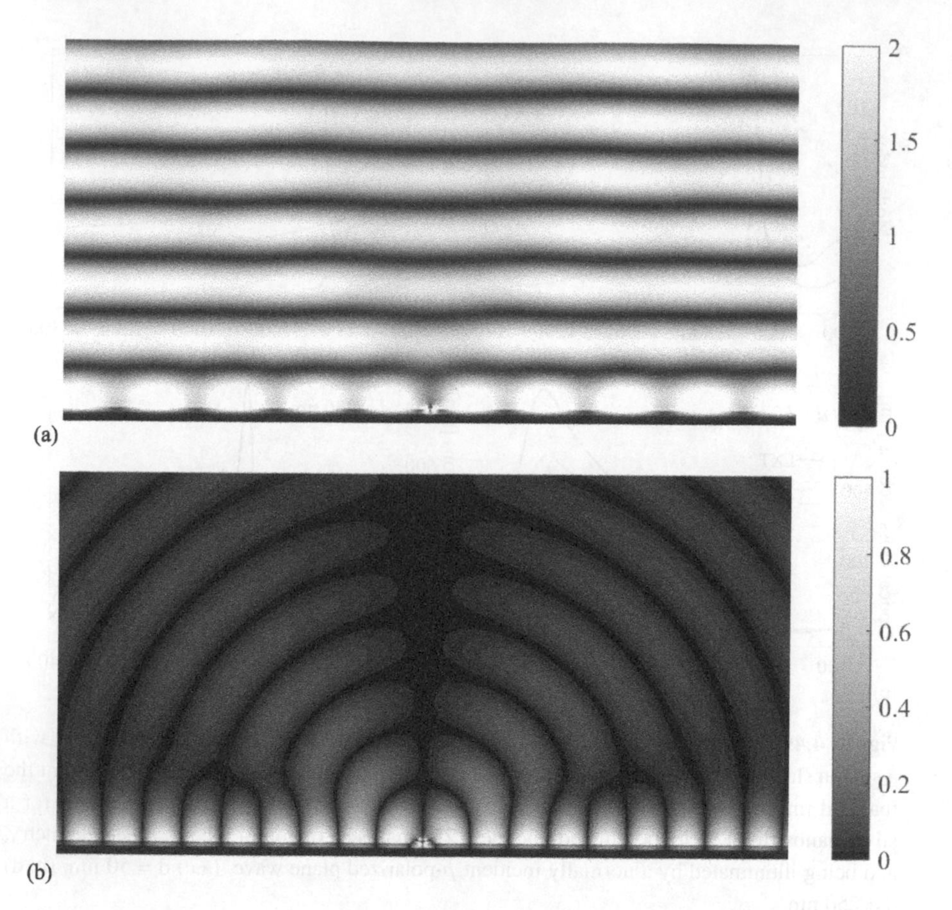

Figure 4.50 Examples of field plots for a silver nanostrip of width 60 nm and thickness 25 nm placed on a 50-nm silicon-on-silver waveguide geometry. The nanostrip is illuminated by normally incident p-polarized light of wavelength 1200 nm. (a) $|\mathbf{E}|$, and (b) $|\text{Real}\{E_y\}|$.

terfere with the reference field, and a contribution from the out-of-plane scattered field is also barely noticeable. Fig 4.50(b) shows instead the absolute value of the real part of the y-component of the electric field $|\text{Real}\{E_y\}|$. The reference field does not contribute to this field component, which thus purely contains contributions from the scattered field. In this case the out-of-plane scattered field and the right- and left-propagating SPP waves are more clearly observed.

Field plots of $|\text{Real}\{H_y(\mathbf{r})\}|$ are shown in Fig. 4.51 for the same geometry but for s polarization. The field component being shown purely contains scattered fields, since the normally incident field and thus also the reference field do not have any magnetic field polarized along the y-axis. The wavelengths 1015 nm and 800 nm are considered in Fig. 4.51(a) and (b), respectively. It is known from Fig. 4.48 that the

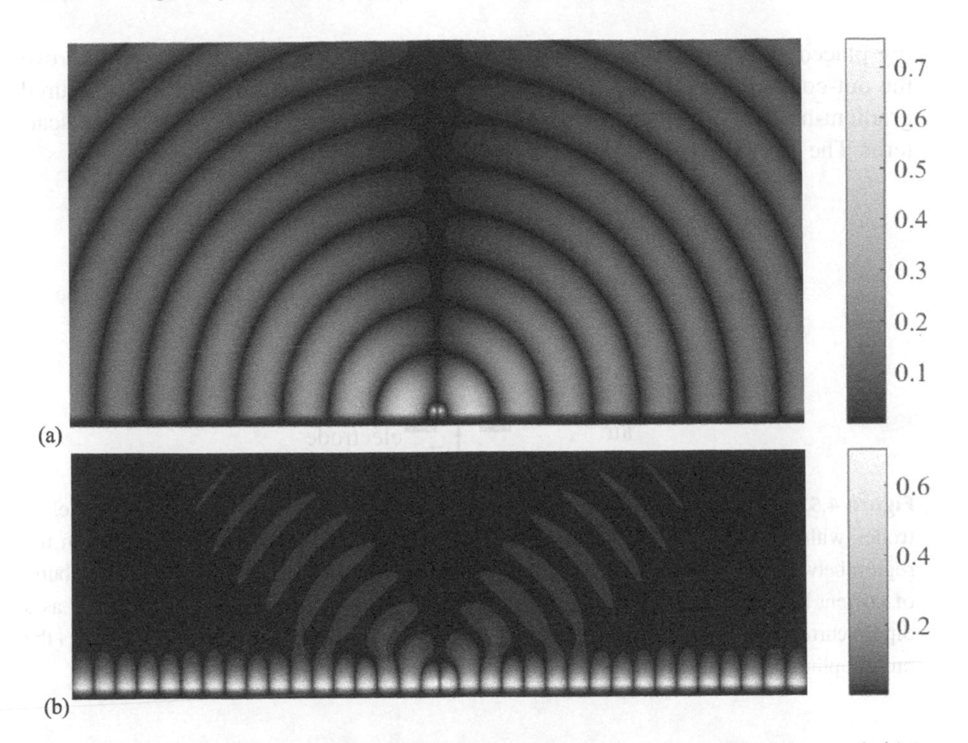

Figure 4.51 Examples of $|\text{Real}\{H_y(\mathbf{r})\}|$ for a silver nanostrip of width 60 nm and thickness 25 nm placed on a 50-nm-silicon-on-silver waveguide geometry being illuminated by normally incident s-polarized light of wavelength (a) 1015 nm, and (b) 800 nm.

structure does not support any guided modes for the wavelength 1015 nm. The scattered field is thus dominated by out-of-plane propagating cylindrical waves moving away from the nanostrip. The wavelength 800 nm is, on the other hand special, in the sense that the scattering cross section for scattering into a guided mode is much stronger than that for scattering into out-of-plane propagating modes (see Fig. 4.48). This can also be observed in the field plot in Fig. 4.51(b) where the out-of-plane scattering is clearly weak compared to the excitation of the guided modes propagating along the a:Si-waveguide.

4.2.8 EXAMPLE: MICROSTRUCTURED GRADIENT-INDEX LENS FOR THZ PHOTOCONDUCTIVE ANTENNAS

In this section we will consider again a three-layer reference geometry with a scatterer but for a situation where the reference field is the field generated by a point source. Here we will be interested in calculating the power emitted in different directions instead of optical cross sections. The geometry that will be considered as an example is that of a point source that emits radiation at the frequency 1 THz being placed at the bottom or backside of a semiconductor slab. In order to improve the out-coupling of light from the upper or front-side of the slab, a microstructured gradient-index (GRIN) lens is placed on the front-side. The GRIN lens is the scatterer. The geometry is illustrated in Fig. 4.52.

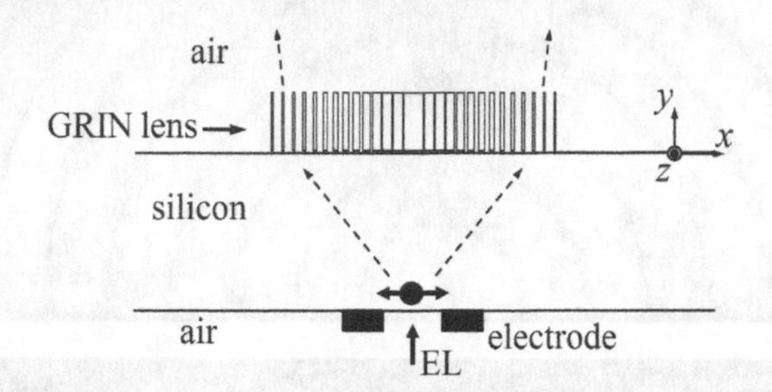

Figure 4.52 Schematic of a microstructured gradient-index lens on a silicon slab. Two electrodes with a bias voltage across are placed on the backside of the silicon slab. When the region between electrodes is illuminated by a short-pulsed excitation laser (EL), a short burst of current between electrodes generates radiation at THz frequencies. This is modeled as a dipole current source at the backside of the silicon slab. The lens collimates and increases the out-coupling of the radiation.

The problem originates from research into generation of THz signals via a photoconductive effect based on placing two closely spaced electrodes with a bias voltage across on the backside of the slab. By illuminating the gap between the electrodes with ultra-short laser pulses then, due to absorption, there will be photo-generated carriers in the semiconductor. Due to the bias voltage this results in a burst of current flowing between the electrodes. The current stops after a time determined by the recombination of photogenerated carriers. In order to obtain spectral components for the generated current in the THz frequency range the semiconductor must be prepared with a distribution of defects leading to fast recombination of photogenerated carriers. In this section we will only consider this problem from an optical point of view and simply treat the current distribution as that of an oscillating electric dipole. The reference field is then the field generated by the dipole in the reference structure (the three-layer geometry without the GRIN lens). In this section we will be interested in obtaining far-field radiation patterns and field distributions.

For the details of the geometry we follow Ref. [87] and use a 400-μm-thick silicon slab and a microstructured silicon GRIN lens with ridges of height 100 μm designed with the restrictions that the maximum lens width (diameter) should not exceed 600 μm, there should be one ridge for each 25 μm and the width of ridges should be between 2.5 μm and 19.5 μm in order to avoid very thin gaps and very thin ridges. Other examples of GRIN lenses have been presented in, for example, Refs. [88, 89]. The main idea is that for the actual structure the lens will consist of silicon rings such that in, e.g., the *xy* plane, the field can be considered as roughly *p* polarized, and in the *yz*-plane it can be considered as roughly *s* polarized. Here we will consider fields propagating in the xy-plane with a point source polarized either along the x-axis (p polarization) or along the z-axis (s polarization).

The beginning and end of each ridge of the GRIN lens along the x-axis is given in the following table:

Start of ridge (μm):	−283	−259	−235.5	−212	−189	−166	−143	−120
End of ridge (μm):	−275	−250	−225	−200	−175	−150	−125	−100
Start of ridge (μm):	−96.5	−72.5	−47.5	−22.5	25	50	75	100
End of ridge (μm):	−75	−50	−25	22.5	47.5	72.5	96.5	120
Start of ridge (μm):	125	150	175	200	225	250	275	
End of ridge (μm):	143	166	189	212	235.5	259	283	

All corners of ridges are rounded by a 2-nm radius. As can be seen from the table, and also from Fig. 4.52 which shows exactly this GRIN lens design, the ridges are thinner and the air gaps between ridges are wider near the ends of the lens. The fraction of high-refractive index material is thus small near the ends of the lens while it is large in the center. This means that the average refractive index is large in the center and decreases toward the ends. This gives the same effect as a normal lens where instead the thickness of the lens is varied. In both cases the phase shift a wave experiences as it propagates through the lens will be large in the center of the lens and smaller near the ends. The microstructured GRIN lens will also give rise to some finite-size effects since the width of ridges and gaps are still a reasonable fraction of the wavelength in silicon. The free-space wavelength at 1 THz is $\lambda_0 = 300$ μm. The gaps and ridge widths are thus much smaller than the free-space wavelength. However, in silicon where we will use the approximate value for the refractive index of $n_2 = n_{L2} = 3.5$, the corresponding wavelength is $\lambda = \lambda_0/3.5 = 86$ μm, and then the considered ridge widths are actually not much smaller than the wavelength in this material. We can thus expect that the GRIN lens improves the out-coupling through the front of the lens and that the radiation is to some extent collimated. However, finite-size effects may also lead to somewhat complicated radiation patterns.

We will first consider a z oriented dipole at position $\mathbf{r}_0 = \hat{\mathbf{x}}x_0 + \hat{\mathbf{y}}y_0$. This dipole orientation will generate s-polarized radiation. The polarization density being considered is given by

$$\mathbf{P}(\mathbf{r},t) = \int_\omega \mathbf{P}(\mathbf{r},\omega)e^{-i\omega t}d\omega, \tag{4.331}$$

where

$$\mathbf{P}(\mathbf{r},\omega) = \hat{\mathbf{z}}p(\omega)\delta(\mathbf{r}-\mathbf{r}_0). \tag{4.332}$$

The corresponding current density is thus given by

$$\mathbf{J}(\mathbf{r},t) = \frac{\partial \mathbf{P}}{\partial t} \tag{4.333}$$

leading to

$$\mathbf{J}(\mathbf{r},\omega) = -i\hat{\mathbf{z}}\omega p(\omega)\delta(\mathbf{r}-\mathbf{r}_0). \tag{4.334}$$

We will consider only a single frequency ω, and the argument ω is suppressed in the following. For the dipole in the reference geometry (no GRIN lens), the wave equation reduces to

$$\left(\nabla^2 + k_0^2 \varepsilon_{\text{ref}}(\mathbf{r})\right) E_0(\mathbf{r}) = -\omega^2 \mu_0 p \delta(\mathbf{r} - \mathbf{r}_0). \tag{4.335}$$

The dipole field is strongly related to the Green's function g of the reference geometry being a solution of

$$\left(\nabla^2 + k_0^2 \varepsilon_{\text{ref}}(\mathbf{r})\right) g(\mathbf{r}, \mathbf{r}') = -\delta(\mathbf{r} - \mathbf{r}'). \tag{4.336}$$

The Green's function must be the one that satisfies appropriate boundary conditions of fields propagating away from the source point $\mathbf{r}'$ for positions outside the film. The reference electric field is thus given by

$$E_0(\mathbf{r}) = \omega^2 \mu_0 p g(\mathbf{r}, \mathbf{r}_0). \tag{4.337}$$

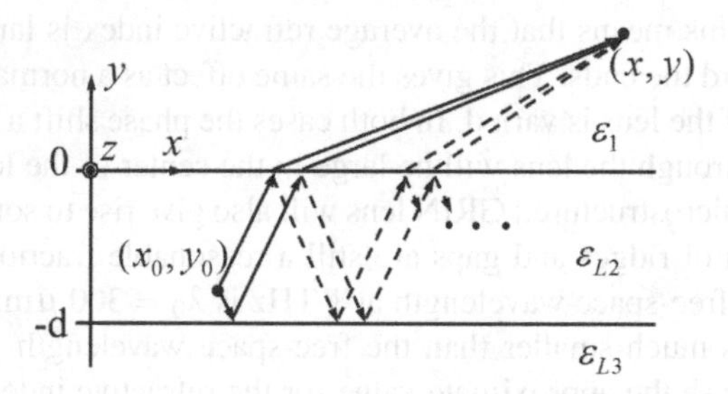

Figure 4.53 Schematic of emission from a dipole at $\mathbf{r}_0 = \hat{\mathbf{x}}x_0 + \hat{\mathbf{y}}y_0$ inside a film. The resulting field at $\mathbf{r}$ can be described as the sum of contributions originating from multiple reflections inside the film.

The emission from a dipole in the middle layer of a three-layer geometry is illustrated in Fig. 4.53. The Green's function for the case of $\mathbf{r}_0$ placed inside the layer $(0 > y_0 > -d)$ can be obtained by starting with the Green's function for the source placed in a homogeneous dielectric with dielectric constant ε_{L2}

$$g^{(d)}(\mathbf{r}, \mathbf{r}') = \frac{i}{2\pi} \int_{k_x=0}^{\infty} \frac{\cos(k_x[x - x'])e^{ik_{y,L2}|y-y'|}}{k_{y,L2}} dk_x. \tag{4.338}$$

The source emits waves that propagate both upward and downward. These waves will then be multiply reflected inside the film, with part of the waves being transmitted into the upper layer, and part being transmitted into the lower layer (not shown), for each round-trip of propagation inside the film. The resulting sum of contributions to the field at $\mathbf{r}$ can be evaluated using

$$\sum_{n=0}^{\infty} x^n = \frac{1}{1-x}, \quad |x| < 1. \tag{4.339}$$

An alternative approach is to simply, for each k_x in Eq. (4.338), add an upward propagating wave and a downward propagating wave inside the layer and construct waves outside the layer that propagate away from the layer, and then to use the electromagnetics boundary conditions to determine the coefficients in front of each term. These approaches both lead to the Green's function

$$g(\mathbf{r},\mathbf{r}_0) = \frac{i}{2\pi}\int_{k_x=0}^{\infty} \frac{\cos(k_x[x-x_0])f(y,y_0,k_x)}{k_{y,L2}\left(1 - r_{L2,1}r_{L2,L3}e^{2ik_{y,L2}d}\right)}dk_x, \tag{4.340}$$

where

$$f(y,y_0,k_x) = \begin{cases} A(k_x,y_0)e^{ik_{y1}y}e^{-ik_{y,L2}y_0}t_{L2,1}, & y>0 \\ A(k_x,y_0)B(k_x,y)e^{ik_{y,L2}(y-y_0)}, & 0>y>y_0 \\ A(k_x,y)B(k_x,y_0)e^{ik_{y,L2}(y_0-y)}, & y_0>y>-d \\ B(k_x,y_0)e^{ik_{y,L2}(d+y_0)}t_{L2,L3}e^{-ik_{y,L3}(y+d)}, & y<-d \end{cases},$$

$$0 > y_0 > -d, \tag{4.341}$$

and

$$A(k_x,y) = \left(1 + r_{L2,L3}e^{2ik_{y,L2}(d+y)}\right), \tag{4.342}$$

$$B(k_x,y) = \left(1 + r_{L2,1}e^{-2ik_{y,L2}y}\right). \tag{4.343}$$

The coefficients $r_{i,j}$ and $t_{i,j}$ are the single-interface reflection and transmission coefficients for an interface between layers i and j and for the polarization being considered. The coefficients are functions of k_x. In order to avoid integration near the poles of the integrand, i.e., values of k_x where the reference structure supports guided modes, the integration along the real axis can be replaced by an integration path going into the complex plane keeping a distance to the poles. This is similar to what has been suggested elsewhere in this book for evaluating the Green's function of a layered reference structure.

Now consider the case of placing the source at $y_0 = -d$ and $x_0 = 0$. Then,

$$E_0(\mathbf{r}) = \omega^2\mu_0 p\frac{i}{2\pi}\int_{k_x=0}^{\infty} \frac{\cos(k_x x)t_{L2,L3}^{(s)}t_{L2,1}^{(s)}e^{ik_{y,L2}d}e^{ik_{y1}y}}{k_{y,L2}\left(1 - r_{L2,1}^{(s)}r_{L2,L3}^{(s)}e^{2ik_{y,L2}d}\right)}dk_x, \quad y>0, \tag{4.344}$$

$$E_0(\mathbf{r}) = \omega^2\mu_0 p\frac{i}{2\pi}\int_{k_x=0}^{\infty} \frac{\cos(k_x x)t_{L2,L3}^{(s)}\left(e^{ik_{y,L2}(y+d)} + r_{L2,1}^{(s)}e^{ik_{y,L2}(d-y)}\right)}{k_{y,L2}\left(1 - r_{L2,1}^{(s)}r_{L2,L3}^{(s)}e^{2ik_{y,L2}d}\right)}dk_x,$$

$$0 > y > -d, \tag{4.345}$$

and

$$E_0(\mathbf{r}) = \omega^2\mu_0 p\frac{i}{2\pi}\int_{k_x=0}^{\infty} \frac{\cos(k_x x)t_{L2,L3}^{(s)}\left(1 + r_{L2,1}^{(s)}e^{2ik_{y,L2}d}\right)e^{-ik_{y,L3}(d+y)}}{k_{y,L2}\left(1 - r_{L2,1}^{(s)}r_{L2,L3}^{(s)}e^{2ik_{y,L2}d}\right)}dk_x,$$

$$y < -d, \tag{4.346}$$

where it was used that $t_{L2,L3} = 1 + r_{L2,L3}$. The magnetic field can be obtained straightforwardly from the electric field.

The out-of-plane far field for the upper half-plane ($y > 0$) can be derived from (4.344) by using

$$\mathbf{r} = \hat{\mathbf{x}} r\cos\theta + \hat{\mathbf{y}} r\sin\theta, \quad 0 < \theta < \pi, \tag{4.347}$$

and

$$k_x = k_0 n_1 \cos\alpha, \quad \text{and} \quad k_{y,1} = k_0 n_1 \sin\alpha. \tag{4.348}$$

The upper integration limit can be replaced with $k_0 n_1$, and then a procedure is followed that is similar to that explained in Sec. 4.2.1 for calculating far-field expressions of Green's functions.

The resulting out-of-plane far field for the upper half-plane is then given by

$$E_0^{(\mathrm{ff})}(r,\theta) = \omega^2 \mu_0 p \frac{e^{i\pi/4}}{4} \sqrt{\frac{2}{\pi k_0 n_1}} \frac{e^{ik_0 n_1 r}}{\sqrt{r}} \frac{t_{L2,1}^{(s)} t_{L2,L3}^{(s)} e^{ik_{y,L2} d}}{1 - r_{L2,1}^{(s)} r_{L2,L3}^{(s)} e^{2ik_{y,L2} d}} \frac{k_{y1}}{k_{y,L2}},$$
$$0 < \theta < \pi, \tag{4.349}$$

where the transmission functions $t_{L2,1}^{(s)}$ and $t_{L2,L3}^{(s)}$, and $k_{y,i}$, that are all functions of k_x should now take the argument $k_x = k_0 n_1 \cos\theta$, which makes them functions of θ.

The corresponding out-of-plane far field for the lower half-plane ($y < -d$) is similarly given by

$$E_0^{(\mathrm{ff})}(r,\theta) = \omega^2 \mu_0 p \frac{e^{i\pi/4}}{4} \sqrt{\frac{2}{\pi k_0 n_{L3}}} \frac{e^{ik_0 n_{L3} r}}{\sqrt{r}} \frac{t_{L2,L3}^{(s)} \left(1 + r_{L2,1}^{(s)} e^{2ik_{y,L2} d}\right) e^{-ik_{y,L3} d}}{1 - r_{L2,1}^{(s)} r_{L2,L3}^{(s)} e^{2ik_{y,L2} d}} \frac{k_{y,L3}}{k_{y,L2}},$$
$$0 > \theta > -\pi, \tag{4.350}$$

where the transmission functions $t_{L2,1}^{(s)}$ and $t_{L2,L3}^{(s)}$, and $k_{y,i}$, that are again functions of k_x should now take the argument $k_x = k_0 n_{L3} \cos\theta$.

The reference far field can be used together with the integral equations to obtain the total far field in the upper or lower half-plane in the following way:

$$E^{(\mathrm{ff})}(\mathbf{r}) = E_0^{(\mathrm{ff})}(\mathbf{r}) - \oint_{C_1} \left\{ g^{(\mathrm{ff})}(\mathbf{r},\mathbf{r}')\hat{\mathbf{n}}' \cdot \nabla' E(\mathbf{r}') - E(\mathbf{r}')\hat{\mathbf{n}}' \cdot \nabla' g^{(\mathrm{ff})}(\mathbf{r},\mathbf{r}') \right\} dl', \tag{4.351}$$

where the integral is taken over the outer surface of the GRIN lens (C_1), and the far-field Green's function is given by (4.185). The time-averaged power emitted into the upper half-plane within the angles θ_1 and θ_2 is given by

$$P = \frac{1}{2}\sqrt{\frac{\varepsilon_0}{\mu_0}} n_1 \int_{\theta=\theta_1}^{\theta_2} |E^{(\mathrm{ff})}(r,\theta)|^2 r d\theta, \tag{4.352}$$

where it has been assumed that n_1 is positive and real. The corresponding differential power, or radiation pattern, is given by

$$\frac{dP}{d\theta} = \frac{1}{2} n_1 \sqrt{\frac{\varepsilon_0}{\mu_0}} |E^{(\mathrm{ff})}(r,\theta)|^2 r. \tag{4.353}$$

Similar expressions apply for the lower half-plane in which case n_1 should be replaced by n_{L3}. For the structure of interest in this section $n_1 = n_{L3}$.

The far-field radiation pattern for the considered geometry with a dipole emitting radiation at 1 THz is shown in Fig. 4.54(a). This result was obtained using the GF-SIEM with polynomial order $m = 2$, and where each ridge was divided into 3 sections for each corner, 10 sections for each straight side along y (vertical sides) and 4 sections for each straight side along x (top and bottom). There are 23 ridges and thus this leads to a total of 920 sections with two sampling points per section. A similar calculation with polynomial order $m = 1$ does not differ much from the result with $m = 2$. For comparison is shown the radiation when the lens is removed. The GRIN lens radiation pattern is normalized relative to the radiation pattern with no lens for the direction along the y-axis ($\theta = \pi/2$). The radiation going in this particular direction is thus increased by a factor of approximately 9.5 due to the presence of the lens. The radiation pattern also has some significantly large side-lobes. Most of the out-of-plane propagating radiation is seen to be going into the upper half-plane.

The magnitude of the electric field in a region surrounding the dipole and the GRIN lens is shown in Fig. 4.54(b). The field distribution is seen to agree well with the radiation pattern. Notice that a large number of guided modes are excited that propagate inside the 400-μm silicon film, which lead to a complicated interference pattern inside the film. The figures shown here differ slightly from those in Ref. [87] in the way that in [87] most of the reflections from the lower interface of the silicon slab were not included. In the results shown in this book, all of the multiple reflections are included, which leads to some minor differences in radiation patterns and field profiles.

We now turn to consider the case of an x polarized electric dipole in the middle layer $L2$ given by

$$\mathbf{P}(\mathbf{r}) = \hat{\mathbf{x}} p \delta(\mathbf{r} - \mathbf{r}_0), \tag{4.354}$$

where again the argument ω has been suppressed. In this case the wave equation for the electric field is given by

$$-\nabla \times \nabla \times \mathbf{E}(\mathbf{r}) + k_0^2 \varepsilon(\mathbf{r}) \mathbf{E}(\mathbf{r}) = -\omega^2 \mu_0 p \hat{\mathbf{x}} \delta(\mathbf{r} - \mathbf{r}_0). \tag{4.355}$$

Now assume that the dipole is in a homogeneous dielectric with dielectric constant ε_{L2}. In that case the reference electric field that is a solution of (4.355) is given by

$$\mathbf{E}(\mathbf{r}) = \left[\left(\mathbf{I} + \frac{1}{k_0^2 \varepsilon_{L2}} \nabla\nabla \right) g^{(d)}(\mathbf{r}, \mathbf{r}_0) \right] \cdot (\omega^2 \mu_0 p \hat{\mathbf{x}}). \tag{4.356}$$

The corresponding magnetic field is then given by

$$\mathbf{H}(\mathbf{r}) = \hat{\mathbf{z}} H(\mathbf{r}) = i\omega p \hat{\mathbf{x}} \times \nabla g^{(d)}(\mathbf{r}, \mathbf{r}_0) = i\omega p \hat{\mathbf{z}} \frac{\partial g(\mathbf{r}, \mathbf{r}_0)}{\partial y}. \tag{4.357}$$

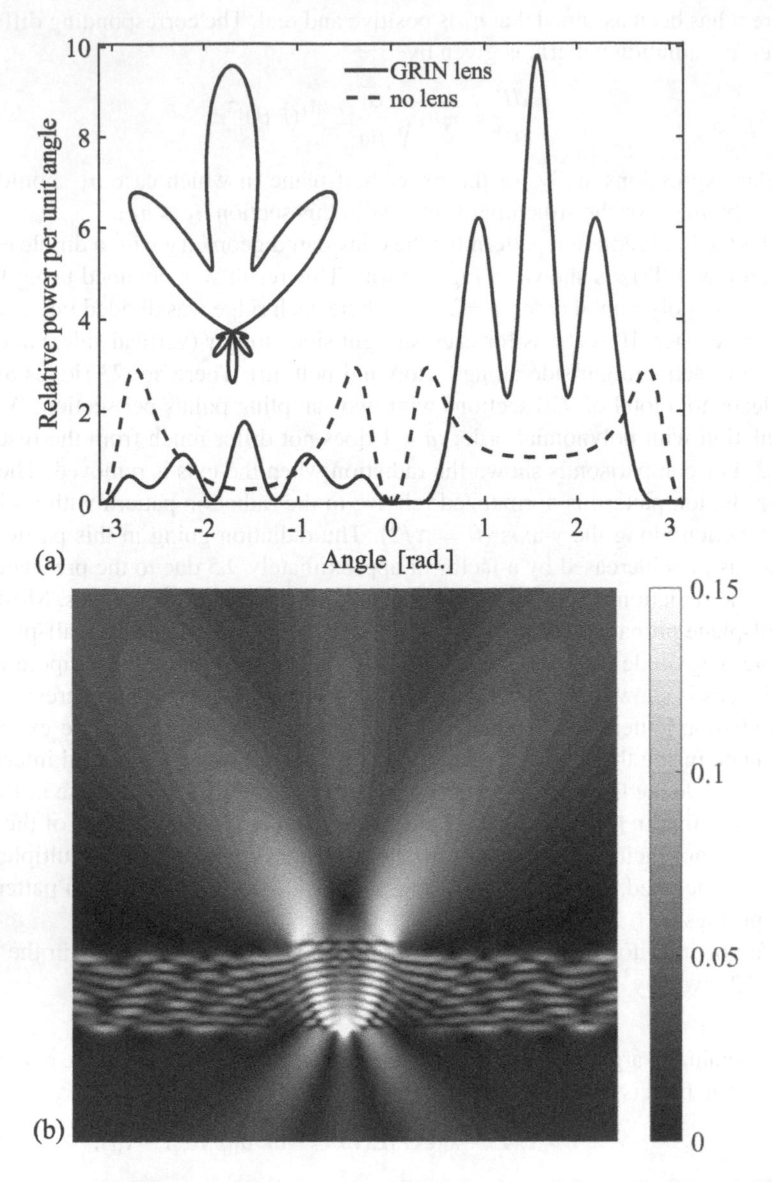

Figure 4.54 (a) Radiation pattern for a z oriented dipole emitting radiation at 1 THz for the configuration in Fig. 4.52 with a 400-μm silicon slab and a microstructured GRIN lens. Also shown is the radiation pattern for the case without the GRIN lens. Inset: Polar plot of the GRIN lens radiation pattern. (b) Magnitude of electric field.

We thus find

$$H(\mathbf{r}) = -i\frac{\omega p}{2\pi}\frac{y-y_0}{|y-y_0|}\int_{k_x=0}^{\infty}\cos(k_x[x-x_0])e^{ik_{y,L2}|y-y_0|}dk_x, \tag{4.358}$$

where the factor $(y-y_0)/|y-y_0|$ is either 1 or -1, depending on whether $y>y_0$ or $y<y_0$.

Now, since the dipole is actually sitting in a layered geometry, then similar to the analysis for the z polarized dipole, terms must be added for each plane wave such that the electromagnetics boundary conditions are satisfied. Following the same approaches as for the z polarized dipole, and considering the limit of $y_0=-d$, the reference magnetic field for the three-layer geometry is given by

$$H_0(\mathbf{r}) = -i\frac{\omega p}{2\pi}\int_{k_x=0}^{\infty} \frac{\cos(k_x[x-x_0])(1-r_{L2,L3}^{(p)})t_{L2,1}^{(p)}e^{ik_{y,L2}d}e^{ik_{y1}y}}{1-r_{L2,1}^{(p)}r_{L2,L3}^{(p)}e^{2ik_{y,L2}d}}dk_x,\ y>0, \quad (4.359)$$

$$H_0(\mathbf{r}) = -i\frac{\omega p}{2\pi}\int_{k_x=0}^{\infty} \frac{\cos(k_x[x-x_0])(1-r_{L2,L3}^{(p)})(e^{ik_{y,L2}(d+y)}+r_{L2,1}^{(p)}e^{ik_{y,L2}(d-y)})}{1-r_{L2,1}^{(p)}r_{L2,L3}^{(p)}e^{2ik_{y,L2}d}}dk_x,$$
$$0>y>-d. \quad (4.360)$$

and

$$H_0(\mathbf{r}) = -i\frac{\omega p}{2\pi}\int_{k_x=0}^{\infty} \frac{\cos(k_x[x-x_0])(-1+r_{L2,1}^{(p)}e^{2ik_{y,L2}d})t_{L2,L3}^{(p)}e^{-ik_{y,L3}(d+y)}}{1-r_{L2,1}^{(p)}r_{L2,L3}^{(p)}e^{2ik_{y,L2}d}}dk_x,$$
$$y<-d. \quad (4.361)$$

For this case the reference field in the far field for $y>0$ and $x_0=0$ can be expressed as

$$H_0^{(\text{ff})}(r,\theta) = -\omega p\frac{e^{i\pi/4}}{4}\sqrt{\frac{2}{\pi k_0 n_1}}\frac{e^{ik_0 n_1 r}}{\sqrt{r}}\frac{\left(1-r_{L2,L3}^{(p)}\right)t_{L2,1}^{(p)}k_{y1}e^{ik_{y,L2}d}}{1-r_{L2,1}^{(p)}r_{L2,L3}^{(p)}e^{2ik_{y,L2}d}},$$
$$0<\theta<\pi, \quad (4.362)$$

and again the reflection coefficients are functions of θ via $k_x=k_0 n_1\cos\theta$.

The reference far field in the lower half-plane ($y<-d$) and $x_0=0$ can be expressed as

$$H_0^{(\text{ff})}(r,\theta) = -\omega p\frac{e^{i\pi/4}}{4}\sqrt{\frac{2}{\pi k_0 n_{L3}}}\frac{e^{ik_0 n_{L3} r}}{\sqrt{r}}\frac{\left(-1+r_{L2,1}^{(p)}e^{2ik_{y,L2}d}\right)t_{L2,L3}^{(p)}k_{y,L3}e^{-ik_{y,L3}d}}{1-r_{L2,1}^{(p)}r_{L2,L3}^{(p)}e^{2ik_{y,L2}d}},$$
$$0>\theta>-\pi. \quad (4.363)$$

The total far field is then given by

$$H^{(\text{ff})}(\mathbf{r}) = H_0^{(\text{ff})}(\mathbf{r}) - \oint_{C_1}\left\{g^{(\text{ff})}(\mathbf{r},\mathbf{r}')\hat{\mathbf{n}}'\cdot\nabla' H(\mathbf{r}') - H(\mathbf{r}')\hat{\mathbf{n}}'\cdot\nabla' g^{(\text{ff})}(\mathbf{r},\mathbf{r}')\right\}dl'. \quad (4.364)$$

The time-averaged power emitted into the upper half-plane within the angles θ_1 and θ_2 is now given by

$$P = \frac{1}{2}\sqrt{\frac{\mu_0}{\varepsilon_0}}\frac{1}{n_1}\int_{\theta=\theta_1}^{\theta_2} |H^{(\mathrm{ff})}(r,\theta)|^2 r d\theta, \tag{4.365}$$

where it has been assumed that n_1 is positive and real. The differential power, or radiation pattern, on the other hand is then given by

$$\frac{dP}{d\theta} = \frac{1}{2}\frac{1}{n_1}\sqrt{\frac{\mu_0}{\varepsilon_0}}|H^{(\mathrm{ff})}(r,\theta)|^2 r, \quad 0 < \theta < \pi. \tag{4.366}$$

The corresponding differential power for the lower half-plane is on the other hand given by

$$\frac{dP}{d\theta} = \frac{1}{2}\frac{1}{n_{L3}}\sqrt{\frac{\mu_0}{\varepsilon_0}}|H^{(\mathrm{ff})}(r,\theta)|^2 r, \quad -\pi < \theta < 0. \tag{4.367}$$

The far-field radiation pattern for this dipole orientation (x) is shown in Fig. 4.55(a). The radiation pattern is compared with the case without the lens, and also here the lens increases the emission in the forward direction ($\theta = \pi/2$). In this case it is increased by a factor of approximately 5.3 which is less compared with s polarization but still significant. Note that here the same lens as for s polarization was used. In Ref. [87] a marginally different lens design was used for s and p polarization in order to optimize for these cases separately. A polar plot of the GRIN lens radiation pattern is shown as an inset.

The electric field can be obtained from the magnetic field. The magnitude of the electric field for this dipole orientation (x) is shown in Fig. 4.55(b). Since the electric field may jump across boundaries the slits in the GRIN lens are clearly visible. The calculated field plot samples the electric field at 10-μm steps along x and y except for the region of the GRIN lens where the electric field is sampled at 2-μm steps in order to capture the slits. The magnitude of the magnetic field (not shown) looks similar at large distances from the silicon film and lens.

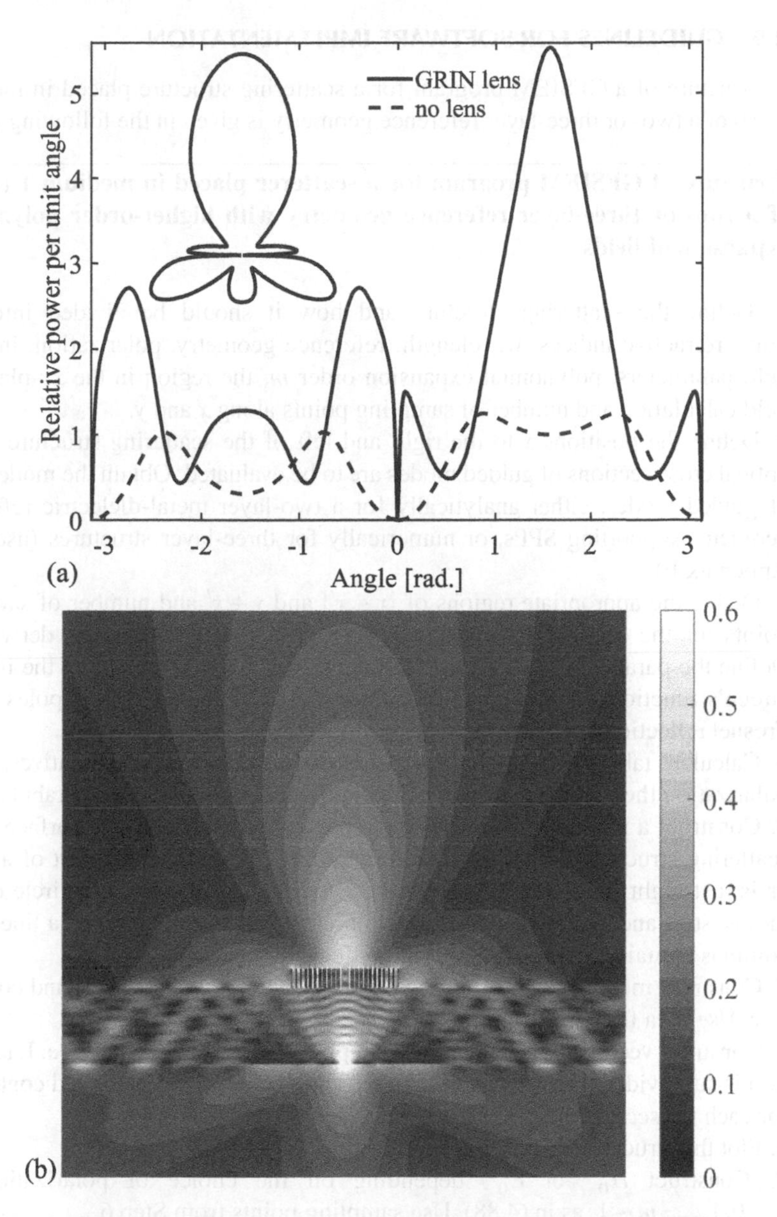

Figure 4.55 Radiation pattern for an x oriented dipole emitting radiation at 1 THz for the configuration in Fig. 4.52 with a 400-μm silicon slab and a microstructured GRIN lens. Also shown is the radiation pattern for the case without the GRIN lens. Inset: Polar plot of the GRIN lens radiation pattern. (b) Magnitude of electric field.

4.2.9 GUIDELINES FOR SOFTWARE IMPLEMENTATION

The structure of a GFSIEM program for a scattering structure placed in medium 1 ($y > 0$) of a two- or three-layer reference geometry is given in the following textbox:

Structure of GFSIEM program for a scatterer placed in medium 1 ($y > 0$) of a two- or three-layer reference geometry with higher-order polynomial expansion of fields

1. Define the scattering structure and how it should be divided into sections, refractive indices, wavelength, reference geometry, polarization, incident field parameters, polynomial expansion order m, the region in the xy-plane for field calculation and number of sampling points along x and y.
2. Define the positions x to the right and left of the scattering structure where optical cross sections of guided modes are to be evaluated. Obtain the mode index of guided modes, either analytically for a two-layer metal-dielectric reference geometry supporting SPPs, or numerically for three-layer structures (use, e.g., Appendix F).
3. Define the appropriate regions of $|x-x'|$ and $y+y'$ and number of sampling points for the calculation of the indirect Green's function and its derivatives. Define the parameters E_L and E_H for taking the integrals governing the indirect Green's function into the complex plane to avoid integration near poles of the Fresnel reflection coefficient related to guided modes.
4. Calculate tables with the indirect Green's function and its derivatives. For p polarization the non-singular part of the indirect Green's function is tabulated.
5. Construct a matrix where each row represents a section of the surface of the scattering structure. The first column identifies if the section is part of a circle or is a straight line. For a circle the next 6 columns describe the circle center, radius, start and end angles, and direction of the normal vector. For a line, the 6 columns contain the start and end of the line and the normal vector.
6. Construct matrices with sampling points $\mathbf{s}_i^{(v)}$. Let rows refer to i, and columns to v. Use data from Step 5.
7. Construct vectors describing a sub-sampling of the scatterer surface. Each section is subdivided into $N_s \cdot m$ subsections, and vectors are constructed containing for each subsection the positions, normal vector, and length.
8. Plot the structure for a visual check.
9. Construct $\overline{H}_0^{(v)}$ or $\overline{E}_0^{(v)}$ depending on the choice of polarization for $v = 0, 1, \ldots, m-1$, as in (4.88). Use sampling points from Step 6.
10. Observe which sections are neighbors by comparing positions of ends of sections and construct the matrices $\overline{\overline{D}}_H$ and $\overline{\overline{D}}_\phi$.

Structure of GFSIEM program: continued

11. Construct the matrices $\overline{\overline{A}}^{(u,v,v')}$ and $\overline{\overline{B}}^{(u,v,v')}$. Use (4.91) and (4.227). Note that for $u = 1$ in these expressions the total Green's function $g^{(d)} + g^{(i)}$ must be applied. For s polarization, $g^{(i)}$ is non-singular and can be constructed directly using the tables from Step 4. For p polarization the contribution from $g^{(i)}$ must be divided into that from the nonsingular part using tables from Step 4 and a contribution from a singular analytic part that can be treated similar to $g^{(d)}$. For the matrix elements $A_{i,j}^{(u,v,v')}$ and $B_{i,j}^{(u,v,v')}$ loop over $v = 0 : m-1$, $v' = 0 : m$, $i = 1 : N$, $j = 1 : N$. If $|\mathbf{s}_i^{(v)} - \mathbf{s}(t)|$ is larger than a threshold for all t on j, then calculate matrix elements as a simple summation over $N_s \cdot m$ terms using data from Step 4. If not, apply an analytic expression for the integrand and integration limits in an integration routine (e.g. Gauss-Kronrod). Give special attention when $\mathbf{s}(t)$ crosses the point $\mathbf{s}_i^{(v)}$.
12. Construct and solve the matrix equation (4.96) using data from Steps 9-11.
13. In the case of a plane-wave incident field, calculate absorption, extinction, guided-mode scattering, differential scattering and scattering cross sections. The absorption cross section can be calculated using (4.97). Other cross sections can be calculated using expressions from Sec. 2.3.2 as detailed in the examples (Sec. 4.2.4 - 4.2.8). Construct vectors with H (or E) and ϕ_1 at positions from Step 7. This can now be used to construct cross sections in a simple way as a summation over $N \cdot N_s \cdot m$ terms. Alternatively, use rigorous integration. In the case of a point-source or other given current source, calculate the far-field radiation patterns.
14. Calculate fields. Loop over all positions $\mathbf{r}$ where the field should be calculated. Determine if $\mathbf{r}$ is inside or outside the scatterer using (4.137). Calculate the field at $\mathbf{r}$ using the procedure in Sec. 4.2.3. For the part of this procedure with spatial integration use simple summation based on data from Steps 7 and 13 if the distance to the scatterer is beyond a threshold. Otherwise use rigorous integration.
15. Wrap the whole program in a for-loop scanning a wavelength interval.
16. Plot the optical cross sections versus wavelength. Plot the calculated fields.

A few notes:

1. For the calculation of the indirect Green's function, the relevant interval for $|x - x'|$ is from 0 to the total width of the scattering structure. The relevant interval for $y + y'$ is from $2 \times y_{\min}$ to $2 \times y_{\max}$, where $y_{\min}$ and $y_{\max}$ are the smallest and largest y on the scattering structure, respectively.
2. E_L/k_0 must be larger than the mode index (real part) of any guided mode.
3. The total cross section of guided modes can, except for lossy structures, be obtained as the difference between the total extinction and total out-of-plane scattering cross section. Direct calculation of guided-mode cross sections is still useful for comparison with indirectly calculated cross sections as a numerical check.

4.2.10 EXERCISES

1. Consider the expression (4.168) and change the sign of η'. Thus, consider instead the limit $\eta' \to 0^-$. Then carry out the integral using residue calculus and show that instead of the expression (4.157) you will get the complex conjugate of (4.157). In particular this means that the integrand will be proportional to $\exp(-ik_{y1}^*|y-y'|)$ instead of $\exp(+ik_{y1}|y-y'|)$. The resulting Green's function is suitable when using the sign convention for the time factor $\exp(+i\omega t)$ since then, the Green's function is equivalent to a cylindrical wave propagating away from $\mathbf{r}'$. With our sign convention, $\exp(-i\omega t)$, the resulting Green's function propagates toward $\mathbf{r}'$ and thus does not satisfy the radiating boundary condition.

2. Derive the expression (4.170) using ordinary Fresnel reflection theory as described in Sec. 2.2 and in basic optics text books, e.g. [38].

3. Derive the far-field Green's function for transmission through a layered structure (4.189).

4. Derive the asymptotic expressions (4.173) and (4.174).

5. Derive the expression for the pole $k_{x,\mathrm{p}}$ in (4.195) and the corresponding coefficient $A(k_{x,\mathrm{p}})$ in (4.196) for a single metal-dielectric interface. The expression for the pole is also the surface plasmon polariton (SPP) dispersion relation.

6. The Green's function for a three-layer reference geometry and for the case of the source point $\mathbf{r}'$ placed in the middle layer is given in Eq. (4.340). Derive this expression by starting out with the plane-wave expansion of the direct Green's function for the material in the middle layer, and then add all additional contributions that occur due to multiple reflections at the upper and lower interfaces. Write the result as a sum that can be evaluated using (4.339). Inspiration for how to construct such a sum can be found in Sec. 4.2.3 and Sec. 2.2.1.

7. Calculate numerically and plot the Green's function for a single dielectric-dielectric interface that does not support guided modes and for a single metal-dielectric interface or a three-layer geometry that does support guided modes. Experiment with choice of structure and wavelength. Compare, in each case, the result with the sum of far-field expressions of the Green's function (out-of-plane far field and guided-mode far field) and observe that there is agreement when the observation point $\mathbf{r}$ is at large distances from the source point $\mathbf{r}'$.

8. Follow the guidelines in Sec. 4.2.9 and develop a GFSIEM code capable of calculating optical cross sections and fields for a metal nanostrip placed on a dielectric substrate.

9. Further develop the code from problem **8** to be able to handle a two-layer dielectric-metal geometry that supports guided modes (SPPs) and calculate the related optical cross sections of SPP excitation. It is advantageous here that the mode-index n_p is related to the pole from problem **5** as $n_\mathrm{p} = k_{x,\mathrm{p}}/k_0$ and is known analytically.

10. Further develop the code from problem **9** to be able to handle a three-layer geometry that supports guided modes and calculate the related optical cross sections of guided modes. Demonstrate, for a situation with materials that are not lossy, that the guided-mode cross sections are equivalent to the difference between extinction and out-of-plane scattering cross sections.

4.3 PERIODIC STRUCTURES

In this section we will consider the Green's function surface integral equation method for structures that are periodic along the x-axis. As examples of this general type of geometry a finite photonic crystal is shown in Fig. 4.56(a) and a periodic array of grooves in Fig. 4.56(b). The photonic crystal is an example of a structure where each region with dielectric constant ε_2 is a closed region. For the array of grooves the region with dielectric constant ε_2 is open. These two types of situations must be treated slightly differently.

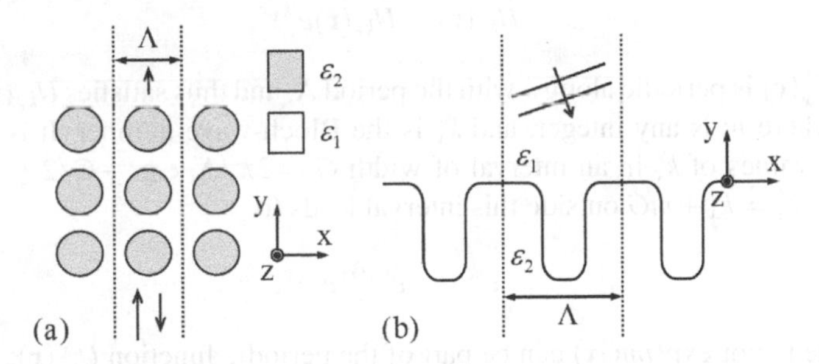

Figure 4.56 (a) Schematic of photonic crystal. Regions with dielectric constant ε_2 are closed. (b) Schematic of a periodic array of grooves in a surface. The region with dielectric constant ε_2 is open.

The photonic crystal consists of a periodic array of circular cylinders with dielectric constant ε_2 on a square lattice in a background with dielectric constant ε_1. The cylinders could also have been placed on, for example, a hexagonal lattice. The structure is fully periodic along the x-axis with a period Λ, while there is only a finite number of layers of cylinders along the y-axis. We can thus consider illuminating the photonic crystal from below, in which case there will be a reflection from the crystal and a transmission through the crystal, as illustrated by arrows in the figure. Photonic crystals are also known as photonic bandgap structures, and are characterized by frequency intervals (bandgaps) where no propagating electromagnetic modes exist in the periodic structure [12, 13, 14, 15]. Similar to electronic bandgap structures (usually semiconductors) it is possible to trap light at frequencies in the bandgap to point-like or line-like defects in the periodic structure. Here, such defects form waveguides and cavities for light, while corresponding defects in electronic crystals form quantum dots, quantum wires, and quantum wells.

For the periodic array of grooves in a surface with dielectric constant ε_2 we will consider both the case of grooves in a dielectric, in which case the groove array can have an anti-reflection effect, and a groove array in a metal, in which case the grooves can function as narrow-band optical resonators or as broadband absorbers depending on the groove geometry [65, 76, 71, 72, 73].

4.3.1 BLOCH WAVES

Due to the periodicity along the x-axis the fields can be expanded in Bloch waves. For p polarization the magnetic field can thus be expressed as

$$H(\mathbf{r}) = \sum_{k_x} H_{k_x}(\mathbf{r}), \tag{4.368}$$

where the sum represents both a sum over discrete k_x or alternatively an integral over a continuum of k_x. The Bloch waves in the expansion are of the form

$$H_{k_x}(\mathbf{r}) = U_{k_x}(\mathbf{r}) e^{ik_x x}, \tag{4.369}$$

where $U_{k_x}(\mathbf{r})$ is periodic along x with the period Λ, and thus satisfies $U_{k_x}(\mathbf{r}) = U_{k_x}(\mathbf{r} + \hat{\mathbf{x}} m\Lambda)$, where m is any integer, and k_x is the Bloch-wave number. It is sufficient to consider values of k_x in an interval of width $G = 2\pi/\Lambda$, e.g. $-G/2 \leq k_x \leq +G/2$, since any $k_x' = k_x + mG$ outside this interval leads to

$$e^{ik_x' x} = e^{imGx} e^{ik_x x}, \tag{4.370}$$

where the factor $\exp(imGx)$ can be part of the periodic function $U_{k_x}(\mathbf{r})$.

In the case of an incident field or reference field being a plane wave, i.e.,

$$H_0(\mathbf{r}) = A e^{ik_x x} e^{\pm ik_{y1} y}, \tag{4.371}$$

the field is of the form (4.368) with only a single term in the sum, or only a single value of k_x. Thus, k_x in (4.371) determines in that case the relevant Bloch-wave number to be used in the calculation. For more general incident fields containing several k_x it should be noted that field components with different k_x are uncoupled. In that case it will be necessary to carry out a calculation for each k_x separately, and then carry out the sum (4.368). In this section we will consider only the case of a single incident plane wave, and it will thus be sufficient to consider a single k_x.

4.3.2 GREEN'S FUNCTION FOR PERIODIC STRUCTURES

The Green's function surface integral equations for periodic structures, and a specific Bloch-wave number k_x, only differ from the case of structures that are not periodic by the Green's function being different. For periodic structures, the Green's function must satisfy periodic boundary conditions along x rather than a radiating boundary condition. For a homogeneous material with dielectric constant ε_u, and with the requirement of a periodic boundary condition along x defined by the Bloch-wave number, the Green's function must satisfy the equation

$$\left(\nabla^2 + k_0^2\varepsilon_u\right) g(\mathbf{r},\mathbf{r}') = -\delta(\mathbf{r}-\mathbf{r}') \tag{4.372}$$

as before, but now with the periodic boundary condition

$$g(\mathbf{r}+\hat{\mathbf{x}}m\Lambda,\mathbf{r}') = g(\mathbf{r},\mathbf{r}')e^{ik_x\Lambda m}, \quad m = 0,\pm1,\pm2,... \tag{4.373}$$

In addition the Green's function must satisfy the radiating boundary condition such that $g(\mathbf{r},\mathbf{r}')$ describes a field that propagates away from the position $\mathbf{r}'$ along y.

The Green's function can be constructed using a complete set of eigenmodes E_λ to the operator in Eq. (4.372) that satisfies the periodic boundary condition. The eigenmodes must thus satisfy

$$\left(\nabla^2 + k_0^2\varepsilon_u\right) E_\lambda = \lambda E_\lambda \tag{4.374}$$

with the Bloch boundary condition

$$E_\lambda(\mathbf{r}+\hat{\mathbf{x}}m\Lambda) = E_\lambda(\mathbf{r})e^{ik_x\Lambda m}. \tag{4.375}$$

These requirements lead to the following eigenmodes

$$E_{n,k_y}(\mathbf{r}) = e^{i(k_x+nG)x}e^{ik_y y}, \tag{4.376}$$

where the corresponding eigenvalue is given by

$$\lambda_{n,k_y} = k_0^2\varepsilon_u - (k_x+nG)^2 - k_y^2. \tag{4.377}$$

The eigenfunctions satisfy the orthogonality relation

$$\int_{x=0}^{\Lambda}\int_{y=-\infty}^{\infty} E_{n,k_y}(\mathbf{r})\left(E_{n',k_y'}(\mathbf{r})\right)^* dxdy = 2\pi\Lambda\delta_{n,n'}\delta(k_y-k_y'). \tag{4.378}$$

The Green's function may now be constructed using the mode-expansion method as follows:

$$g(\mathbf{r},\mathbf{r}') = -\sum_n \int_{k_y=-\infty}^{\infty} \frac{E_{n,k_y}(\mathbf{r})\left(E_{n,k_y}(\mathbf{r})\right)^*}{2\pi\Lambda\lambda_{n,k_y}} dk_y. \tag{4.379}$$

In order to obtain the Green's function that satisfies also the radiating boundary condition along y, a very small imaginary term is added to the eigenvalue in the denominator, and the integral is then carried out using residue calculus. This leads to the final expression for the Green's function, which is also the main result of this subsection:

Green's function for a medium with dielectric constant ε_u and periodic boundary condition along x. Case of period Λ and Bloch-wave number k_x (2D):

$$g_u(\mathbf{r},\mathbf{r}') = \frac{i}{2\Lambda}\sum_n \frac{e^{i(k_x+nG)(x-x')}e^{ik_{y,u,n}|y-y'|}}{k_{y,u,n}}, \tag{4.380}$$

where $G = 2\pi/\Lambda$, $k_{y,u,n} = \sqrt{k_0^2\varepsilon_u - (k_x+nG)^2}$ with $\text{Imag}\{k_{y,u,n}\} \geq 0$.

The periodic Green's function is singular for $\mathbf{r} \to \mathbf{r}'$ similar to the non-periodic case. This can be seen by using the fact that in this limit, the summation can be approximated with an integral. Thus, by using $\Delta k_x = G = 2\pi/\Lambda$ we find

$$g_u(\mathbf{r},\mathbf{r}') \approx \frac{i}{4\pi}\int_{k_x} \frac{e^{ik_x(x-x')}e^{ik_{y,u}|y-y'|}}{k_{y,u}}dk_x, \tag{4.381}$$

which is approximately the same as the non-periodic homogeneous-medium Green's function. While the singular part of the Green's function is the same as for the non-periodic case, the non-singular imaginary part of the Green's function is different in the limit $\mathbf{r} \to \mathbf{r}'$. The singularity must be dealt with in numerical calculations similar to the non-periodic case. The similarity between Green's functions for the periodic and non-periodic cases may also be used to verify that by applying the left-hand-side operator in Eq. (4.372) to the periodic Green's function the result will be minus the identity operator, i.e., the right-hand-side operator in Eq. (4.372).

It is also possible to construct a Green's function for a layered reference structure using similar principles as were used for the non-periodic case. In the case of $y > 0$ and $y' > 0$, the Green's function for a layered reference structure can be obtained by adding a reflection term for each $k_x + nG$. Thus,

$$g(\mathbf{r},\mathbf{r}') = g^{(d)}(\mathbf{r},\mathbf{r}') + g^{(i)}(\mathbf{r},\mathbf{r}'),\ \ y, y' > 0, \tag{4.382}$$

where $g^{(d)}(\mathbf{r},\mathbf{r}') = g_1(\mathbf{r},\mathbf{r}')$ and

$$g^{(i)}(\mathbf{r},\mathbf{r}') = \frac{i}{2\Lambda}\sum_n \frac{e^{i(k_x+nG)(x-x')}e^{ik_{y,u,n}(y+y')}r^{(v)}(k_x+nG)}{k_{y,u,n}}, \tag{4.383}$$

where here $v = s$ or p depending on the polarization being considered. For the specific periodic structures of interest in this section it is sufficient to use only the simpler Green's functions in Eq. (4.380). This will be explained in the next subsection.

4.3.3 GFSIEM FOR PERIODIC STRUCTURES

An example of a simple structure which is periodic along x with the period Λ is shown in Fig. 4.57, where surfaces and normal vectors have been included for the following discussion of the Green's function surface integral equations. The structure is

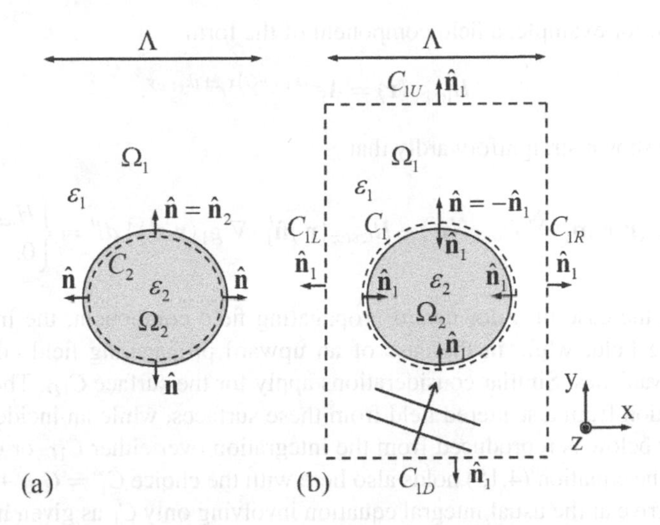

Figure 4.57 Schematic of one period of a periodic structure with surfaces and normal vectors for calculating (a) the field inside the scatterer in region Ω_2, and (b) outside the scatterer in region Ω_1.

divided into two regions Ω_1 and Ω_2 with dielectric constants ε_1 and ε_2, respectively. In this example the scatterer surface is closed.

The Green's function integral equations from the non-periodic case still apply, except that the Green's function must be replaced with the periodic Green's function for the Bloch-wave number k_x. Thus, considering, for example, the magnetic field for p polarization, the magnetic field at any position $\mathbf{r} \in \Omega_2$ can be obtained from the integral equation (4.7), where the integral must be taken over the dashed curve C_2 in Fig. 4.57(a), and the periodic Green's function $g_2(\mathbf{r}, \mathbf{r}')$ must be used.

Considering instead a position $\mathbf{r} \in \Omega_1$ the magnetic field can be obtained from a similar surface integral equation as Eq. (4.10) except that here the periodic Green's function $g_1(\mathbf{r}, \mathbf{r}')$ for the Bloch-wave number k_x and dielectric constant ε_1 must be used. The appropriate surfaces to be used in the integral equation in this case are shown in Fig. 4.57(b). The surface C_1 is as usual the outer surface of the scatterer. However, the outer surface C_1^{∞} in Eq. (4.10) should now be replaced with the surface $C_{1L}+C_{1R}+C_{1U}+C_{1D}$. Here, a rectangular surface has been chosen where L, R, U, D, refer to the left, right, upper and lower sides of the rectangle. Now, we may note that the integral along the two lines C_{1L} and C_{1R} will cancel each other due to the periodic boundary conditions in the field and Green's function, since these lines are separated by exactly one period along x. We may thus use $C_1^{\infty} = C_{1U} + C_{1D}$. In addition, it can be shown that there will only be contributions from the integrals along the upper and lower lines C_{1U} and C_{1L} for field components propagating toward the scatterer.

Consider, for example, a field component of the form

$$H_{\text{test}\pm}(\mathbf{r}) = Ae^{i(k_x+nG)x}e^{\pm ik_{y1,n}y}. \tag{4.384}$$

It can be shown straightforwardly that

$$\int_{C_{1U}} \{g_1(\mathbf{r},\mathbf{r}')\hat{\mathbf{n}}_1' \cdot \nabla' H_{\text{test}\pm}(\mathbf{r}') - H_{\text{test}\pm}(\mathbf{r}')\hat{\mathbf{n}}_1' \cdot \nabla' g_1(\mathbf{r},\mathbf{r}')\}\, dl' = \begin{cases} H_{\text{test}-}(\mathbf{r}), & - \\ 0, & + \end{cases}. \tag{4.385}$$

Thus, in the case of a downward propagating field component, the integral reproduces the field, while in the case of an upward propagating field component the integral vanishes. Similar considerations apply for the surface C_{1D}. There is thus no contribution from a scattered field from these surfaces, while an incident field from above or below is reproduced from the integration over either C_{1U} or C_{1D}. To summarize, the equation (4.15) holds also here with the choice $C_1^\infty = C_{1U} + C_{1D}$, and we finally arrive at the usual integral equation involving only C_1 as given in Eq. (4.16).

Now consider the case of a periodic array of grooves being illuminated from above as illustrated in Fig. 4.56(b). As illustrated in Fig. 4.58 the region above the grooves may be referred to as Ω_1 and the region below as Ω_2. Since the incident field is

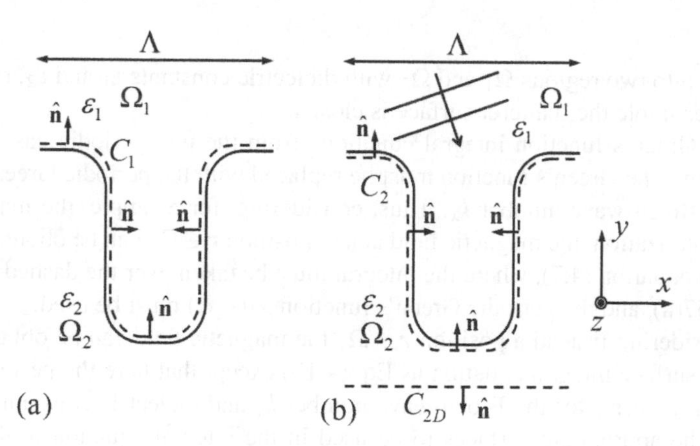

Figure 4.58 Schematic of one period of a periodic array of grooves in a surface with surfaces and normal vectors for calculating (a) the field in region Ω_1, and (b) the field in region Ω_2.

incident from above, the field at any position in Ω_1 can be obtained by using the integral equation

$$H(\mathbf{r}) = H_0(\mathbf{r}) - \int_{C_1} \{g_1(\mathbf{r},\mathbf{r}')\hat{\mathbf{n}}' \cdot \nabla' H(\mathbf{r}') - H(\mathbf{r}')\hat{\mathbf{n}}' \cdot \nabla' g_1(\mathbf{r},\mathbf{r}')\}\, dl', \quad \mathbf{r} \in \Omega_1, \tag{4.386}$$

where the term $H_0(\mathbf{r})$ is related to an integral over a surface similar to C_{1U} in Fig. 4.57. The only difference from before is that the surface C_1 is not closed in the usual sense, but instead, in a periodic sense. Here C_1 refers to the structure surface within one period along x.

Considering instead a position $\mathbf{r} \in \Omega_2$ the field is in principle given by a surface integral over a surface surrounding $\mathbf{r}$, where again integrals over vertical sides separated by one period Λ along x cancel each other. In addition, since all field components at the surface C_{2D} are propagating downward the integral over this surface vanishes too. Thus, the field in Ω_2 is given by

$$H(\mathbf{r}) = \int_{C_2} \{g_2(\mathbf{r},\mathbf{r}')\hat{\mathbf{n}}' \cdot \nabla' H(\mathbf{r}') - H(\mathbf{r}')\hat{\mathbf{n}}' \cdot \nabla' g_2(\mathbf{r},\mathbf{r}')\}\, dl', \quad \mathbf{r} \in \Omega_2, \tag{4.387}$$

where again the surface C_2 is not closed in the usual sense. By applying the electromagnetics boundary conditions across the surface between the regions Ω_1 and Ω_2 the integral equations can be solved similar to the case of a scatterer in free space. It is convenient that it is possible to use the simple Green's functions with no built-in layered structure.

4.3.4 DERIVATIVES OF PERIODIC GREEN'S FUNCTION AND TABULATION

The periodic Green's functions (4.380) must be calculated numerically. They can, for example, be tabulated on a grid of values for $x-x'$ and $|y-y'|$ that can then be used subsequently for interpolation to find the Green's function for other values. Here it can be useful to subtract the singular part of the Green's function when building the table such that a smooth and slowly varying function is tabulated rather than a singular function. We may thus tabulate

$$g_{u,\text{Table}}(\mathbf{r},\mathbf{r}') = g_u(\mathbf{r},\mathbf{r}') - f(\mathbf{r},\mathbf{r}'), \tag{4.388}$$

where $f(\mathbf{r},\mathbf{r}')$ is a relatively simple analytic function with the same periodic array of singularities as g_u. The subtraction of f will thus cancel the singularities of $g_u(\mathbf{r},\mathbf{r}')$ occurring whenever $x-x' = m\Lambda$ and $y-y' = 0$.

A suitable function that works for singularity subtraction with periodic boundary conditions is

$$f(\mathbf{r},\mathbf{r}') = \frac{i}{4} H_0^{(1)}(k_0 n_u R_B(\mathbf{r},\mathbf{r}'))e^{ik_x(x-x')}, \tag{4.389}$$

where

$$R_B(\mathbf{r},\mathbf{r}') \equiv \sqrt{(y-y')^2 + \frac{4}{G^2}(\sin(G(x-x')/2))^2}. \tag{4.390}$$

For the tabulation of the derivatives of the Green's function the following derivatives of f are also useful

$$\frac{\partial f(\mathbf{r},\mathbf{r}')}{\partial x'} = \frac{-i}{4} H_1^{(1)}(k_0 n_u R_B(\mathbf{r},\mathbf{r}'))k_0 n_u \frac{\partial R_B(\mathbf{r},\mathbf{r}')}{\partial x'} - ik_x f(\mathbf{r},\mathbf{r}'), \tag{4.391}$$

$$\frac{\partial f(\mathbf{r},\mathbf{r}')}{\partial y'} = \frac{-i}{4} H_1^{(1)}(k_0 n_u R_B(\mathbf{r},\mathbf{r}'))k_0 n_u \frac{\partial R_B(\mathbf{r},\mathbf{r}')}{\partial y'}, \tag{4.392}$$

where

$$\frac{\partial R_B(\mathbf{r},\mathbf{r}')}{\partial x'} = \frac{-\sin(G(x-x')/2)\cos(G(x-x')/2)}{R_B(\mathbf{r},\mathbf{r}')}\frac{2}{G}, \tag{4.393}$$

$$\frac{\partial R_B(\mathbf{r},\mathbf{r}')}{\partial y'} = \frac{-(y-y')}{R_B(\mathbf{r},\mathbf{r}')}. \tag{4.394}$$

We may thus also tabulate

$$\frac{\partial g_{u,\text{Table}}(\mathbf{r},\mathbf{r}')}{\partial x'} = \frac{\partial g_u(\mathbf{r},\mathbf{r}')}{\partial x'} - \frac{\partial f(\mathbf{r},\mathbf{r}')}{\partial x'}, \tag{4.395}$$

and

$$\frac{\partial g_{u,\text{Table}}(\mathbf{r},\mathbf{r}')}{\partial y'} = \frac{\partial g_u(\mathbf{r},\mathbf{r}')}{\partial y'} - \frac{\partial f(\mathbf{r},\mathbf{r}')}{\partial y'}. \tag{4.396}$$

We note that the only difference when changing the sign of $y-y'$ is that the sign of $\partial g_u/\partial y'$ will also change. It is thus sufficient to tabulate only for the case of $y-y' \geq 0$.

The partial derivatives of the periodic Green's functions are given by

$$\frac{\partial g_u(\mathbf{r},\mathbf{r}')}{\partial x'} = \frac{i}{2\Lambda}\sum_n \frac{e^{i(k_x+nG)(x-x')}e^{ik_{y,u,n}|y-y'|}}{k_{y,u,n}}(-i)(k_x+nG), \tag{4.397}$$

$$\frac{\partial g_u(\mathbf{r},\mathbf{r}')}{\partial y'} = \frac{i}{2\Lambda}\sum_n e^{i(k_x+nG)(x-x')}e^{ik_{y,u,n}|y-y'|}(-i)\text{sign}\{y-y'\}. \tag{4.398}$$

When calculating (4.380), (4.397) and (4.398) it is necessary to include enough terms (values of n) that the summand practically vanishes for the remaining terms that are ignored, which can happen due to the term $\exp(ik_{y,u,n}|y-y'|)$ for $|y-y'|>0$. Clearly, more terms are needed when $|y-y'|$ is small compared with the case when $|y-y'|$ is large. The case $|y-y'|=0$ can be handled by using a very small but non-zero value of $|y-y'|$. After subtracting the singular term the resulting non-singular function being tabulated will differ very little from the case of using exactly $|y-y'| = 0$ and the approximation being made is thus acceptable.

Now, except that the Green's functions g_1 and g_2 are different, the approach for calculating the fields is also the same as for the case of a scatterer in a homogeneous dielectric for the non-periodic case. For a calculation of all field components, we also need the second derivatives of the Green's function given in the following:

$$\frac{\partial^2 g_{u,\text{Table}}(\mathbf{r},\mathbf{r}')}{\partial x'\partial x} = \frac{\partial^2 g_u(\mathbf{r},\mathbf{r}')}{\partial x'\partial x} - \frac{\partial^2 f(\mathbf{r},\mathbf{r}')}{\partial x'\partial x}, \tag{4.399}$$

$$\frac{\partial^2 g_{u,\text{Table}}(\mathbf{r},\mathbf{r}')}{\partial x'\partial y} = \frac{\partial^2 g_u(\mathbf{r},\mathbf{r}')}{\partial x'\partial y} - \frac{\partial^2 f(\mathbf{r},\mathbf{r}')}{\partial x'\partial y}, \tag{4.400}$$

$$\frac{\partial^2 g_u(\mathbf{r},\mathbf{r}')}{\partial y'\partial x} = \frac{\partial^2 g_u(\mathbf{r},\mathbf{r}')}{\partial x'\partial y}, \tag{4.401}$$

$$\frac{\partial^2 g_{u,\mathrm{Table}}(\mathbf{r},\mathbf{r}')}{\partial y'\partial y} = \frac{\partial^2 g_u(\mathbf{r},\mathbf{r}')}{\partial y'\partial y} - \frac{\partial^2 f(\mathbf{r},\mathbf{r}')}{\partial y'\partial y}, \tag{4.402}$$

where, again, the singular part of derivatives has been subtracted. Here,

$$\frac{\partial^2 g_u(\mathbf{r},\mathbf{r}')}{\partial x'\partial x} = \frac{i}{2\Lambda}\sum_n \frac{e^{i(k_x+nG)(x-x')}e^{ik_{y,u,n}|y-y'|}}{k_{y,u,n}}(k_x+nG)^2, \tag{4.403}$$

$$\frac{\partial^2 g_u(\mathbf{r},\mathbf{r}')}{\partial x'\partial y} = \frac{i}{2\Lambda}\sum_n e^{i(k_x+nG)(x-x')}e^{ik_{y,u,n}|y-y'|}(k_x+nG)\mathrm{sign}\{y-y'\}, \tag{4.404}$$

$$\frac{\partial^2 g_u(\mathbf{r},\mathbf{r}')}{\partial y'\partial y} = \frac{i}{2\Lambda}\sum_n e^{i(k_x+nG)(x-x')}e^{ik_{y,u,n}|y-y'|}k_{y,u,n}. \tag{4.405}$$

An alternative analytic function with a suitable singularity that also works with periodic boundary conditions is given by

$$f(\mathbf{r},\mathbf{r}') = \frac{-1}{4\pi}\ln\left((G(y-y')/2)^2 + (\sin(G(x-x')/2))^2\right)e^{ik_x(x-x')}. \tag{4.406}$$

The other choice (4.389) depends on u but also captures the region near the singularity better for each u.

4.3.5 CALCULATING THE FIELDS

In this subsection we will assume that the fields on C_1 and C_2 (field and normal derivative) have been calculated by solving the integral equations. Using the periodic GFSIEM, the field can then be obtained at any other position in the regions Ω_1 and Ω_2 by using the integral equations and those fields, similar to the non-periodic case.

We shall thus instead here focus on the cases where the y-coordinate is above any scatterer or groove, or below any scatterer or groove. At those positions the scattered field will strictly propagate upward or downward, and the angular spectrum representation is applicable. If we now let $y_{\max}$ denote the maximum y-coordinate of any scatterer, then for $y > y_{\max}$ all scattered field components will propagate upward. It is thus possible to express the total fields in the form

$$H(\mathbf{r}) = H_0(\mathbf{r}) + \sum_n \tilde{H}_n^+ e^{i(k_x+nG)x}e^{ik_{y1,n}(y-y_{\max})}, \quad y > y_{\max}, \tag{4.407}$$

$$E_x(\mathbf{r}) = E_{x,0}(\mathbf{r}) + \frac{i}{\omega\varepsilon_0\varepsilon_1}\sum_n ik_{y1,n}\tilde{H}_n^+ e^{i(k_x+nG)x}e^{ik_{y1,n}(y-y_{\max})}, \quad y > y_{\max}, \tag{4.408}$$

$$E_y(\mathbf{r}) = E_{y,0}(\mathbf{r}) + \frac{-i}{\omega\varepsilon_0\varepsilon_1}\sum_n i(k_x+nG)\tilde{H}_n^+ e^{i(k_x+nG)x}e^{ik_{y1,n}(y-y_{\max})}, \quad y > y_{\max}, \tag{4.409}$$

where the Fourier components of the field are given by

$$\tilde{H}_n^+ = \frac{-i}{2\Lambda}\frac{1}{k_{y1,n}}\oint_{C_1} e^{-i(k_x+nG)x'}e^{ik_{y1,n}(y_{\max}-y')}$$

$$\times \left(\phi(\mathbf{r}') - H(\mathbf{r}')\hat{\mathbf{n}}' \cdot [\hat{\mathbf{x}}(-i(k_x + nG)) + \hat{\mathbf{y}}(-ik_{y1,n})]\right) dl'. \tag{4.410}$$

For values of y being slightly larger than $y_{\max}$, e.g. $y > y_{\max} + \Delta y$, only a limited number of terms is needed in the summation due to the exponential decrease with y of $\exp(ik_{y1,n}(y - y_{\max}))$ for those n where $k_{y1,n}$ is imaginary.

Similarly, if we let $y_{\min}$ denote the minimum y-coordinate of any scatterer, then for smaller values of y, all scattered field components will propagate downward. In the region $y < y_{\min}$ the dielectric constant will be ε_u, where $u = 1$ for the photonic crystal structure, and $u = 2$ for the groove array in a dielectric or in a metal.

For the photonic crystal, the field is given by

$$H(\mathbf{r}) = H_0(\mathbf{r}) + \sum_n \tilde{H}_n^- e^{i(k_x+nG)x} e^{ik_{y1,n}(y_{\min}-y)}, \quad y < y_{\min}, \tag{4.411}$$

$$E_x(\mathbf{r}) = E_{x,0}(\mathbf{r}) + \frac{i}{\omega\varepsilon_0\varepsilon_1}\sum_n (-ik_{y1,n})\tilde{H}_n^- e^{i(k_x+nG)x} e^{ik_{y1,n}(y_{\min}-y)}, \quad y < y_{\min}, \tag{4.412}$$

$$E_y(\mathbf{r}) = E_{y,0}(\mathbf{r}) + \frac{-i}{\omega\varepsilon_0\varepsilon_1}\sum_n i(k_x+nG)\tilde{H}_n^- e^{i(k_x+nG)x} e^{ik_{y1,n}(y_{\min}-y)}, \quad y < y_{\min}, \tag{4.413}$$

with the Fourier components

$$\tilde{H}_n^- = \frac{-i}{2\Lambda}\frac{1}{k_{y1,n}} \oint_{C_1} e^{-i(k_x+nG)x'} e^{ik_{y1,n}(y'-y_{\min})}$$

$$\times \left(\phi(\mathbf{r}') - H(\mathbf{r}')\hat{\mathbf{n}}' \cdot [\hat{\mathbf{x}}(-i(k_x + nG)) + \hat{\mathbf{y}}(ik_{y1,n})]\right) dl'. \tag{4.414}$$

In the case of the groove array, the fields are instead given by

$$H(\mathbf{r}) = \sum_n \tilde{H}_n^- e^{i(k_x+nG)x} e^{ik_{y2,n}(y_{\min}-y)}, \quad y < y_{\min}, \tag{4.415}$$

$$E_x(\mathbf{r}) = \frac{i}{\omega\varepsilon_0\varepsilon_2}\sum_n (-ik_{y2,n})\tilde{H}_n^- e^{i(k_x+nG)x} e^{ik_{y2,n}(y_{\min}-y)}, \quad y < y_{\min}, \tag{4.416}$$

$$E_y(\mathbf{r}) = \frac{-i}{\omega\varepsilon_0\varepsilon_2}\sum_n i(k_x+nG)\tilde{H}_n^- e^{i(k_x+nG)x} e^{ik_{y2,n}(y_{\min}-y)}, \quad y < y_{\min}, \tag{4.417}$$

with the Fourier coefficients

$$\tilde{H}_n^- = \frac{i}{2\Lambda}\frac{1}{k_{y2,n}} \oint_{C_1} e^{-i(k_x+nG)x'} e^{ik_{y2,n}(y'-y_{\min})}$$

$$\times \left(\frac{\varepsilon_2}{\varepsilon_1}\phi(\mathbf{r}') - H(\mathbf{r}')\hat{\mathbf{n}}' \cdot [\hat{\mathbf{x}}(-i(k_x + nG)) + \hat{\mathbf{y}}(ik_{y2,n})]\right) dl'. \tag{4.418}$$

In the case of s-polarized light the corresponding expressions can be obtained by making the substitutions $H \to E$ and vice versa, $i/\omega\varepsilon_u \to -i/\omega\mu_0$, and $\varepsilon_2/\varepsilon_1 \to 1$.

4.3.6 CALCULATING REFLECTION AND TRANSMISSION

Reflection and transmission for the case of illuminating the photonic crystal [Fig. 4.56(a)] with a p-polarized plane wave from below will be considered first. The reference field is thus given by

$$H_0(\mathbf{r}) = Ae^{ik_x x}e^{+ik_{y1,0}y}, \tag{4.419}$$

where it is assumed that $-k_0 n_1 < k_x < +k_0 n_1$ such that $k_{y1,0}$ is real and positive. The power in the incident field crossing a line from $x = 0$ to $x = \Lambda$ at a fixed value of y is given by

$$P_i = \frac{1}{2}\int_{x=0}^{x=\Lambda} \text{Real}\{\mathbf{E}_0(\mathbf{r}) \times (\mathbf{H}_0(\mathbf{r}))^*\} \cdot \hat{\mathbf{y}} dx = \frac{1}{2}|H_0|^2 \frac{k_{y1,0}}{\omega\varepsilon_0\varepsilon_1}. \tag{4.420}$$

Note that it is assumed that ε_1 is a real number.

The reflected field, or backward scattered field, as a result of scattering by the photonic crystal, is given by

$$H_r(\mathbf{r}) = \sum_n \tilde{H}_n^- e^{i(k_x+nG)x} e^{ik_{y1,n}(y_{\min}-y)}, \quad y < y_{\min}, \tag{4.421}$$

where the Fourier coefficients $\tilde{H}_n^-$ were defined in the previous section. The power in the total reflected field is given by

$$P_r = \frac{1}{2}\int_{x=0}^{x=\Lambda} \text{Real}\{\mathbf{E}_r(\mathbf{r}) \times (\mathbf{H}_r(\mathbf{r}))^*\} \cdot (-\hat{\mathbf{y}}) dx = \sum_n \frac{1}{2}|\tilde{H}_n^-|^2 \frac{\text{Real}\{k_{y1,n}\}}{\omega\varepsilon_0\varepsilon_1}, \tag{4.422}$$

where, although $\mathbf{H}_r$ and $\mathbf{E}_r$ are both described by a sum over their discrete Fourier components, the result for the power is a single sum since cross terms cancel upon integration. Note that any term where $k_{y1,n}$ is purely imaginary will not contribute. The total reflectance is thus given by

$$R_{\text{total}} = \frac{P_r}{P_i} = \sum_n R_n, \tag{4.423}$$

where

$$R_n = \frac{|\tilde{H}_n^-|^2 \text{Real}\{k_{y,1,n}\}}{|H_0|^2 k_{y1,0}} \tag{4.424}$$

is the fraction of power reflected into the n-th reflection diffraction order. R_0 is the specular reflectance.

The transmitted power is, on the other hand, given by

$$P_t = \frac{1}{2}\int_{x=0}^{x=\Lambda} \text{Real}\{\mathbf{E}(\mathbf{r}) \times (\mathbf{H}(\mathbf{r}))^*\} \cdot \hat{\mathbf{y}} dx$$

$$= \sum_n \frac{1}{2}|Ae^{ik_{y1,0}y_{\max}}\delta_{n,0} + \tilde{H}_n^+|^2 \frac{\text{Real}\{k_{y1,n}\}}{\omega\varepsilon_0\varepsilon_1}, \quad y > y_{\max}. \tag{4.425}$$

The total transmittance can thus be written as

$$T_{\text{total}} = \frac{P_t}{P_i} = \sum_n T_n, \tag{4.426}$$

where

$$T_n = \frac{|Ae^{ik_{y1,0}y_{\max}}\delta_{n,0} + \tilde{H}_n^+|^2 \text{Real}\{k_{y1,n}\}}{|H_0|^2 k_{y1,0}}. \tag{4.427}$$

Here T_n represents the fraction of incident power which is scattered into the n-th transmission diffraction order, and T_0 is the zero-order transmittance.

In the case of the groove array in a dielectric or a metal, a plane wave being incident from the top is considered. Thus,

$$H_0(\mathbf{r}) = Ae^{ik_x x}e^{-ik_{y1,0}y}. \tag{4.428}$$

Here, the n-th reflectance order is given by

$$R_n = \frac{|\tilde{H}_n^+|^2 \text{Real}\{k_{y,1,n}\}}{|H_0|^2 k_{y1,0}}. \tag{4.429}$$

In the case of a dielectric (ε_2 being real and positive), the n-th transmittance order is given by

$$T_n = \frac{|\tilde{H}_n^-|^2 \text{Real}\{k_{y2,n}\}}{|H_0|^2 k_{y1,0}} \frac{\varepsilon_1}{\varepsilon_2}. \tag{4.430}$$

Note that this time in the transmission expression there is no component related to the incident field since we are not considering positions in the region $u = 1$.

In the case of s polarization the incident electric field (z-component) will instead be on the form

$$E_0(\mathbf{r}) = Ae^{ik_x x}e^{\pm ik_{y1,0}y}, \tag{4.431}$$

and the reflected field, or backward scattered field, as a result of scattering by the photonic crystal, is given by

$$E_r(\mathbf{r}) = \sum_n \tilde{E}_n^- e^{i(k_x+nG)x} e^{ik_{y1,n}(y_{\min}-y)}, \quad y < y_{\min}. \tag{4.432}$$

The power in the incident field is now given by

$$P_i = \frac{1}{2}|E_0|^2 \frac{\text{Real}\{k_{y1,0}\}}{\omega\mu_0}, \tag{4.433}$$

and the power in the total reflected field is given by

$$P_r = \sum_n \frac{1}{2}|\tilde{E}_n^-|^2 \frac{\text{Real}\{k_{y1,n}\}}{\omega\mu_0}. \tag{4.434}$$

Thus,

$$R_n = \frac{|\tilde{E}_n^-|^2 \text{Real}\{k_{y,1,n}\}}{|E_0|^2 k_{y1,0}}. \tag{4.435}$$

Similarly, for the photonic crystal

$$T_n = \frac{|Ae^{ik_{y1,0}y_{\max}}\delta_{n,0} + \tilde{E}_n^+|^2 \mathrm{Real}\{k_{y1,n}\}}{|E_0|^2 k_{y1,0}}, \tag{4.436}$$

and for the groove array

$$T_n = \frac{|\tilde{E}_n^-|^2 \mathrm{Real}\{k_{y2,n}\}}{|E_0|^2 k_{y1,0}}. \tag{4.437}$$

4.3.7 MULTILAYER PERIODIC STRUCTURES

In this subsection the GFSIEMs will be considered for multilayer periodic structures. It is possible to obtain relations between the fields on the first and last interfaces, such that the numerical problem can be cast in a form which is no larger than that for a single interface.

A general example of an N-layer periodic structure is shown in Fig. 4.59. The structure is divided into N regions $\Omega_1, \Omega_2, \ldots \Omega_N$ with dielectric constants $\varepsilon_1, \varepsilon_2 \ldots$, ε_N, respectively. At an infinitesimal distance on each side of a physical boundary is placed a surface. For example, the surface $C_{1,2}$ is placed in the region Ω_1 at an infinitesimal distance from the physical boundary between regions Ω_1 and Ω_2. The surface $C_{2,1}$ is placed in Ω_2 at an infinitesimal distance from the same physical boundary. Similar surfaces $C_{2,3}$ and $C_{3,2}$ are placed on either side of the physical boundary between Ω_2 and Ω_3 etc. The normal vector at the boundary between Ω_1 and Ω_2 pointing into Ω_1 is referred to as $\hat{\mathbf{n}}_1$. The normal vector pointing into Ω_2 at the boundary between Ω_2 and Ω_3 is referred to as $\hat{\mathbf{n}}_2$, etc. The regions are ordered such that Ω_1 is placed above Ω_2, which is again placed above Ω_3 and so on.

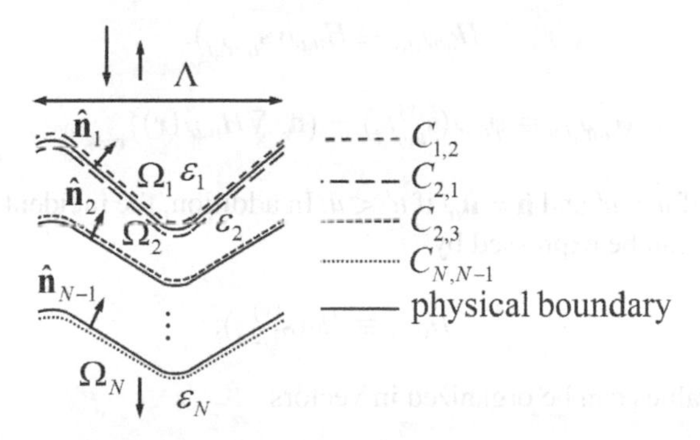

Figure 4.59 Schematic of a periodic N-layer structure with definitions of normal vectors for each surface and surfaces placed infinitesimally on either side of physical boundaries.

The following integral equations apply to an N-layer geometry, where one equation is formulated for each of the N regions in terms of the surfaces and normal vectors described in Fig. 4.59:

$$H(\mathbf{r}) = H_0(\mathbf{r}) - \oint_{C_{1,2}} \{g_1(\mathbf{r},\mathbf{r}')\phi_{1,2}(\mathbf{r}') - H_{1,2}(\mathbf{r}')\hat{\mathbf{n}}_1' \cdot \nabla' g_1(\mathbf{r},\mathbf{r}')\} dl', \quad \mathbf{r} \in \Omega_1, \tag{4.438}$$

$$H(\mathbf{r}) = + \oint_{C_{u,u-1}} \{g_u(\mathbf{r},\mathbf{r}')\phi_{u,u-1}(\mathbf{r}') - H_{u,u-1}(\mathbf{r}')\hat{\mathbf{n}}_{u-1}' \cdot \nabla' g_u(\mathbf{r},\mathbf{r}')\} dl' -$$

$$\oint_{C_{u,u+1}} \{g_u(\mathbf{r},\mathbf{r}')\phi_{u,u+1}(\mathbf{r}') - H_{u,u+1}(\mathbf{r}')\hat{\mathbf{n}}_u' \cdot \nabla' g_u(\mathbf{r},\mathbf{r}')\} dl', \quad \mathbf{r} \in \Omega_u, \quad u = 1, \ldots, N-1 \tag{4.439}$$

$$H(\mathbf{r}) = + \oint_{C_{N,N-1}} \{g_N(\mathbf{r},\mathbf{r}')\phi_{N,N-1}(\mathbf{r}') - H_{N,N-1}(\mathbf{r}')\hat{\mathbf{n}}_{N-1}' \cdot \nabla' g_N(\mathbf{r},\mathbf{r}')\} dl', \quad \mathbf{r} \in \Omega_N. \tag{4.440}$$

In these equations, $H_{u,u'}$ refers to the magnetic field on $C_{u,u'}$ while $\phi_{u,u'}$ is the normal derivative of the magnetic field on $C_{u,u'}$ corresponding to the normal vector $\hat{\mathbf{n}}_u$ if $u < u'$ and $\hat{\mathbf{n}}_{u'}$ if $u' < u$. H_0 is, as usual, the incident field in Ω_1.

We are now going to express the integral equations in matrix form by using the polynomial expansion of fields with order m similar to previous sections. The position on each of the surfaces $C_{u,u'}$ is described by the parametrization $\mathbf{s}_{u,u'}(t)$ with $0 \leq t \leq L_{u,u'}$ such that $L_{u,u'}$ is the total length of surface $C_{u,u'}$ within one period of the structure along x, and t is the distance along the surface. The surface $C_{u,u'}$ is divided into $N_{u,u'}$ sections, and each section is assigned $m+1$ sampling points with $\mathbf{s}_{u,u',i}^{(0)}$ representing the start of section i and $\mathbf{s}_{u,u',i}^{(m)}$ representing the end of section i, and $\mathbf{s}_{u,u',i}^{(v)}$ for $v = 1, \ldots, m-1$ representing sampling points that are in-between.

The total fields at sampling points can be expressed as

$$H_{u,u',i,v} \equiv H_{u,u'}(\mathbf{s}_{u,u',i}^{(v)}), \tag{4.441}$$

$$\phi_{u,u',i,v} \equiv \phi_{u,u'}(\mathbf{s}_{u,u',i}^{(v)}) = \left(\hat{\mathbf{n}} \cdot \nabla H_{u,u'}(\mathbf{r})\right)_{\mathbf{r}=\mathbf{s}_{u,u',i}^{(v)}}, \tag{4.442}$$

where $\hat{\mathbf{n}} = \hat{\mathbf{n}}_u$ if $u < u'$ and $\hat{\mathbf{n}} = \hat{\mathbf{n}}_{u'}$ if $u' < u$. In addition, the incident field at sampling points on $C_{1,2}$ can be expressed by

$$H_{0,i,v} \equiv H_0(\mathbf{s}_{1,2,i}^{(v)}). \tag{4.443}$$

The field values can be organized in vectors

$$\overline{H}_{u,u'}^{(v)} = \left[H_{u,u',1,v} \quad H_{u,u',2,v} \quad \ldots \quad H_{u,u',N_{u,u'},v}\right]^T, \tag{4.444}$$

$$\overline{\phi}_{u,u'}^{(v)} = \left[\phi_{u,u',1,v} \quad \phi_{u,u',2,v} \quad \ldots \quad \phi_{u,u',N_{u,u'},v}\right]^T, \tag{4.445}$$

$$\overline{H}_0^{(v)} = \left[H_{0,1,v} \quad H_{0,2,v} \quad \ldots \quad H_{0,N_{1,2},v}\right]^T. \tag{4.446}$$

It will be convenient to further build total vectors for each surface

$$\overline{H}_{u,u'} = \left[\left(\overline{H}^{(0)}_{u,u'}\right)^T \; \left(\overline{H}^{(1)}_{u,u'}\right)^T \; \dots \; \left(\overline{H}^{(m-1)}_{u,u'}\right)^T \right]^T, \tag{4.447}$$

$$\overline{\phi}_{u,u'} = \left[\left(\overline{\phi}^{(0)}_{u,u'}\right)^T \; \left(\overline{\phi}^{(1)}_{u,u'}\right)^T \; \dots \; \left(\overline{\phi}^{(m-1)}_{u,u'}\right)^T \right]^T, \tag{4.448}$$

$$\overline{H}_{0} = \left[\left(\overline{H}^{(0)}_{0}\right)^T \; \left(\overline{H}^{(1)}_{0}\right)^T \; \dots \; \left(\overline{H}^{(m-1)}_{0}\right)^T \right]^T. \tag{4.449}$$

These may be supplemented by boundary conditions for each surface on the form

$$\overline{H}^{(m)}_{u,u'} = \overline{\overline{D}}_{u,u',H}\overline{H}^{(0)}_{u,u'}. \tag{4.450}$$

$$\overline{\phi}^{(m)}_{u,u'} = \overline{\overline{D}}_{u,u',\phi}\overline{\phi}^{(0)}_{u,u'}. \tag{4.451}$$

These boundary conditions on matrix form must simply describe that $H_{u,u'}$ is continuous on $C_{u,u'}$ and also satisfies the periodic boundary condition. For a smooth surface with no sharp corners the same applies for $\phi_{u,u'}$.

Now, if we let the position $\mathbf{r}$ approach the sampling points on the surfaces $C_{1,2}$ and $C_{N,N-1}$ the integral equations (4.438) and (4.440) can be formulated as

$$\overline{H}^{(v)}_{1,2} + \sum_{v'=0}^{m} \left\{ \overline{\overline{A}}^{(1,2,v,v')}\overline{\phi}^{(v')}_{1,2} - \overline{\overline{B}}^{(1,2,v,v')}\overline{H}^{(v')}_{1,2} \right\} = \overline{H}^{(v)}_{0}, \quad v = 0,1,\dots,m-1, \tag{4.452}$$

$$\overline{H}^{(v)}_{N,N-1} - \sum_{v'=0}^{m} \left\{ \overline{\overline{A}}^{(N,N-1,v,v')}\overline{\phi}^{(v')}_{N,N-1} - \overline{\overline{B}}^{(N,N-1,v,v')}\overline{H}^{(v')}_{N,N-1} \right\} = \overline{0}, \quad v = 0,1,\dots,m-1, \tag{4.453}$$

where

$$A^{(u,u',v,v')}_{ij} = P\int_t g_u\left(\mathbf{s}^{(v)}_{u,u',i}, \mathbf{s}_{u,u'}(t)\right) f^{(m,v')}_{u,u',j}(t)dt, \tag{4.454}$$

$$B^{(u,u',v,v')}_{ij} = P\int_t \left(\hat{\mathbf{n}}' \cdot \nabla' g_u(\mathbf{s}^{(v)}_{u,u',i}, \mathbf{r}')\right)_{\mathbf{r}'=\mathbf{s}_{u,u'}(t)} f^{(m,v')}_{u,u',j}(t)dt$$
$$+\delta_{i,j}\delta_{v,v'}\frac{1}{2}\left(\delta_{u,u'-1} - \delta_{u,u'+1}\right). \tag{4.455}$$

Again, $\hat{\mathbf{n}} = \hat{\mathbf{n}}_u$ if $u < u'$ and $\hat{\mathbf{n}} = \hat{\mathbf{n}}_{u'}$ if $u' < u$. The functions $f^{(m,v)}_{u,u',i}$ are given by

$$f^{(m,v)}_{u,u',i}(t) = \begin{cases} f^{(m,v)}\left(\frac{t - t^{(s,i)}_{u,u'}}{t^{(e,i)}_{u,u'} - t^{(s,i)}_{u,u'}}\right), & t^{(s,i)}_{u,u'} < t < t^{(e,i)}_{u,u'} \\ 0, & \text{otherwise} \end{cases}, \tag{4.456}$$

where the general polynomial expansion functions $f^{(m,v)}$ are defined in Eq. (4.32) with $x_0 = 0$ and $x_m = 1$. The position along the surface $C_{u,u'}$ referred to as $t^{(s,i)}_{u,u'}$ represents the start of section i on $C_{u,u'}$, i.e., $\mathbf{s}^{(0)}_{u,u',i} = \mathbf{s}(t^{(s,i)}_{u,u'})$, and $t^{(e,i)}_{u,u'}$ represents the end of section i on $C_{u,u'}$, i.e., $\mathbf{s}^{(m)}_{u,u',i} = \mathbf{s}(t^{(e,i)}_{u,u'})$. In the last half of the expression (4.455) it has been assumed for simplicity that the surface is smooth with no sharp corners.

Note that for the expressions (4.452) and (4.453) there is one matrix equation for each v. By collecting these equations and applying the boundary conditions (4.450) and (4.451), grand matrix equations can be constructed of the form

$$\left(\overline{\overline{I}} - \overline{\overline{B}}^{(1,2)}\right)\overline{H}_{1,2} + \overline{\overline{A}}^{(1,2)}\overline{\phi}_{1,2} = \overline{H}_0, \tag{4.457}$$

$$\left(\overline{\overline{I}} + \overline{\overline{B}}^{(N,N-1)}\right)\overline{H}_{N,N-1} - \overline{\overline{A}}^{(N,N-1)}\overline{\phi}_{N,N-1} = \overline{0}, \tag{4.458}$$

where the matrices $\overline{\overline{A}}^{(u,u')}$ and $\overline{\overline{B}}^{(u,u')}$ follow the same structure as the matrices in Eqs. (4.94) and (4.95).

We now turn to consider the integral equation for the region Ω_2 given in (4.439). In this case the position $\mathbf{r}$ may approach infinitesimally either the sampling points on the surface $C_{2,1}$ or on the surface $C_{2,3}$. This leads to the following two matrix equations

$$\overline{H}^{(v)}_{2,1} - \sum_{v'=0}^{m}\left\{\overline{\overline{A}}^{(2,1,v,v')}\overline{\phi}^{(v')}_{2,1} - \overline{\overline{B}}^{(2,1,v,v')}\overline{H}^{(v')}_{2,1}\right\} +$$
$$\sum_{v'=0}^{m}\left\{\overline{\overline{A}}^{(2,3)\to(2,1),(v,v')}\overline{\phi}^{(v')}_{2,3} - \overline{\overline{B}}^{(2,3)\to(2,1),(v,v')}\overline{H}^{(v')}_{2,3}\right\} = \overline{0},\ \ v = 0,1,\ldots,m-1, \tag{4.459}$$

$$\overline{H}^{(v)}_{2,3} + \sum_{v'=0}^{m}\left\{\overline{\overline{A}}^{(2,3,v,v')}\overline{\phi}^{(v')}_{2,3} - \overline{\overline{B}}^{(2,3,v,v')}\overline{H}^{(v')}_{2,3}\right\} -$$
$$\sum_{v'=0}^{m}\left\{\overline{\overline{A}}^{(2,1)\to(2,3),(v,v')}\overline{\phi}^{(v')}_{2,1} - \overline{\overline{B}}^{(2,1)\to(2,3),(v,v')}\overline{H}^{(v')}_{2,1}\right\} = \overline{0},\ \ v = 0,1,\ldots,m-1. \tag{4.460}$$

The matrices that couple a surface to itself have already been described above in (4.454) and (4.455). The other matrices $\overline{\overline{A}}^{(u,u')\to(u,u''),(v,v')}$ and $\overline{\overline{B}}^{(u,u')\to(u,u''),(v,v')}$ describe the coupling between the two different surfaces $C_{u,u'}$ and $C_{u,u''}$. These matrices have the matrix elements

$$A^{(u,u')\to(u,u''),(v,v')}_{ij} = \int_t g_u\left(\mathbf{s}^{(v)}_{u,u'',i}, \mathbf{s}_{u,u'}(t)\right) f^{(m,v')}_{u,u',j}(t)dt, \tag{4.461}$$

$$B^{(u,u')\to(u,u''),(v,v')}_{ij} = \int_t \left(\hat{\mathbf{n}}'\cdot\nabla' g_u(\mathbf{s}^{(v)}_{u,u'',i}, \mathbf{r}')\right)_{\mathbf{r}'=\mathbf{s}_{u,u'}(t)} f^{(m,v')}_{u,u',j}(t)dt, \tag{4.462}$$

and $\hat{\mathbf{n}} = \hat{\mathbf{n}}_{u'}$ if $u' < u''$ and $\hat{\mathbf{n}} = \hat{\mathbf{n}}_u$ if $u'' < u'$.

By applying boundary conditions (4.450) and (4.451) and collecting equations for each v into grand matrix equations we obtain matrix equations of the form

$$\left(\overline{\overline{I}} + \overline{\overline{B}}^{(2,1)}\right)\overline{H}_{2,1} - \overline{\overline{A}}^{(2,1)}\overline{\phi}_{2,1} - \overline{\overline{B}}^{(2,3)\to(2,1)}\overline{H}_{2,3} + \overline{\overline{A}}^{(2,3)\to(2,1)}\overline{\phi}_{2,3} = \overline{0}, \tag{4.463}$$

$$\overline{\overline{B}}^{(2,1)\to(2,3)}\overline{H}_{2,1} - \overline{\overline{A}}^{(2,1)\to(2,3)}\overline{\phi}_{2,1} + \left(\overline{\overline{I}} - \overline{\overline{B}}^{(2,3)}\right)\overline{H}_{2,3} + \overline{\overline{A}}^{(2,3)}\overline{\phi}_{2,3} = \overline{0}. \tag{4.464}$$

Consider a situation where the number of layers N is large. Then it will be highly inefficient to just combine all the equations into a grand matrix equation due to the large number of unknowns. Instead we shall now seek to write the equations in a form which is no larger than that of having just a single interface corresponding to 2 layers even if N is a large number. We are going to construct a matrix relation that couples the fields (H and ϕ) on the very first interface $C_{1,2}$ with the corresponding fields on the very last interface $C_{N,N-1}$. Those fields will thus be coupled through a somewhat complicated boundary condition in much the same way as the fields on $C_{1,2}$ and $C_{2,1}$ are coupled. Such a relation can be constructed from relations for how the fields are coupled across individual interfaces and across individual layers.

The matrix equations that govern how fields are coupled across interfaces are given from the electromagnetics boundary conditions for how H and ϕ on one and the other side of a physical boundary are related:

$$\overline{H}_{i,i+1} = \overline{H}_{i+1,i}, \tag{4.465}$$

$$\frac{1}{\varepsilon_i}\overline{\phi}_{i,i+1} = \frac{1}{\varepsilon_{i+1}}\overline{\phi}_{i+1,i}. \tag{4.466}$$

These relations can be organized in matrix form as

$$\begin{bmatrix} \overline{H}_{i,i+1} \\ \overline{\phi}_{i,i+1} \end{bmatrix} = \overline{\overline{T}}_{i,i+1} \begin{bmatrix} \overline{H}_{i+1,i} \\ \overline{\phi}_{i+1,i} \end{bmatrix} = \begin{bmatrix} \overline{\overline{I}} & \overline{\overline{0}} \\ \overline{\overline{0}} & \frac{\varepsilon_i}{\varepsilon_{i+1}}\overline{\overline{I}} \end{bmatrix} \begin{bmatrix} \overline{H}_{i+1,i} \\ \overline{\phi}_{i+1,i} \end{bmatrix}. \tag{4.467}$$

The matrix $\overline{\overline{T}}_{i,i+1}$ thus relates H and ϕ on $C_{i,i+1}$ to H and ϕ on $C_{i+1,i}$.

Now consider again the equations (4.463) and (4.464) but re-arranged as a relation that gives H and ϕ on $C_{2,1}$ in terms of H and ϕ on $C_{2,3}$:

$$\begin{bmatrix} \overline{H}_{2,1} \\ \overline{\phi}_{2,1} \end{bmatrix} = \overline{\overline{T}}_2 \begin{bmatrix} \overline{H}_{2,3} \\ \overline{\phi}_{2,3} \end{bmatrix}, \tag{4.468}$$

with

$$\overline{\overline{T}}_2 = \begin{bmatrix} \left(\overline{\overline{I}} + \overline{\overline{B}}^{(2,1)}\right) & -\overline{\overline{A}}^{(2,1)} \\ \overline{\overline{B}}^{(2,1)\to(2,3)} & -\overline{\overline{A}}^{(2,1)\to(2,3)} \end{bmatrix}^{-1} \begin{bmatrix} \overline{\overline{B}}^{(2,3)\to(2,1)} & -\overline{\overline{A}}^{(2,3)\to(2,1)} \\ -\left(\overline{\overline{I}} - \overline{\overline{B}}^{(2,3)}\right) & -\overline{\overline{A}}^{(2,3)} \end{bmatrix}. \tag{4.469}$$

Now, for an N-layer structure it follows that

$$\begin{bmatrix} \overline{H}_{1,2} \\ \overline{\phi}_{1,2} \end{bmatrix} = \overline{\overline{T}}_{1,2}\overline{\overline{T}}_2\overline{\overline{T}}_{2,3}\ldots\overline{\overline{T}}_{N-1}\overline{\overline{T}}_{N-1,N} \begin{bmatrix} \overline{H}_{N,N-1} \\ \overline{\phi}_{N,N-1} \end{bmatrix} = \begin{bmatrix} \overline{\overline{S}}_{1,1} & \overline{\overline{S}}_{1,2} \\ \overline{\overline{S}}_{2,1} & \overline{\overline{S}}_{2,2} \end{bmatrix} \begin{bmatrix} \overline{H}_{N,N-1} \\ \overline{\phi}_{N,N-1} \end{bmatrix}. \tag{4.470}$$

This is now the boundary condition that relates the fields at $C_{1,2}$ and $C_{N,N-1}$. By direct inversion this can also be formulated as

$$\begin{bmatrix} \overline{H}_{N,N-1} \\ \overline{\phi}_{N,N-1} \end{bmatrix} = \begin{bmatrix} \overline{\overline{C}}_{1,1} & \overline{\overline{C}}_{1,2} \\ \overline{\overline{C}}_{2,1} & \overline{\overline{C}}_{2,2} \end{bmatrix} \begin{bmatrix} \overline{H}_{1,2} \\ \overline{\phi}_{1,2} \end{bmatrix}, \tag{4.471}$$

where

$$\begin{bmatrix} \overline{\overline{C}}_{1,1} & \overline{\overline{C}}_{1,2} \\ \overline{\overline{C}}_{2,1} & \overline{\overline{C}}_{2,2} \end{bmatrix} = \begin{bmatrix} \overline{\overline{S}}_{1,1} & \overline{\overline{S}}_{1,2} \\ \overline{\overline{S}}_{2,1} & \overline{\overline{S}}_{2,2} \end{bmatrix}^{-1}. \tag{4.472}$$

This boundary condition can then be combined with the equations (4.457) and (4.458). This leads finally to a matrix equation of the same size as for a single interface, namely

$$\begin{bmatrix} \overline{\overline{M}}_{11} & \overline{\overline{M}}_{12} \\ \overline{\overline{M}}_{21} & \overline{\overline{M}}_{22} \end{bmatrix} \begin{bmatrix} \overline{H}_{1,2} \\ \overline{\phi}_{1,2} \end{bmatrix} = \begin{bmatrix} \overline{H}_0 \\ \overline{0} \end{bmatrix}, \tag{4.473}$$

where

$$\overline{\overline{M}}_{11} = \overline{\overline{I}} - \overline{\overline{B}}^{(1,2)}, \tag{4.474}$$

$$\overline{\overline{M}}_{12} = \overline{\overline{A}}^{(1,2)}, \tag{4.475}$$

$$\overline{\overline{M}}_{21} = \left(\overline{\overline{I}} + \overline{\overline{B}}^{(N,N-1)}\right)\overline{\overline{C}}_{11} - \overline{\overline{A}}^{(N,N-1)}\overline{\overline{C}}_{21}, \tag{4.476}$$

$$\overline{\overline{M}}_{22} = \left(\overline{\overline{I}} + \overline{\overline{B}}^{(N,N-1)}\right)\overline{\overline{C}}_{12} - \overline{\overline{A}}^{(N,N-1)}\overline{\overline{C}}_{22}. \tag{4.477}$$

Once the equation (4.473) has been solved and the fields $H_{1,2}$ and $\phi_{1,2}$ on $C_{1,2}$ have been determined, then the corresponding fields on other surfaces can be straightforwardly obtained via the inverse of the matrices $\overline{\overline{T}}_{(u)}$ and $\overline{\overline{T}}_{(u,u')}$.

A few final words are in order for this subsection. Consider a situation where, e.g., Ω_{N-1} is a metal layer which is so thick that it is not penetrable, and there is effectively no coupling between the fields on $C_{N-1,N-2}$ and $C_{N-1,N}$. In that case the matrices that describe coupling between surfaces $C_{N-1,N-2}$ and $C_{N-1,N}$ will vanish. The required inversion of certain matrices is thus no longer numerically feasible. A straightforward solution then is to simply treat the geometry as a layered structure with $N-1$ layers. On the other hand, if the layer Ω_{N-1} is less thick so that there is a non-negligible coupling between the two surfaces then there is no problem. Also, it is not a problem if the last medium Ω_N is not penetrable as this is the last medium and there are no coupling matrices to any surfaces farther down. An iterative method where one layer is added at a time, and which will not become numerically unstable if there are thick impenetrable layers, will be described in Section 4.3.8.

4.3.8 TRANSFER-MATRIX METHOD FOR LARGE STRUCTURES

In this section we will consider a transfer-matrix method that can be used for large structures with multiple scatterers. It will be assumed that the structures are periodic along x with period Λ, and along y they can be naturally divided into a number of regions. The idea now is to construct transfer matrices for each region based on the GFSIEM, such that an entire structure consisting of many regions can be modeled by constructing a transfer matrix for the entire structure based on transfer matrices for the individual regions. In Sec. 4.3.7 a transfer-matrix method was considered for multilayer structures. In that case the surfaces of a multilayer structure were the natural interfaces between otherwise unstructured regions. For the situation considered in this section there is no physical surface between regions but instead there is a scattering structure inside each region. In addition, in Sec. 4.3.7 the transfer matrices were based on connecting the field and its normal derivative on different interfaces. In this section the transfer matrices will instead connect upward and downward propagating field components on one and the other side of a region. This has the advantage that as more regions are added to a structure, the total transfer matrix can be built up iteratively by considering the behavior of forward propagating fields being multiply reflected. In the case of evanescent fields that vanish at the other end of a region this will not pose any problems, while some care had to be exercised in such cases with the method presented in Sec. 4.3.7.

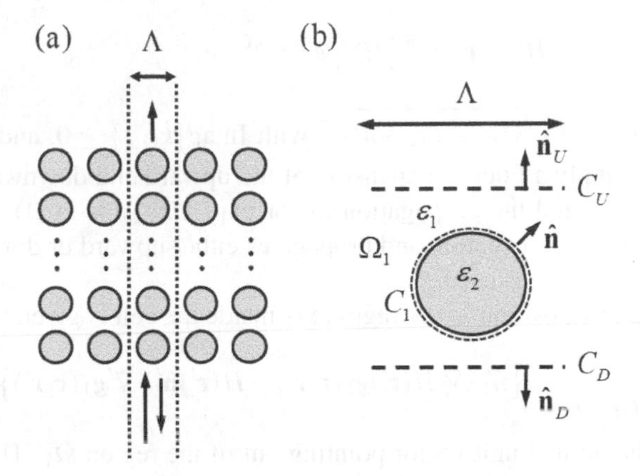

Figure 4.60 (a) Schematic of a photonic crystal with periodicity along x and a finite but large length along y. (b) Schematic of one cell along y of the photonic crystal with cell surfaces and normal vectors included for the following discussion.

Note that as far as the iterative approach is concerned, the method presented in this section uses similar principles as the Fourier-Modal-Method (FMM) [8, 9, 10, 11]. However, with the FMM, the structure would always be approximated with a stair-cased geometry, and especially round structures would require many matrices. Here, matrices will be constructed using the GFSIEM, which easily handles an entire scatterer with a round surface, and the scatterer will thus not be approximated with a stair-cased structure.

An example of a periodic structure that is naturally divided into regions along y is the photonic crystal structure shown in Fig. 4.60(a). The structure has a number of rods along y, and it is thus natural to divide the structure into regions along y that each contain one rod (within one period along x) as illustrated in Fig. 4.60(b).

The structure can in principle be modeled using the GFSIEM for periodic structures presented in Section 4.3. However, for large structures with many rods along y it should be noted that in the case of N rods along y, the memory requirements will scale as N^2, the time for constructing the matrix equation will scale as N^2, and the time for matrix inversion will scale as N^3. For a large N this becomes prohibitive both in terms of calculation time and required computer memory. The transfer-matrix method being presented in this section will greatly reduce the calculation time and memory requirements for such a structure.

We will consider again the example of one cell of a multi-region structure shown in Fig. 4.60(b). The dashed lines C_U and C_D represent the cell boundaries that are parallel with the x-axis and defined by $y = y_U$ or $y = y_D$.

For a Bloch wave with a specific k_x the field at the upper ($A = U$) and lower ($A = D$) interfaces of the cell can be expressed in the form

$$H_A(\mathbf{r}) = H_A^{(+)}(\mathbf{r}) + H_A^{(-)}(\mathbf{r}), \tag{4.478}$$

where

$$H_A^{(\pm)}(\mathbf{r}) = \sum_n \tilde{H}_{A,n}^{(\pm)} e^{i(k_x+nG)x} e^{\pm ik_{y1,n}(y-y_A)}, \tag{4.479}$$

and as usual $k_{y1,n} = \sqrt{k_0^2\varepsilon_1 - (k_x + nG)^2}$ with $\text{Imag}\{k_{y1,n}\} \geq 0$, and $G = 2\pi/\Lambda$. At $y = y_A$ this is simply a Fourier expansion of the upward and downward propagating field components, and the propagation factor $\exp(\pm ik_{y1,n}(y - y_A))$ ensures that the field satisfies the wave equation and propagates either upward or downward depending on whether $+$ or $-$ is used.

The field at any position in the region Ω_1 inside the cell is given by the integral

$$H(\mathbf{r}) = \int_{C_U+C_D+C_1} \{\hat{\mathbf{n}}_1' \cdot \nabla' H(\mathbf{r}') g_1(\mathbf{r},\mathbf{r}') - H(\mathbf{r}')\hat{\mathbf{n}}_1' \cdot \nabla' g_1(\mathbf{r},\mathbf{r}')\} \, dl', \tag{4.480}$$

where $\hat{\mathbf{n}}_1$ is the normal unit vector pointing out of the region Ω_1. Thus $\hat{\mathbf{n}}_1 = -\hat{\mathbf{n}}$ on the outer scatterer surface C_1, and it is equal to $\hat{\mathbf{n}}_U$ or $\hat{\mathbf{n}}_D$ on the surfaces C_U and C_D, respectively.

It can be shown by straightforward application of the periodic Green's function (4.380) that the part of the integral over the surfaces C_U and C_D reduces to the fields propagating into the cell

$$H_0(\mathbf{r}) \equiv H_D^{(+)}(\mathbf{r}) + H_U^{(-)}(\mathbf{r}). \tag{4.481}$$

With this definition we find the usual form of the integral equation

$$H(\mathbf{r}) = H_0(\mathbf{r}) - \oint_{C_1} \{\phi(\mathbf{r}') g_1(\mathbf{r},\mathbf{r}') - H(\mathbf{r}')\hat{\mathbf{n}}' \cdot \nabla' g_1(\mathbf{r},\mathbf{r}')\} \, dl', \tag{4.482}$$

where here $\hat{\mathbf{n}}$ is as usual the normal unit vector pointing out of the scatterer and thus into region Ω_1.

We will consider again the GFSIEM with a polynomial expansion of fields on C_1 with polynomial order m. The incident field $H_0(\mathbf{r})$ must then be determined at the sampling points on the scatterer surface C_1. H_0 must thus be determined at the positions $\mathbf{s}_i^{(v)}$, where i refers to a section on C_1, and v refers to one of m sampling points on that section. Clearly,

$$H_{0,i}^{(v)} \equiv H_0(\mathbf{s}_i^{(v)}) = \sum_n \tilde{H}_{D,n}^{(+)} e^{i(k_x+nG)x_i^{(v)}} e^{+ik_{y1,n}(y_i^{(v)} - y_D)}$$

$$+\sum_n \tilde{H}_{U,n}^{(-)} e^{i(k_x+nG)x_i^{(v)}} e^{-ik_{y1,n}(y_i^{(v)}-y_U)}. \tag{4.483}$$

The coefficients $H_{0,i}^{(v)}$ are now as usual collected in a vector

$$\overline{H}_0^{(v)} = \left[H_{0,1}^{(v)} \quad H_{0,2}^{(v)} \quad \dots \quad H_{0,N}^{(v)} \right]^T, \tag{4.484}$$

and a grand vector is constructed for the incident field as

$$\overline{H}_0 = \left[\left(\overline{H}_0^{(0)}\right)^T \quad \left(\overline{H}_0^{(1)}\right)^T \quad \dots \quad \left(\overline{H}_0^{(m-1)}\right)^T \right]^T. \tag{4.485}$$

The coefficients $\tilde{H}_{D,n}^{(+)}$ and $\tilde{H}_{U,n}^{(-)}$ are also collected in vectors $\overline{H}_D^{(+)}$ and $\overline{H}_U^{(-)}$, respectively. The vectors must be truncated such that only a finite range of n are considered, e.g. $-N_A \leq n \leq +N_A$. The equation (4.483) can now be expressed in the matrix form

$$\overline{H}_0 = \overline{\overline{C}}^{(U)} \overline{H}_U^{(-)} + \overline{\overline{C}}^{(D)} \overline{H}_D^{(+)}, \tag{4.486}$$

where the elements of the matrices $\overline{\overline{C}}^{(U)}$ and $\overline{\overline{C}}^{(D)}$ are chosen in accordance with Eq. (4.483).

The magnetic field at the same sampling points on the scatterer surface $H_i^{(v)}$ and its normal derivative $\phi_i^{(v)}$ are also organized as usual in vectors

$$\overline{H}^{(v)} = \left[H_1^{(v)} \quad H_2^{(v)} \quad \dots \quad H_N^{(v)} \right]^T, \tag{4.487}$$

$$\overline{\phi}^{(v)} = \left[\phi_1^{(v)} \quad \phi_2^{(v)} \quad \dots \quad \phi_N^{(v)} \right]^T, \tag{4.488}$$

and grand vectors are formed

$$\overline{H} = \left[\left(\overline{H}^{(0)}\right)^T \quad \left(\overline{H}^{(1)}\right)^T \quad \dots \quad \left(\overline{H}^{(m-1)}\right)^T \right]^T, \tag{4.489}$$

$$\overline{\phi} = \left[\left(\overline{\phi}^{(0)}\right)^T \quad \left(\overline{\phi}^{(1)}\right)^T \quad \dots \quad \left(\overline{\phi}^{(m-1)}\right)^T \right]^T. \tag{4.490}$$

The total field and its normal derivative at the scatterer surface C_1 can now be obtained in the usual way from a matrix relation on the form

$$\begin{bmatrix} \overline{H} \\ \overline{\phi} \end{bmatrix} = \overline{\overline{M}}_S \overline{H}_0, \tag{4.491}$$

where the matrix $\overline{\overline{M}}_S$ will be similar to the left-hand part of the inverse of the matrix in (4.96), except that matrix elements should be calculated using the periodic Green's functions. This is entirely similar to Section 4.3, where $\overline{H}$ and $\overline{\phi}$ are also obtained from $\overline{H}_0$ through a matrix relation.

After obtaining H and ϕ on the scatterer surface, it now follows from Eq. (4.482) that for positions $\mathbf{r}$ on C_U the total field is given by

$$H(\mathbf{r}) = H_U^{(-)}(\mathbf{r}) + H_D^{(+)}(\mathbf{r}) - \frac{i}{2\Lambda}\sum_n e^{i(k_x+nG)x} e^{+ik_{y1,n}y}$$

$$\times \int_{C_1} \frac{e^{-i(k_x+nG)x'} e^{-ik_{y1,n}y'}}{k_{y1,n}} \left(\phi(\mathbf{r}') + i\hat{\mathbf{n}}' \cdot (\hat{\mathbf{x}}[k_x+nG] + \hat{\mathbf{y}}k_{y1,n})\right) dl'. \tag{4.492}$$

This follows from (4.482) by inserting the expression for the periodic Green's function g_1 given in Eq. (4.380) and using that $y - y' = |y - y'| > 0$ since C_U is placed above the scatterer.

At C_U the total field can also be written as

$$H(\mathbf{r}) = H_U^{(+)}(\mathbf{r}) + H_U^{(-)}(\mathbf{r}). \tag{4.493}$$

We thus find

$$H_U^{(+)}(\mathbf{r}) = H_D^{(+)}(\mathbf{r}) - \frac{i}{2\Lambda}\sum_n e^{i(k_x+nG)x} e^{ik_{y1,n}y} \int_t \frac{e^{-i(k_x+nG)x(t)} e^{-ik_{y1,n}y(t)}}{k_{y1,n}}$$

$$\times \sum_{i=1}^{N}\sum_{v=0}^{m} \left(\phi_i^{(v)} + i\hat{\mathbf{n}}(t) \cdot (\hat{\mathbf{x}}(k_x+nG) + \hat{\mathbf{y}}k_{y1,n}) H_i^{(v)}\right) f_i^{(v)}(t) dt, \tag{4.494}$$

where as usual $\mathbf{s}(t) = \hat{\mathbf{x}}x(t) + \hat{\mathbf{y}}y(t)$ describes the surface of C_1 with t representing the distance along C_1 from a starting point, and $\hat{\mathbf{n}}(t)$ is the corresponding surface normal vector.

Note that

$$H_U^{(+)}(\mathbf{r}) = \sum_n \tilde{H}_{U,n}^{(+)} e^{i(k_x+nG)x} e^{ik_{y1,n}(y-y_U)}, \tag{4.495}$$

and

$$H_D^{(+)}(\mathbf{r}) = \sum_n \tilde{H}_{D,n}^{(+)} e^{i(k_x+nG)x} e^{ik_{y1,n}(y-y_D)} = \sum_n \tilde{H}_{D,n}^{(+)} P_n e^{i(k_x+nG)x} e^{ik_{y1,n}(y-y_U)}, \tag{4.496}$$

where

$$P_n \equiv e^{ik_{y1,n}(y_U-y_D)}. \tag{4.497}$$

The scattered part of the field at C_U can be expressed as

$$H_{\mathrm{sc}}(\mathbf{r}) = \sum_n e^{i(k_x+nG)x} e^{ik_{y1,n}(y-y_U)} \sum_{i=1}^{N}\sum_{v=0}^{m} \left(\phi_i^{(v)} E_{n,i,v}^{(U)} + H_i^{(v)} F_{n,i,v}^{(U)}\right), \tag{4.498}$$

where

$$E_{n,i,v}^{(U)} \equiv -\frac{i}{2\Lambda}\int_t \frac{e^{-i(k_x+nG)x(t)} e^{ik_{y1,n}(y_U-y(t))}}{k_{y1,n}} f_i^{(v)}(t) dt, \tag{4.499}$$

$$F_{n,i,v}^{(U)} \equiv -\frac{i}{2\Lambda}\int_t \frac{e^{-i(k_x+nG)x(t)} e^{ik_{y1,n}(y_U-y(t))}}{k_{y1,n}} i\hat{\mathbf{n}}(t) \cdot (\hat{\mathbf{x}}(k_x+nG) + \hat{\mathbf{y}}k_{y1,n}) f_i^{(v)}(t) dt. \tag{4.500}$$

The expression (4.494) thus becomes

$$\tilde{H}_{U,n}^{(+)} = \tilde{H}_{D,n}^{(+)} P_n + \sum_{i=1}^{N} \sum_{v=0}^{m} \left\{ \phi_i^{(v)} E_{n,i,v}^{(U)} + H_i^{(v)} F_{n,i,v}^{(U)} \right\}. \tag{4.501}$$

This can also be expressed in matrix form as

$$\overline{H}_U^{(+)} = \overline{\overline{P}}\, \overline{H}_D^{(+)} + \sum_{v=0}^{m} \left\{ \overline{\overline{E}}^{(U,v)} \overline{\phi}^{(v)} + \overline{\overline{F}}^{(U,v)} \overline{H}^{(v)} \right\}, \tag{4.502}$$

where the matrices $\overline{\overline{P}}$, $\overline{\overline{E}}^{(U,v)}$, and $\overline{\overline{F}}^{(U,v)}$ are constructed in accordance with Eq. (4.501).

This can be further written as

$$\overline{H}_U^{(+)} = \overline{\overline{P}}\, \overline{H}_D^{(+)} + \overline{\overline{M}}_{EF}^{(U)} \begin{bmatrix} \overline{H} \\ \overline{\phi} \end{bmatrix}, \tag{4.503}$$

where

$$\overline{\overline{M}}_{EF}^{(A)} = \left[\overline{\overline{F}}^{(A)} \quad \overline{\overline{E}}^{(A)} \right], \tag{4.504}$$

with

$$\overline{\overline{F}}^{(A)} = \left[\left(\overline{\overline{F}}^{(A,0)} + \overline{\overline{F}}^{(A,m)} \overline{\overline{D}}_H \right) \quad \overline{\overline{F}}^{(A,1)} \quad \ldots \quad \overline{\overline{F}}^{(A,m-1)} \right], \tag{4.505}$$

$$\overline{\overline{E}}^{(A)} = \left[\left(\overline{\overline{E}}^{(A,0)} + \overline{\overline{E}}^{(A,m)} \overline{\overline{D}}_\phi \right) \quad \overline{\overline{E}}^{(A,1)} \quad \ldots \quad \overline{\overline{E}}^{(A,m-1)} \right]. \tag{4.506}$$

Again, A represents either U or D.

If we now combine Eqs. (4.503), (4.491) and (4.486) we find

$$\overline{H}_U^{(+)} = \overline{\overline{P}}\, \overline{H}_D^{(+)} + \overline{\overline{M}}_{EF}^{(U)} \overline{\overline{M}}_S \left[\overline{\overline{C}}^{(U)} \quad \overline{\overline{C}}^{(D)} \right] \begin{bmatrix} \overline{H}_U^{(-)} \\ \overline{H}_D^{(+)} \end{bmatrix}. \tag{4.507}$$

It can be shown in a similar way that

$$\tilde{H}_{D,n}^{(-)} = \tilde{H}_{U,n}^{(-)} P_n + \sum_{i=1}^{N} \sum_{v=0}^{m} \left\{ \phi_i^{(v)} E_{n,i,v}^{(D)} + H_i^{(v)} F_{n,i,v}^{(D)} \right\}, \tag{4.508}$$

where the matrix elements $E_{n,i,v}^{(D)}$ and $F_{n,i,v}^{(D)}$ are given by

$$E_{n,i,v}^{(D)} \equiv -\frac{i}{2\Lambda} \int_t \frac{e^{-i(k_x+nG)x(t)} e^{ik_{y1,n}(y(t)-y_D)}}{k_{y1,n}} f_i^{(v)}(t) dt, \tag{4.509}$$

$$F_{n,i,v}^{(D)} \equiv -\frac{i}{2\Lambda} \int_t \frac{e^{-i(k_x+nG)x(t)} e^{ik_{y1,n}(y(t)-y_D)}}{k_{y1,n}} i\hat{\mathbf{n}}(t) \cdot (\hat{\mathbf{x}}(k_x+nG) - \hat{\mathbf{y}} k_{y1,n}) f_i^{(v)}(t) dt. \tag{4.510}$$

This can also be expressed in matrix form as

$$\overline{H}_D^{(-)} = \overline{\overline{P}}\, \overline{H}_U^{(-)} + \sum_{v=0}^{m} \left\{ \overline{\overline{E}}^{(D,v)} \overline{\phi}^{(v)} + \overline{\overline{F}}^{(D,v)} \overline{H}^{(v)} \right\}, \tag{4.511}$$

where the matrices $\overline{\overline{P}}$, $\overline{\overline{D}}^{(U,\nu)}$, and $\overline{\overline{F}}^{(D,\nu)}$ are constructed in accordance with Eq. (4.511), and this can be further written as

$$\overline{H}_D^{(-)} = \overline{\overline{P}}\,\overline{H}_U^{(-)} + \overline{\overline{M}}_{EF}^{(D)} \begin{bmatrix} \overline{H} \\ \overline{\phi} \end{bmatrix}. \tag{4.512}$$

By combining (4.507) and (4.512), a scattering matrix relation is obtained

$$\begin{bmatrix} \overline{H}_U^{(+)} \\ \overline{H}_D^{(-)} \end{bmatrix} = \begin{bmatrix} \overline{\overline{R}}_U & \overline{\overline{T}}_{UD} \\ \overline{\overline{T}}_{DU} & \overline{\overline{R}}_D \end{bmatrix} \begin{bmatrix} \overline{H}_U^{(-)} \\ \overline{H}_D^{(+)} \end{bmatrix}, \tag{4.513}$$

where

$$\overline{\overline{R}}_U = \overline{\overline{M}}_{EF}^{(U)} \overline{\overline{M}}_S \overline{\overline{C}}^{(U)}, \tag{4.514}$$

$$\overline{\overline{T}}_{UD} = \left(\overline{\overline{P}} + \overline{\overline{M}}_{EF}^{(U)} \overline{\overline{M}}_S \overline{\overline{C}}^{(D)} \right), \tag{4.515}$$

$$\overline{\overline{T}}_{DU} = \left(\overline{\overline{P}} + \overline{\overline{M}}_{EF}^{(D)} \overline{\overline{M}}_S \overline{\overline{C}}^{(U)} \right), \tag{4.516}$$

$$\overline{\overline{R}}_D = \overline{\overline{M}}_{EF}^{(D)} \overline{\overline{M}}_S \overline{\overline{C}}^{(D)}. \tag{4.517}$$

The matrix $\overline{\overline{T}}_{UD}$ governs transmission from C_U to C_D, $\overline{\overline{R}}_U$ governs reflection of a wave being incident into the single layer at C_U, $\overline{\overline{T}}_{DU}$ governs transmission from C_D to C_U, and $\overline{\overline{R}}_D$ governs reflection of a wave being incident into the layer at C_D.

For further discussion we will label the transmission and reflection matrices for a single cell i as $\overline{\overline{T}}_{UD,i}$, $\overline{\overline{T}}_{DU,i}$, $\overline{\overline{R}}_{U,i}$, $\overline{\overline{R}}_{D,i}$. These matrices represent the total transmission and reflection matrices for a structure consisting of a single cell such as the one shown in Fig. 4.61(a). In the case of a single plane wave being incident from below

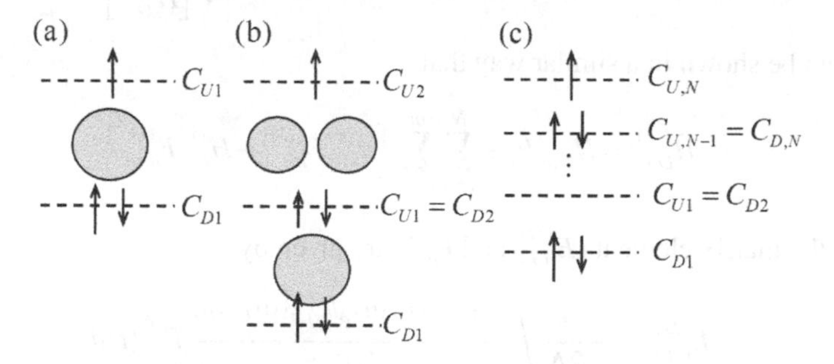

Figure 4.61 (a) Schematic of a structure consisting of a single cell with boundaries C_{D1} and C_{U1}. (b) Schematic of a structure consisting of two cells with boundaries C_{D1} and C_{U1} for cell 1, and boundaries C_{D2} and C_{U2} for cell 2. (c) Schematic of a structure consisting of N cells.

we may use

$$\tilde{H}_{D1,n}^{(+)} = \delta_{n,0} A, \tag{4.518}$$

where A is the amplitude of the plane wave. Thus, only a single element in the vector $\overline{H}_{D,1}^{(+)}$ is non-zero. In this situation with no incident waves from above, $\overline{H}_{U,1}^{(-)} = \overline{0}$. The Fourier coefficients governing transmitted and reflected fields for a 1-cell geometry can now be obtained as

$$\overline{H}_{U1}^{(+)} = \overline{\overline{T}}_{UD,1}\overline{H}_{D,1}^{(+)}, \tag{4.519}$$

$$\overline{H}_{D1}^{(-)} = \overline{\overline{R}}_{D,1}\overline{H}_{D,1}^{(+)}. \tag{4.520}$$

Clearly, $\overline{\overline{T}}_{UD,1}$ governs transmission from C_{D1} to C_{U1}, and $\overline{\overline{R}}_{D,1}$ governs the reflection of a wave propagating into the cell at C_{D1}.

In the case of a structure with two cells labeled 1 and 2 as illustrated in Fig. 4.61(b) we must take into account multiple reflections between the cells. It is illustrated in Fig. 4.61(b) that the two cells need not be the same. The transmission through the total structure from the interface C_{D1} to the interface C_{U2} is given by an infinite series describing all propagation paths through the structure

$$\overline{\overline{T}}_{U2,D1} = \overline{\overline{T}}_{UD,2}\overline{\overline{T}}_{UD,1} + \overline{\overline{T}}_{UD,2}\overline{\overline{R}}_{U,1}\overline{\overline{R}}_{D,2}\overline{\overline{T}}_{UD,1} + \overline{\overline{T}}_{UD,2}\left(\overline{\overline{R}}_{U,1}\overline{\overline{R}}_{D,2}\right)^2\overline{\overline{T}}_{UD,1} + \dots \tag{4.521}$$

By evaluating the series analytically we find

$$\overline{\overline{T}}_{U2,D1} = \overline{\overline{T}}_{UD,2}\left(\overline{\overline{I}} - \overline{\overline{R}}_{U,1}\overline{\overline{R}}_{D,2}\right)^{-1}\overline{\overline{T}}_{UD,1}. \tag{4.522}$$

The reflection coefficient for the total structure can be obtained in a similar way as

$$\overline{\overline{R}}_{U2,D1} = \overline{\overline{R}}_{D1} + \overline{\overline{T}}_{DU,1}\left(\overline{\overline{I}} - \overline{\overline{R}}_{D2}\overline{\overline{R}}_{U1}\right)^{-1}\overline{\overline{R}}_{D2}\overline{\overline{T}}_{UD,1}. \tag{4.523}$$

The transmission and reflection the other way through the two-cell structure will be given instead by

$$\overline{\overline{T}}_{D1,U2} = \overline{\overline{T}}_{DU,1}\left(\overline{\overline{I}} - \overline{\overline{R}}_{D,2}\overline{\overline{R}}_{U,1}\right)^{-1}\overline{\overline{T}}_{DU,2}. \tag{4.524}$$

$$\overline{\overline{R}}_{D1,U2} = \overline{\overline{R}}_{U2} + \overline{\overline{T}}_{UD,2}\left(\overline{\overline{I}} - \overline{\overline{R}}_{U1}\overline{\overline{R}}_{D2}\right)^{-1}\overline{\overline{R}}_{U1}\overline{\overline{T}}_{DU,2}. \tag{4.525}$$

For a symmetric structure

$$\overline{\overline{T}}_{D1,U2} = \overline{\overline{T}}_{U2,D1}, \quad \overline{\overline{R}}_{D1,U2} = \overline{\overline{R}}_{U2,D1}. \tag{4.526}$$

For the N layer geometry illustrated in Fig. 4.61(c), the reflection and transmission matrices can be obtained in terms of those for the $(N-1)$-layer geometry using again multiple reflections between layer N and now the stack of the $N-1$ other layers. Thus,

$$\overline{\overline{T}}_{UN,D1} = \overline{\overline{T}}_{UD,N}\left(\overline{\overline{I}} - \overline{\overline{R}}_{D1,UN-1}\overline{\overline{R}}_{D,N}\right)^{-1}\overline{\overline{T}}_{UN-1,D1}, \tag{4.527}$$

$$\overline{\overline{R}}_{UN,D1} = \overline{\overline{R}}_{UN-1,D1} + \overline{\overline{T}}_{D1,UN-1}\left(\overline{\overline{I}} - \overline{\overline{R}}_{D,N}\overline{\overline{R}}_{D1,UN-1}\right)^{-1}\overline{\overline{R}}_{DN}\overline{\overline{T}}_{UN-1,D1}, \tag{4.528}$$

$$\overline{\overline{T}}_{D1,UN} = \overline{\overline{T}}_{D1,UN-1}\left(\overline{\overline{I}} - \overline{\overline{R}}_{D,N}\overline{\overline{R}}_{D1,UN-1}\right)^{-1}\overline{\overline{T}}_{DU,N}, \tag{4.529}$$

$$\overline{\overline{R}}_{D1,UN} = \overline{\overline{R}}_{U,N} + \overline{\overline{T}}_{UD,N}\left(\overline{\overline{I}} - \overline{\overline{R}}_{D1,UN-1}\overline{\overline{R}}_{D,N}\right)^{-1}\overline{\overline{R}}_{D1,UN-1}\overline{\overline{T}}_{DU,N}. \tag{4.530}$$

The above expressions can now be applied iteratively for setting up transmission and reflection for geometries with many layers. In the case of identical layers, as is the case for the photonic crystal, it is sufficient to construct transmission and reflection matrices for a single layer and then use these iteratively to find reflection and transmission for any number of layers. Otherwise single-layer matrices must be calculated for those layers that are different, and then the same iterative approach as before applies for the construction of reflection and transmission for multi-layer geometries.

Transmission and reflection of a plane wave may be calculated by using

$$\overline{H}_{D1}^{(+)} = \begin{bmatrix} 0 & \dots & 0 & 1 & 0 & \dots & 0 \end{bmatrix}^T, \tag{4.531}$$

where all elements are 0 except for the element number $N_A + 1$. This is equivalent to using the incident field $H_D^{(+)}(\mathbf{r}) = e^{ik_x x}e^{ik_{y1,0}(y-y_D)}$. The Fourier coefficients of the transmitted and reflected fields for an N-layer structure are now obtained as

$$\overline{H}_{U,N}^{(+)} = \overline{\overline{T}}_{UN,D1}\overline{H}_{D,1}^{(+)}, \tag{4.532}$$

$$\overline{H}_{D1}^{(-)} = \overline{\overline{R}}_{UN,D1}\overline{H}_{D,1}^{(+)}. \tag{4.533}$$

The n-th-order transmittance is then given by

$$T_n = \frac{|\tilde{H}_{U,N,n}^{(+)}|^2 \mathrm{Real}\left\{k_{y1,n}\right\}}{k_{y1,0}}, \tag{4.534}$$

and the n-th-order reflectance is given by

$$R_n = \frac{|\tilde{H}_{D,1,n}^{(-)}|^2 \mathrm{Real}\left\{k_{y1,n}\right\}}{k_{y1,0}}. \tag{4.535}$$

Here, the dielectric constant on both sides of the layered structure is the same, and the amplitude of the incident plane wave is unity. Also note that here the Fourier coefficients $\tilde{H}_{U,N,n}^{(+)}$ and $\tilde{H}_{D,1,n}^{(-)}$ represent the total field and not just the scattered field.

We may also be interested in calculating the field inside each cell. The inside of a cell j can be treated as a single isolated structure in terms of the incident fields $\overline{H}_{D,j}^{(+)}$ and $\overline{H}_{U,j}^{(-)}$. For the first layer ($j = 1$) in a geometry with N layers, the first of those fields ($\overline{H}_{D,1}^{(+)}$) is known as simply the given incident field from below. Taking into account multiple reflections the incident field at C_{U1} is given by

$$\overline{H}_{U,1}^{(-)} = \sum_{n=0}^{\infty}\left(\overline{\overline{R}}_{UN,D2}\overline{\overline{R}}_{DU,1}\right)^n \overline{\overline{R}}_{UN,D2}\overline{\overline{T}}_{UD,1}\overline{H}_{D,1}^{(+)}$$

$$= \left(\overline{\overline{I}} - \overline{\overline{R}}_{UN,D2}\overline{\overline{R}}_{DU,1}\right)^{-1}\overline{\overline{R}}_{UN,D2}\overline{\overline{T}}_{UD,1}\overline{H}_{D,1}^{(+)}. \quad (4.536)$$

For other layers we find

$$\overline{H}_{D,j}^{(+)} = \left(\overline{\overline{I}} - \overline{\overline{R}}_{D1,Uj-1}\overline{\overline{R}}_{UN,Dj}\right)^{-1}\overline{\overline{T}}_{Uj-1,D1}\overline{H}_{D,1}^{(+)}, \quad (4.537)$$

$$\overline{H}_{U,j}^{(-)} = \left(\overline{\overline{I}} - \overline{\overline{R}}_{UN,Dj+1}\overline{\overline{R}}_{D1,Uj}\right)^{-1}\overline{\overline{R}}_{UN,Dj+1}\overline{\overline{T}}_{Uj,D1}\overline{H}_{D,1}^{(+)}. \quad (4.538)$$

Using these fields it is possible to construct the incident fields on the surface C_1 of the scatterer in cell j by using the matrices $\overline{\overline{C}}_U$ and $\overline{\overline{C}}_D$ for cell j. The field and its normal derivative can then be obtained on C_1 by using the matrix $\overline{\overline{M}}_S$ for cell j, and we can then finally use the integral (4.482) for calculating the field in Ω_1. The field and normal derivative just inside the scatterer surface is obtained from the electromagnetics boundary conditions, and then the field inside is obtained using another integral relation for this region.

For the other polarization there is not much difference. H should be replaced by E, and the matrix $\overline{\overline{M}}_S$ will be different due to different boundary conditions across the scatterer surface between regions Ω_1 and Ω_2.

4.3.9 EXAMPLE: PHOTONIC CRYSTAL

In this section we consider an example of transmission and reflection for s-polarized light being normally incident ($k_x = 0$) on a finite photonic crystal as in Fig. 4.56(a). The photonic crystal consists of dielectric rods with dielectric constant $\varepsilon_2 = 8.9$ in air. The rods are arranged on a square lattice with period Λ, and the radius of the rods is given by $a = 0.2\ \Lambda$. The structure is fully periodic along the x-axis while the number of rods along the y-axis is finite. Transmission and reflection spectra for 1, 2, 3 and 4 rods along the y-axis have been calculated using the periodic GFSIEM using polynomial order $m = 2$ and $N = 40$ sections per rod, and the results are presented in Fig. 4.62. Transmission and reflection have been calculated as a function of the normalized frequency Λ/λ. The reflection and transmission spectra become more complicated as the number of rods increases due to multiple scattering and the resulting interference. However, for a specific frequency range from approximately 0.27 to 0.44 a reflection peak builds up such that for many layers there is almost total reflection. In addition a reflection peak also builds up with an increasing number of rods at a frequency interval around $\Lambda/\lambda = 0.6$. These frequency intervals can be understood from a band diagram of the photonic crystal.

The band diagram of the photonic crystal is related to the properties of the photonic crystal being fully periodic along both the x and y axes. For a moment we are thus going to consider the fully periodic dielectric function satisfying

$$\varepsilon(x,y) = \varepsilon(x + n_x\Lambda, y + n_y\Lambda) \quad (4.539)$$

for any integer n_x and n_y. The solutions for the electric field for such a geometry can be expressed as Bloch waves of the form

$$E(\mathbf{r}) = U_{\mathbf{k},n}(\mathbf{r})e^{i\mathbf{k}\cdot\mathbf{r}}, \quad (4.540)$$

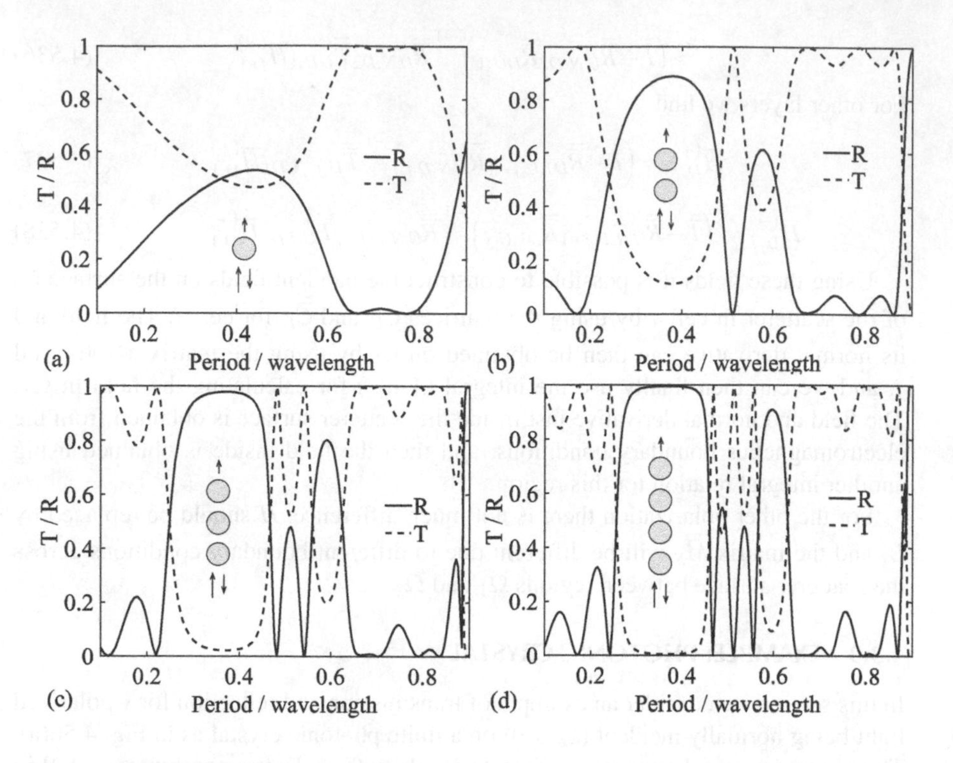

Figure 4.62 Transmission and reflection spectra for light being normally incident on (a) 1, (b) 2, (c) 3, and (d) 4 layers of a photonic crystal with dielectric rods ($\varepsilon = 8.9$) in air having radius $r = 0.2\,\Lambda$.

where $\mathbf{k} = \hat{x}k_x + \hat{y}k_y$ is the Bloch-wave vector.

The electric field on this form must as usual satisfy the wave equation

$$\left(\nabla^2 + k_0^2\varepsilon(\mathbf{r})\right)E(\mathbf{r}) = 0, \tag{4.541}$$

which can here be thought of as an eigenvalue-problem where E is the eigenfunction and k_0^2 is the eigenvalue. By requiring that the field should be of the form (4.540) with a specific Bloch-wave vector $\mathbf{k}$, the resulting eigenvalues when solving (4.541) will become functions of $\mathbf{k}$. The band index n in (4.540) is used to distinguish between different solutions for the same $\mathbf{k}$. It is common practice to investigate the eigenvalues on the boundary of the 1st irreducible Brillouin zone shown in the inset in Fig. 4.63. By exploiting symmetry considerations all Bloch-mode solutions can be obtained from the solutions with $\mathbf{k}$ within the irreducible Brillouin zone. The path along the boundary can be described as $\Gamma \to X \to M \to \Gamma$ = $(0,0) \to (G/2,0) \to (G/2,G/2) \to (0,0)$. Fig. 4.63 shows $\Lambda/\lambda = k_0\Lambda/2\pi$ as a function of $\mathbf{k}$ along this path. The band diagram was calculated using plane-wave-expansion theory as described in Appendix G.

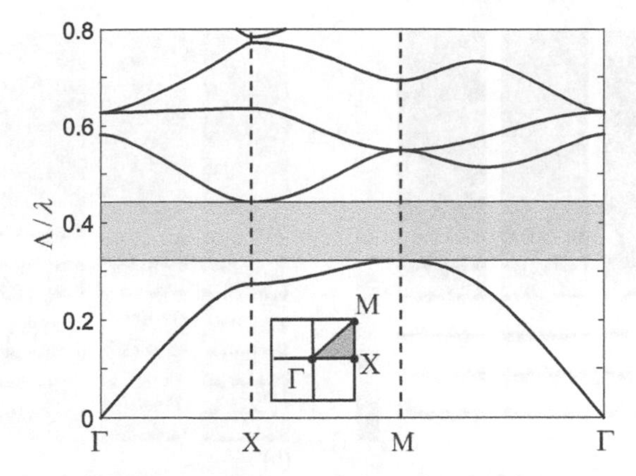

Figure 4.63 Band diagram of photonic crystal with dielectric rods of radius $a = 0.2\Lambda$ and dielectric constant $\varepsilon_2 = 8.9$ arranged on a square lattice with lattice constant Λ. Light is *s* polarized and is restricted to propagation in the *xy*-plane.

The gray-shaded region in Fig. 4.63 indicates the photonic bandgap which covers the frequency interval $0.32 \leq \Lambda/\lambda \leq 0.44$. For frequencies within the bandgap no propagating modes exist in the fully periodic structure, meaning that no Bloch modes exist with a real **k**. Evanescent or exponentially decreasing modes with a complex **k** do exist but these modes will not penetrate far into the photonic crystal. The reflection peak that appeared in Fig. 4.62 with an increasing number of rods covered a wider frequency interval than the bandgap. This is because light was normally incident which in the band diagram corresponds to **k** being on the line from Γ to X, and for this part of the band diagram there are no propagating modes within the frequency interval $0.27 \leq \Lambda/\lambda \leq 0.44$ explaining the actual width of the observed reflection peak. There is no complete bandgap at the frequency $\Lambda/\lambda = 0.6$. However, in the band diagram along the path from Γ to X there are also no propagating modes for the interval $0.58 \leq \Lambda/\lambda \leq 0.63$, which explains why a reflection peak builds up around this frequency with an increasing number of rods along *y*.

A few examples of electric field distributions for the photonic crystal are shown in Fig. 4.64. The case of frequency $\Lambda/\lambda = 0.38$ is considered in Fig. 4.64(a) where the different field plots correspond to having 1, 2, 3 and 4 rods along the *y*-axis. Below the photonic crystal an interference pattern is observed between the incident and the reflected waves. No such interference is present on the transmission side where light only propagates upward. The field strength on the transmission side drops rapidly with an increasing number of rods. The corresponding field distributions for the frequency $\Lambda/\lambda = 0.5$ are shown in Fig. 4.64(b). This frequency is not in any bandgap and light can be transmitted.

By introducing line and point defects into the photonic crystal structure it is possible to create waveguides and cavities where light can be trapped in a line-like or point-like region for frequencies within the bandgap. We will consider an example

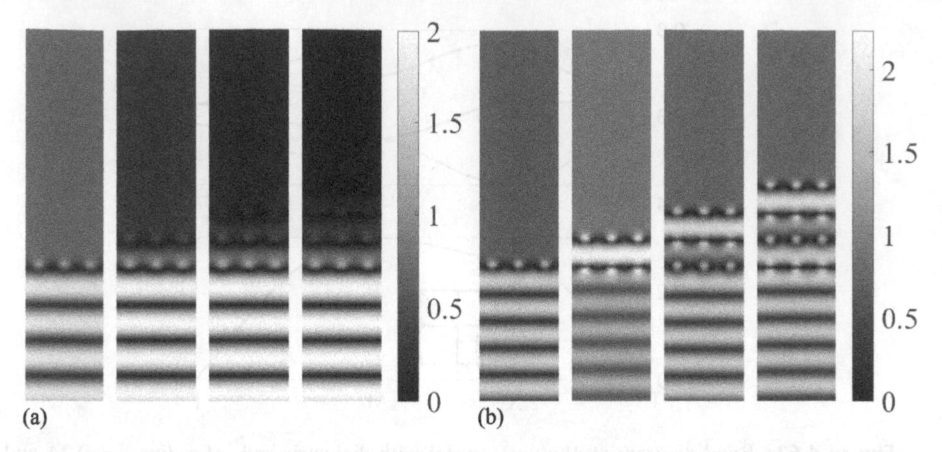

Figure 4.64 Electric field magnitude for s-polarized light being normally incident on a photonic crystal with 1, 2, 3 and 4 rods along the y-axis. (a) $\Lambda/\lambda = 0.38$, and (b) $\Lambda/\lambda = 0.5$.

of a photonic crystal waveguide (PC-WG) in the following.

The properties of a PC-WG can to a large extent be understood from a banddiagram for a fully periodic structure (along both x and y). We will consider a periodic structure where waveguides are periodically repeated along the x-axis. The period along the x-axis will be $\Lambda_x = 7\Lambda$, where Λ is the period of the simple photonic crystal and is now also the period along y. There are only 6 rods along the x-axis within the distance Λ_x corresponding to a photonic crystal where every 7th column of rods has been removed which creates a periodic array of waveguides. The unit cell of the PC-WG is shown as an inset in Fig. 4.65. The band diagram is thus based on a fully periodic dielectric function satisfying here

$$\varepsilon(x,y) = \varepsilon(x+n_x\Lambda_x, y+n_y\Lambda) \tag{4.542}$$

for any integer n_x and n_y. The solutions for the electric field for such a geometry can still be expressed as Bloch waves in the form (4.540) but now with

$$U_{\mathbf{k},n}(x+n_x\Lambda_x, y+n_y\Lambda) = U_{\mathbf{k},n}(x,y). \tag{4.543}$$

The function $U_{\mathbf{k},n}$ is thus still periodic with the same period as the structure.

While it might be desirable that there should be only one waveguide and not a periodic array, then the case of a single waveguide can be approximated well by using a large Λ_x, such that there is negligible coupling between the neighbor waveguides. In that case the part of the band diagram related to the properties of the waveguide will be unaffected by the periodicity along x.

The band diagram uses $k_x = 0$ and considers the bands for a varying $k_y \in [0;\pi/\Lambda]$ $(G_y = 2\pi/\Lambda)$. Notice that there is a band in the region of the bandgap of a simple photonic crystal. This band corresponds to guided modes of the photonic crystal waveguide. The photonic crystal itself does not support any propagating modes in

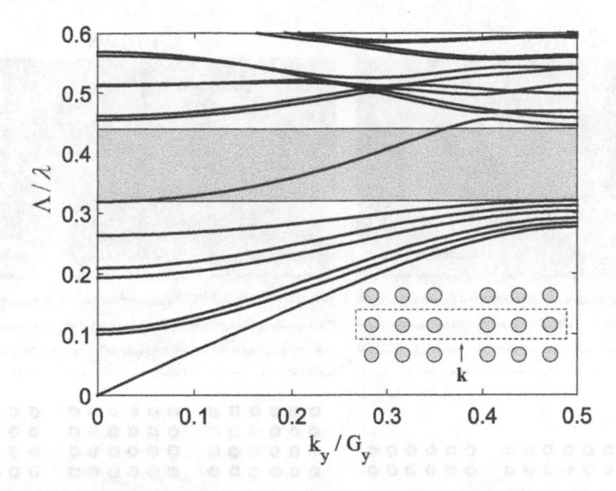

Figure 4.65 Band diagram of photonic crystal waveguide. The unit cell shown as an inset (marked with a dashed line) is repeated periodically. The unit cell contains 6 dielectric rods of radius $a = 0.2\ \Lambda$ and dielectric constant $\varepsilon_2 = 8.9$. The unit cell height is $\Lambda_y = \Lambda$, and the width is $\Lambda_x = 7\Lambda$. Light is *s* polarized and is restricted to propagation in the *xy*-plane. The Bloch-wave vector has $k_x = 0$.

the bandgap, and the modes seen here must thus be localized to the defect region or the waveguide in the otherwise periodic structure. Notice that now because the unit cell is much larger, the number of bands is also larger. For example, the first band of the simple band diagram in Fig. 4.63 is now turned into 6 bands. If there were no waveguide and a unit cell would thus contain 7 rods, then there would be 7 bands instead. In the limit where $\Lambda_x \rightarrow \infty$ there would be a continuum of bands above and below the bandgap. However, regarding the band in the bandgap of the simple photonic crystal, there will still be only one band.

In an ordinary index-guiding waveguide, light would be trapped in the wave-guide by the mechanism of total internal reflection (TIR). This then requires that the refractive index of the waveguide core is higher than the refractive index of the surrounding medium. For the photonic crystal waveguide considered here, the refractive index in the waveguide region is lower than the average refractive index of the surrounding photonic crystal structure, and the mechanism of light guiding is different and is due to an interference phenomenon.

A few examples of plots of the electric field magnitude for a photonic crystal waveguide with two or four rows of rods being illuminated from below with a plane wave of frequency $\Lambda/\lambda = 0.4$ are shown in Fig. 4.66. These plots illustrate that light can be confined to the PC-WGs for a frequency in the bandgap. At the considered frequency the PC-WG supports a single guided mode which can be used to transmit light through the photonic crystal. Below the PC waveguide an interference pattern is seen due to interference between the incident field and the downward reflected field. Most of the light is reflected. However, some light is also seen to be localized to the

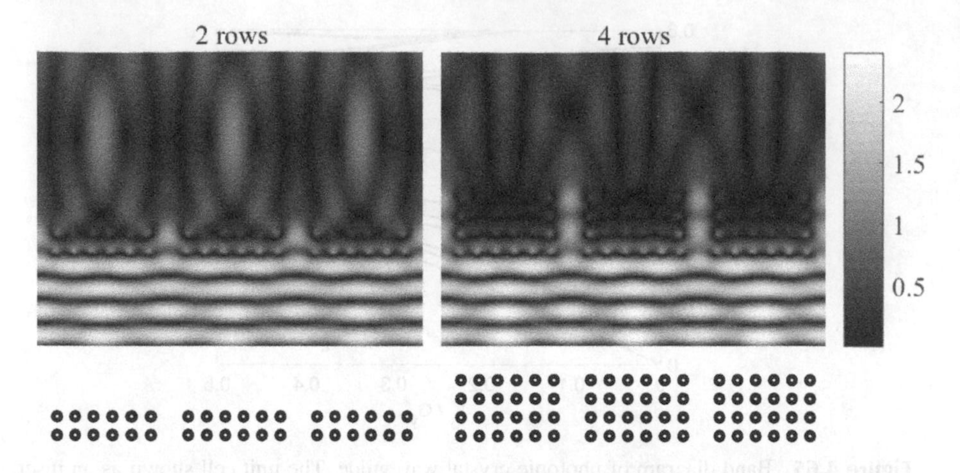

Figure 4.66 Magnitude of electric field for a periodic array of photonic crystal wave-guides with 2 rows of rods or 4 rows of rods. The period along x is 7Λ, where Λ is the usual period of a simple photonic crystal, and within the length 7Λ along x there are only 6 rods such that a waveguide has been formed.

waveguides. Especially for the case of 4 rows of rods it is visible that light is strongly localized to the waveguide region and cannot penetrate much into the surrounding photonic crystal. In the latter calculation there are 24 rods in the unit cell. The surface of each rod was divided into 20 sections and polynomial order $m = 2$ was applied. This is equivalent to using 960 sampling points in the periodic GFSIEM.

4.3.10 EXAMPLE: ANTI-REFLECTIVE GROOVE ARRAY IN A DIELECTRIC

In this section the periodic GFSIEM will be applied to dielectric surfaces that are structured with a periodic array of grooves for anti-reflective purposes [90]. The anti-reflective effect may arise for rectangular grooves due to multiple reflection interference, and for tapered grooves due to a gradual transition from the outside material to the dielectric material. Similar effects are possible with a series of planar thin-film layers of appropriately varying refractive index and thickness. However, a material with the optimum refractive index may not be available, and multiple materials are then needed. With the microstructured surfaces, only one material is needed.

Anti-reflection is important in optical systems with many lens surfaces, as even though only a few percent of the light is lost due to reflection at each interface, the total reflection loss from several lenses may be considerably large. It is thus desirable to make the surfaces anti-reflective by either depositing a thin-film geometry or by microstructuring the surface. Anti-reflection is also highly important for, e.g., solar cells [91], since, if there is no anti-reflective coating or microstructure, then due to the high refractive index of semiconductors, there will easily be a 30% reflection loss from the reflection at the first interface alone. In the case of polymer optics, one possibility for fabricating anti-reflective surface microstructures is to use hot embossing where the microstructure is imprinted into the surface [92, 93, 94]. Other examples of anti-reflective microstructures that have been fabricated include nanowire, nanocone, and nanopyramid arrays [95, 96, 97, 98]. Periodic arrays of microstructures with an antireflective effect can also be found in nature in, for example, moth eyes [99].

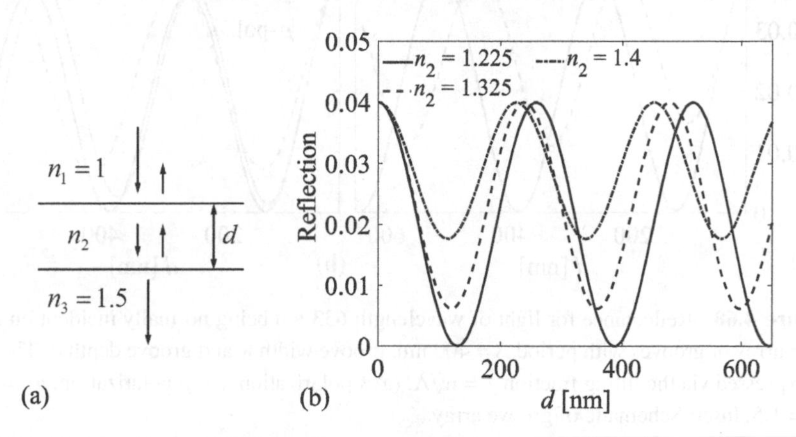

Figure 4.67 (a) Schematic of light being normally incident on a three-layer structure with a thin-film layer of refractive index n_2 and thickness d being sandwiched between media with refractive indices $n_1 = 1$ and $n_3 = 1.5$. (b) Reflection versus film thickness d for the wavelength 633 nm.

The anti-reflective effect of microstructured surfaces can be understood to a large extent from principles of planar thin-film anti-reflective structures. Consider first the simple three-layer geometry illustrated in Fig. 4.67(a) with a thin film of thickness d and refractive index n_2 being placed between semi-infinite media with refractive indices $n_1 = 1$ and $n_3 = 1.5$. According to Eq. (2.54) the reflection of light being incident from medium 1 will vanish when two conditions are satisfied, namely

$$r_{1,2}^{(u)} = r_{2,3}^{(u)}, \tag{4.544}$$

and

$$e^{ik_{y,2}d} = -1. \tag{4.545}$$

For loss-less dielectrics and normally incident light this translates into the requirements

$$n_2 = \sqrt{n_1 n_3}, \tag{4.546}$$

and

$$d = \frac{\lambda}{4n_2}(1+2m), \quad m = 0, 1, 2, \ldots \tag{4.547}$$

The reflectance as a function of the film thickness d is shown in Fig. 4.67(b) for three different film refractive indices n_2, where the refractive index $n_2 = \sqrt{1.5} =$

1.225 results in vanishing reflection for certain thicknesses d given in (4.547), while vanishing reflection cannot be achieved for other film refractive indices.

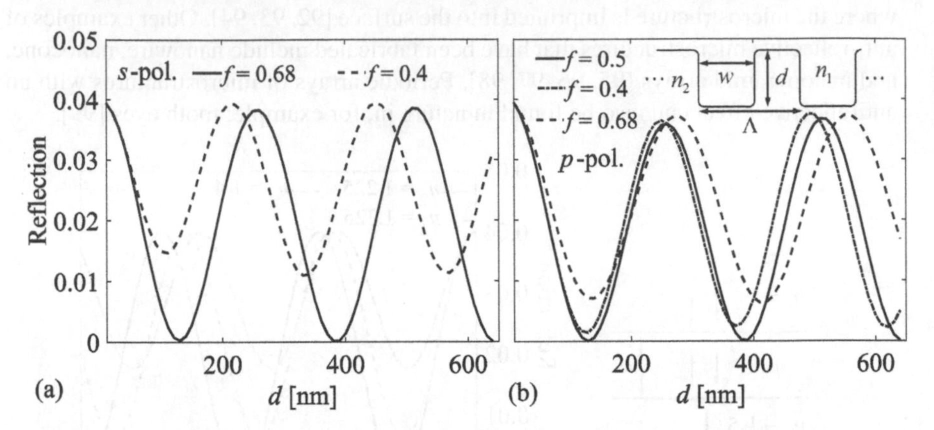

Figure 4.68 Reflectance for light of wavelength 633 nm being normally incident on a periodic array of grooves with period $\Lambda = 400$ nm, groove width w and groove depth d. The width is expressed via the filling fraction $f = w/\Lambda$. (a) s polarization. (b) p polarization. $n_1 = 1$, and $n_2 = 1.5$. Inset: Schematic of groove array.

A similar anti-reflective effect can be obtained by periodically microstructuring a surface with a groove of width w and depth d within each period of width Λ. A schematic of the rectangular groove array is shown in the inset in Fig. 4.68(b). Now there are only two materials with refractive indices n_1 and n_2 but the region with a microstructure of thickness d can function as an effective medium with an effective refractive index lying somewhere between the refractive indices n_1 and n_2. The effective refractive index will depend on the groove width w, the period Λ, the wavelength λ, and the polarization of light. In the limit of $\Lambda \to 0$ or $\lambda >> \Lambda$ the effective refractive index can be obtained from simple effective-medium theory [100, 101, 102] and is given from just n_1, n_2, the filling fraction w/Λ, and the polarization:

$$n_{\text{eff}}^{(s)} = \sqrt{\varepsilon_1 w/\Lambda + \varepsilon_2(1 - w/\Lambda)}, \quad \Lambda/\lambda \to 0, \tag{4.548}$$

$$n_{\text{eff}}^{(p)} = 1/\sqrt{(1/\varepsilon_1)w/\Lambda + (1/\varepsilon_2)(1 - w/\Lambda)}, \quad \Lambda/\lambda \to 0. \tag{4.549}$$

The expression for s polarization can be obtained by noting that the electric field is continuous across interfaces, and thus from the expression for the displacement field

$$\mathbf{D} = \varepsilon_0\mathbf{E} + \mathbf{P} = \varepsilon_0\varepsilon(\mathbf{r})\mathbf{E} \tag{4.550}$$

it can be observed that the average response or average polarization is equivalent to the response of a homogeneous medium with a dielectric constant equal to the geometrical average of the dielectric constant. For p polarization the displacement field is continuous across interfaces, and in this case the inverse of the effective dielectric constant is obtained as the geometric average of the inverse of the dielectric constant using similar considerations [102]. As the period Λ increases or the wavelength λ decreases, the electric field tends to be trapped inside the high-refractive index material, which leads to a higher effective refractive index than the long-wavelength expressions (4.548) and (4.549). This is a waveguide effect which has, for example, been discussed in [103].

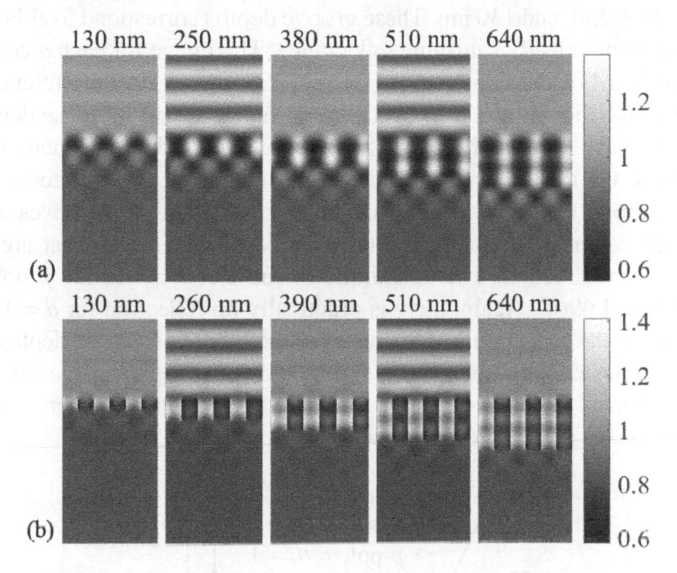

Figure 4.69 Electric field magnitude for light of the wavelength 633 nm being normally incident on a periodic array of rectangular grooves with period Λ = 400 nm and different groove depths d indicated in the figure. (a) groove width w = 272 nm (filling factor 0.68) and s polarization. (b) groove width w = 200 nm (filling factor 0.5) and p polarization

The reflectance for light being normally incident on a periodic array of rectangular grooves calculated using the periodic GFSIEM is shown in Fig. 4.68(a) and (b) for s polarization and p polarization, respectively. The period of the groove array is Λ = 400 nm, and the wavelength is 633 nm. Corners are rounded with a 2-nm radius. The refractive index of the structured surface is $n_2 = 1.5$ and above the surface there is air $n_1 = 1$. The effective refractive index may now be controlled via the filling fraction $f \equiv w/\Lambda$. For s polarization it is seen that for $f = 0.68$ and groove depths d = 130 nm, 380 nm, and 640 nm, the reflection practically vanishes. These groove depths are close to the film thicknesses in Fig. 4.67 that lead to vanishing reflection for a homogeneous film with refractive index $\sqrt{n_1 n_2}$. For another filling factor $f = 0.4$ the reflection does not vanish for any groove depth. This is equivalent to the requirement (4.544) not being satisfied due to a non-optimum effective refractive index. While the filling fraction 0.68 was optimum for s-polarized light, this is not the case for p-polarized light where instead $f = 0.5$ is optimum as can be seen from Fig. 4.68(b). This is a consequence of the effective refractive index being different for the two polarizations.

The calculation was based on polynomial order $m = 2$ and 30 sections for each vertical line, 20 sections for each horizontal line, and 3 sections for each corner. Thus, a total of 112 sections being equivalent to 224 sampling points. The magnitude of the electric field is shown in Fig. 4.69(a) for the case of normally incident s-polarized light for a groove of width $w = 272$ nm and different groove depths $d =$ 640, 510, 380, 250, and 130 nm. These groove depths correspond to either maximum reflectance or practically vanishing reflectance. The image for each d corresponds to a width of 3Λ. For the depths $d = 510$ and 250 nm a clear interference pattern is observed above the grooves as the reflectance is maximum for these depths. For the other depths very little interference is seen above the grooves as there is practically no reflection. On the transmission side below the grooves an interference pattern is observable along the x-axis which vanishes farther below the grooves as this is the interference pattern as a result of interference between waves that are evanescent along the $(-)y$ direction. The magnitude of the electric field for p polarization is shown in Fig. 4.69(b). Again, there is practically no reflection for $d = 130$, 390 and 640 nm, while the reflection is maximum for the other considered depths. Notice the jump in the electric field magnitude across the vertical part of the groove surface, which is expected from the condition that here εE_x is conserved across the interface leading to a jump in the field magnitude.

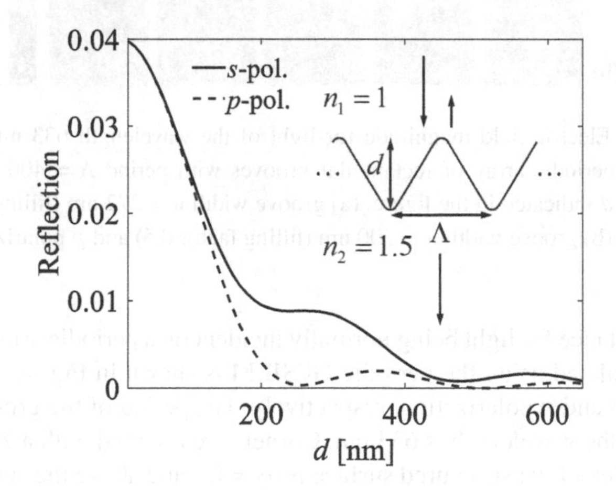

Figure 4.70 Reflectance for light of wavelength 633 nm being normally incident on a periodic array of V-grooves with period $\Lambda = 400$ nm and groove depth d. Corners are rounded by a corner radius of 2 nm. Inset: Schematic of groove array.

Another approach to obtain a microstructured surface with an anti-reflection effect is to use tapered grooves. An example of a tapered groove that is sometimes referred to as a V-groove due to its shape is shown in the inset in Fig. 4.70. The V-groove shown here can be constructed from two straight lines and two parts of small circles connected such that there are no sharp corners. The V-groove surface is thus completely described from the period Λ, the groove depth d, and the corner radius a. Since now the fraction of high-refractive-index material increases from the top of the groove to the bottom of the groove, this structure is roughly equivalent to having a graded-index profile from top to bottom starting from $n = n_1$ and ending at $n = n_2$.

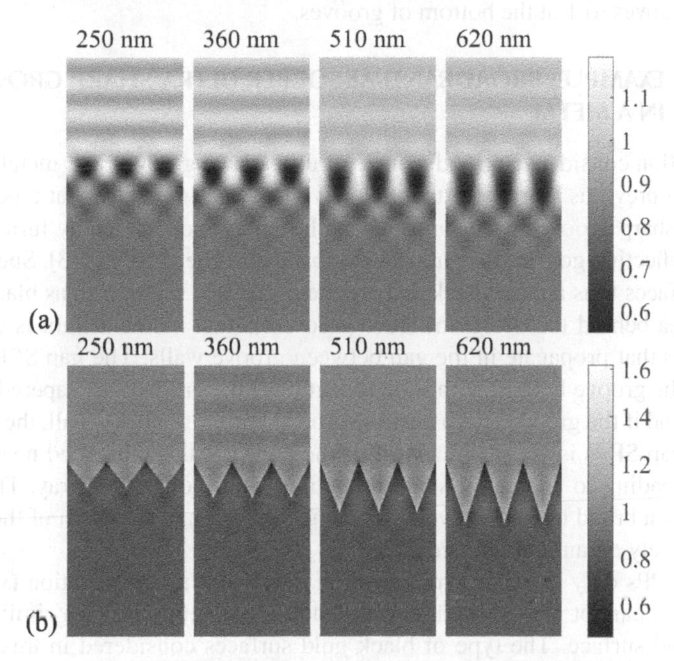

Figure 4.71 Electric field magnitude for a periodic array of V grooves in a glass surface ($n_2 = 1.5$) for different groove depths d (indicated in the figure) being illuminated by normally incident light of the wavelength 633 nm. (a) s polarization, and (b) p polarization.

Fig. 4.70 shows a calculation of the reflectance for light being normally incident on a V-groove characterized by the period Λ = 400 nm, corner radius a = 2 nm, $n_1 = 1$, $n_2 = 1.5$, and a range of groove depths d. The calculation was made using the periodic GFSIEM with 6 sections per corner and 20 sections per straight line, and polynomial order $m = 2$. This is equivalent to 104 sampling points. In the numerical calculation of both reflectance R and transmittance T the deviation $|R+T-1|$ is less than 0.0001. When doubling the number of sections the change in reflectance appeared only on the sixth decimal.

It can be observed that with increasing groove depth d the reflectance for p-polarization decreases faster than the reflectance for s polarization. However, if the groove depth is large enough, a good anti-reflection effect can be obtained for both polarizations at the same time. There are still some interference-related oscillations in the reflectance versus d similar to Fig. 4.68. They are just much weaker here and are dampened with increasing d.

The electric field magnitude for a few selected depths of V grooves is shown in

Fig. 4.71. Contrary to the rectangular grooves, here there is no clearly observable interference pattern in the region between the top and the bottom of the grooves. This is an effect of the gradual and slow change from one medium to the other by gradually increasing the filling factor of high-refractive index material from 0 at the top of grooves to 1 at the bottom of grooves.

4.3.11 EXAMPLE: BROADBAND-ABSORBER ULTRA-SHARP GROOVE ARRAY IN A METAL

This section considers a periodic array of ultra-sharp grooves in a metal. It has been shown in previous work both theoretically and experimentally that a periodic array of ultra-sharp grooves with subwavelength period and depth may turn a shiny and highly reflecting gold surface into a broadband absorber [71, 72, 73]. Such structured gold surfaces thus appear black and are therefore also referred to as black gold. The main idea behind the design of these grooves is that incident light is coupled into gap SPPs that propagate in the gap between groove walls. The gap SPPs propagate toward the groove bottom with little reflection if the grooves are tapered sufficiently slowly, and if the groove width near the groove bottom is ultra-small, the propagation loss of gap SPPs is very large. Most of the power is thus absorbed near the groove bottom leading to an overall small reflection from the groove array. This principle works for a broad band of wavelengths since the main mechanism of the absorption does not rely on an interference effect.

Gap SPPs only exist for p polarization. For the other polarization (s) broadband absorption cannot be achieved, and instead the reflection will be similar to that of a flat gold surface. The type of black gold surfaces considered in this section can thus be used as a polarizer that works in reflection mode. The main advantage of such polarizers is that an ultra-short laser pulse (e.g. 10 femtoseconds) will not be stretched noticeably when it is reflected by a black gold surface [104] as there will be practically no dispersion. It is thus possible to make a simple polarizer for ultra-short pulses that do not require any dispersion compensation afterward. Black metal polarizers based on the same principle but for other metals have been studied theoretically in Ref. [105].

For simplicity, the grooves considered in this section are slightly simpler than those investigated in previous work [71, 72, 73] since there is no flat plateau between neighbor grooves. The groove array is shown in Fig. 4.72 and is defined by the bottom groove width $w = 2\delta$, the groove depth d, the period Λ, and the top groove corner radius a. The groove surface within one period is composed of 4 parts of circles stitched together such that there are no sharp corners. The bottom circle with radius δ is a half-circle, which results in a groove with parallel groove walls near the groove bottom. The half-circle is described by

$$\mathbf{s}(\theta) = \hat{\mathbf{x}}\delta\cos\theta + \hat{\mathbf{y}}(-d+\delta+\delta\sin\theta), \quad \theta \in [-\pi;0]. \tag{4.551}$$

The part of a circle at the top of a groove is described by

$$\mathbf{s}(\theta) = \hat{\mathbf{x}}\left(\frac{\Lambda}{2}+a\cos\theta\right) + \hat{\mathbf{y}}(-a+a\sin\theta), \quad \theta \in [\pi-\alpha;\alpha], \tag{4.552}$$

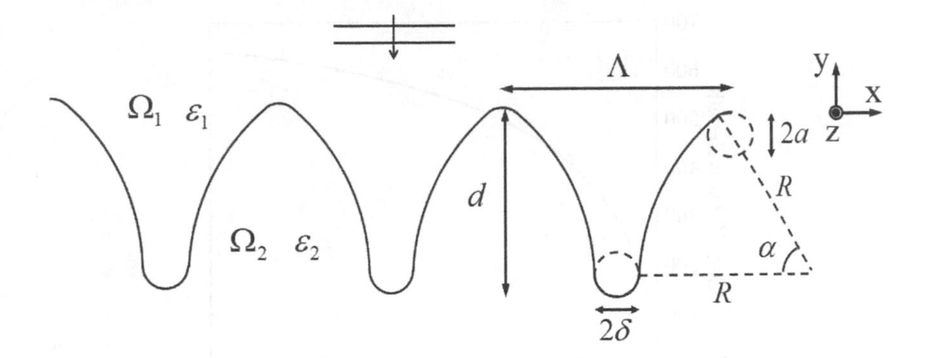

Figure 4.72 Schematic of ultra-sharp groove array.

and the two circles that describe the main part of the groove walls are described by

$$\mathbf{s}(\theta) = \hat{\mathbf{x}}(\delta + R + R\cos\theta) + \hat{\mathbf{y}}(-d + \delta + R\sin\theta), \quad \theta \in [\pi - \alpha; \pi], \tag{4.553}$$

and

$$\mathbf{s}(\theta) = \hat{\mathbf{x}}(-\delta - R + R\cos\theta) + \hat{\mathbf{y}}(-d + \delta + R\sin\theta), \quad \theta \in [0; \alpha]. \tag{4.554}$$

By requiring that all parts of circles are smoothly connected with no sharp corners, R and α are uniquely defined from

$$R(\alpha) = \frac{-a + d - \delta + a\sin\alpha}{\sin\alpha}, \tag{4.555}$$

and

$$f(\alpha) = \frac{\Lambda}{2} - a\cos\alpha - \delta - R(\alpha)(1 - \cos\alpha) = 0. \tag{4.556}$$

The equation (4.556) can be solved numerically for α. The corresponding radius R is then obtained from (4.555), and finally one period of the groove is then defined by (4.551)-(4.554).

The magnetic field of a gap SPP (p polarization) propagating in an air gap between parallel metal surfaces can be expressed in the form

$$H_{\text{Gap-SPP}}(x,y) = f(x)e^{\pm ik_0 n_m y}, \tag{4.557}$$

where n_m is the mode index. The mode index of a gap SPP in an air gap between two gold surfaces is shown versus the gap width for the wavelength 800 nm in Fig. F.3 (in Appendix F). In the limit of very small gaps, the mode index increases drastically as the gap width decreases. This means that the effective wavelength of the gap SPP defined as $\lambda_{\text{gap-SPP}} = \lambda_0/\text{Real}\{n_m\}$ will be very small at the bottom of an ultra-sharp groove due to the very small gap width there. The gap-SPP wavelength versus gap-width for the wavelength 800 nm is shown in Fig. 4.73.

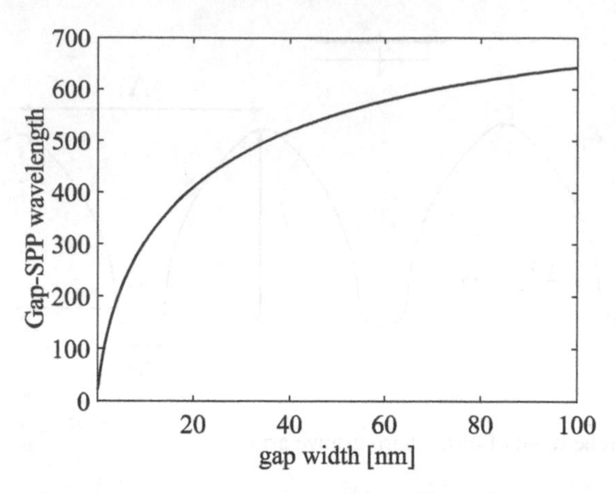

Figure 4.73 Gap-SPP wavelength versus gap width for an air gap between two gold surfaces. The free-space wavelength is 800 nm. The smallest gap width considered in the figure is 0.3 nm.

In order to avoid reflection as the gap SPP propagates into the groove, the mode-index should not change much over a distance of one gap-SPP wavelength [72]. For large groove widths, e.g. 100 nm, the gap-SPP mode index will only change by a small percentage if the groove width is reduced by a factor of 2. However, if the groove width is very small, e.g. 1 nm, the gap-SPP mode index will be roughly doubled if the groove width is reduced by a factor of 2. Fast tapering is thus possible near the groove top without it leading to significant reflection, while tapering must be very slow near the groove bottom. This is the type of groove design being considered.

The discretization of the groove surface is non-trivial. As a general rule of thumb at least 10 sampling points per wavelength are needed. However, the gap-SPP wavelength differs substantially from the top to the bottom of the groove due to the tapering of the groove width as shown in Fig. 4.73. In the extreme case of a groove with a bottom gap width of $w = 0.3$ nm the gap-SPP wavelength is approximately 22 nm, while at the groove top the gap-SPP wavelength easily exceeds 600 nm. If we follow the rule of thumb, then this translates into the requirement of one sampling point per every approximately 2 nm near the groove bottom, while approximately one sampling point is needed per every 60 nm near the groove top. It is clearly not preferable to use equidistant sampling on the main part of the groove walls. Equidistant sampling will require too many sampling points before sufficient sampling density is reached near the bottom. A simple approach to make a non-equidistant sampling is to exploit the fact that for the ultra-small gaps between the groove walls, the gap-SPP mode index is approximately inversely proportional to the gap width. The gap-SPP wavelength is thus approximately proportional to the groove width. The length of a section may then be required to not exceed a certain factor F times the smallest gap width related to that section, and there should also be an upper limit on the length of a section L_{max}. The sampling is then determined from the chosen factor and the chosen maximum section length.

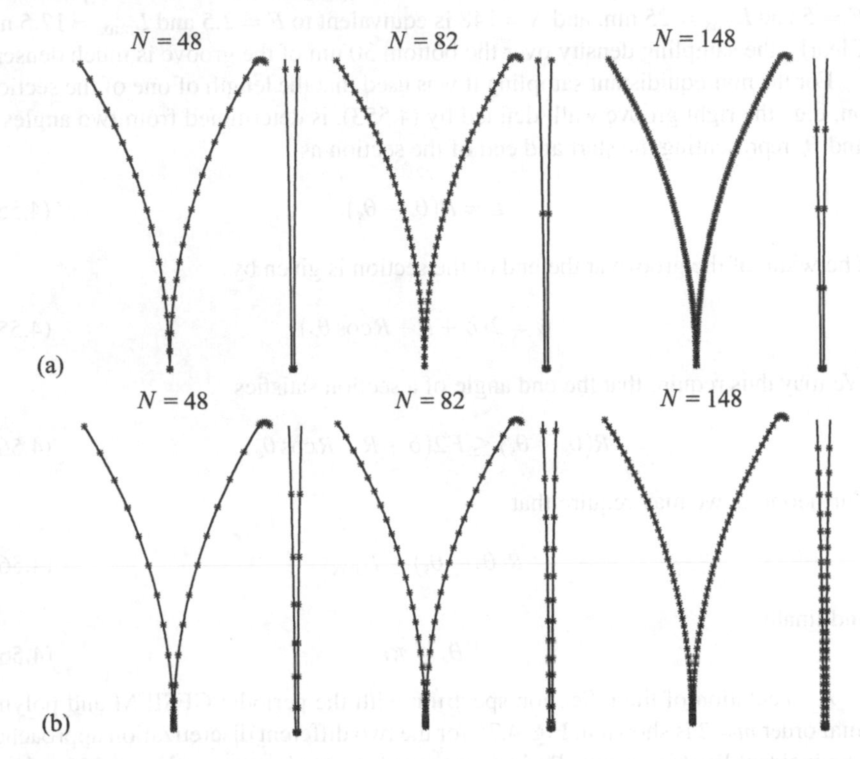

Figure 4.74 (a) Equidistant discretization of the two parts of circles of radius R of the ultra-sharp groove shown in Fig. 4.72. (b) Non-equidistant discretization of the two circles. Six sections are used for each of the two small circles at the top and bottom of the groove of radius a and δ. The remaining $N-12$ sections are used on the circles of radius R. The asterisks mark the beginning and end of a section. Both the entire groove and the bottom 30 nm of the groove (along y) are shown.

The discretization into sections is considered in Fig. 4.74 for a groove of width $\Lambda = 300$ nm, groove depth $d = 500$ nm, upper corner radius $a = 10$ nm, and bottom groove width $w = 0.3$ nm. The beginning and end of each section is marked with an asterisk. Each of the small corners at the groove top and bottom are divided into 6 sections. The remaining $N-12$ sections are used on the two parts of circles with radius R that constitute the main part of the groove walls. The case of equidistant sampling on these parts of circles is considered in Fig. 4.74(a) for $N = 48$, 82, and 148 sections. For each N the bottom 30 nm of the groove is also shown. It is seen that over the bottom 30 nm, even for $N = 148$ there are less than 4 sections on the two parts of circles of radius R, which is not sufficient. The alternative discretization approach with more sections near the groove bottom and less near the top is shown in Fig. 4.74(b). The same number of sections are considered as with equidistant sampling. Here, $N = 48$ is equivalent to $F = 10$ and $L_{\max} = 50$ nm, $N = 82$ is equivalent to $F = 5$ and $L_{\max} = 25$ nm, and $N = 148$ is equivalent to $F = 2.5$ and $L_{\max} = 12.5$ nm. Clearly, the sampling density over the bottom 30 nm of the groove is much denser.

For the non-equidistant sampling it was used that the length of one of the sections on, e.g., the right groove wall, defined by (4.553), is determined from two angles θ_s and θ_e representing the start and end of the section as

$$L = R(\theta_e - \theta_s). \tag{4.558}$$

The width of the groove at the end of the section is given by

$$w = 2(\delta + R + R\cos\theta_e). \tag{4.559}$$

We may thus require that the end angle of a section satisfies

$$R(\theta_e - \theta_s) \leq F2(\delta + R + R\cos\theta_e). \tag{4.560}$$

Furthermore, we may require that

$$R(\theta_e - \theta_s) \leq L_{\max}, \tag{4.561}$$

and finally

$$\theta_e \leq \pi. \tag{4.562}$$

A calculation of the reflection spectrum with the periodic GFSIEM and polynomial order $m = 2$ is shown in Fig. 4.75 for the two different discretization approaches. The incident light is a normally incident p-polarized plane wave. Fig. 4.75(a) shows

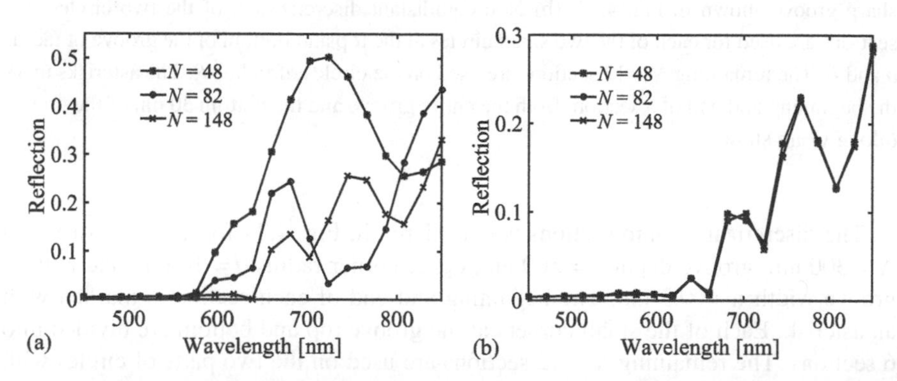

Figure 4.75 (a) Reflection of ultra-sharp groove calculated with equidistant discretization of the two circles that constitute the main part of the groove surface for different number of sections N. (b) Reflection of ultra-sharp groove calculated with non-equidistant discretization of the two circles for different number of sections N. The discretization in each case is shown in Fig. 4.74.

the calculated reflectance spectra for the case of equidistant sampling corresponding to the discretization shown in Fig. 4.74(a). It is clear that convergence is not reached. The result fluctuates substantially with the number of sections N. Fig. 4.75(b) shows the calculated reflectance spectra for the case of the non-equidistant sampling shown in Fig. 4.74(b). There is a very small change when going from $N = 48$ to $N = 82$, and no observable change when going from $N = 82$ to 148. The result obtained with $N = 82$ can be considered converged, and even the case of $N = 48$ appears to be reasonably good as well.

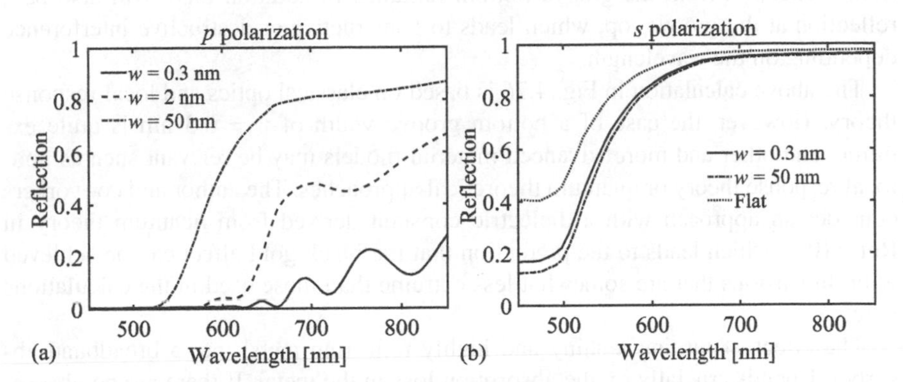

Figure 4.76 Reflection from periodic ultra-sharp groove array for different bottom groove widths $w = 2\delta$. (a) p polarization. (b) s polarization. $\Lambda = 300$ nm, $d = 500$ nm, and $a = 10$ nm. The grooves are illuminated by a normally incident plane wave.

The reflection spectra for the periodic array of ultra-sharp grooves in gold calculated using the periodic GFSIEM is shown in Fig. 4.76 for different bottom groove widths $w = 2\delta = 0.3$ nm, 2 nm, and 50 nm. The grooves are characterized by the period $\Lambda = 300$ nm, groove depth $d = 500$ nm, and upper groove corner radius $a = 10$ nm. The grooves are illuminated by normally incident light. The cases of p polarization and s polarization are considered in Fig. 4.76(a) and (b), respectively. The reflection from a flat gold surface with no grooves is also shown in Fig. 4.76(b). Notice that for the smallest considered wavelength, the reflectivity from a flat gold surface is only approximately 40%, and the reflectivity is much higher at the other end of the considered wavelength range approaching 98% at the wavelength 850 nm. For s polarization the presence of grooves reduces the reflectivity by more than a factor of two for the shortest considered wavelengths, but at the other end of the spectrum, the reflectivity is only reduced by a few percent. As there are no gap SPPs for s polarization then light is not able to penetrate very far down into the grooves.

Due to the higher losses for short wavelengths, for p polarization the reflectivity is very small for wavelengths below 550 nm. For the longer wavelengths the incident light can excite the gap SPPs that are bound to the region between the groove walls and will propagate toward the groove bottom. For the smallest bottom groove widths w, the reflection is smaller because the propagation losses of the gap SPPs is quite large, and more so as w decreases. For the largest bottom width $w = 50$ nm the propagation losses near the groove bottom are not very large and most of the gap SPP will be reflected at the groove bottom leading to an overall high reflectivity for the longer considered wavelengths. As w decreases toward the limit of an ultra-small bottom width corresponding to the width of one gold atom ($w = 0.3$ nm), the reflectivity drops below 30% for all considered wavelengths and is substantially smaller for the visible wavelength range. This explains why such a structured gold surface turns black in appearance [71]. Also notice the oscillations in the reflection spectra. Some reflection from the groove bottom remains. In addition there will also be a reflection at the groove top, which leads to constructive or destructive interference depending on the wavelength.

The above calculation in Fig. 4.76 is based on classical optics and local-response theory. However, the case of a bottom groove width of $w = 0.3$ nm is quite extreme, and other and more advanced material models may be relevant such as non-local response theory or quantum theoretical approaches. The author and co-workers consider an approach with a dielectric constant derived from quantum theory in Ref. [106], which leads to the prediction that the black-gold effect can be achieved with dimensions that are somewhat less extreme than those used in the calculations here.

The effect of turning a shiny and highly reflecting metal into a broadband absorber depends crucially on the absorption loss in the metal. If there are no absorption losses, or Ohmic losses, then the reflection will always be 100%. For silver the losses are smaller than for gold, and it has been shown theoretically that if silver is used instead of gold, then the reflection will generally be higher, while if chromium, nickel, platinum or palladium are used, the reflection will generally be smaller [72]. In the latter case this means that the grooves in, for example, nickel do not have to be nearly as sharp in order to make a black metal [73]. On the other hand, the absorption for s polarization is higher as well, and a polarizer for ultra-short pulses based on nickel will absorb part of the light for the polarization that should not be eliminated [105].

4.3.12 GUIDELINES FOR SOFTWARE IMPLEMENTATION

Structure of GFSIEM program for a periodic structure

1. Define the structure and how it should be divided into sections, refractive indices, wavelength, polarization, incident field parameters, polynomial expansion order m, the region in the xy-plane for field calculation, and the number of sampling points along x and y.
2. Calculate tables with the periodic Green's functions g_1 and g_2 with singularity subtraction and their derivatives (4.388), (4.395) and (4.396).
3. Construct a matrix where each row represents a section of the surface of the structure within one period. The first column identifies if the section is part of a circle or is a straight line. For a circle the next 6 columns describe the circle center, radius, start and end angles, and direction of the normal vector. For a line the 6 columns contain the start and end of the line and the normal vector.
4. Construct matrices with sampling points $\mathbf{s}_i^{(v)}$. Let rows refer to i, and columns to v. Use data from Step 3.
5. Construct vectors describing a sub-sampling of the scatterer surface. Each section is subdivided into $N_s \cdot m$ subsections, and vectors are constructed containing for each subsection the positions, normal vector, and length.
6. Plot the structure for a visual check.
7. Construct $\overline{H}_0^{(v)}$ or $\overline{E}_0^{(v)}$ depending on the choice of polarization for $v = 0, 1, \ldots, m-1$, as in (4.88). Use sampling points from Step 4.
8. Observe which sections are neighbors (in a periodic sense) by comparing positions of ends of sections and construct the matrices $\overline{\overline{D}}_H$ and $\overline{\overline{D}}_\phi$. Note that the matrices may contain values such as $\exp(\pm ik_x\Lambda)$.
9. Construct the matrices $\overline{\overline{A}}^{(u,v,v')}$ and $\overline{\overline{B}}^{(u,v,v')}$. Use (4.91) and (4.92) with the periodic Green's functions. Use the tables from Step 2 and add the contribution from the subtracted analytical (and singular) function. For the matrix elements $A_{i,j}^{(u,v,v')}$ and $B_{i,j}^{(u,v,v')}$, loop over $v = 0 : m-1$, $v' = 0 : m$, $i = 1 : N$, $j = 1 : N$. If $|\mathbf{s}_i^{(v)} - \mathbf{s}(t)|$ is larger than a threshold (in a periodic sense) for all t on j, then calculate matrix elements as a simple summation over $N_s \cdot m$ terms using data from Step 5. If not, apply an integration routine (e.g. Gauss-Kronrod) for the analytic part of the Green's functions. Give special attention when $\mathbf{s}(t)$ crosses the point $\mathbf{s}_i^{(v)}$.
10. Construct and solve the matrix equation (4.96) using data from Steps 7-9.
11. Calculate the reflection and transmission using the approach in Sec. 4.3.6. The procedure is slightly different for the photonic crystal and the grooves.
12. Calculate the fields. Loop over all positions $\mathbf{r}$ where the field should be calculated. Determine if $\mathbf{r}$ is inside or outside the scatterer. Calculate the field at $\mathbf{r}$ using the procedure in Sec. 4.2.3.
13. Wrap the whole program in a for-loop scanning a wavelength interval or a groove depth interval.
14. Plot the reflectance and transmittance versus wavelength or groove depth. Plot the calculated fields.

A few notes:

1. For the calculation of the periodic Green's functions, the relevant interval for $x-x'$ is from $-\Lambda$ to Λ. The code can in some cases be simplified if a slightly larger region is used since then it will not be necessary to apply the periodicity boundary condition of the Green's functions. The relevant interval for $|y-y'|$ is from 0 to the total height of the structure. In practice the smallest value for $|y-y'|$ considered is slightly larger than 0. The case of $|y-y'|=0$ can be evaluated by interpolation.
2. For loss-less structures the total sum of transmission and reflection must equal unity, which can be used as a numerical check.
3. Note that there may be particular regions of a structure that require many sections such as is, e.g., the case for an ultra-sharp groove in metal.
4. For the photonic crystal and photonic crystal-based structure, there are only circular structure surfaces. When determining if $\mathbf{r}$ is inside, it is thus possible to simply check for each circle if $(x-x_0)^2+(y-y_0)^2<r_0^2$, where (x_0,y_0) is the center of the circle and r_0 is the radius. For the grooves, one approach is to, e.g., count how many structure surfaces will be crossed along a line along (-)y from (x,y) to $(x,-\infty)$. If this is an even number, then (x,y) is outside, and otherwise it is inside the structure.
5. For simplicity, the magnetic field components for s polarization or the electric field components for p polarization may be obtained from calculations of the electric and magnetic field, respectively, and numerical differentiation.

4.3.13 EXERCISES

1. Prove (4.385) by inserting (4.384) and (4.380) and carrying out the integral.

2. Calculate a table with the value of (4.388) for a range of $x-x'$ and $|y-y'|$ both with and without the subtraction of the singularity. Plot the real part of the table versus $x-x'$ and $|y-y'|$ and observe that the singularity subtraction procedure results in a smooth and well-behaved function. Do the same for the derivatives of the Green's function in (4.395) and (4.396).

3. Prove the expressions for the incident power (4.420), reflected power (4.422), and the transmitted power (4.425), by carrying out the integrals. Note that although $\mathbf{H}$ and $\mathbf{E}$ are both expressed as a Fourier sum the resulting expression only contains a single sum since cross terms cancel out upon integration of the Poynting vector flux.

4. Follow the guidelines in Sec. 4.3.12 and develop a GFSIEM code capable of calculating reflection and transmission for photonic crystals and periodic groove arrays, and which is capable of calculating field distributions.

5. Consider the transmission spectra in Fig. 4.62 and the band diagram in Fig. 4.63. Calculate similar transmission spectra and band diagrams for rods with a slightly increased radius or rods with a slightly higher refractive index. What is the effect on the bandgap?

6. Prove the expressions (4.546) and (4.547) for the optimum parameters for antireflection with a single thin film.

7. Use the optimum fill-factors for a rectangular groove array as found in Fig. 4.68 in the long-wavelength-limit effective-medium-theory refractive indices (4.548) and (4.549). Show that the resulting refractive index values are smaller than the thin-film-theory expression (4.546). Explain this difference.

8. Consider the anti-reflective periodic groove arrays in Sec. 4.3.10 and carry out calculations with the periodic GFSIEM for similar structures with a varying period. Consider especially the case of a period that slightly exceeds the wavelength ($\Lambda > \lambda$), and observe the effect of the presence of higher-order diffraction on total reflection (into all diffraction orders). It is usually preferable that the period is smaller than the wavelength for normal incidence, and even smaller for oblique incidence. In addition, if the anti-reflection effect should be applied to a lens, then higher-order diffraction is also not desirable on the transmission side. How small should the period be in order to avoid higher-order diffraction in the transmitted light?

9. The surface of a V-groove may be constructed by stitching together two parts of circles with radius a and two straight lines such that there are no sharp corners. The two parts of circles are placed at the top and bottom of the groove and rounds the corners. For a given choice of period Λ, groove depth d, and a, show that the part of a circle in the top of the groove can be described by

$$\mathbf{r} = \mathbf{r}_c + a(\hat{\mathbf{x}}\cos\alpha + \hat{\mathbf{y}}\sin\alpha), \quad \alpha_0 \leq \alpha \leq \pi - \alpha_0, \tag{4.563}$$

where $\mathbf{r}_c$ is the center of the circle, and α_0 is given by

$$\frac{\Lambda}{2}\cos\alpha_0 + (2a - d)\sin\alpha_0 - 2a = 0. \tag{4.564}$$

The equation (4.564) can be solved numerically. The center of the bottom groove is displaced from the top groove by $\Lambda/2$ along the x-axis and by $d - 2a$ along the (-)y-axis. The straight lines connect end points on the two parts of circles such that a smooth surface with no sharp corners is obtained.

10. Carry out calculations of reflection for black metals based on silver and nickel using the periodic GFSIEM. Show that in one case the reflection increases while in the other case it decreases as compared with gold.

5 Area integral equation method for 2D scattering problems

This chapter is concerned with electric-field area integral equations for 2D scattering problems. These methods follow closely the physical principle that the material of the scatterer is polarized due to an incident field, and the incident field at a local position is the sum of an external field and radiation from all polarized parts of the scatterer itself. The polarization at any position inside the scatterer is thus partly driven by the fields originating from polarization currents at all positions inside the scatterer, and thus all parts of the scatterer are coupled.

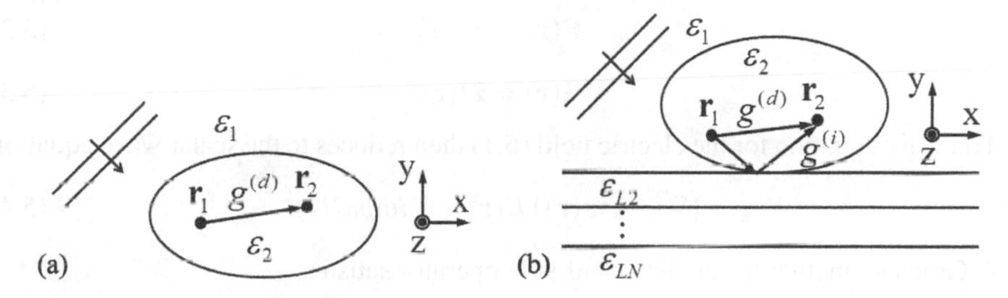

Figure 5.1 (a) Schematic of a scatterer in a homogeneous background being illuminated by a plane wave. (b) Schematic of a scatterer placed on a layered reference structure being illuminated by a plane wave.

A schematic of a scatterer placed in a homogeneous background and being illuminated by a plane wave is illustrated in Fig. 5.1(a). The total electric field $\mathbf{E}(\mathbf{r}_2)$ at the position $\mathbf{r}_2$ is the sum of the incident plane-wave field and the radiation from polarization currents at any other position inside the scatterer. The contribution to the field at $\mathbf{r}_2$ from polarization currents at $\mathbf{r}_1$ is governed by a direct Green's function $g^{(d)}(\mathbf{r}_2, \mathbf{r}_1)$. The polarization at $\mathbf{r}_2$ is given as $\mathbf{P}(\mathbf{r}_2) = \varepsilon_0 (\varepsilon(\mathbf{r}_2) - \varepsilon_{\text{ref}}) \mathbf{E}(\mathbf{r}_2)$ and is thus driven or induced by the field $\mathbf{E}(\mathbf{r}_2)$. Similarly, $\mathbf{P}(\mathbf{r}_1) = \varepsilon_0 (\varepsilon(\mathbf{r}_1) - \varepsilon_{\text{ref}}) \mathbf{E}(\mathbf{r}_1)$, which is the polarization at $\mathbf{r}_1$, depends on a field that partly originates from polarization currents at $\mathbf{r}_2$. Thus, $\mathbf{P}(\mathbf{r}_2)$ depends on $\mathbf{P}(\mathbf{r}_1)$, and in turn $\mathbf{P}(\mathbf{r}_1)$ depends on $\mathbf{P}(\mathbf{r}_2)$. All positions inside the scatterer are coupled in this way, and it is required that the field or polarization inside the scatterer is calculated self-consistently.

When the scatterer is placed on a layered reference geometry as illustrated in Fig. 5.1(b), the two positions $\mathbf{r}_1$ and $\mathbf{r}_2$ are not only directly coupled but there is also indirect coupling via reflection from the layered structure governed by the indirect part of the Green's function $g^{(i)}(\mathbf{r}_2,\mathbf{r}_1)$.

5.1 GREEN'S FUNCTION INTEGRAL EQUATIONS

The starting point is the wave equation for the electric field

$$-\nabla\times\nabla\times\mathbf{E}(\mathbf{r})+k_0^2\varepsilon(\mathbf{r})\mathbf{E}(\mathbf{r})=-i\omega\mu_0\mathbf{J}(\mathbf{r}), \tag{5.1}$$

where the electric field $\mathbf{E}$ and the current density $\mathbf{J}$ only vary in the xy-plane, i.e., $\mathbf{r}=\hat{\mathbf{x}}x+\hat{\mathbf{y}}y$. The fields, currents, and the dielectric constant are thus invariant along z. The electric field can be divided into two independent components corresponding to either s or p polarization, where the electric field is either polarized along the z-axis or in the xy-plane. Integral equations will be obtained in the following for solving the wave equation for s and p polarization, respectively.

5.1.1 *S* POLARIZATION

For s polarization the field and the current density are polarized along the z-axis, i.e.,

$$\mathbf{E}(\mathbf{r})=\hat{\mathbf{z}}E(\mathbf{r}), \tag{5.2}$$

$$\mathbf{J}(\mathbf{r})=\hat{\mathbf{z}}J(\mathbf{r}). \tag{5.3}$$

The wave equation for the electric field (5.1) then reduces to the scalar wave equation

$$\left(\nabla^2+k_0^2\varepsilon(\mathbf{r})\right)E(\mathbf{r})=-i\omega\mu_0J(\mathbf{r}). \tag{5.4}$$

A Green's function for the left-hand-side operator satisfies

$$\left(\nabla^2+k_0^2\varepsilon(\mathbf{r})\right)g(\mathbf{r},\mathbf{r}')=-\delta(\mathbf{r}-\mathbf{r}'). \tag{5.5}$$

In terms of the Green's function, a solution to (5.4) can be obtained as the following area integral

$$E(\mathbf{r})=i\omega\mu_0\int g(\mathbf{r},\mathbf{r}')J(\mathbf{r}')d^2r'. \tag{5.6}$$

The appropriate solution here is the one that corresponds to using the Green's function that satisfies the radiating boundary condition, since then the field solution will describe a field that propagates away from the current density.

For a current density in free space, the Green's function satisfying the radiating boundary condition is given as

$$g^{(\mathrm{d})}(\mathbf{r},\mathbf{r}')=\frac{i}{4}H_0^{(1)}(k_0|\mathbf{r}-\mathbf{r}'|), \tag{5.7}$$

and the integral (5.6) can be straightforwardly carried out.

Now consider the situation where the Green's function satisfying the radiating boundary condition is known for a more general reference geometry, i.e.,

$$\left(\nabla^2+k_0^2\varepsilon_{\mathrm{ref}}(\mathbf{r})\right)g(\mathbf{r},\mathbf{r}')=-\delta(\mathbf{r}-\mathbf{r}'). \tag{5.8}$$

A field solution for the reference geometry satisfies the equation

$$\left(\nabla^2 + k_0^2 \varepsilon_{\mathrm{ref}}(\mathbf{r})\right) E_0(\mathbf{r}) = -i\omega\mu_0 J(\mathbf{r}). \tag{5.9}$$

The field E_0 may, e.g., simply be a plane wave, in which case $J(\mathbf{r})$ is zero in the region of interest, or E_0 may be the field generated by a given current density in the region of interest. If the current density is even a δ-function, then E_0 is in fact proportional to the Green's function with $\mathbf{r}'$ at the source position.

The equation (5.4) may now be rewritten as

$$\left(\nabla^2 + k_0^2 \varepsilon_{\mathrm{ref}}(\mathbf{r})\right) E(\mathbf{r}) = -k_0^2 \left(\varepsilon(\mathbf{r}) - \varepsilon_{\mathrm{ref}}(\mathbf{r})\right) E(\mathbf{r}) - i\omega\mu_0 J(\mathbf{r}). \tag{5.10}$$

By subtracting (5.9) we find

$$\left(\nabla^2 + k_0^2 \varepsilon_{\mathrm{ref}}(\mathbf{r})\right) \left(E(\mathbf{r}) - E_0(\mathbf{r})\right) = -k_0^2 \left(\varepsilon(\mathbf{r}) - \varepsilon_{\mathrm{ref}}(\mathbf{r})\right) E(\mathbf{r}), \tag{5.11}$$

and a solution for the electric field may then be obtained by solving the following area integral equation:

Green's function area integral equation for *s* polarization:

$$E(\mathbf{r}) = E_0(\mathbf{r}) + \int g(\mathbf{r},\mathbf{r}') k_0^2 \left(\varepsilon(\mathbf{r}') - \varepsilon_{\mathrm{ref}}(\mathbf{r}')\right) E(\mathbf{r}') d^2 r'. \tag{5.12}$$

It should be noticed that the field can be obtained at all positions from the field at those positions where $(\varepsilon(\mathbf{r}') - \varepsilon_{\mathrm{ref}}(\mathbf{r}')) \neq 0$. This means that it is sufficient at first to consider (5.12) for the positions where the reference structure has been modified, and once the field has been obtained here, it is straightforwardly given at all other positions from the same integral equation.

5.1.2 *P* POLARIZATION

In the case of p polarization, the electric field and the corresponding component of the current density are polarized in the xy-plane, and the fields are invariant along the z-axis, i.e.,

$$\mathbf{E}(\mathbf{r}) = \hat{\mathbf{x}} E_x(\mathbf{r}) + \hat{\mathbf{y}} E_y(\mathbf{r}), \tag{5.13}$$

$$\mathbf{J}(\mathbf{r}) = \hat{\mathbf{x}} J_x(\mathbf{r}) + \hat{\mathbf{y}} J_y(\mathbf{r}). \tag{5.14}$$

The wave equation for the electric field is then still of the form (5.1) with the only difference that $\mathbf{E}$ and $\mathbf{J}$ do not have a z-component. We may rewrite the vector wave equation by using

$$-\nabla \times \nabla = -\nabla\nabla \cdot + \nabla^2, \tag{5.15}$$

which leads to

$$\left(-\nabla\nabla \cdot + \nabla^2 + k_0^2 \varepsilon(\mathbf{r})\right) \mathbf{E}(\mathbf{r}) = -i\omega\mu_0 \mathbf{J}(\mathbf{r}). \tag{5.16}$$

Here, $\nabla\nabla$ is a dyadic operator. It works as follows

$$\nabla\nabla f(\mathbf{r}) = \nabla\left(\nabla f(\mathbf{r})\right), \tag{5.17}$$

and

$$\nabla\nabla\cdot\mathbf{A}(\mathbf{r}) = \nabla(\nabla\cdot\mathbf{A}(\mathbf{r})). \tag{5.18}$$

A dyadic Green's tensor for the left-hand-side operator in the wave equation satisfies

$$\left(-\nabla\nabla\cdot+\nabla^2+k_0^2\varepsilon(\mathbf{r})\right)\mathbf{G}(\mathbf{r},\mathbf{r}') = -\mathbf{I}\delta(\mathbf{r}-\mathbf{r}'), \tag{5.19}$$

where the identity dyadic is given by

$$\mathbf{I} = \hat{\mathbf{x}}\hat{\mathbf{x}}+\hat{\mathbf{y}}\hat{\mathbf{y}}, \tag{5.20}$$

and the dyadic Green's tensor is given by

$$\mathbf{G}(\mathbf{r},\mathbf{r}') = \hat{\mathbf{x}}\hat{\mathbf{x}}G_{xx}(\mathbf{r},\mathbf{r}')+\hat{\mathbf{x}}\hat{\mathbf{y}}G_{xy}(\mathbf{r},\mathbf{r}')+\hat{\mathbf{y}}\hat{\mathbf{x}}G_{yx}(\mathbf{r},\mathbf{r}')+\hat{\mathbf{y}}\hat{\mathbf{y}}G_{yy}(\mathbf{r},\mathbf{r}'). \tag{5.21}$$

The identity operator reproduces a vector. This works as follows

$$\mathbf{I}\cdot\mathbf{A} = \mathbf{A}\cdot\mathbf{I} = \mathbf{A}\cdot(\hat{\mathbf{x}}\hat{\mathbf{x}}+\hat{\mathbf{y}}\hat{\mathbf{y}}) = (\mathbf{A}\cdot\hat{\mathbf{x}})\,\hat{\mathbf{x}}+(\mathbf{A}\cdot\hat{\mathbf{y}})\,\hat{\mathbf{y}} = \mathbf{A}. \tag{5.22}$$

This last equality follows straightforwardly by inserting $\mathbf{A} = \hat{\mathbf{x}}A_x+\hat{\mathbf{y}}A_y$.

For a current density in free space the dyadic Green's tensor is given as

$$\mathbf{G}^{(\mathrm{d})}(\mathbf{r},\mathbf{r}') = \left(\mathbf{I}+\frac{1}{k_0^2}\nabla\nabla\right)g^{(\mathrm{d})}(\mathbf{r},\mathbf{r}'), \tag{5.23}$$

where $g^{(\mathrm{d})}$ is given in (5.7), and here a solution for the electric field that satisfies (5.16) can be expressed as the following area integral

$$\mathbf{E}(\mathbf{r}) = i\omega\mu_0\int\mathbf{G}^{(d)}(\mathbf{r},\mathbf{r}')\cdot\mathbf{J}(\mathbf{r}')d^2r'. \tag{5.24}$$

Due to the choice of $\mathbf{G}^{(d)}(\mathbf{r},\mathbf{r}')$, the solution satisfies the radiating boundary condition.

Now consider a general reference structure given by the dielectric function $\varepsilon_{\mathrm{ref}}(\mathbf{r})$. The field $\mathbf{E}_0$ generated by a current density in this structure must satisfy

$$\left(-\nabla\nabla\cdot+\nabla^2+k_0^2\varepsilon_{\mathrm{ref}}(\mathbf{r})\right)\mathbf{E}_0(\mathbf{r}) = -i\omega\mu_0\mathbf{J}(\mathbf{r}). \tag{5.25}$$

The dyadic Green's tensor for the reference structure satisfies the equation

$$\left(-\nabla\nabla\cdot+\nabla^2+k_0^2\varepsilon_{\mathrm{ref}}(\mathbf{r})\right)\mathbf{G}(\mathbf{r},\mathbf{r}') = -\mathbf{I}\delta(\mathbf{r}-\mathbf{r}'), \tag{5.26}$$

and a solution to (5.25) for the reference field $\mathbf{E}_0$ can thus be obtained as the following area integral

$$\mathbf{E}_0(\mathbf{r}) = i\omega\mu_0\int\mathbf{G}(\mathbf{r},\mathbf{r}')\cdot\mathbf{J}(\mathbf{r}')d^2r'. \tag{5.27}$$

If the Green's tensor is chosen that satisfies the radiating boundary condition, then the resulting reference field will be propagating away from the current distribution and thus also satisfy the radiating boundary condition. It may also be the case that

$\mathbf{E}_0$ is not generated by sources in the region of interest, but is instead, in that region, a homogeneous solution of (5.25).

If the same current density is placed in another structure with dielectric function $\varepsilon(\mathbf{r})$ being a modification of the reference structure, the total field must satisfy the equation

$$\left(-\nabla\nabla\cdot+\nabla^2+k_0^2\varepsilon(\mathbf{r})\right)\mathbf{E}(\mathbf{r}) = -i\omega\mu_0\mathbf{J}(\mathbf{r}). \tag{5.28}$$

This equation can be rewritten as

$$\left(-\nabla\nabla\cdot+\nabla^2+k_0^2\varepsilon_{\text{ref}}(\mathbf{r})\right)\mathbf{E}(\mathbf{r}) = -i\omega\mu_0\mathbf{J}(\mathbf{r}) - k_0^2\left(\varepsilon(\mathbf{r})-\varepsilon_{\text{ref}}(\mathbf{r})\right)\mathbf{E}(\mathbf{r}), \tag{5.29}$$

and when this is combined with (5.25) we obtain

$$\left(-\nabla\nabla\cdot+\nabla^2+k_0^2\varepsilon_{\text{ref}}(\mathbf{r})\right)\left(\mathbf{E}(\mathbf{r})-\mathbf{E}_0(\mathbf{r})\right) = -k_0^2\left(\varepsilon(\mathbf{r})-\varepsilon_{\text{ref}}(\mathbf{r})\right)\mathbf{E}(\mathbf{r}). \tag{5.30}$$

Now it follows that the total electric field satisfies the Green's function area integral equation:

Green's function area integral equation for p polarization:

$$\mathbf{E}(\mathbf{r}) = \mathbf{E}_0(\mathbf{r}) + \int \mathbf{G}(\mathbf{r},\mathbf{r}')k_0^2\cdot\left(\varepsilon(\mathbf{r}')-\varepsilon_{\text{ref}}(\mathbf{r}')\right)\mathbf{E}(\mathbf{r}')d^2r'. \tag{5.31}$$

Again, it follows that once the field has been calculated in the region where $(\varepsilon(\mathbf{r})-\varepsilon_{\text{ref}}(\mathbf{r}))\neq 0$, then the field can be obtained straightforwardly at all other positions using (5.31).

5.2 DISCRETIZATION WITH SQUARE-SHAPED ELEMENTS

In this section discretization of a scatterer with square-shaped area elements is considered. An example of a circular scatterer with dielectric constant ε_2 in a reference structure with dielectric constant ε_1 is illustrated in Fig. 5.2. An area that includes the scatterer is divided into a number of square-shaped area elements. The idea now is to represent the scatterer by assigning a single dielectric constant to each element. As illustrated in Fig. 5.2 this leads to a stair-cased representation of the scatterer surface. Elements near the boundary may contain both some of the material with dielectric constant ε_1 and some of the material with dielectric constant ε_2. The presence of both materials can be taken into account by assigning an appropriate averaging of the dielectric constant over the cell.

For s polarization, where the electric field is along the z-axis and thus parallel with all interfaces, it is appropriate to assign the following geometric average to cell i [102]:

$$\varepsilon_{\|,i} = \frac{1}{\Delta_i}\int_i \varepsilon(\mathbf{r})d^2r, \tag{5.32}$$

where Δ_i is the area of cell i, and the integration is over cell i.

For p polarization, the electric field may have both a component which is parallel to the interface, and a component which is perpendicular to the interface. The effective dielectric constant should be different for these two field directions. An example

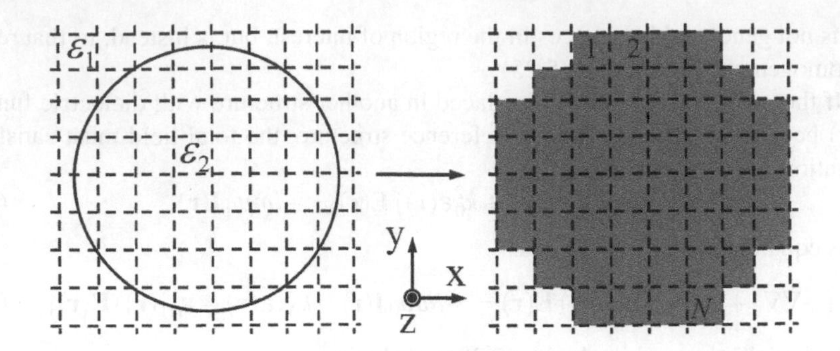

Figure 5.2 Schematic of discretization of a circular object with square-shaped elements.

of a cell that should describe a part of a structure with a curved interface is illustrated in Fig. 5.3. The average normal vector $\hat{\mathbf{n}}$ of a surface that can be described as part of

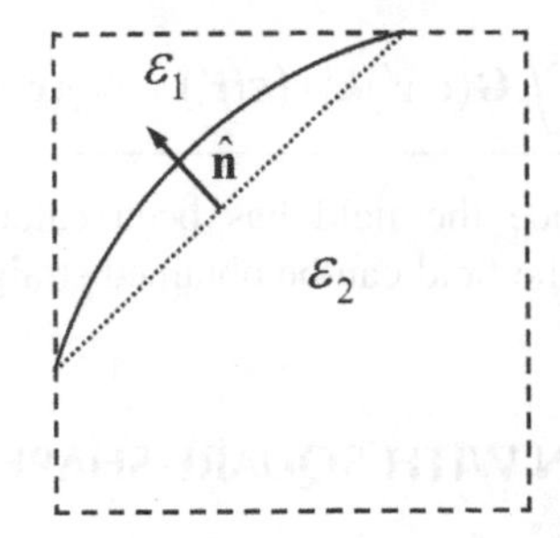

Figure 5.3 Schematic of a square-shaped cell with an interface between two media.

a circle can be constructed by first finding the vector between the two end points of the part of the circle within the cell, and then constructing a perpendicular unit vector. For the component of the field which is parallel to the interface the appropriate effective dielectric constant is still given by Eq. (5.32). For the field component in the direction of $\hat{\mathbf{n}}$ the appropriate effective dielectric constant is given as [102]:

$$\frac{1}{\varepsilon_{\perp,i}} = \frac{1}{\Delta_i} \int_i \frac{1}{\varepsilon(\mathbf{r})} d^2 r. \tag{5.33}$$

A tensor expression for the dielectric constant of the cell may now be constructed as

$$\varepsilon_i = \varepsilon_{\perp,i} \hat{\mathbf{n}}\hat{\mathbf{n}} + \varepsilon_{\parallel,i} \left(\mathbf{I} - \hat{\mathbf{n}}\hat{\mathbf{n}} \right). \tag{5.34}$$

This is based on the fact that $(\hat{\mathbf{n}}\hat{\mathbf{n}}) \cdot \mathbf{E}_i = \hat{\mathbf{n}} (\hat{\mathbf{n}} \cdot \mathbf{E}_i)$ is the component of $\mathbf{E}_i$ along $\hat{\mathbf{n}}$ times the vector $\hat{\mathbf{n}}$. Similarly, $(\mathbf{I} - \hat{\mathbf{n}}\hat{\mathbf{n}}) \cdot \mathbf{E}_i$ is the component of $\mathbf{E}_i$ which is perpendicular to $\hat{\mathbf{n}}$ times a unit vector in the direction perpendicular to $\hat{\mathbf{n}}$. The dielectric tensor should be applied to the electric field in the following way:

$$\varepsilon_i \cdot \mathbf{E}_i = \left(\varepsilon_{\perp,i} - \varepsilon_{\parallel,i}\right)\hat{\mathbf{n}}\left(\hat{\mathbf{n}} \cdot \mathbf{E}_i\right) + \varepsilon_{\parallel,i}\mathbf{E}_i. \tag{5.35}$$

5.3 DISCRETIZATION WITH TRIANGULAR ELEMENTS

In this section discretization of a scatterer with triangle-shaped elements and linear expansion functions is considered. An example of how to divide a circular scatterer into triangles is shown in Fig. 5.4. The representation of the outer surface is now nearly circular and there is no stair-casing.

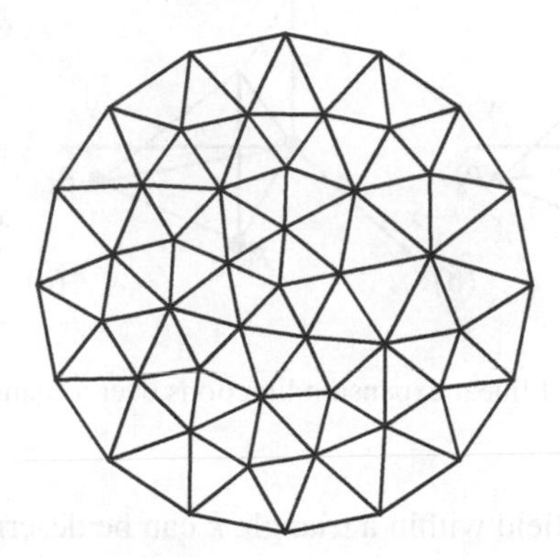

Figure 5.4 Example of triangular mesh for a circular scatterer.

The triangular mesh consists of a large number of triangles that are each identified from three corner points. Corner points may be shared by several triangles. The mesh may thus be described as a set of points, and each triangle is then identified as three connected points. The points may be organized in a matrix $\overline{\overline{P}}$ of the form

$$\overline{\overline{P}} = \begin{bmatrix} x_1 & x_2 & \dots & x_N \\ y_1 & y_2 & \dots & y_N \end{bmatrix}, \tag{5.36}$$

where $\mathbf{r}_i = \hat{\mathbf{x}}x_i + \hat{\mathbf{y}}y_i$ is the position of point i, and N is the total number of points. The coordinates of point i are thus represented by column i.

The triangles may then be represented by a connectivity matrix $\overline{\overline{T}}$ of the form

$$\overline{\overline{T}} = \begin{bmatrix} i_1^{(1)} & i_2^{(1)} & \dots & i_M^{(1)} \\ i_1^{(2)} & i_2^{(2)} & \dots & i_M^{(2)} \\ i_1^{(3)} & i_2^{(3)} & \dots & i_M^{(3)} \\ u_1 & u_2 & \dots & u_M \end{bmatrix}, \tag{5.37}$$

where the first three elements of column k represent which three points are connected to form triangle k, and the fourth element represents which domain the triangle belongs to. The latter index can be used to identify the dielectric constant in the region of a triangle if a structure is divided into different domains with different dielectric constants. Thus, if $\left[i_k^{(1)}\ i_k^{(2)}\ i_k^{(3)}\ u_k\right]^T = \left[2\ 7\ 5\ 3\right]^T$ then the points 2, 7, and 5 form triangle k, and triangle k is in domain 3. M is the total number of triangles. The matrices $\overline{\overline{P}}$, $\overline{\overline{T}}$, and an additional matrix $\overline{\overline{E}}$ representing outer edges, can be obtained from the MATLAB® *pdetool* or another implementation of, e.g., the Delaunay algorithm.

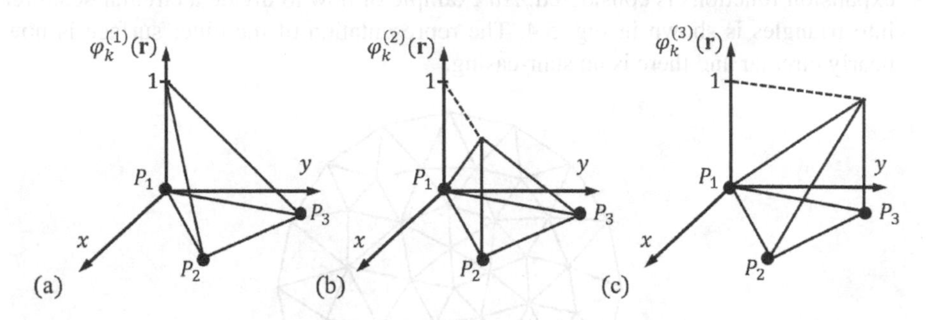

Figure 5.5 Schematic of linear expansion functions over a triangle k.

A linearly varying field within a triangle k can be described from three functions $\varphi_k^{(i)}(\mathbf{r})$, $i = 1,2,3$, illustrated in Fig. 5.5. The expansion functions are characterized by being linear functions within the triangle and zero outside. The triangle has three corner points P_i with positions $\mathbf{r}_k^{(i)}$, $i = 1,2,3$, and the expansion function $\varphi_k^{(i)}(\mathbf{r})$ has the property that it equals 1 at point P_i and equals zero at the other corner points. This can be expressed as

$$\varphi_k^{(i)}(\mathbf{r}_k^{(j)}) = \delta_{i,j}. \tag{5.38}$$

For other positions the expansion functions can be expressed as

$$\varphi_k^{(i)}(\mathbf{r}) = \begin{cases} a_k^{(i)} + b_k^{(i)} x + c_k^{(i)} y, & \mathbf{r} \in \Omega_k \\ 0, & \text{otherwise} \end{cases}, \tag{5.39}$$

where Ω_k represents the region of the triangle, and the coefficients $a_k^{(i)}$, $b_k^{(i)}$, and $c_k^{(i)}$ can be obtained from (5.38) by solving

$$\begin{bmatrix} 1 & x_k^{(1)} & y_k^{(1)} \\ 1 & x_k^{(2)} & y_k^{(2)} \\ 1 & x_k^{(3)} & y_k^{(3)} \end{bmatrix} \begin{bmatrix} a_k^{(i)} \\ b_k^{(i)} \\ c_k^{(i)} \end{bmatrix} = \begin{bmatrix} \delta_{i,1} \\ \delta_{i,2} \\ \delta_{i,3} \end{bmatrix}. \tag{5.40}$$

Instead of using the coordinates x and y alternative coordinates u and v may be introduced such that a position within the triangle k is given from

$$\mathbf{r}_k(u,v) = \mathbf{r}_k^{(1)} + \left(\mathbf{r}_k^{(2)} - \mathbf{r}_k^{(1)}\right) u + \left(\mathbf{r}_k^{(3)} - \mathbf{r}_k^{(1)}\right) v, \quad 0 \le u+v \le 1, \quad u,v \ge 0. \tag{5.41}$$

For integration purposes the differential area on triangle k may then be expressed as

$$d^2r = \left| \frac{\partial \mathbf{r}_k(u,v)}{\partial u} \times \frac{\partial \mathbf{r}_k(u,v)}{\partial v} \right| dudv = 2\Delta_k dudv, \tag{5.42}$$

where Δ_k is the area of triangle k, which can be obtained as

$$\Delta_k = \frac{1}{2}\left|(x_k^{(2)} - x_k^{(1)})(y_k^{(3)} - y_k^{(1)}) - (y_k^{(2)} - y_k^{(1)})(x_k^{(3)} - x_k^{(1)})\right|. \tag{5.43}$$

In terms of the new coordinates the expansion functions can be expressed simply as

$$\varphi_k^{(i)}(u,v) = \begin{cases} 1-u-v, & i=1 \\ u, & i=2\,, \quad 0 \le u+v \le 1. \\ v, & i=3 \end{cases} \tag{5.44}$$

Once the transformation has been made from (x,y) to (u,v), the expansion functions are independent of the triangle number k. The expansion functions in terms of u and v are shown in Fig. 5.6

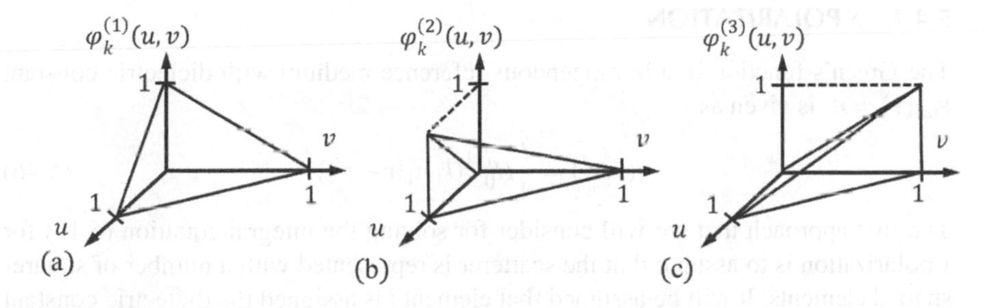

Figure 5.6 Schematic of linear expansion functions over a triangle k.

If we consider s polarization the electric field within the region of the triangular mesh may now be represented as

$$E(\mathbf{r}) = \sum_{k=1}^{M}\sum_{j=1}^{3} \varphi_k^{(j)}(\mathbf{r})a_k^{(j)}, \tag{5.45}$$

where $a_k^{(j)}$ is the value of the electric field at corner j of the triangle k. However, since the same point may be shared by several triangles, and the field must be continuous, several of the coefficients $a_k^{(j)}$ must assume the same value. An example of one point being shared by seven triangles is shown in Fig. 5.7, where the net effect of one specific value of the field at the point is that seven of the coefficients $a_k^{(j)}$ assume the same value, and the resulting function governed by this value is shown in the figure as a total of seven triangle expansion functions with the same value in the point.

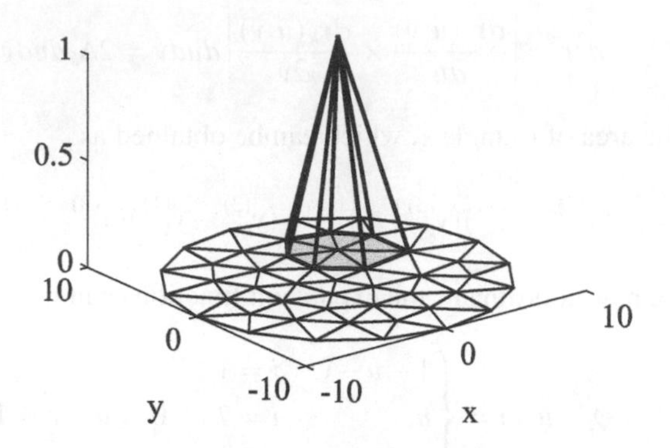

Figure 5.7 Example of total function related to one point in a triangular mesh.

5.4 SCATTERER IN A HOMOGENEOUS MEDIUM

5.4.1 *S* POLARIZATION

The Green's function in a homogeneous reference medium with dielectric constant $\varepsilon_{\mathrm{ref}}(\mathbf{r}) = \varepsilon_1$ is given as

$$g(\mathbf{r},\mathbf{r}') = \frac{i}{4} H_0^{(1)}(k_0 n_1 |\mathbf{r}-\mathbf{r}'|). \tag{5.46}$$

The first approach that we will consider for solving the integral equation (5.12) for *s* polarization is to assume that the scatterer is represented with a number of square-shaped elements. It will be assumed that element i is assigned the dielectric constant ε_i, and that the electric field in element i is constant and given by E_i. The field in element i will be evaluated in the center of the element at the position $\mathbf{r}_i$. The reference field in element i is given as $E_{0,i} \equiv E_0(\mathbf{r}_i)$.

With these assumptions the integral equation (5.12) can be expressed in discrete form as

$$E_i = E_{0,i} + \sum_j g_{ij} k_0^2 (\varepsilon_j - \varepsilon_{\mathrm{ref}}) E_j, \tag{5.47}$$

where

$$g_{ij} = P \int_j g(\mathbf{r}_i, \mathbf{r}') d^2 r'. \tag{5.48}$$

Here, the integral is over discretization element j, and P refers to the principal value integral meaning that for $i = j$ the point $\mathbf{r}' = \mathbf{r}_i$ is excluded from the integral. This is perfectly acceptable as the singularity of g is only a logarithmic singularity.

In a simple approach we may use the approximations

$$g_{ij} \approx \begin{cases} g(\mathbf{r}_i, \mathbf{r}_j)\Delta_j, & i \neq j \\ \frac{i}{2(ka)^2}\left[kaH_1^{(1)}(ka) + i\frac{2}{\pi} \right] \Delta_i, & i = j \end{cases}, \tag{5.49}$$

where $k = k_0 n_1$, $a = \sqrt{\Delta_j/\pi}$, and Δ_j is the area of element j. The analytic expression for $i = j$ is obtained from (5.48) by assuming that the element $i = j$ is circular shaped with a radius a chosen such that the circle has the same area as the corresponding square-shaped element. With this assumption the integral can be evaluated analytically [25]. Note that $g_{i,i}$ vanishes in the limit $\Delta_i \to 0$.

We now introduce the following vectors

$$\overline{E} = \begin{bmatrix} E_1 & E_2 & \ldots & E_N \end{bmatrix}^T, \tag{5.50}$$

$$\overline{E}_0 = \begin{bmatrix} E_{0,1} & E_{0,2} & \ldots & E_{0,N} \end{bmatrix}^T. \tag{5.51}$$

In addition we introduce the matrix $\overline{\overline{g}}$ with the matrix elements g_{ij}, and a dielectric constant matrix $\overline{\overline{\varepsilon}}$ being a diagonal matrix with the diagonal elements ε_i.

The discretized integral equation (5.47) can now be written in matrix form as

$$\left(\overline{\overline{I}} - \overline{\overline{g}} k_0^2 \left(\overline{\overline{\varepsilon}} - \overline{\overline{I}} \varepsilon_{\text{ref}}\right)\right) \overline{E} = \overline{E}_0, \tag{5.52}$$

and the total field inside the scatterer is obtained as

$$\overline{E} = \left(\overline{\overline{I}} - \overline{\overline{g}} k_0^2 \left(\overline{\overline{\varepsilon}} - \overline{\overline{I}} \varepsilon_{\text{ref}}\right)\right)^{-1} \overline{E}_0. \tag{5.53}$$

When the field has been calculated inside the scatterer using this expression, the field can be obtained at any other position by using the integral equation (5.12).

The scattered far field can be obtained by using the far-field Green's function for a homogeneous reference medium (4.21), which leads to

$$E_{\text{scat}}^{(\text{ff})}(r,\theta) = \frac{1}{4}\sqrt{\frac{2}{\pi k r}} e^{ikr} e^{i\pi/4} \int e^{-ik\hat{\mathbf{r}}\cdot\mathbf{r}'} k_0^2 \left(\varepsilon(\mathbf{r}') - \varepsilon_{\text{ref}}\right) E(\mathbf{r}') d^2 r'. \tag{5.54}$$

Here, $\mathbf{r} = \hat{\mathbf{x}} r\cos\theta + \hat{\mathbf{y}} r \sin\theta$, $r = |\mathbf{r}|$, and $\hat{\mathbf{r}} = \mathbf{r}/r$. If the incident field is a plane wave the scattered far field can be used to obtain, for example, the extinction and scattering cross sections.

Within the approximation considered here, with square-shaped elements and constant fields in elements, the far field can be calculated using

$$E_{\text{scat}}^{(\text{ff})}(r,\theta) = \frac{1}{4}\sqrt{\frac{2}{\pi k r}} e^{ikr} e^{i\pi/4} \sum_j e^{-ik\hat{\mathbf{r}}\cdot\mathbf{r}_j} k_0^2 \left(\varepsilon_j - \varepsilon_{\text{ref}}\right) E_j \Delta_j. \tag{5.55}$$

If the reference field is generated by a distribution of currents in the region of the scatterer, the total far field propagating away from the region of the scatterer is given as $E^{(\text{ff})}(\mathbf{r}) = E_{\text{scat}}^{(\text{ff})}(r,\theta) + E_0^{(\text{ff})}(r,\theta)$, where

$$E_0^{(\text{ff})}(r,\theta) = i\omega\mu_0 \frac{1}{4}\sqrt{\frac{2}{\pi k r}} e^{ikr} e^{i\pi/4} \int e^{-ik\hat{\mathbf{r}}\cdot\mathbf{r}'} J(\mathbf{r}') d^2 r'. \tag{5.56}$$

Here, the far field $E^{(\text{ff})}(\mathbf{r})$ can be used to calculate the radiation pattern and by integration the total radiated power. The total power emitted by a point source may also be related to the field itself at the position of the point source, which will be discussed further in the example in Sec. 5.8.

In order to illustrate the convergence of the method, the case of a plane wave of wavelength 700 nm propagating along the x-axis and being incident on a circular cylinder with dielectric constant $\varepsilon_2 = 12$ and radius a = 200 nm is considered in Fig. 5.8. Calculations are shown for the magnitude of the electric field as obtained with the GFAIEM using square-shaped cells of different widths: 400 nm / 11, 400 nm / 21, and 400 nm / 41. The expression (5.49) was applied for the Green's function. The diameter of the cylinder is thus discretized in 11, 21 or 41 elements. Also

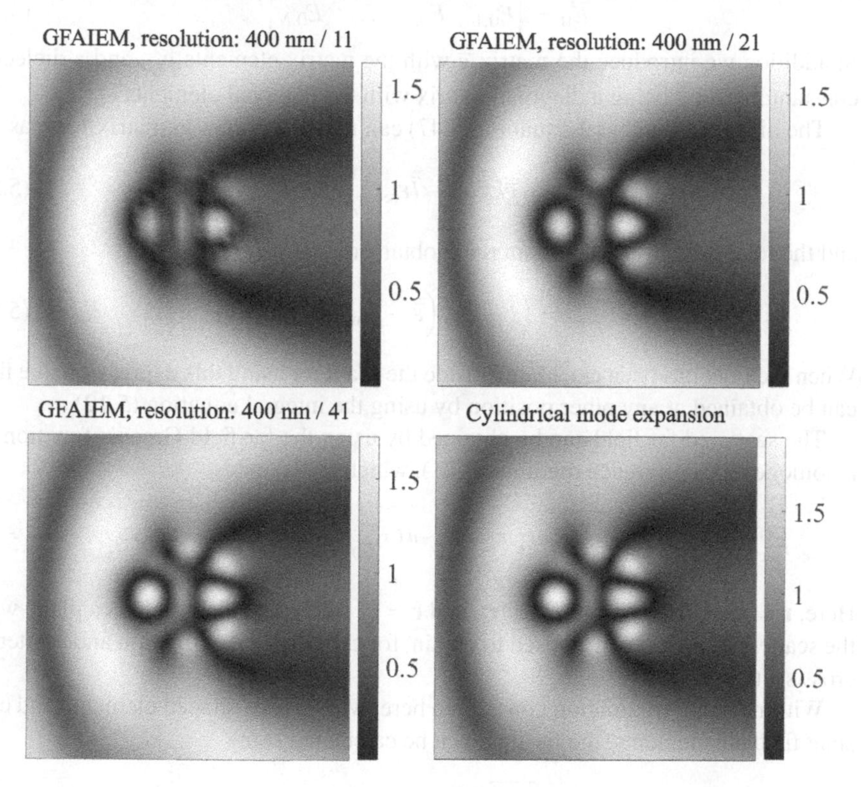

Figure 5.8 Field magnitude calculated with the GFAIEM using square-shaped elements for a plane wave of wavelength 700 nm being incident on a cylinder with dielectric constant $\varepsilon_2 = 12$ and radius a = 200 nm in a background material with dielectric constant $\varepsilon_1 = 1$. Calculations are shown based on square-shaped elements of width 400 nm / 11, 400 nm / 21, and 400 nm / 41, respectively. Also shown is the exact analytic result obtained with cylindrical mode-expansion theory (Appendix D).

shown is the exact analytic result obtained using the cylindrical mode-expansion theory presented in Appendix D. As a rule of thumb, at least 10 elements are needed per wavelength in the medium, which is here equivalent to 20 elements across the cylinder. The cases of 21 or 41 elements across the diameter look reasonably good.

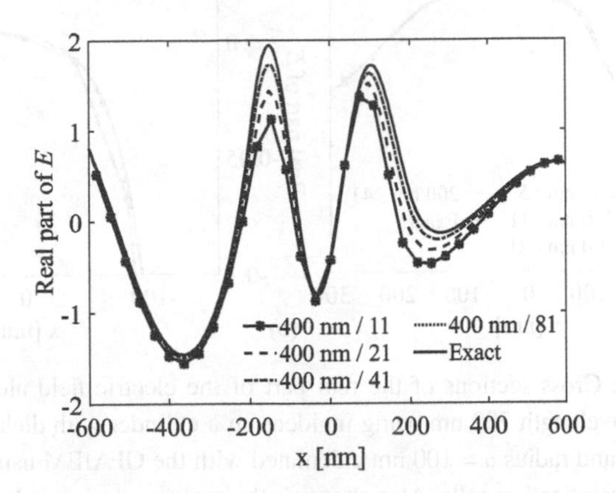

Figure 5.9 Cross sections of the real part of the electric field along the x-axis for the same situation as in Fig. 5.8 using different sizes of square-shaped discretization cells. Also shown is the exact analytic result.

In order to give a more precise comparison with the analytic result, cross sections of the real part of the field along the x-axis are shown in Fig. 5.9 for different sizes of discretization cells. Again, also the exact result is shown for comparison. The field appears to converge toward the exact analytic result as the discretization cell size decreases. For the case of 81 elements across the diameter, the numerical result practically coincides with the exact result to the extent that the curve for the numerical result is not distinguishable from the exact result on the scale of the figure. Note that if (5.48) is used instead of the approximation (5.49), a quite similar result is obtained. In that case though, the result for the side length of elements of 400 nm / 81 will be barely visible.

One more example of the convergence is presented in Fig. 5.10 for a plane wave of wavelength 700 nm being incident on a silver cylinder with $\varepsilon_2 = -22.99 + i0.395$ and radius a = 100 nm. The dielectric constant of silver was obtained from [40]. Although the dielectric constant is quite large, and with a quite large negative real part, the GFAIEM has no problem in handling this case for s polarization. The results using discretization cells of size 200 nm /21 and 200 nm / 41 coincide with the exact result on the scale of the figure. This is a significant difference compared with p polarization. In the next subsection we shall see that when using square-shaped elements in the GFAIEM for a metal cylinder for p polarization, the convergence is not acceptable, and better discretization approaches are required.

We now consider the alternative strategy of using a triangular mesh as in Fig. 5.4 described by N points, and M triangles.

The electric field at point i is given by

$$E_i = E(\mathbf{r}_i), \tag{5.57}$$

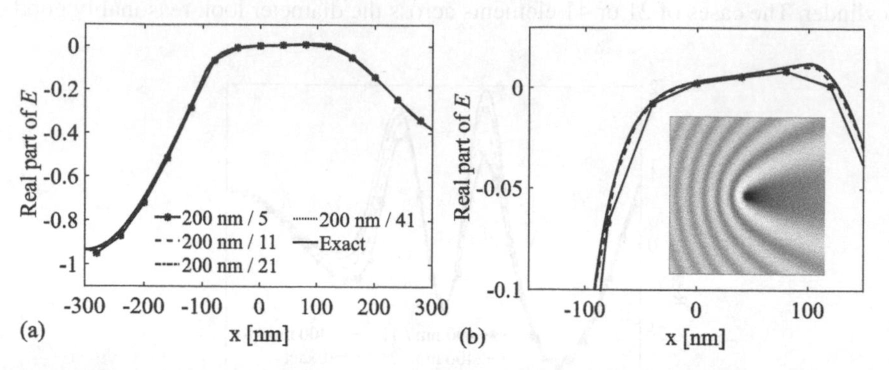

Figure 5.10 (a) Cross sections of the real part of the electric field along the x-axis for a plane wave of wavelength 700 nm being incident on a cylinder with dielectric constant $\varepsilon_2 = -22.99 + i0.395$ and radius $a = 100$ nm calculated with the GFAIEM using different sizes of square-shaped discretization cells. Also shown is the exact analytic result. (b) Zoom-in on the result in (a). Inset: magnitude of electric field in a 2000 nm × 2000 nm window.

and the corresponding reference field is given as

$$E_{0,i} = E_0(\mathbf{r}_i). \tag{5.58}$$

The integral equation may now be discretized by using the expansion of the field inside the scatterer using (5.45) and point-matching:

$$E_i = E_{0,i} + \sum_{k=1}^{M} \sum_{j=1}^{3} A_{i,j}^{(k)} a_k^{(j)}, \quad i = 1, 2, \ldots, N, \tag{5.59}$$

where

$$A_{i,j}^{(k)} = \int g(\mathbf{r}_i, \mathbf{r}') k_0^2 \left(\varepsilon(\mathbf{r}') - \varepsilon_{\text{ref}}\right) \varphi_k^{(j)}(\mathbf{r}') d^2 r' \tag{5.60}$$

are the coefficients of a matrix $\overline{\overline{A}}^{(k)}$ with N rows and 3 columns.

The following vectors are now introduced

$$\overline{E} = \begin{bmatrix} E_1 & E_2 & \ldots & E_N \end{bmatrix}^T, \tag{5.61}$$

$$\overline{E}_0 = \begin{bmatrix} E_{0,1} & E_{0,2} & \ldots & E_{0,N} \end{bmatrix}^T, \tag{5.62}$$

and

$$\overline{a}^{(k)} = \begin{bmatrix} a_k^{(1)} & a_k^{(2)} & a_k^{(3)} \end{bmatrix}^T. \tag{5.63}$$

The coefficients $a_k^{(j)}$ are related to the field coefficients E_i through yet another matrix relation

$$\overline{a}^{(k)} = \overline{\overline{P}}^{(k)} \overline{E}, \tag{5.64}$$

where $\overline{\overline{P}}^{(k)}$ is a matrix with N columns and 3 rows. In each row of $\overline{\overline{P}}^{(k)}$, all elements are equal to 0 except for one element that equals 1. The element that equals one is found at column $i_k^{(1)}$, $i_k^{(2)}$, and $i_k^{(3)}$ for rows 1, 2, and 3, respectively. The three elements that equal 1 are thus identified from column k of the matrix $\overline{\overline{T}}$ (See Eq. 5.37).

The discretized integral equation may now finally be expressed as

$$\overline{E} = \overline{E}_0 + \left(\sum_{k=1}^{M} \overline{\overline{A}}^{(k)} \overline{\overline{P}}^{(k)} \right) \overline{E}, \tag{5.65}$$

where the total matrix within the parantheses will be of size $N \times N$. Note that it is not very efficient to carry out the matrix products $\overline{\overline{A}}^{(k)}\overline{\overline{P}}^{(k)}$. For each k just add the three columns of $\overline{\overline{A}}^{(k)}$ to three different columns determined from column k of $\overline{\overline{T}}$ in a matrix of size $N \times N$.

For the numerical integration in (5.60) rigorous integration may be applied for triangles that are close to $\mathbf{r}_i$, while for triangles farther away, this will be too time consuming. In the latter case the integration can be carried out approximately by first introducing

$$f(\mathbf{r}, \mathbf{r}') \equiv g(\mathbf{r}, \mathbf{r}') k_0^2 \left(\varepsilon(\mathbf{r}') - \varepsilon_{\text{ref}}(\mathbf{r}') \right), \tag{5.66}$$

and thus for $\mathbf{r}'$ on triangle k the following approximation applies:

$$f(\mathbf{r}_i, \mathbf{r}') \approx f(\mathbf{r}_i, \mathbf{r}_k^{(1)}) \varphi_k^{(1)}(\mathbf{r}') + f(\mathbf{r}_i, \mathbf{r}_k^{(2)}) \varphi_k^{(2)}(\mathbf{r}') + f(\mathbf{r}_i, \mathbf{r}_k^{(3)}) \varphi_k^{(3)}(\mathbf{r}'),$$
$$\mathbf{r}' \in \Omega_k. \tag{5.67}$$

The coefficients $f(\mathbf{r}_i, \mathbf{r}_k^{(j)})$ can be obtained by evaluating f at three positions. The following integral over triangle k

$$I^{(j)} = \int_{\Omega_k} f(\mathbf{r}_i, \mathbf{r}') \varphi_k^{(j)}(\mathbf{r}') d^2 r' \tag{5.68}$$

may now be carried out analytically by using

$$\int_{u=0}^{1} \int_{v=0}^{1-u} \varphi_k^{(i)}(u,v) \varphi_k^{(j)}(u,v) du dv = \frac{1}{24} (1 + \delta_{i,j}). \tag{5.69}$$

Thus,

$$I^{(j)} = \frac{1}{12} \Delta_k \left(f(\mathbf{r}_k^{(1)})(1 + \delta_{j,1}) + f(\mathbf{r}_k^{(2)})(1 + \delta_{j,2}) + f(\mathbf{r}_k^{(3)})(1 + \delta_{j,3}) \right), \tag{5.70}$$

where Δ_k is the area of triangle k given in (5.43).

Once the field has been calculated at all points in the mesh for the scatterer, which shall now be referred to as the inner mesh, the field can be calculated on another mesh covering a region outside the scatterer (outer mesh). It is preferable that the mesh points for the outer mesh on the scatterer boundary coincide with mesh points for the inner mesh. An example of an inner and outer mesh is shown in Fig. 5.11(a) for a

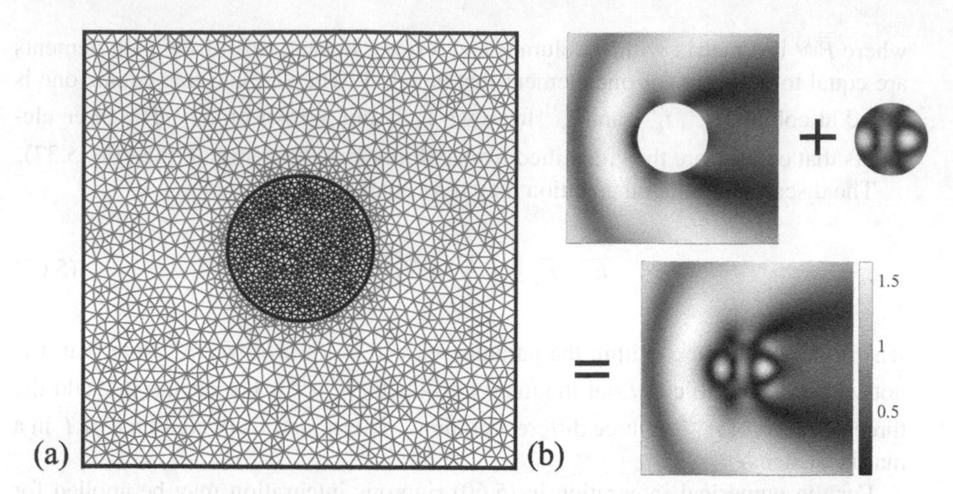

Figure 5.11 (a) Two triangular meshes for the inside of a circular scatterer and a region outside, respectively, with a factor of two difference in the maximum side length of triangles. (b) Calculation of electric field magnitude inside a circular scatterer of radius $a = 200$ nm and dielectric constant $\varepsilon_2 = 12$ being illuminated by a plane wave of wavelength 700 nm obtained using the inner mesh, the calculation of the field outside the scatterer using the outer mesh, and the fields combined in one plot.

circular scatterer of radius $a = 200$ nm. Here, the maximum side length of triangles in the inner mesh is approximately 20 nm, while the maximum side length of triangles in the outer mesh is twice as large. The field within the scatterer calculated using the inner mesh is shown in Fig. 5.11(b) for the case of the scatterer having dielectric constant $\varepsilon_2 = 12$ being illuminated by a plane wave of wavelength 700 nm. Also shown is the field calculated on the outer mesh, and the combined fields.

The field on all mesh points on the outer mesh can be obtained from the result for the field inside the scatterer using the Green's function integral equation. Consider a position of one of the outer mesh points $\mathbf{r}$. The field at this position can be obtained by first defining

$$f_i \equiv f(\mathbf{r}, \mathbf{r}_i). \tag{5.71}$$

If the position $\mathbf{r}$ coincides with one of the mesh points $\mathbf{r}_i$ then f_i will be infinity. In that case the already calculated field E_i at this position is assigned to $E(\mathbf{r})$.

Otherwise, for $\mathbf{r}'$ on triangle k

$$f(\mathbf{r}, \mathbf{r}') \approx f_{T_{k,1}} \varphi_k^{(1)}(\mathbf{r}') + f_{T_{k,2}} \varphi_k^{(2)}(\mathbf{r}') + f_{T_{k,3}} \varphi_k^{(3)}(\mathbf{r}'), \quad \mathbf{r}' \in \Omega_k, \tag{5.72}$$

where $T_{k,i}$ refers to the element in column k and row i of the matrix $\overline{\overline{T}}$. The electric field on this triangle can be approximated in a similar way as

$$E(\mathbf{r}') \approx E_{T_{k,1}} \varphi_k^{(1)}(\mathbf{r}') + E_{T_{k,2}} \varphi_k^{(2)}(\mathbf{r}') + E_{T_{k,3}} \varphi_k^{(3)}(\mathbf{r}'). \tag{5.73}$$

By using these approximations in the integral equation the field at position **r** is obtained as

$$E(\mathbf{r}) = E_0(\mathbf{r}) + \frac{1}{12}\sum_{k=1}^{M} \Delta_k \left(\overline{f}^{(k)}\right)^T \overline{\overline{B}}\,\overline{E}^{(k)}, \tag{5.74}$$

where

$$\overline{f}^{(k)} = \left[f_{T_{k,1}} \quad f_{T_{k,2}} \quad f_{T_{k,3}}\right]^T, \tag{5.75}$$

$$\overline{E}^{(k)} = \left[E_{T_{k,1}} \quad E_{T_{k,2}} \quad E_{T_{k,3}}\right]^T, \tag{5.76}$$

and

$$\overline{\overline{B}} = \begin{bmatrix} 2 & 1 & 1 \\ 1 & 2 & 1 \\ 1 & 1 & 2 \end{bmatrix}. \tag{5.77}$$

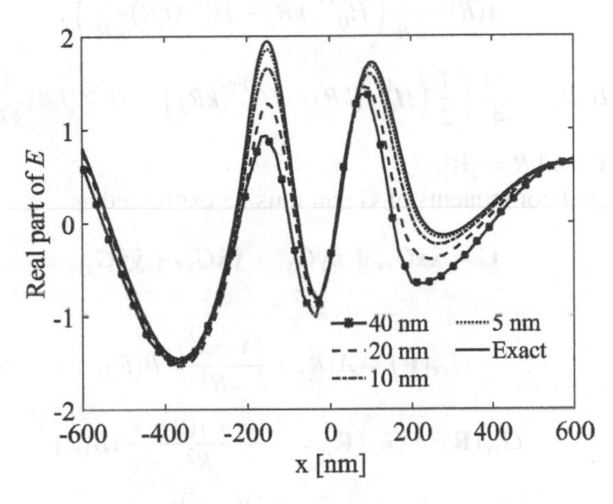

Figure 5.12 Cross sections of the real part of the electric field along the x-axis for the same situation as in Fig. 5.11 with different maximum side lengths of triangles in the inner mesh, and twice the maximum triangle side length in the outer mesh. Also shown is the exact analytic result.

Cross sections of the real part of the electric field for the same scatterer using an inner triangular mesh with different maximum side length of triangles and an outer mesh with twice the maximum triangle side length is shown in Fig. 5.12. The result can be compared with Fig. 5.9 showing a similar calculation based on square-shaped discretization elements. The convergence is quite similar using either type of discretization element.

5.4.2 *P* POLARIZATION

The dyadic Green's tensor for a homogeneous reference medium with dielectric constant $\varepsilon_{\text{ref}}(\mathbf{r}) = \varepsilon_1$ is given as

$$\mathbf{G}(\mathbf{r},\mathbf{r}') = \left(\mathbf{I} + \frac{1}{k^2}\nabla\nabla\right)\frac{i}{4}H_0^{(1)}(k|\mathbf{r}-\mathbf{r}'|), \tag{5.78}$$

where $k = k_0 n_1$.

This leads to the following analytic expression for the Green's tensor

$$\mathbf{G}(\mathbf{r},\mathbf{r}') = \mathbf{I}A(R) + \frac{\mathbf{RR}}{R^2}B(R), \tag{5.79}$$

where

$$A(R) = \frac{i}{4}\left(H_0^{(1)}(kR) - H_1^{(1)}(kR)\frac{1}{kR}\right), \tag{5.80}$$

$$B(R) = \frac{-i}{4}\left(\frac{1}{2}\left(H_0^{(1)}(kR) - H_2^{(1)}(kR)\right) - H_1^{(1)}(kR)\frac{1}{kR}\right), \tag{5.81}$$

with $\mathbf{R} = \mathbf{r} - \mathbf{r}'$ and $R = |\mathbf{R}|$.

The individual components of $\mathbf{G}$ can thus be expressed as

$$\mathbf{G} = \hat{\mathbf{x}}\hat{\mathbf{x}}G_{xx} + \hat{\mathbf{x}}\hat{\mathbf{y}}G_{xy} + \hat{\mathbf{y}}\hat{\mathbf{x}}G_{yx} + \hat{\mathbf{y}}\hat{\mathbf{y}}G_{yy}, \tag{5.82}$$

with

$$G_{xx}(\mathbf{R}) = A(R) + \frac{(x-x')^2}{R^2}B(R), \tag{5.83}$$

$$G_{xy}(\mathbf{R}) = G_{yx}(\mathbf{R}) = \frac{(x-x')(y-y')}{R^2}B(R), \tag{5.84}$$

$$G_{yy}(\mathbf{R}) = A(R) + \frac{(y-y')^2}{R^2}B(R). \tag{5.85}$$

The first approach that will be considered for solving the integral equation for p polarization (5.31) is to assume that the scatterer is represented with a number of square-shaped elements. It will be assumed that element i is assigned the dielectric constant tensor

$$\boldsymbol{\varepsilon}_i = \hat{\mathbf{x}}\hat{\mathbf{x}}\varepsilon_{xx,i} + \hat{\mathbf{x}}\hat{\mathbf{y}}\varepsilon_{xy,i} + \hat{\mathbf{y}}\hat{\mathbf{x}}\varepsilon_{yx,i} + \hat{\mathbf{y}}\hat{\mathbf{y}}\varepsilon_{yy,i},$$

and that the electric field in element i is constant and given by $\mathbf{E}_i = \hat{\mathbf{x}}E_{x,i} + \hat{\mathbf{y}}E_{y,i}$. The field in element i will be evaluated in the center of the element at the position $\mathbf{r}_i$. The reference field in element i is given as $\mathbf{E}_{0,i} \equiv \mathbf{E}_0(\mathbf{r}_i) = \hat{\mathbf{x}}E_{0,x,i} + \hat{\mathbf{y}}E_{0,y,i}$.

With these assumptions, the integral equation (5.31) can be expressed in discrete form as

$$\mathbf{E}_i = \mathbf{E}_{0,i} + \sum_j \mathbf{G}_{ij} \cdot k_0^2(\boldsymbol{\varepsilon}_j - \mathbf{I}\varepsilon_{\text{ref}}) \cdot \mathbf{E}_j, \tag{5.86}$$

where

$$\mathbf{G}_{ij} = \int_j \mathbf{G}(\mathbf{r}_i,\mathbf{r}')d^2r'. \tag{5.87}$$

For $i = j$ it should be noted that the integrand is highly singular, and it is not sufficient to just exclude the point $\mathbf{r}' = \mathbf{r}_i$ from the integration. A method of handling the singularity is to convert the area integral to a surface integral over a surface placed away from the singular point [107, 108]. The first step is to apply a modified version of (5.78) with differential operators applied to the primed coordinates, leading to

$$\int_i \mathbf{G}(\mathbf{r}_i, \mathbf{r}')d^2r' = \int_i \left(\mathbf{I} + \frac{1}{k^2}\nabla'\nabla'\right) g^{(d)}(\mathbf{r}_i, \mathbf{r}')d^2r'. \tag{5.88}$$

The scalar Green's function satisfies

$$g^{(d)}(\mathbf{r}, \mathbf{r}') = -\frac{1}{k^2}\nabla^2 g^{(d)}(\mathbf{r}, \mathbf{r}') - \frac{1}{k^2}\delta(\mathbf{r} - \mathbf{r}'), \tag{5.89}$$

which is inserted into (5.88). Gauss's law is then applied, which leads to

$$\int_i \mathbf{G}(\mathbf{r}_i, \mathbf{r}')d^2r' = -\frac{1}{k^2}\mathbf{I} + \frac{1}{k^2}\int_{\partial i} \left(\hat{\mathbf{n}}'\nabla' - \mathbf{I}\hat{\mathbf{n}}'\cdot\nabla'\right) g^{(d)}(\mathbf{r}_i, \mathbf{r}')dl', \tag{5.90}$$

where ∂i refers to the boundary of element i.

In the limit of very small elements, the integral (5.90) does not vanish but instead it approaches a constant value

$$\mathbf{G}_{ii} \approx -\frac{1}{k^2}\frac{1}{2}\mathbf{I}. \tag{5.91}$$

This means that the radiation generated by the polarization of element i gives a very significant contribution to the field in the same element, which again partly induces the polarization in that element. Note that the expression (5.91) applies to a square-shaped element but not necessarily to elements of other shapes. A slightly better approximation for $\mathbf{G}_{ii}$ is given as

$$\mathbf{G}_{ii} \approx -\frac{1}{k^2}\frac{1}{2}\mathbf{I} + \frac{1}{2}\frac{i}{2(ka)^2}\left[kaH_1^{(1)}(ka) + i\frac{2}{\pi}\right]\Delta_i, \tag{5.92}$$

where $\pi a^2 = \Delta_i$ is the area of element i. Note that the second term on the right-hand side vanishes for $\Delta_i \to 0$. This expression is derived in exercise 5.

An approximation of the Green's tensor elements is thus given as

$$\mathbf{G}_{ij} \approx \begin{cases} \mathbf{G}(\mathbf{r}_i, \mathbf{r}_j)\Delta_j, & i \neq j \\ -\frac{1}{k^2}\frac{1}{2}\mathbf{I}, & i = j \end{cases}. \tag{5.93}$$

We may now introduce the vectors

$$\overline{E}_\alpha = \left[E_{\alpha,1} \quad E_{\alpha,2} \quad \ldots \quad E_{\alpha,N}\right]^T, \tag{5.94}$$

$$\overline{E}_{0,\alpha} = \left[E_{0,\alpha,1} \quad E_{0,\alpha,2} \quad \ldots \quad E_{0,\alpha,N}\right]^T, \tag{5.95}$$

where $\alpha = x$ or y. We also introduce the matrices $\overline{\overline{G}}_{\alpha,\beta}$ with the matrix elements $G_{\alpha,\beta,i,j} = \hat{\alpha}\cdot\mathbf{G}_{ij}\cdot\hat{\beta}$, where α and β can be equal to x or y, and the dielectric constant matrices $\overline{\overline{\varepsilon}}_{\alpha,\beta}$ as diagonal matrices with diagonal elements $\varepsilon_{\alpha,\beta,i} = \hat{\alpha}\cdot\varepsilon_i\cdot\hat{\beta}$.

The equation (5.86) can now be expressed in matrix form as

$$\left(\overline{\overline{I}} - \begin{bmatrix} \overline{\overline{G}}_{xx} & \overline{\overline{G}}_{xy} \\ \overline{\overline{G}}_{yx} & \overline{\overline{G}}_{yy} \end{bmatrix} k_0^2 \left(\begin{bmatrix} \overline{\overline{\varepsilon}}_{xx} & \overline{\overline{\varepsilon}}_{xy} \\ \overline{\overline{\varepsilon}}_{yx} & \overline{\overline{\varepsilon}}_{yy} \end{bmatrix} - \overline{\overline{I}}\varepsilon_{\text{ref}} \right)\right) \begin{bmatrix} \overline{E}_x \\ \overline{E}_y \end{bmatrix} = \begin{bmatrix} \overline{E}_{0,x} \\ \overline{E}_{0,y} \end{bmatrix}. \tag{5.96}$$

After solving (5.96) for the field inside the scatterer, the field at all other positions outside the scatterer can be obtained from the integral equation (5.31).

In order to calculate the scattered far field the far-field Green's tensor must be obtained. This can be obtained straightforwardly by inserting the far-field expression for $g^{(d)}$ given in (4.21) into (5.78), i.e.,

$$\mathbf{G}^{(\text{ff},d)}(\mathbf{r},\mathbf{r}') = \left(\mathbf{I} + \frac{1}{k^2}\nabla\nabla\right)\frac{1}{4}\sqrt{\frac{2}{\pi kr}}e^{ikr}e^{i\pi/4}e^{-ik\hat{\mathbf{r}}\cdot\mathbf{r}'}, \tag{5.97}$$

which leads to

$$\mathbf{G}^{(\text{ff},d)}(\mathbf{r},\mathbf{r}') = \hat{\theta}\hat{\theta}\frac{1}{4}\sqrt{\frac{2}{\pi kr}}e^{ikr}e^{i\pi/4}e^{-ik\hat{\mathbf{r}}\cdot\mathbf{r}'}, \tag{5.98}$$

where $\mathbf{r} = \hat{\mathbf{x}}r\cos\theta + \hat{\mathbf{y}}r\sin\theta$, $\hat{\mathbf{r}} = \mathbf{r}/r$, and $\hat{\theta} = -\hat{\mathbf{x}}\sin\theta + \hat{\mathbf{y}}\cos\theta$. Any term that decays faster with r than $1/\sqrt{r}$ has been ignored.

The scattered far field can thus be expressed as

$$\mathbf{E}_{\text{sc}}^{(\text{ff})}(r,\theta) = \hat{\theta}\left(\frac{1}{4}\sqrt{\frac{2}{\pi kr}}e^{ikr}e^{i\pi/4}\int e^{-ik\hat{\mathbf{r}}\cdot\mathbf{r}'}k_0^2\left(\varepsilon(\mathbf{r}') - \varepsilon_{\text{ref}}\right)\hat{\theta}\cdot\mathbf{E}(\mathbf{r}')d^2r'\right). \tag{5.99}$$

Within the approximation considered here with square-shaped elements and constant fields in elements, this can also be expressed as

$$\mathbf{E}_{\text{sc}}^{(\text{ff})}(r,\theta) = \hat{\theta}\left(\frac{1}{4}\sqrt{\frac{2}{\pi kr}}e^{ikr}e^{i\pi/4}\sum_j e^{-ik\hat{\mathbf{r}}\cdot\mathbf{r}_j}k_0^2\hat{\theta}\cdot(\varepsilon_j - \varepsilon_{\text{ref}})\cdot\mathbf{E}_j\Delta_j\right). \tag{5.100}$$

If the reference field is generated by a distribution of currents in the region of the scatterer, the total far field propagating away from the scatterer region is the sum of $\mathbf{E}_{\text{sc}}^{(\text{ff})}(r,\theta)$ and $\mathbf{E}_0^{(\text{ff})}(r,\theta)$, where

$$\mathbf{E}_0^{(\text{ff})}(r,\theta) = i\omega\mu_0\int\mathbf{G}^{(\text{ff},d)}(\mathbf{r},\mathbf{r}')\cdot\mathbf{J}(\mathbf{r}')d^2r'. \tag{5.101}$$

The total far field can be used to calculate the radiation pattern and the total radiated power.

A few examples of the convergence of the method are given in the following. In Fig. 5.13 the convergence of the GFAIEM is considered for a p-polarized plane wave of wavelength 700 nm, which is propagating along the x-axis and is incident on a circular cylinder with dielectric constant $\varepsilon_2 = 12$ and radius $a = 150$ nm. Calculations are shown for the magnitude of the electric field as obtained with the GFAIEM using square-shaped cells of widths 300 nm / 11 and 300 nm / 41. The exact result obtained with cylindrical-mode-expansion theory is also shown. The expression (5.93) has been applied for the Green's function.

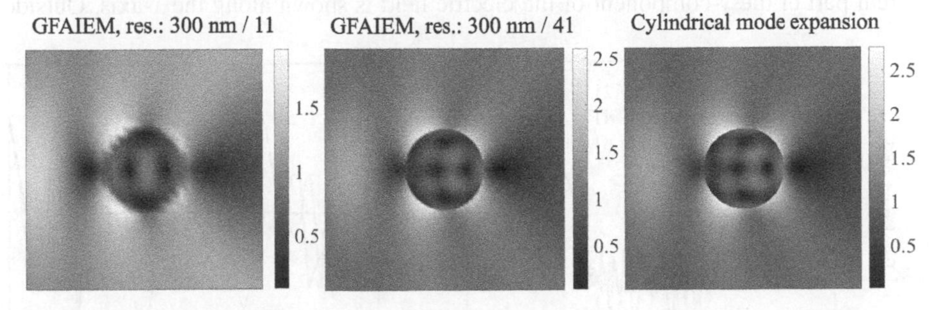

Figure 5.13 Field magnitude calculated with the GFAIEM using square-shaped elements for a p-polarized plane wave of wavelength 700 nm being incident on a cylinder with dielectric constant $\varepsilon_2 = 12$ and radius $a = 150$ nm in a background material with dielectric constant $\varepsilon_1 = 1$. Calculations are shown based on square-shaped elements of width 300 nm / 11 and 300 nm / 41. Also shown is the exact analytic result obtained with cylindrical mode-expansion theory.

Cross sections of the y-component of the electric field along the x-axis as obtained with the GFAIEM using square-shaped elements of different widths are shown in Fig. 5.14 along with the exact analytic result. It appears that the field converges

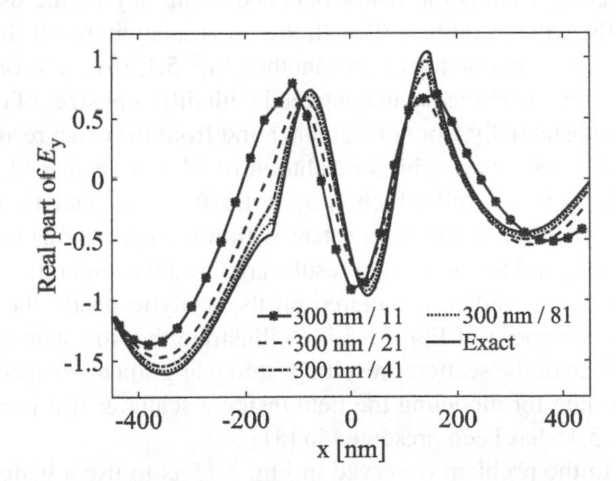

Figure 5.14 Cross sections of the real part of the y-component of the electric field along the x-axis for the same situation as in Fig. 5.13 using different sizes of discretization cells. Also shown is the exact analytic result.

reasonably well toward the analytic result as the size of discretization cells decreases. This example shows that the simple GFAIEM can handle a relatively large dielectric constant if it is positive.

The GFAIEM is considered in Fig. 5.15 for the case of a p-polarized plane wave of wavelength 700 nm being incident on a silver cylinder of radius 10 nm. Again, the real part of the y-component of the electric field is shown along the x-axis. Outside

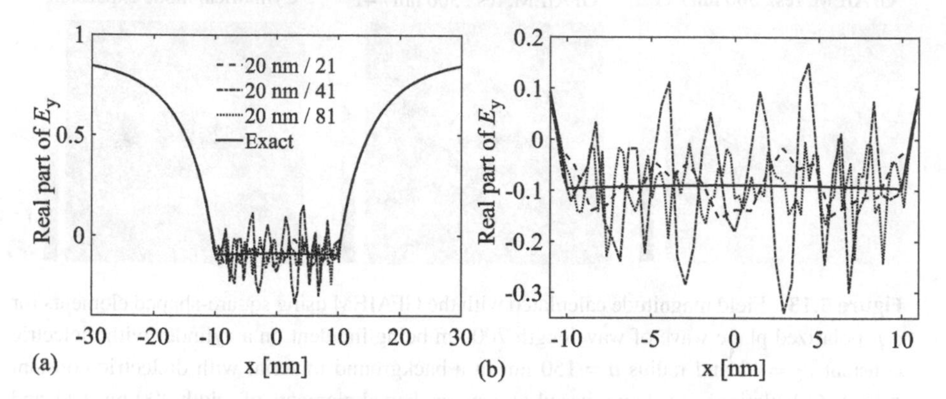

Figure 5.15 (a) Cross sections of the real part of the y-component of the electric field along the x-axis for a plane wave of wavelength 700 nm being incident on a cylinder with dielectric constant $\varepsilon_2 = -22.99 + i0.395$ and radius 10 nm calculated with the GFAIEM using different sizes of square-shaped discretization cells. Also shown is the exact analytic result. (b) Zoom-in on the result in (a).

the cylinder ($|x| > 10$ nm) the result obtained using any of the used sizes of discretization cells matches quite well with the exact analytic result. In that region the different curves are right on top of one another. Fig. 5.15(b) is a zoom-in on the field inside the cylinder. Here, the results obtained with different sizes of discretization elements differ substantially from one another and from the exact result. Although the cylinder is very small, even a high resolution of 81 square-shaped elements across the cylinder leads to a result which is quite far from the analytic result. The analytic result is smooth and almost constant within the cylinder as is expected in the electrostatic limit, and the numerical results are rapidly oscillating, and although the oscillations are approximately centered on the analytic result, the oscillations are quite large. The purpose of Fig. 5.15 is to illustrate that the stair-cased representation of the surface of the scatterer resulting from using square-shaped elements is not always appropriate for modeling the field inside a scatterer if it is a metal. A result similar to Fig. 5.15 has been presented in [51].

A remedy to the problem observed in Fig. 5.15 is to use a better representation of the scatterer surface, which is possible by discretizing the scatterer into triangular elements similar to Fig. 5.11. We will thus consider the discretized integral equation (5.86) once more but for triangular elements. Thus, $\mathbf{r}_i$ now represents the center of triangle i, and Δ_i is the area of triangle i. The three corners of triangle i are referred to as $\mathbf{r}_i^{(u)}$, $u = 1,2,3$. The tensor $\mathbf{G}_{ij}$ in (5.87) should now be obtained by integrating over triangle j. Thus, for $i \neq j$

$$\mathbf{G}_{ij} = \int_{u=0}^{1}\int_{v=0}^{1-u} \mathbf{G}(\mathbf{r}_i, \left\{\mathbf{r}_j^{(1)} + (\mathbf{r}_j^{(2)} - \mathbf{r}_j^{(1)})u + (\mathbf{r}_j^{(3)} - \mathbf{r}_j^{(1)})v\right\})2\Delta_j du dv. \quad (5.102)$$

If the distance between elements i and j is sufficient for the Green's tensor to vary approximately linearly across element j, the following approximation is applicable:

$$\mathbf{G}_{ij} \approx \frac{1}{3}\left(\mathbf{G}(\mathbf{r}_i, \mathbf{r}_j^{(1)}) + \mathbf{G}(\mathbf{r}_i, \mathbf{r}_j^{(2)}) + \mathbf{G}(\mathbf{r}_i, \mathbf{r}_j^{(3)})\right)\Delta_j. \quad (5.103)$$

The tensor $\mathbf{G}_{ii}$ can be calculated by using the expression (5.90). Here, when integrating over the boundary of element i the outward normal vector is needed. For example, when integrating along the straight line of the triangle from $\mathbf{r}_i^{(1)}$ to $\mathbf{r}_i^{(2)}$, the normal vector can be obtained as

$$\hat{\mathbf{n}} = \pm\hat{\mathbf{z}} \times \frac{\mathbf{r}_i^{(2)} - \mathbf{r}_i^{(1)}}{|\mathbf{r}_i^{(2)} - \mathbf{r}_i^{(1)}|}, \quad (5.104)$$

where $+$ is used if $\hat{\mathbf{z}} \cdot \left((\mathbf{r}_i^{(2)} - \mathbf{r}_i^{(1)}) \times (\mathbf{r}_i - \mathbf{r}_i^{(1)})\right) < 0$, and $-$ is used if $\hat{\mathbf{z}} \cdot \left((\mathbf{r}_i^{(2)} - \mathbf{r}_i^{(1)}) \times (\mathbf{r}_i - \mathbf{r}_i^{(1)})\right) > 0$. In addition, the integral over the line from $\mathbf{r}_i^{(1)}$ to $\mathbf{r}_i^{(2)}$ can be carried out by using the coordinate transformation

$$\mathbf{r}' = \mathbf{r}_i^{(1)} + \left(\mathbf{r}_i^{(2)} - \mathbf{r}_i^{(1)}\right)u, \quad (5.105)$$

and thus $dl' = |\mathbf{r}_i^{(2)} - \mathbf{r}_i^{(1)}|du$.

The silver cylinder of radius 10 nm from Fig. 5.15 is considered once more but now the calculation is carried out using the triangle-shaped elements. The result for different numbers of triangular elements within the cylinder is shown in Fig. 5.16. From Fig. 5.16(a) it is clear that with only N = 76 triangles, the field is already very close to the exact analytic result. The oscillations in the field inside the cylinder that occurred when using square-shaped elements have practically disappeared. A zoom-in on the field inside the cylinder is shown in Fig. 5.16(b). For the cross section of the field along the x-axis, the field assigned to a given position is the field within the triangle that the position belongs to. By assigning the field in this way the field becomes piecewise constant. The calculation is clearly closer to the analytic result as the number of triangles increases. It was found in [51] that when using square-shaped elements inside the cylinder, and edge elements that match the surface profile, the oscillations in the field inside the cylinder also disappear. It thus appears that the representation of the scatterer surface is essential for modeling of a metal scatterer. The GFAIEM with triangle-shaped cells has been used for modeling of resonant plasmonic scatterers of different shapes in, for example, Refs. [109] using the method presented in [110]. Sharp corners may themselves support a plasmonic resonant mode and must sometimes be rounded in order to obtain convergence [109].

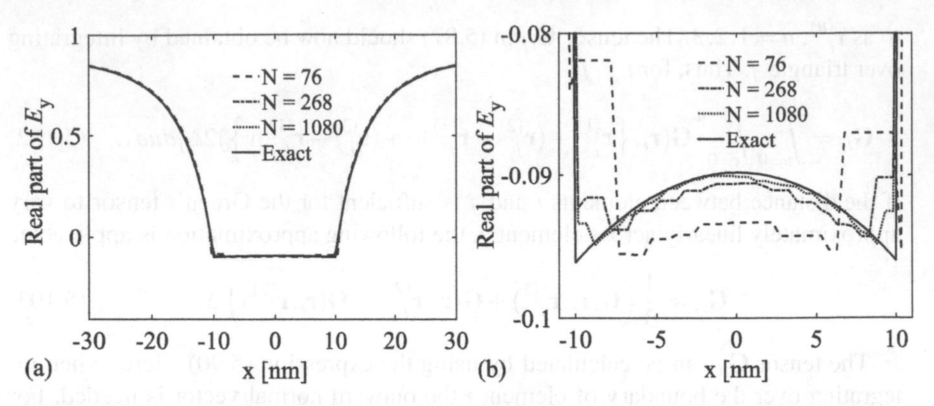

Figure 5.16 (a) Cross sections of the real part of the y-component of the electric field along the x-axis for a plane wave of wavelength 700 nm being incident on a cylinder with dielectric constant $\varepsilon_2 = -22.99 + i0.395$ and radius 10 nm calculated with the GFAIEM using different numbers of triangle-shaped discretization cells. Also shown is the exact analytic result. (b) Zoom-in on the result in (a).

With the assumption of a constant field $\mathbf{E}_2$ inside the circular cylinder, as must be the case in the quasi-static limit, the field in the integrand in (5.31) can be moved outside the integral. For a position $\mathbf{r}$ inside the cylinder the integral equation can then be expressed as

$$\mathbf{E}_2 = \mathbf{E}_0 + \int_{\text{cylinder}} \mathbf{G}(\mathbf{r},\mathbf{r}')d^2r' \cdot \left(k_0^2(\varepsilon_2 - \varepsilon_1)\mathbf{E}_2\right). \tag{5.106}$$

For a circular cylinder in the quasi-static limit, the integral reduces to the same value as $\mathbf{G}_{ii}$ for a square-shaped area element, and the integral equation thus reduces to

$$\mathbf{E}_2 = \mathbf{E}_0 - \frac{1}{2}\frac{(\varepsilon_2 - \varepsilon_1)}{\varepsilon_1}\mathbf{E}_2, \tag{5.107}$$

which can be easily solved as

$$\mathbf{E}_2 = \frac{2\varepsilon_1}{\varepsilon_1 + \varepsilon_2}\mathbf{E}_0. \tag{5.108}$$

In the example in Fig. 5.16 the incident field is given at the cylinder position as $\mathbf{E}_0 \approx \hat{\mathbf{y}}$ (unity amplitude) and (5.108) leads to the real part of the y-component being approximately equal to -0.091, which is approximately the value of the field obtained inside the cylinder in Fig. 5.16.

5.5 SCATTERER ON OR NEAR PLANAR SURFACES

5.5.1 *S* POLARIZATION

The main difference between a scatterer in a homogeneous reference medium as in Fig. 5.1(a), and a scatterer on or near a layered reference structure as in Fig. 5.1(b), is that the reference dielectric function $\varepsilon_{\text{ref}}(\mathbf{r})$ in (5.12) now depends on the position $\mathbf{r}$. Furthermore, the Green's function must now also satisfy (5.8) corresponding to the spatially varying reference medium.

The Green's function for s polarization and a layered reference structure has already been given in Sec. 4.2.1, and there it has been discussed how it can be calculated numerically, and far-field expressions governing both out-of-plane radiation (4.185) and (4.189), and excitation of guided modes for a three-layer structure (4.213) have also been derived.

The far fields are obtained by inserting the far-field Green's functions into the integral equation (5.12). The out-of-plane scattered far field going into the upper half-plane is thus given by

$$E_{\text{sc}}(\mathbf{r}) = \frac{e^{ik_1 r}}{\sqrt{r}} \frac{e^{i\frac{\pi}{4}}}{4} \sqrt{\frac{2}{\pi k_1}} \int e^{-ik_1 \cos\theta x'} \left(e^{-ik_1 \sin\theta y'} + r^{(s)}(\theta) e^{ik_1 \sin\theta y'} \right)$$

$$\times k_0^2 \left(\varepsilon(\mathbf{r}') - \varepsilon_{\text{ref}}(\mathbf{r}') \right) E(\mathbf{r}') d^2 r', \quad 0 < \theta < \pi, \tag{5.109}$$

where $k_1 = k_0 n_1$.

The out-of-plane scattered far field going into the lower half-plane is given by

$$E_{\text{sc}}(\mathbf{r}) = \frac{e^{ik_{LN} r}}{\sqrt{r}} \frac{e^{i\frac{\pi}{4}}}{4} \sqrt{\frac{2}{\pi k_{LN}}} \frac{k_{y,LN}}{k_{y,1}} e^{-ik_{y,LN} d} t^{(s)}(\theta) \int e^{-ik_{LN} \cos\theta x'} e^{ik_{y,1} y'}$$

$$\times k_0^2 \left(\varepsilon(\mathbf{r}') - \varepsilon_{\text{ref}}(\mathbf{r}') \right) E(\mathbf{r}') d^2 r', \quad -\pi < \theta < 0, \tag{5.110}$$

where $k_{LN} = k_0 n_{LN}$.

5.5.2 *P* POLARIZATION

The integral equation (5.31) for p polarization for a scatterer on or near a layered structure requires that the dyadic Green's tensor is known. The Green's tensor will be constructed in the following.

The first step is to obtain a plane-wave expansion of the dyadic Green's tensor for p polarization. This can be achieved by inserting the plane-wave expansion for the direct Green's function (4.157) into the expression (5.78). This leads to

$$\mathbf{G}^{(d)}(\mathbf{r}, \mathbf{r}') = \frac{i}{2\pi} \int_0^\infty \left(\hat{\mathbf{x}}\hat{\mathbf{x}} \frac{k_{y,1}^2}{k_1^2} + \hat{\mathbf{y}}\hat{\mathbf{y}} \frac{k_x^2}{k_1^2} \right) \cos(k_x [x - x']) \frac{e^{ik_{y,1}|y-y'|}}{k_{y,1}} dk_x +$$

$$\frac{i}{2\pi} \int_0^\infty (\hat{\mathbf{x}}\hat{\mathbf{y}} + \hat{\mathbf{y}}\hat{\mathbf{x}}) \frac{-ik_x k_{y,1}}{k_1^2} \frac{y - y'}{|y - y'|} \sin(k_x [x - x']) \frac{e^{ik_{y,1}|y-y'|}}{k_{y,1}} dk_x. \tag{5.111}$$

Now consider a two-layer reference structure with

$$\varepsilon_{\text{ref}}(\mathbf{r}) = \begin{cases} \varepsilon_1, & y > 0 \\ \varepsilon_{L2}, & y < 0 \end{cases}. \tag{5.112}$$

The Green's tensor can now be expressed as

$$\mathbf{G}(\mathbf{r},\mathbf{r}') = \begin{cases} \mathbf{G}^{(\mathrm{d})}(\mathbf{r},\mathbf{r}') + \mathbf{G}^{(\mathrm{i})}(\mathbf{r},\mathbf{r}'), & y,y' > 0 \\ \mathbf{G}^{(\mathrm{t})}(\mathbf{r},\mathbf{r}'), & y' > 0,\ y < 0 \end{cases}, \tag{5.113}$$

where $\mathbf{G}^{(\mathrm{i})}$ represents reflection or the indirect part of the Green's tensor, and $\mathbf{G}^{(\mathrm{t})}$ represents the transmitted part. These parts of the Green's tensor will now be derived.

First consider the field

$$\mathbf{E}_{\mathrm{i},x}(\mathbf{r}) = \mathbf{G}^{(\mathrm{d})}(\mathbf{r},\mathbf{r}')\cdot\hat{\mathbf{x}}, \quad y < y'. \tag{5.114}$$

By using (5.111) this can be expressed as the following field being incident on the interface between the two media

$$\mathbf{E}_{\mathrm{i},x}(\mathbf{r}) = \frac{i}{2\pi}\int_0^\infty \left(\hat{\mathbf{x}}\frac{k_{y,1}^2}{k_1^2}\cos(k_x[x-x']) + \hat{\mathbf{y}}\frac{ik_xk_{y,1}}{k_1^2}\sin(k_x[x-x'])\right)\frac{e^{ik_{y,1}(y'-y)}}{k_{y,1}}dk_x. \tag{5.115}$$

For each plane wave in this expression there will be a reflected wave leading to the total reflected field

$$\mathbf{E}_{\mathrm{r},x}(\mathbf{r}) = \frac{i}{2\pi}\int_0^\infty r^{(p)}(k_x)\left(-\hat{\mathbf{x}}\frac{k_{y,1}^2}{k_1^2}\cos(k_x[x-x'])\right.$$
$$\left. + \hat{\mathbf{y}}\frac{ik_xk_{y,1}}{k_1^2}\sin(k_x[x-x'])\right)\frac{e^{ik_{y,1}(y+y')}}{k_{y,1}}dk_x, \tag{5.116}$$

and a transmitted wave leading to the total transmitted field

$$\mathbf{E}_{\mathrm{t},x}(\mathbf{r}) = \frac{i}{2\pi}\int_0^\infty t_{1,L2}^{(p)}(k_x)\frac{\varepsilon_1}{\varepsilon_{L2}}\left(\hat{\mathbf{x}}\frac{k_{y,1}k_{y,L2}}{k_1^2}\cos(k_x[x-x'])\right.$$
$$\left. + \hat{\mathbf{y}}\frac{ik_xk_{y,1}}{k_1^2}\sin(k_x[x-x'])\right)\frac{e^{ik_{y,1}y'}e^{-ik_{y,L2}y}}{k_{y,1}}dk_x. \tag{5.117}$$

The resulting field is thus given by

$$\mathbf{E}_x(\mathbf{r}) = \begin{cases} \mathbf{E}_{i,x}(\mathbf{r}) + \mathbf{E}_{r,x}(\mathbf{r}), & y > 0 \\ \mathbf{E}_{t,x}(\mathbf{r}), & y < 0 \end{cases}. \tag{5.118}$$

The electromagnetics boundary conditions may now be applied across the boundary at $y = 0$, which leads to the usual Fresnel reflection and transmission coefficients

$$r^{(p)}(k_x) = \frac{k_{y,1}\varepsilon_{L2} - k_{y,L2}\varepsilon_1}{k_{y,1}\varepsilon_{L2} + k_{y,L2}\varepsilon_1}, \tag{5.119}$$

and

$$t_{1,L2}^{(p)}(k_x) = 1 + r^{(p)}(k_x). \tag{5.120}$$

In a similar way, another incident field may be constructed as

$$\mathbf{E}_{i,y}(\mathbf{r}) = \mathbf{G}^{(d)}(\mathbf{r},\mathbf{r}')\cdot\hat{\mathbf{y}}, \quad y < y', \tag{5.121}$$

and the corresponding reflected and transmitted fields can be obtained by constructing a reflected and a transmitted wave for each of the incident plane waves using ordinary Fresnel reflection and transmission theory. This leads finally to the following indirect part of the Green's tensor

$$\mathbf{G}^{(i)}(\mathbf{r},\mathbf{r}') = \frac{i}{2\pi}\int_0^\infty r^{(p)}(k_x)\left(-\hat{\mathbf{x}}\hat{\mathbf{x}}\frac{k_{y,1}^2}{k_1^2} + \hat{\mathbf{y}}\hat{\mathbf{y}}\frac{k_x^2}{k_1^2}\right)\cos(k_x[x-x'])\frac{e^{ik_{y,1}(y+y')}}{k_{y,1}}dk_x +$$

$$\frac{i}{2\pi}\int_0^\infty r^{(p)}(k_x)\left(\hat{\mathbf{y}}\hat{\mathbf{x}} - \hat{\mathbf{x}}\hat{\mathbf{y}}\right)\frac{ik_x k_{y,1}}{k_1^2}\sin(k_x[x-x'])\frac{e^{ik_{y,1}(y+y')}}{k_{y,1}}dk_x, \tag{5.122}$$

and to the following transmitted part of the Green's tensor

$$\mathbf{G}^{(t)}(\mathbf{r},\mathbf{r}') =$$

$$\frac{i}{2\pi}\int_0^\infty t_{1,L2}^{(p)}(k_x)\frac{\varepsilon_1}{\varepsilon_{L2}}\left(\hat{\mathbf{x}}\hat{\mathbf{x}}\frac{k_{y,1}k_{y,L2}}{k_1^2} + \hat{\mathbf{y}}\hat{\mathbf{y}}\frac{k_x^2}{k_1^2}\right)\cos(k_x[x-x'])\frac{e^{ik_{y,1}y'}e^{-ik_{y,L2}y}}{k_{y,1}}dk_x +$$

$$\frac{i}{2\pi}\int_0^\infty t_{1,L2}^{(p)}(k_x)\frac{\varepsilon_1}{\varepsilon_{L2}}\left(\hat{\mathbf{y}}\hat{\mathbf{x}} + \hat{\mathbf{x}}\hat{\mathbf{y}}\frac{k_{y,L2}}{k_{y,1}}\right)\frac{ik_x k_{y,1}}{k_1^2}\sin(k_x[x-x'])\frac{e^{ik_{y,1}y'}e^{-ik_{y,L2}y}}{k_{y,1}}dk_x. \tag{5.123}$$

For an N-layer reference geometry with total thickness d of the layers, the reflected (or indirect) part of the Green's tensor will be unchanged except that the Fresnel reflection coefficient should be the one for the multi-layer geometry instead. The transmitted Green's tensor will instead assume the form

$$\mathbf{G}^{(t)}(\mathbf{r},\mathbf{r}') =$$

$$\frac{i}{2\pi}\int_0^\infty t_{1,LN}^{(p)}(k_x)\frac{\varepsilon_1}{\varepsilon_{LN}}\left(\hat{\mathbf{x}}\hat{\mathbf{x}}\frac{k_{y,1}k_{y,LN}}{k_1^2} + \hat{\mathbf{y}}\hat{\mathbf{y}}\frac{k_x^2}{k_1^2}\right)\cos(k_x[x-x'])\frac{e^{ik_{y,1}y'}e^{-ik_{y,LN}(y+d)}}{k_{y,1}}dk_x +$$

$$\frac{i}{2\pi}\int_0^\infty t_{1,LN}^{(p)}(k_x)\frac{\varepsilon_1}{\varepsilon_{LN}}\left(\hat{\mathbf{y}}\hat{\mathbf{x}} + \hat{\mathbf{x}}\hat{\mathbf{y}}\frac{k_{y,LN}}{k_{y,1}}\right)\frac{ik_x k_{y,1}}{k_1^2}\sin(k_x[x-x'])\frac{e^{ik_{y,1}y'}e^{-ik_{y,LN}(y+d)}}{k_{y,1}}dk_x. \tag{5.124}$$

For a three-layer geometry the Green's tensor for positions $\mathbf{r}$ in the layer $L2$ ($0 > y > -d$) and $y' > 0$ is given by

$$\mathbf{G}(\mathbf{r},\mathbf{r}') =$$

$$\frac{i}{2\pi}\int_0^\infty A(k_x)\frac{\varepsilon_1}{\varepsilon_{L2}}\left(\hat{\mathbf{x}}\hat{\mathbf{x}}\frac{k_{y,1}k_{y,L2}}{k_1^2} + \hat{\mathbf{y}}\hat{\mathbf{y}}\frac{k_x^2}{k_1^2}\right)\cos(k_x[x-x'])\frac{e^{ik_{y,1}y'}e^{-ik_{y,L2}y}}{k_{y,1}}dk_x +$$

$$\frac{i}{2\pi}\int_0^\infty A(k_x)\frac{\varepsilon_1}{\varepsilon_{L2}}\left(\hat{\mathbf{y}}\hat{\mathbf{x}} + \hat{\mathbf{x}}\hat{\mathbf{y}}\frac{k_{y,L2}}{k_{y,1}}\right)\frac{ik_x k_{y,1}}{k_1^2}\sin(k_x[x-x'])\frac{e^{ik_{y,1}y'}e^{-ik_{y,L2}y}}{k_{y,1}}dk_x +$$

$$\frac{i}{2\pi}\int_0^\infty B(k_x)\frac{\varepsilon_1}{\varepsilon_{L2}}\left(-\hat{\mathbf{x}}\hat{\mathbf{x}}\frac{k_{y,1}k_{y,L2}}{k_1^2}+\hat{\mathbf{y}}\hat{\mathbf{y}}\frac{k_x^2}{k_1^2}\right)\cos(k_x[x-x'])\frac{e^{ik_{y,1}y'}e^{+ik_{y,L2}y}}{k_{y,1}}dk_x+$$

$$\frac{i}{2\pi}\int_0^\infty B(k_x)\frac{\varepsilon_1}{\varepsilon_{L2}}\left(\hat{\mathbf{y}}\hat{\mathbf{x}}-\hat{\mathbf{x}}\hat{\mathbf{y}}\frac{k_{y,L2}}{k_{y,1}}\right)\frac{ik_xk_{y,1}}{k_1^2}\sin(k_x[x-x'])\frac{e^{ik_{y,1}y'}e^{+ik_{y,L2}y}}{k_{y,1}}dk_x, \quad (5.125)$$

where

$$A(k_x)=\frac{t^{(p)}_{1,L2}(k_x)}{1+r^{(p)}_{1,L2}(k_x)r^{(p)}_{L2,L3}(k_x)e^{2ik_{y,L2}d}}, \quad (5.126)$$

$$B(k_x)=\frac{t^{(p)}_{1,L2}(k_x)e^{2ik_{y,L2}d}r^{(p)}_{L2,L3}(k_x)}{1+r^{(p)}_{1,L2}(k_x)r^{(p)}_{L2,L3}(k_x)e^{2ik_{y,L2}d}}. \quad (5.127)$$

It is also useful to have expressions for the out-of-plane far-field Green's tensor in order to be able to calculate far-field scattering or radiation. We will first consider the indirect part of the Green's tensor in the far field. The first step is to note that in the far field the integration limits in (5.122) can be replaced with integration from 0 to k_1 since the term in the integrand $\exp(ik_{y,1}(y+y'))$ will be vanishingly small in the far field for $k_x > k_1$. The following coordinate transformation is now applied:

$$x=r\cos\theta, \quad y=r\sin\theta, \quad 0<\theta<\pi, \quad (5.128)$$

$$k_x=k_1\cos\alpha, \quad k_{y,1}=k_1\sin\alpha, \quad 0<\alpha<\pi/2, \quad (5.129)$$

in which case $dk_x=-k_{y,1}d\alpha$, and the integration variable is now changed from k_x to α.

Now consider the integral

$$\int_{\alpha=0}^{\pi/2} f(\alpha)\cos(k_x[x-x'])e^{ik_{y,1}(y+y')}d\alpha=$$

$$\int_{\alpha=0}^{\pi/2} f(\alpha)\frac{1}{2}\left(e^{ik_1r\cos(\alpha-\theta)}e^{-ik_xx'}+e^{-ik_1r\cos(\alpha+\theta)}e^{ik_xx'}\right)e^{ik_{y,1}y'}d\alpha. \quad (5.130)$$

Due to $k_1r>>1$ (far field) the terms $\exp(ik_1r\cos(\alpha-\theta))$ and $\exp(-ik_1r\cos(\alpha+\theta))$ will oscillate fast with α except when $\alpha\approx\theta$ and $\alpha\approx\pi-\theta$, respectively. In the ranges of α where the integrand varies fast, the integral vanishes as $k_1r\to\infty$ and these ranges can be ignored in the far field. First consider the case of $0<\theta<\pi/2$. The term $\exp(-ik_1r\cos(\alpha+\theta))$ may then be ignored, and for the other term we may use the approximation

$$e^{ik_1r\cos(\alpha-\theta)}\approx e^{ik_1r}e^{-ik_1r\frac{1}{2}(\alpha-\theta)^2}. \quad (5.131)$$

The integration limits may furthermore be replaced with integration over all α from $-\infty$ to $+\infty$ since the added integration interval is for α with a fast oscillating integrand that vanishes in the limit $k_1r\to\infty$. The resulting integral can be evaluated analytically by replacing α with θ in the slowly varying part of the integrand, and using

$$\int_{\alpha} e^{-i\frac{1}{2}k_1 r(\alpha-\theta)^2} d\alpha = \sqrt{\frac{2\pi}{ik_1 r}}. \tag{5.132}$$

For $\pi/2 < \theta < \pi$ we may use for the second term

$$e^{-ik_1 r\cos(\alpha+\theta)} = e^{ik_1 r\cos(\alpha-(\pi-\theta))} \approx e^{ik_1 r} e^{-ik_1 r\frac{1}{2}(\alpha-(\pi-\theta))^2} \tag{5.133}$$

and discard the first term $\exp(ik_1 r\cos(\alpha-\theta))$. The second term is handled with a similar procedure. These considerations lead to the indirect far-field Green's tensor

$$\mathbf{G}^{(\mathrm{ff},i)}(\mathbf{r},\mathbf{r}') = (-\hat{\mathbf{x}}\sin\theta + \hat{\mathbf{y}}\cos\theta)(\hat{\mathbf{x}}\sin\theta + \hat{\mathbf{y}}\cos\theta) \tag{5.134}$$

$$\times e^{ik_1 r}\sqrt{\frac{2}{\pi k_1 r}}\frac{e^{i\frac{\pi}{4}}}{4} r^{(p)}(k_1\cos\theta) e^{-ik_1\cos\theta x'} e^{ik_{y,1}y'}, \quad 0 < \theta < \pi. \tag{5.135}$$

Similar considerations may be applied to obtain the transmitted far-field Green's tensor

$$\mathbf{G}^{(\mathrm{ff},t)}(\mathbf{r},\mathbf{r}') = (-\hat{\mathbf{x}}\sin\theta + \hat{\mathbf{y}}\cos\theta)\left(\hat{\mathbf{x}}\frac{k_{y,1}k_{LN}}{k_1^2} + \hat{\mathbf{y}}\cos\theta\frac{k_{LN}^2}{k_1^2}\right) e^{ik_1 r}\sqrt{\frac{2}{\pi k_{LN} r}}\frac{e^{i\frac{\pi}{4}}}{4} \tag{5.136}$$

$$\times t_{1,LN}^{(p)}(k_{LN}\cos\theta)\frac{\varepsilon_1}{\varepsilon_{LN}}\frac{k_{y,LN}}{k_{y,1}} e^{-ik_{LN}\cos\theta x'} e^{ik_{y,1}y'} e^{-ik_{y,LN}d}, \quad 0 > \theta > -\pi, \tag{5.137}$$

where here $k_{y,i} = \sqrt{k_0^2\varepsilon_i - k_x^2}$ with $\mathrm{Imag}\{k_{y,i}\} \geq 0$.

In the case of guided modes being supported by the multilayer reference structure, it is also relevant to calculate the scattering into these modes. For this purpose the far-field Green's tensor component being related to guided modes is needed. For the upper half-plane $(y, y' > 0)$ a small value $y + y'$ is of interest, since a small value of y' is necessary for excitation of a guided mode, and a small value of y is necessary for the guided mode to have a non-negligible amplitude since the modes are evanescent along the y-axis. In the far-field approximation of the Green's tensor, a large value of $|x - x'|$ is considered. For the indirect part of the Green's tensor (5.122) we shall now consider the part of the integral for k_x from $k_0 n_1$ to ∞. For a large $|x - x'|$ the part of the integrand $\cos(k_x[x - x'])$ will be rapidly oscillating with k_x, and there will only be a significant contribution from the part of the integral where the remaining part of the integrand is not slowly varying, i.e, from the part of the integral being near a pole of the reflection coefficient $r^{(p)}(k_x)$.

Near a pole $k_{x,p}$ the reflection coefficient can be expressed as

$$r^{(p)}(k_x) \approx \frac{A^{(g)}}{k_x - k_{x,p}}, \tag{5.138}$$

where $A^{(g)}$ can be obtained by following the principles of Sec. 4.2.1 for guided modes. Thus

$$A^{(g)} = \frac{r_{1,L2}^{(p)}(k_{x,p}) + r_{L2,L3}^{(p)}(k_{x,p}) e^{2ik_{y,L2,p}d_{L2}}}{f'(k_{x,p})}, \tag{5.139}$$

where $f'(k_x)$ is given in (4.201). For each pole in the reflection coefficient it can now be shown, using residue calculus similar to Sec. 4.2.1, that there is a far-field contribution to the Green's tensor

$$\mathbf{G}^{(g)}(\mathbf{r},\mathbf{r}') = -\frac{1}{2}\frac{A^{(g)}}{k_{y1,p}}\left(-\hat{\mathbf{x}}\frac{k_{y1,p}}{k_1} + \hat{\mathbf{y}}\frac{k_{x,p}}{k_1}\frac{x-x'}{|x-x'|}\right)\left(\hat{\mathbf{x}}\frac{k_{y1,p}}{k_1} + \hat{\mathbf{y}}\frac{k_{x,p}}{k_1}\frac{x-x'}{|x-x'|}\right)$$
$$\times e^{ik_{x,p}|x-x'|}e^{ik_{y1,p}(y+y')}, \quad y,y' > 0. \tag{5.140}$$

An example of applying the GFAIEM for an organic fiber on a silver film on glass will be presented in Sec. 5.9. The scattered light going into the glass substrate can be used to measure the excitation of surface plasmon polaritons on the air-silver interface due to leakage of SPPs into the substrate at a specific angle. There will also be excitation of SPPs on the glass-silver interface that are guided modes of the reference structure. In addition, second-harmonic generation in the fiber will be considered with the GFAIEM.

5.6 PERIODIC SURFACE MICROSTRUCTURES

In this section the GFAIEM is considered for a structured surface which is periodic along the x-axis. The reference geometry is a planar layered structure with the dielectric constant

$$\varepsilon_{\text{ref}}(\mathbf{r}) = \begin{cases} \varepsilon_1, & y > 0 \\ \varepsilon_{L2}, & y < 0 \end{cases}, \tag{5.141}$$

and a periodic array of scatterers is placed on top of the layered reference geometry. The dielectric constant of the total structure is thus periodic along x with the period Λ such that

$$\varepsilon(\mathbf{r} + \hat{\mathbf{x}}\Lambda) = \varepsilon(\mathbf{r}). \tag{5.142}$$

An example of a layered reference geometry with a periodic array of scatterers is illustrated in Fig. 5.17.

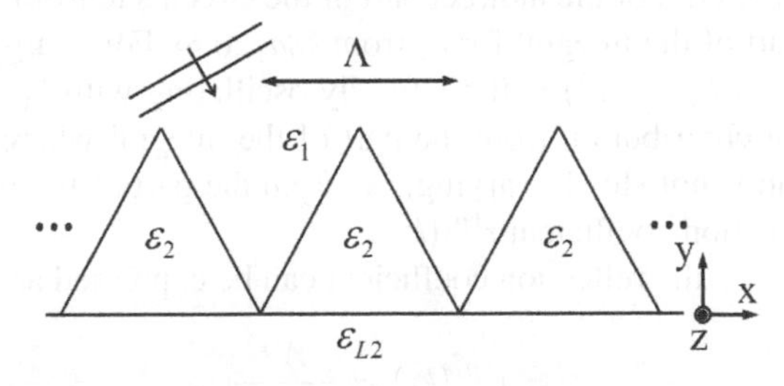

Figure 5.17 Illustration of a layered reference geometry with a periodic array of scatterers on top.

5.6.1 *S* POLARIZATION

For *s* polarization the reference electric field is a solution of

$$\left(\nabla^2 + k_0^2 \varepsilon_{\text{ref}}(\mathbf{r})\right) E_0(\mathbf{r}) = 0. \tag{5.143}$$

We will consider the following reference field that corresponds to illuminating the structure with a plane wave from above:

$$E_0(\mathbf{r}) = \begin{cases} A\left(e^{-ik_{y1}y} + r^{(s)}(k_x)e^{+ik_{y1}y}\right)e^{ik_x x}, & y > 0 \\ At^{(s)}(k_x)e^{-ik_{y,L2}y}, & y < 0 \end{cases}, \tag{5.144}$$

where as usual $r^{(s)}$ and $t^{(s)}$ are the Fresnel reflection and transmission coefficients for the layered reference geometry, and $k_x = k_0 n_1 \cos\theta$, where θ is the angle of propagation relative to the x-axis.

The reference field is a Bloch wave satisfying the periodic boundary condition

$$E_0(\mathbf{r} + \hat{\mathbf{x}}\Lambda) = E_0(\mathbf{r})e^{ik_x\Lambda}, \tag{5.145}$$

and the resulting total field E must satisfy the same periodic boundary condition.

The Green's function for the reference geometry and $y, y' > 0$ has been derived in Sec. 4.3.2. However, in Sec. 4.3, only the periodic Green's function for homogeneous media was needed, whereas here the periodic Green's function for a layered reference geometry is needed instead:

Green's function for a medium with dielectric constant ε_1 for $y > 0$, a layered structure for $y < 0$, and periodic boundary condition along x. Case of period Λ, Bloch-wave number k_x, and $y, y' > 0$:

$$g(\mathbf{r}, \mathbf{r}') = g^{(d)}(\mathbf{r}, \mathbf{r}') + g^{(i)}(\mathbf{r}, \mathbf{r}'), \tag{5.146}$$

with

$$g^{(d)}(\mathbf{r}, \mathbf{r}') = \frac{i}{2\Lambda}\sum_n \frac{e^{i(k_x+nG)(x-x')}e^{ik_{y,1,n}|y-y'|}}{k_{y,1,n}}, \tag{5.147}$$

$$g^{(i)}(\mathbf{r}, \mathbf{r}') = \frac{i}{2\Lambda}\sum_n \frac{e^{i(k_x+nG)(x-x')}e^{ik_{y,1,n}(y+y')}r^{(s)}(k_x+nG)}{k_{y,1,n}}, \tag{5.148}$$

where $G = 2\pi/\Lambda$, $k_{y,1,n} = \sqrt{k_0^2\varepsilon_1 - (k_x + nG)^2}$ with $\text{Imag}\left\{k_{y,1,n}\right\} \geq 0$.

The Green's function must be calculated numerically. It is convenient to split the Green's function into the direct and indirect contributions since $g^{(d)}(\mathbf{r}, \mathbf{r}') = g^{(d)}(x - x', |y - y'|)$ and $g^{(i)}(\mathbf{r}, \mathbf{r}') = g^{(i)}(x - x', y + y')$. The total Green's function only satisfies $g(\mathbf{r}, \mathbf{r}') = g(x - x', y, y')$ and is effectively a function of three coordinates. The direct and indirect Green's functions only need to be tabulated for a two-dimensional space.

The integral equation (5.12) may now be solved using the chosen reference field and Green's function. If we apply the approach of discretizing the structure using square-shaped elements, the integral equation in discrete form reads

$$E_i = E_{0,i} + \sum_j g_{ij} k_0^2 \left(\varepsilon_j - \varepsilon_{\text{ref},j}\right) E_j, \tag{5.149}$$

where

$$g_{ij} = P\int_j g(\mathbf{r}_i, \mathbf{r}')d^2r' = P\int_j g^{(d)}(\mathbf{r}_i, \mathbf{r}')d^2r' + \int_j g^{(i)}(\mathbf{r}_i, \mathbf{r}')d^2r'. \tag{5.150}$$

In (5.149) the field and the structure should only be considered within one period along x. Since $\varepsilon_{\text{ref},j}$ is constant in the region where the scatterer is placed, the discrete integral equation has exactly the same form as for a scatterer in a homogeneous dielectric.

For $i \neq j$ we may use the approximation $g_{ij} \approx g(\mathbf{r}_i, \mathbf{r}_j)\Delta$, where Δ is the area of the discretization elements. For $i = j$, one approach of handling the singularity is to use

$$\int_i g^{(d)}(\mathbf{r}_i, \mathbf{r}')d^2r' = -\frac{1}{k^2} - \frac{1}{k^2}\oint_{\partial i} \hat{\mathbf{n}}' \cdot \nabla g^{(d)}(\mathbf{r}_i, \mathbf{r}')dl', \tag{5.151}$$

where ∂i refers to the surface of element i. Another approach is to use

$$g^{(d)}(\mathbf{r}_i, \mathbf{r}') = \left(g^{(d)}(\mathbf{r}_i, \mathbf{r}') - \frac{i}{4}H_0^{(1)}(k_1|\mathbf{r}_i - \mathbf{r}'|) \right) + \frac{i}{4}H_0^{(1)}(k_1|\mathbf{r}_i - \mathbf{r}'|), \tag{5.152}$$

where $k_1 = k_0 n_1$. The first term is non-singular for $\mathbf{r}' \to \mathbf{r}_i$, and the second term is the Green's function used in Sec. 5.4.1. We may thus use

$$g_{ii} \approx \left(\lim_{\mathbf{r}' \to \mathbf{r}_i} \left(g^{(d)}(\mathbf{r}_i, \mathbf{r}') - \frac{i}{4}H_0^{(1)}(k_1|\mathbf{r}_i - \mathbf{r}'|) \right) + g^{(i)}(\mathbf{r}_i, \mathbf{r}_i) \right)\Delta +$$

$$\frac{i}{2(k_1 a)^2}\left[k_1 a H_1^{(1)}(k_1 a) + i\frac{2}{\pi} \right]\Delta, \tag{5.153}$$

where the last term is the same as the term that was used in (5.49). Note that the limiting value is the one that would be tabulated in Eq. (4.388) in Sec. 4.3.4.

After solving the discretized integral equation, the reflected field for y above the surface structure can be expressed as the Fourier sum

$$E_{\text{r}} = \sum_n A_n e^{i(k_x + nG)x} e^{ik_{y1,n}y}, \tag{5.154}$$

where

$$A_n = \delta_{n,0} A r^{(s)}(k_x) + \int f_n(\mathbf{r}') k_0^2 \left(\varepsilon(\mathbf{r}') - \varepsilon_{\text{ref}}(\mathbf{r}')\right) E(\mathbf{r}')d^2r' \tag{5.155}$$

with

$$f_n(\mathbf{r}) = \frac{i}{2\Lambda}\frac{1}{k_{y1,n}} e^{-i(k_x+nG)x'} \left(e^{-ik_{y1,n}y'} + r^{(s)}(k_x + nG)e^{ik_{y1,n}y'} \right). \tag{5.156}$$

Note that the first term in (5.155) is the reflected part of the reference field E_0. The Fourier sum arises straightforwardly when inserting the periodic Green's function in the integral equation.

The power in the incident field per period along x, where the incident field is the downward propagating part of E_0, is obtained from the downward Poynting vector flux through a line from, e.g., $x=0$ to $x=\Lambda$ with a fixed y. This leads to the incident power

$$P_{\mathrm{i}} = \frac{1}{2}\frac{k_{y1,0}\Lambda}{\omega\mu_0}|A|^2. \tag{5.157}$$

The reflected power is obtained as the upward Poynting vector flux related to the reflected field, and is given as

$$P_{\mathrm{r}} = \sum_n \frac{1}{2}\frac{\mathrm{Real}\{k_{y1,n}\}\Lambda}{\omega\mu_0}|A_n|^2. \tag{5.158}$$

The reflectance $P_{\mathrm{r}}/P_{\mathrm{i}}$ is thus given as

$$R = \sum_n R_n, \tag{5.159}$$

where the nth-order reflectance is given as

$$R_n = \frac{|A_n|^2}{|A|^2}\frac{\mathrm{Real}\{k_{y1,n}\}}{k_{y1}}. \tag{5.160}$$

It has been assumed that ε_1 is purely real and positive, and that $k_{y1} = k_{y1,0}$ is purely real, meaning that the incident field is not evanescent.

The corresponding transmittance is given as

$$T = \sum_n T_n, \tag{5.161}$$

where the nth-order transmittance is

$$T_n = \frac{|B_n|^2}{|A|^2}\frac{\mathrm{Real}\{k_{y2,n}\}}{k_{y1,0}}, \tag{5.162}$$

with

$$B_n = \delta_{n,0}At^{(s)}(k_x) + \int h_n(\mathbf{r}')k_0^2\left(\varepsilon(\mathbf{r}') - \varepsilon_{\mathrm{ref}}(\mathbf{r}')\right)E(\mathbf{r}')d^2r' \tag{5.163}$$

and

$$h_n(\mathbf{r}) = \frac{i}{2\Lambda}\frac{1}{k_{y1,n}}t^{(s)}(k_x+nG)e^{-i(k_x+nG)x'}e^{ik_{y1,n}y'}. \tag{5.164}$$

Here it has been assumed that also ε_{L2} is real and positive.

5.6.2 *P* POLARIZATION

For p polarization the reference electric field is a solution of

$$\left(-\nabla\nabla\cdot+\nabla^2+k_0^2\varepsilon_{\text{ref}}(\mathbf{r})\right)\mathbf{E}_0(\mathbf{r})=\mathbf{0}. \tag{5.165}$$

The reference field corresponding to illuminating the structure with a plane wave from above is given by

$$\mathbf{E}_0(\mathbf{r})=\begin{cases}A\left\{\left(\hat{\mathbf{x}}+\hat{\mathbf{y}}\frac{k_x}{k_{y1}}\right)e^{ik_xx}e^{-ik_{y1}y}+r^{(p)}(k_x)\left(-\hat{\mathbf{x}}+\hat{\mathbf{y}}\frac{k_x}{k_{y1}}\right)e^{ik_xx}e^{ik_{y1}y}\right\}, & y>0\\ At^{(p)}(k_x)\frac{\varepsilon_1}{\varepsilon_{L2}}\left(\hat{\mathbf{x}}\frac{k_{y,L2}}{k_{y1}}+\hat{\mathbf{y}}\frac{k_x}{k_{y1}}\right)e^{ik_xx}e^{-ik_{y,L2}y}, & y<0\end{cases}, \tag{5.166}$$

where $r^{(p)}$ and $t^{(p)}$ are the Fresnel reflection and transmission coefficients for the layered reference geometry.

As for s polarization, the reference field and total field satisfy the Bloch boundary condition. The direct periodic Green's tensor is given as

$$\mathbf{G}^{(d)}(\mathbf{r},\mathbf{r}')=\left(\mathbf{I}+\frac{1}{k_1^2}\nabla\nabla\right)g^{(d)}(\mathbf{r},\mathbf{r}'), \tag{5.167}$$

where the scalar periodic Green's function $g^{(d)}$ is given in (5.147).

This leads to

$$\mathbf{G}^{(d)}(\mathbf{r},\mathbf{r}')=\frac{i}{2\Lambda}\sum_n\frac{e^{i(k_x+nG)(x-x')}e^{ik_{y1,n}|y-y'|}}{k_{y1,n}}\tilde{\mathbf{G}}_n^{(d)}, \tag{5.168}$$

where

$$\tilde{\mathbf{G}}_n^{(d)}=\left\{\hat{\mathbf{x}}\hat{\mathbf{x}}\frac{k_{y1,n}^2}{k_1^2}+\hat{\mathbf{y}}\hat{\mathbf{y}}\frac{(k_x+nG)^2}{k_1^2}-(\hat{\mathbf{x}}\hat{\mathbf{y}}+\hat{\mathbf{y}}\hat{\mathbf{x}})\frac{(k_x+nG)k_{y1,n}}{k_1^2}\frac{y-y'}{|y-y'|}\right\}. \tag{5.169}$$

Now consider the incident field

$$\mathbf{E}_{i,x}(\mathbf{r})=\mathbf{G}^{(d)}(\mathbf{r},\mathbf{r}')\cdot\hat{\mathbf{x}},\quad y<y'. \tag{5.170}$$

From ordinary Fresnel reflection theory and the electromagnetics boundary conditions, this leads to the following expressions for the incident field, reflected field, and transmitted field:

$$\mathbf{E}_{i,x}(\mathbf{r})=\frac{i}{2\Lambda}\sum_n\frac{e^{i(k_x+nG)(x-x')}e^{ik_{y1,n}|y-y'|}}{k_{y1,n}}\left\{\hat{\mathbf{x}}\frac{k_{y1,n}^2}{k_1^2}+\hat{\mathbf{y}}\frac{(k_x+nG)k_{y1,n}}{k_1^2}\right\}, \tag{5.171}$$

$$\mathbf{E}_{r,x}(\mathbf{r})=\frac{i}{2\Lambda}\sum_n r^{(p)}(k_x+nG)\frac{e^{i(k_x+nG)(x-x')}e^{ik_{y1,n}(y+y')}}{k_{y1,n}}$$
$$\times\left\{-\hat{\mathbf{x}}\frac{k_{y1,n}^2}{k_1^2}+\hat{\mathbf{y}}\frac{(k_x+nG)k_{y1,n}}{k_1^2}\right\},\quad y>0 \tag{5.172}$$

$$\mathbf{E}_{t,x}(\mathbf{r}) = \frac{i}{2\Lambda}\sum_n t^{(p)}\frac{\varepsilon_1}{\varepsilon_{L2}}(k_x+nG)\frac{e^{i(k_x+nG)(x-x')}e^{ik_{y1,n}y'}e^{-ik_{y,L2,n}y}}{k_{y1,n}}$$
$$\times\left\{\hat{\mathbf{x}}\frac{k_{y1,n}k_{y,L2,n}}{k_1^2}+\hat{\mathbf{y}}\frac{(k_x+nG)k_{y1,n}}{k_1^2}\right\}, \quad y<0. \tag{5.173}$$

Similar expressions can be made for an incident field given as $\mathbf{E}_{i,y}(\mathbf{r}) = \mathbf{G}^{(d)}(\mathbf{r},\mathbf{r}')\cdot\hat{\mathbf{y}}$.

The Green's tensor can be expressed in terms of the direct, indirect, and the transmitted Green's tensors as in Eq. (5.113). The indirect and the transmitted Green's tensor components may now be constructed using the above considerations as

$$\mathbf{G}^{(i)}(\mathbf{r},\mathbf{r}') = \frac{i}{2\Lambda}\sum_n r^{(p)}(k_x+nG)\frac{e^{i(k_x+nG)(x-x')}e^{ik_{y1,n}(y+y')}}{k_{y1,n}}\tilde{\mathbf{G}}_n^{(i)}, \tag{5.174}$$

where

$$\tilde{\mathbf{G}}_n^{(i)} = \left\{-\hat{\mathbf{x}}\hat{\mathbf{x}}\frac{k_{y1,n}^2}{k_1^2}+\hat{\mathbf{y}}\hat{\mathbf{y}}\frac{(k_x+nG)^2}{k_1^2}+(\hat{\mathbf{y}}\hat{\mathbf{x}}-\hat{\mathbf{x}}\hat{\mathbf{y}})\frac{(k_x+nG)k_{y1,n}}{k_1^2}\right\}, \tag{5.175}$$

and

$$\mathbf{G}^{(t)}(\mathbf{r},\mathbf{r}') = \frac{i}{2\Lambda}\sum_n t^{(p)}(k_x+nG)\frac{\varepsilon_1}{\varepsilon_{L2}}\frac{e^{i(k_x+nG)(x-x')}e^{ik_{y1,n}y'}e^{-ik_{y,L2,n}y}}{k_{y1,n}}\tilde{\mathbf{G}}_n^{(t)}, \tag{5.176}$$

where

$$\tilde{\mathbf{G}}_n^{(t)} = \left\{\hat{\mathbf{x}}\hat{\mathbf{x}}\frac{k_{y1,n}k_{y,L2,n}}{k_1^2}+\hat{\mathbf{y}}\hat{\mathbf{y}}\frac{(k_x+nG)^2}{k_1^2}+\left(\hat{\mathbf{x}}\hat{\mathbf{y}}\frac{k_{y,L2,n}}{k_{y1,n}}+\hat{\mathbf{y}}\hat{\mathbf{x}}\right)\frac{(k_x+nG)k_{y1,n}}{k_1^2}\right\}. \tag{5.177}$$

When the expressions for the Green's tensor are applied in the integral equation, and positions above the scatterer are considered, the resulting reflected field can be expressed as

$$\mathbf{E}_r(\mathbf{r}) = \sum_n \mathbf{E}_{r,n}e^{i(k_x+nG)x}e^{ik_{y1,n}y}, \tag{5.178}$$

where the vector coefficients are given by

$$\mathbf{E}_{r,n} = \frac{i}{2\Lambda}\int\frac{e^{-i(k_x+nG)x'}}{k_{y1,n}}\left(e^{-ik_{y1,n}y'}\tilde{\mathbf{G}}^{(d)}_{n,y>y'}+r^{(p)}(k_x+nG)e^{+ik_{y1,n}y'}\tilde{\mathbf{G}}_n^{(i)}\right)$$
$$\times k_0^2\left(\varepsilon(\mathbf{r}')-\varepsilon_{\mathrm{ref}}(\mathbf{r}')\right)\cdot\mathbf{E}(\mathbf{r}')d^2r' + \delta_{n,0}r^{(p)}(k_x)A\left(-\hat{\mathbf{x}}+\hat{\mathbf{y}}\frac{k_x}{k_{y1}}\right). \tag{5.179}$$

The corresponding magnetic field is given by

$$\mathbf{H}_r(\mathbf{r}) = \frac{-i}{\omega\mu_0}\nabla\times\mathbf{E}_r(\mathbf{r}), \tag{5.180}$$

and this can be used to construct the reflected power per period as

$$P_r = \frac{1}{2}\text{Real}\left\{\int_{x=0}^{\Lambda} \mathbf{E}_r(\mathbf{r}) \times (\mathbf{H}_r(\mathbf{r}))^* dx\right\} = \frac{1}{2}\frac{\Lambda}{\omega\mu_0}\sum_n |\hat{\mathbf{x}}\cdot\mathbf{E}_{r,n}|^2\,\text{Real}\left\{\frac{k_1^2}{k_{y1,n}^*}\right\}. \tag{5.181}$$

It was used in the derivation that $\hat{\mathbf{y}}\cdot\mathbf{E}_{r,n} = -\hat{\mathbf{x}}\cdot\mathbf{E}_{r,n}(k_x+nG)/k_{y1,n}$, which follows from the requirement $\nabla\cdot\mathbf{E}_r(\mathbf{r}) = 0$. As usual there is no contribution for those n where $k_{y1,n}$ is imaginary.

The power in the incident field is correspondingly given by

$$P_i = \frac{1}{2}\frac{\Lambda}{\omega\mu_0}|A|^2\frac{k_1^2}{k_{y1}}, \tag{5.182}$$

and the reflectance is thus

$$R = \sum_n R_n, \tag{5.183}$$

with the reflectance in the nth reflectance diffraction order

$$R_n = \frac{|\hat{\mathbf{x}}\cdot\mathbf{E}_{r,n}|^2\,\text{Real}\left\{\frac{k_{y1}}{k_{y1,n}^*}\right\}}{|A|^2}. \tag{5.184}$$

For positions below the layered structure, the transmitted field can be expressed as

$$\mathbf{E}_t(\mathbf{r}) = \sum_n \mathbf{E}_{t,n}e^{i(k_x+nG)x}e^{-ik_{y,L2,n}y}, \tag{5.185}$$

where the vector coefficients are

$$\mathbf{E}_{t,n} = \frac{i}{2\Lambda}\int t^{(p)}(k_x+nG)\frac{e^{-i(k_x+nG)x'}}{k_{y1,n}}e^{ik_{y1,n}y'}\tilde{\mathbf{G}}_n^{(t)}k_0^2\left(\varepsilon(\mathbf{r}')-\varepsilon_{\text{ref}}(\mathbf{r}')\right)\cdot\mathbf{E}(\mathbf{r}')d^2r'$$
$$+\delta_{n,0}t^{(p)}(k_x)\frac{\varepsilon_1}{\varepsilon_{L2}}A\left(\hat{\mathbf{x}}\frac{k_{y,L2}}{k_{y1}}+\hat{\mathbf{y}}\frac{k_x}{k_{y1}}\right). \tag{5.186}$$

The transmitted power per period is

$$P_t = \frac{1}{2}\text{Real}\left\{\int_{x=0}^{\Lambda}\mathbf{E}_t(\mathbf{r})\times(\mathbf{H}_t(\mathbf{r}))^* dx\right\}$$
$$= \frac{1}{2}\frac{\Lambda}{\omega\mu_0}\sum_n |\hat{\mathbf{x}}\cdot\mathbf{E}_{t,n}|^2\,\text{Real}\left\{\frac{k_0^2\varepsilon_{L2}}{k_{y,L2,n}^*}\right\}. \tag{5.187}$$

There is no contribution for those n where $k_{y,L2,n}$ is imaginary.

The transmittance $T = P_t/P_r$ is thus

$$T = \sum_n T_n, \tag{5.188}$$

with the nth-order transmittance

$$T_n = \frac{|\hat{\mathbf{x}}\cdot\mathbf{E}_{t,n}|^2}{|A|^2}\frac{\varepsilon_{L2}}{\varepsilon_1}\text{Real}\left\{\frac{k_{y1}}{k_{y,L2,n}^*}\right\}. \tag{5.189}$$

5.7 FAST ITERATIVE FFT-BASED APPROACH FOR LARGE STRUCTURES

The previous matrix approaches considered in this chapter for solving the GFAIEM will have a matrix that scales as $N \times N$ for N discretization elements, and the time for the matrix inversion will scale as N^3. The memory requirements and the calculation time therefore increase very fast with the size of the numerical problem N. If a single scattering object is discretized in approximately 100 elements along each direction x and y, this already amounts to $N = 10000$. The matrix in the discretized form of the GFAIEM will thus contain 10^8 elements, and the calculation time will scale as 10^{12}. This is still possible, but as N increases further, another method is needed that scales better with N.

Notice that for a scatterer in a homogeneous reference medium, the Green's function used for s polarization and the Green's tensor used for p polarization satisfy

$$g^{(d)}(\mathbf{r},\mathbf{r}') = g^{(d)}(\mathbf{r}-\mathbf{r}'), \tag{5.190}$$

and

$$\mathbf{G}^{(d)}(\mathbf{r},\mathbf{r}') = \mathbf{G}^{(d)}(\mathbf{r}-\mathbf{r}'). \tag{5.191}$$

Due to this property, the integrals in the Green's function integral equations become two-dimensional convolution integrals of the form

$$E(x,y) = \int f(x-x',y-y')h(x',y')dx'dy'. \tag{5.192}$$

This type of integral can be calculated fast by using the Fast Fourier Transformation algorithm. The integral equation can then be solved using iterative methods rather than matrix inversion. In these methods the memory requirements will scale as N, and the calculation time may ideally scale as $N \log N$. This will be explained in the following.

The two-dimensional Fourier transform of the function E in (5.192) is defined as

$$\tilde{E}(k_x,k_y) = \int E(x,y)e^{-ik_x x}e^{-ik_y y}dxdy, \tag{5.193}$$

and the inverse Fourier transform is then given as

$$E(x,y) = \frac{1}{(2\pi)^2}\int \tilde{E}(k_x,k_y)e^{ik_x x}e^{ik_y y}dk_x dk_y. \tag{5.194}$$

When the two-dimensional Fourier transformation is applied to the right-hand side in (5.192), it can be shown that

$$\tilde{E}(k_x,k_y) = \tilde{f}(k_x,k_y)\tilde{h}(k_x,k_y). \tag{5.195}$$

The time-consuming convolution integral in real space thus becomes a simple product in reciprocal space. The convolution integral (5.192) may thus be carried out by first Fourier-transforming $f(x,y)$ and $h(x,y)$, carrying out the product of the two functions in reciprocal space, and then applying the inverse Fourier transform to the product as

$$E(x,y) = \frac{1}{(2\pi)^2}\int \tilde{f}(k_x,k_y)\tilde{h}(k_x,k_y)e^{ik_x x}e^{ik_y y}dk_x dk_y. \tag{5.196}$$

In order for this approach to be useful, the Fourier transform and inverse Fourier transform should be calculated using a fast method. This can be achieved by using the Fast Fourier Transform (FFT) algorithm. The FFT must be applied to a discrete form of the convolution integral, and a discrete form of the convolution integral in the GFAIEM is obtained when using square-shaped discretization elements. If (5.192) is converted to a sum where (x,y) is defined on N points, then for each discrete (x,y) the resulting sum is a sum over N terms, thus, carrying out the sum for all discrete (x,y) scales as N^2. However, the FFT and the inverse FFT scale as $N\log N$, and the product in (5.195) only scales as N. The steps required for the convolution integral in discrete form thus scales as $N\log N$, which is much better than N^2.

Consider the GFAIEM for s polarization (5.12) with square-shaped discretization elements. In order to convert the integral equation to a discrete convolution we now choose to represent each discrete position in the xy-plane using two discrete coordinates i_x and i_y. The corresponding position is given by

$$\mathbf{r}_{i_x,i_y} \equiv \hat{\mathbf{x}}(x_0 + i_x\Delta_x) + \hat{\mathbf{y}}(y_0 + i_y\Delta_y), \tag{5.197}$$

where $1 \leq i_x \leq N_x$ and $1 \leq i_y \leq N_y$ leading to a total number of discrete positions $N = N_x N_y$.

The value of the total field and reference field at the same position is also identified using the two discrete coordinates as

$$E_{i_x,i_y} \equiv E(\mathbf{r}_{i_x,i_y}), \tag{5.198}$$

$$E_{0,i_x,i_y} \equiv E_0(\mathbf{r}_{i_x,i_y}). \tag{5.199}$$

The discrete value of the dielectric constant and the Green's function are furthermore defined as

$$\varepsilon_{i_x,i_y} \equiv \varepsilon(\mathbf{r}_{i_x,i_y}), \tag{5.200}$$

and

$$g_{i_x-j_x,i_y-j_y} \equiv \begin{cases} \frac{i}{4}H_0^{(1)}(k|\mathbf{r}_{i_x,i_y} - \mathbf{r}_{j_x,j_y}|)\Delta_x\Delta_y, & i_x \neq j_x \text{ or } i_y \neq j_y \\ \frac{i}{2(ka)^2}\left[kaH_1^{(1)}(ka) + i\frac{2}{\pi}\right]\Delta_x\Delta_y, & i_x = j_x \text{ and } i_y = j_y \end{cases}, \tag{5.201}$$

where $\pi a^2 = \Delta_x\Delta_y$ and $k = k_0 n_1$. This is just the expression (5.49) adapted to using two indices for the position.

The integral equation in discrete form may now be expressed as

$$E_{i_x,i_y} - \sum_{j_x,j_y} g_{i_x-j_x,i_y-j_y} k_0^2 \left(\varepsilon_{j_x,j_y} - \varepsilon_{\text{ref}}\right) E_{j_x,j_y} = E_{0,i_x,i_y}, \tag{5.202}$$

which is in the form of a discrete convolution. The discrete convolution relies here on the use of square-shaped or rectangular-shaped elements, and this form does not directly arise if other more general shapes of elements are used.

It is convenient to organize the values of the fields in the square-shaped discretization cells E_{i_x,i_y} and E_{0,i_x,i_y} in matrices $\overline{\overline{E}}$ and $\overline{\overline{E}}_0$, whereas the values were organized in a vector in the discrete GFAIEM in (5.52). The discrete values of the dielectric constant may also be organized in a matrix $\overline{\overline{\varepsilon}}$. Here, this is a full matrix where each matrix element corresponds to the dielectric constant at one of the discrete positions i_x and i_y. The values $g_{i_x-j_x,i_y-j_y}$ may also be organized in a matrix $\overline{\overline{g}}$ with $g_{i_x-j_x,i_y-j_y}$ in matrix element $i_x - j_x + Nx, i_y - j_y + Ny$.

We may now define the operator $\hat{C}$ such that (5.202) is given as

$$\hat{C}\overline{\overline{E}} = \overline{\overline{E}}_0. \tag{5.203}$$

The first step in applying the operator $\hat{C}$ is to apply elementwise multiplication in real space of $\overline{\overline{E}}$ and $k_0^2\left(\overline{\overline{\varepsilon}} - \overline{\overline{1}}\varepsilon_{\text{ref}}\right)$, where $\overline{\overline{1}}$ is a matrix with the value 1 at all positions. This can also be formulated as

$$\overline{\overline{E}}' = k_0^2\left(\overline{\overline{\varepsilon}} - \overline{\overline{1}}\varepsilon_{\text{ref}}\right) .* \overline{\overline{E}}, \tag{5.204}$$

where .* means that element i_x, i_y of the resulting matrix is obtained by multiplying together element i_x, i_y in one matrix and element i_x, i_y in the other matrix. The matrix containing the values of the Green's function $\overline{\overline{g}}$ and the matrix $\overline{\overline{E}}'$ are now Fourier-transformed using the two-dimensional FFT. The resulting matrices are multiplied together in reciprocal space, and the inverse two-dimensional FFT is then applied to the product. The application of the operator $\hat{C}$ can now be formulated

$$\hat{C}\overline{\overline{E}} = \overline{\overline{E}} - IFFT\left(FFT\left(\overline{\overline{g}}\right) .* FFT\left(k_0^2\left(\overline{\overline{\varepsilon}} - \overline{\overline{1}}\varepsilon_{\text{ref}}\right) .* \overline{\overline{E}}\right)\right). \tag{5.205}$$

Note that the matrix $\overline{\overline{g}}$ is of size $(2N_x - 1) \times (2N_y - 1)$ where, for example, $\overline{\overline{E}}'$ is of size $N_x \times N_y$. Before applying the FFT to $\overline{\overline{E}}'$, this matrix should be zero-padded to have the same size as $\overline{\overline{g}}$. After the $IFFT$, the appropriate submatrix of size $N_x \times N_y$ must be taken. The zero-padding and the extraction of a submatrix are implicit in the expression (5.205).

Now consider a scatterer placed on or near a surface. The Green's function no longer assumes the simple form (5.190). However, for positions above the surface, the Green's function can be split into two terms in the form

$$g(\mathbf{r},\mathbf{r}') = g^{(d)}(x-x', y-y') + g^{(i)}(x-x', y+y'). \tag{5.206}$$

The integral in the GFAIEM can thus be split into two separate convolution integrals, one containing $g^{(\text{d})}$, and one containing $g^{(\text{i})}$. When the integral equation is converted to discrete form, an equation similar to (5.203) is obtained with another operator $\hat{C}$. However, the operation $\hat{C}\overline{\overline{E}}$ still scales as $N\log N$ when carried out using the FFT/IFFT procedure for the two discrete convolutions.

The idea now is to solve the equation (5.203) by using an iterative method where we start out with a guess for the field $\overline{\overline{E}}$, which is then iteratively improved until (5.203) is satisfied within a given convergence criteria. One iterative approach is the conjugate gradient algorithm described in Appendix B. Another is the generalized minimal residual algorithm (see for example Refs. [111, 112]).

The conjugate gradient algorithm requires that the Hermitian conjugate of the operator $\hat{C}$ is known, which is expressed as $\hat{C}^{\dagger}$. For the case of the scatterer in a homogeneous reference medium the operation $\overline{\overline{E}}' = \hat{C}^{\dagger}\overline{\overline{E}}$ is equivalent to

$$E'_{i_x,i_y} = E_{i_x,i_y} - k_0^2 \left(\varepsilon_{i_x,i_y} - \varepsilon_{\text{ref}}\right)^* \sum_{j_x,j_y} g^*_{j_x-i_x,j_y-i_y} E_{j_x,j_y}. \tag{5.207}$$

For the homogeneous reference medium, $g_{j_x-i_x,j_y-i_y} = g_{i_x-j_x,i_y-j_y}$, and the equivalent of (5.205) becomes

$$\hat{C}^{\dagger}\overline{\overline{E}} = \overline{\overline{E}} - k_0^2 \left(\overline{\overline{\varepsilon}} - \overline{\overline{1}}\varepsilon_{\text{ref}}\right)^* .* IFFT\left(FFT\left(\overline{\overline{g}}^*\right) .* FFT\left(\overline{\overline{E}}\right)\right). \tag{5.208}$$

The generalized minimum residual algorithm on the other hand does not require the Hermitian conjugate operator.

5.8 EXAMPLE: PURCELL FACTOR OF EMITTER IN A PHOTONIC CRYSTAL

The rate of spontaneous emission from a quantum emitter depends on the optical properties of the surrounding structure. This can be quantified by the Purcell factor that gives the ratio between the emission for the emitter in a given structure relative to the emission from the same emitter placed in a reference structure such as free space. It should be assumed that the structure is only changed at such a distance from the emitter that the electronic properties of the emitter will be unaffected.

In the original expression $f = 3Q\lambda^3/4\pi^2 V$ given by Purcell [31], a cavity that supports one mode is considered, where the cavity has the quality factor Q and mode volume V, and the emitter emits radiation tuned to the cavity mode with the wavelength λ. According to this expression the rate of emission can be increased by many orders of magnitude if the cavity has a high quality factor and is small at the same time. In this section the term *Purcell factor* will be used to generally describe the enhancement of emission from a quantum emitter. The Purcell factor can be modeled classically in the weak-coupling limit by noting that the enhancement in the rate of emission from a quantum emitter is the same as the enhancement in the emission from a classical dipole emitter when placed in the same structure relative to being placed in the reference structure. This follows from the Fermi golden rule [32, 18] and that in both cases the emission is proportional to the local density of optical states (LDOS).

The enhancement or inhibition of spontaneous emission has been demonstrated experimentally for Rydberg atoms between parallel metallic surfaces [113, 114], for cyclotron radiation from electrons in a microwave cavity [115], for quantum dots in a monolithic optical microcavity with distributed Bragg reflector mirrors in a micropillar geometry [116], and for emitters placed near planar metal and dielectric surfaces or between such surfaces [117, 118, 119, 120, 121, 122, 123, 124, 125]. The fluorescence from a molecule in the vicinity of a metal nanoparticle or an optical antenna has also been studied experimentally [126, 127, 128], and spontaneous emission from an emitter in the vicinity of a photonic wire (optical fiber) or plasmonic wire geometry has also been studied including the coupling to the guided modes of the wire [129, 130]. A collection of works on both experiments and theory on spontaneous emission in various microcavities can be found in the books [33, 34].

The rate of spontaneous emission can be rigorously forbidden in a photonic crystal of infinite extent [131] as there are no propagating modes that the emitter can couple to for frequencies in the photonic bandgap. For a photonic crystal of finite extent the emitter can still couple to some modes that are propagating outside the crystal and decaying exponentially into the crystal, in which case the emission is significantly suppressed but no longer zero. On the other hand if an emitter is placed in a photonic crystal cavity, then such a cavity can be extremely small and thus according to Purcell's enhancement factor, the emission enhancement can be orders of magnitude for that reason alone. As the photonic crystals can be based on dielectric materials the quality factor can also be very high at the same time. Some of the work on the control of spontaneous emission by photonic crystals and other nanocavities has been reviewed in Ref. [132].

In this section the GFAIEM is applied to calculate the Purcell factor for a dipole placed in a finite photonic crystal and in a photonic crystal cavity. In the first case

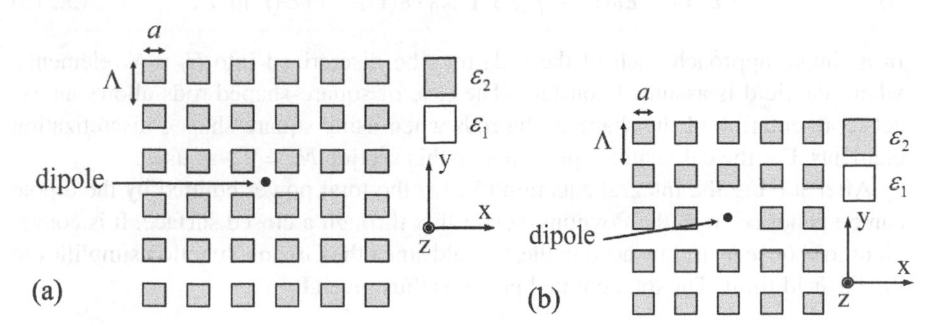

Figure 5.18 (a) Schematic of a dipole placed in the center of a finite photonic crystal with 6×6 square-shaped rods with side length a and dielectric constant ε_2 placed on a square lattice in a medium with dielectric constant ε_1. (b) Schematic of a dipole placed in the center of a photonic crystal cavity with $5 \times 5 - 1$ rods.

the emission from the dipole is suppressed, and in the latter case it may be strongly enhanced. The case of a dipole in a photonic crystal with $2N \times 2N$ square-shaped rods is illustrated in Fig. 5.18(a) for $2N = 6$. The rods have the dielectric constant $\varepsilon_2 = 12$ and are placed on a square lattice with period Λ in vacuum ($\varepsilon_{\text{ref}} = \varepsilon_1 = 1$). The side-length of the rods is $a = \Lambda/3$. The dipole is placed in the center of the array of rods and thus in vacuum for $2N \times 2N$ rods. We will assume that it is a two-dimensional geometry with light propagating only in the xy-plane and that light is s polarized. The electric field is thus polarized along the z-axis. We will also consider the case of a dipole in the center of a photonic crystal cavity with $(2N+1) \times (2N+1) - 1$ rods as illustrated in Fig. 5.18(b) for $2N+1 = 5$. The cavity is obtained by starting out with a $(2N+1) \times (2N+1)$ array of rods in a photonic crystal and then removing the center rod. The dipole is placed in the center of the cavity and is thus again located in vacuum.

The oscillating dipole can be described as having the following current density

$$\mathbf{J}(\mathbf{r}) = \hat{\mathbf{z}}J(\mathbf{r}) = -i\omega\varepsilon_0\delta(\mathbf{r}). \tag{5.209}$$

The field generated by this current density, if it is placed in vacuum, can be obtained from the Green's function integral approach as

$$E_0(\mathbf{r}) = i\omega\mu_0 \int g(\mathbf{r},\mathbf{r}')J(\mathbf{r}')d^2r' = k_0^2 g(\mathbf{r},\mathbf{0}), \tag{5.210}$$

where the Green's function is given by

$$g(\mathbf{r},\mathbf{r}') = \frac{i}{4}H_0^{(1)}(k_0|\mathbf{r}-\mathbf{r}'|). \tag{5.211}$$

The total field generated by the dipole placed not in vacuum but in the photonic crystal can now be obtained by solving the integral equation

$$E(\mathbf{r}) = E_0(\mathbf{r}) + \int g(\mathbf{r},\mathbf{r}')k_0^2\left(\varepsilon(\mathbf{r}') - 1\right)E(\mathbf{r}')d^2r'. \tag{5.212}$$

In a simple approach each of the rods may be discretized into $N_r \times N_r$ elements, where the field is assumed constant. The case of square-shaped rods allows an exact representation of the shape of the rods when using square-shaped discretization elements. For the calculations presented in this section $N_r = 9$ was used.

After solving the integral equation (5.212) the total power emitted by the dipole can be obtained from the Poynting-vector flux through a closed surface. It is convenient to choose a circle placed in the far field since the Green's function simplifies in the far-field limit. The total emitted power is thus given by

$$P = \frac{1}{2}\oint \text{Real}\left\{\mathbf{E}\times\mathbf{H}^*\right\}\cdot\hat{\mathbf{n}}dl \propto \int_{\theta=0}^{2\pi} |E(r,\theta)|^2 r d\theta, \tag{5.213}$$

where the far field is given by

$$E(r,\theta) = \frac{i}{4}e^{-i\pi/4}e^{ik_0 r}\sqrt{\frac{2}{\pi k_0 r}}k_0^2\left(1 + \int e^{-ik_0(x'\cos\theta + y'\sin\theta)}\left(\varepsilon(\mathbf{r}') - 1\right)E(\mathbf{r}')d^2r'\right), \tag{5.214}$$

and $\mathbf{r} = \hat{\mathbf{x}}r\cos\theta + \hat{\mathbf{y}}r\sin\theta$, and $\hat{\mathbf{n}}$ is the normal vector of the closed surface.

Alternatively, the emitted power can be obtained from the imaginary part of the field at the dipole location as follows

$$P \propto \text{Imag}\left\{E(\mathbf{0})\right\}. \tag{5.215}$$

This expression can be obtained by starting with the left-hand part of (5.213) and instead using a closed surface surrounding the dipole in the near-field. The surface integral is then rewritten into an area integral by using Gauss's law and then applying $\nabla \cdot (\mathbf{E} \times \mathbf{H}^*) = \mathbf{H}^* \cdot \nabla \times \mathbf{E} - \mathbf{E} \cdot \nabla \times \mathbf{H}^*$. By further using Maxwell's equations and the expression for the current density (5.209), the resulting integral can be readily carried out and leads to (5.215).

Now consider the Green's function for the total structure $\tilde{g}$ satisfying the equation

$$\left(\nabla^2 + k_0^2 \varepsilon(\mathbf{r})\right) \tilde{g}(\mathbf{r}, \mathbf{r}') = -\delta(\mathbf{r} - \mathbf{r}'), \tag{5.216}$$

and which also satisfies the radiating boundary condition. The electric field obtained from (5.212) is in fact the solution for a point source in the total structure and is related to the Green's function as

$$E(\mathbf{r}) = k_0^2 \tilde{g}(\mathbf{r}, \mathbf{0}). \tag{5.217}$$

Thus, the emission from the point source can also be expressed as

$$P \propto \mathrm{Imag}\left\{\tilde{g}(\mathbf{0}, \mathbf{0})\right\}. \tag{5.218}$$

In general the emission from a point source at any other position $\mathbf{r}$ is given as $P \propto \mathrm{Imag}\left\{\tilde{g}(\mathbf{r}, \mathbf{r})\right\}$.

The power that is emitted when the dipole is placed in the free space reference structure is denoted P_0. The fraction P/P_0 is thus the Purcell factor.

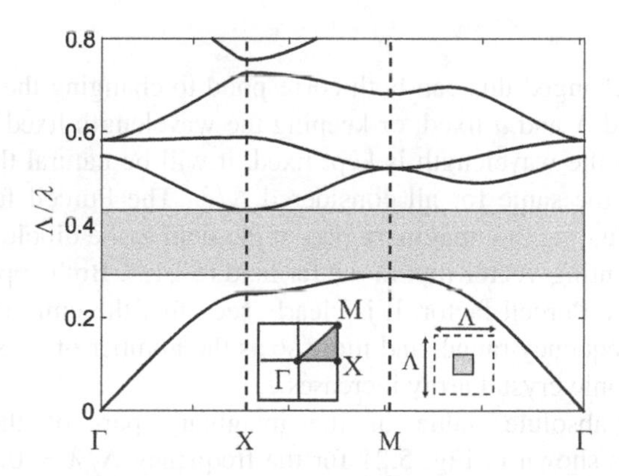

Figure 5.19 Band diagram for a photonic crystal with square-shaped rods with side length a and dielectric constant ε_2 placed on a square lattice in a medium with dielectric constant ε_1 and period Λ. The unit cell and the path on the irreducible Brillouin zone are shown as insets.

The band diagram for the fully periodic photonic crystal structure is shown in Fig. 5.19. This figure was calculated using the method in Appendix G but with Fourier-expansion coefficients for the structure that corresponds to square-shaped rods instead of round rods. It can be noticed that the structure has a bandgap for $0.29 \leq \Lambda/\lambda \leq 0.43$. For the calculation of the Purcell factor we will consider a range of normalized frequencies $\Lambda/\lambda \in [0;0.5]$ that thus include the bandgap of the photonic crystal. The Purcell factor versus the frequency is shown in Fig. 5.20. When

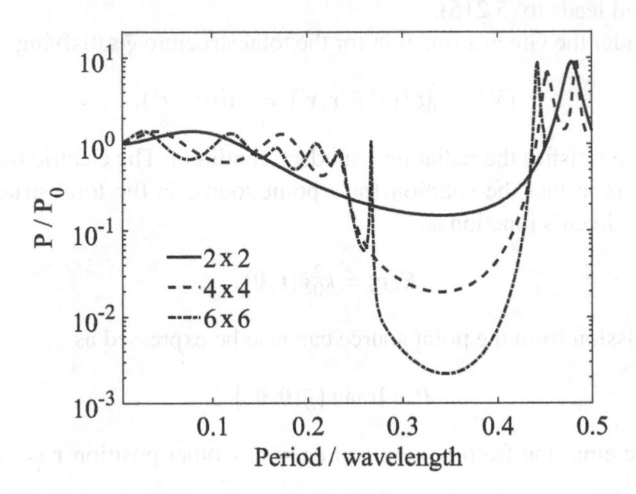

Figure 5.20 Purcell factor for a dipole emitter placed in the center of a finite photonic crystal with square-shaped rods of dielectric constant $\varepsilon_2 = 12$ placed on a square lattice with period Λ in a background with dielectric constant $\varepsilon_1 = 1$.

the frequency is changed this can both correspond to changing the wavelength and keeping the period Λ and a fixed, or keeping the wavelength fixed and changing Λ (and $a = \Lambda/3$). If the wavelength is kept fixed, it will be natural that the dielectric constants remain the same for all considered Λ/λ. The Purcell factor was calculated by both evaluating the imaginary part of the field at the dipole position (5.215) and using the Poynting-vector flux in the far field (5.213). Both approaches give the same result for the Purcell factor. It is clearly seen that the emission is suppressed in the bandgap-frequency range, and more so as the number of rods $2N \times 2N$ in the surrounding photonic crystal array increases.

Plots of the absolute value of the imaginary part of the electric field ($|\text{Imag}\{E(\mathbf{r})\}|$) is shown in Fig. 5.21 for the frequency $\Lambda/\lambda = 0.35$ and different array sizes. The field is normalized relative to the imaginary part of the reference field at the dipole position $\text{Imag}\{E_0(\mathbf{0})\}$. The value of the normalized imaginary part of the field in the center of the field plots is thus also equivalent to the Purcell factor. Apart from that, the field plots represent the field magnitude at a snap-shot in time, and outside the photonic crystal the field is seen to propagate away from the photonic crystal.

We now turn to consider the dipole placed in the photonic crystal cavity as illustrated in Fig. 5.18(b). The band diagram for a unit cell with $5 \times 5 - 1$ rods being

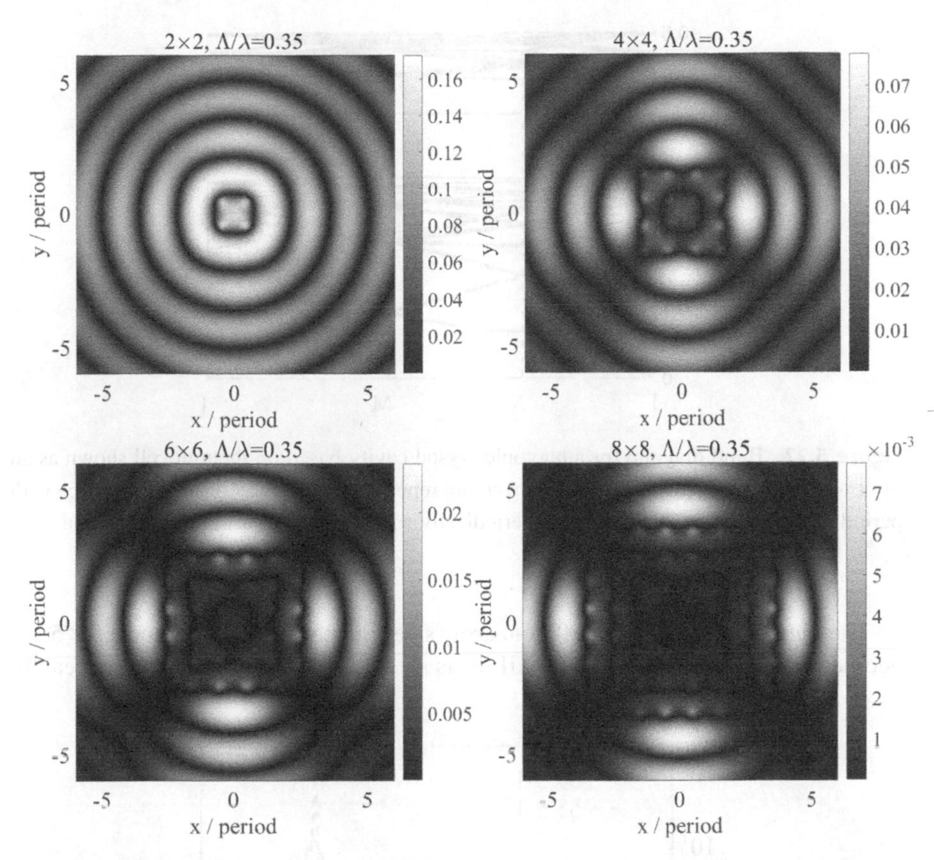

Figure 5.21 $|\text{Imag}\{E(\mathbf{r})\}|$ for a dipole emitter placed in a finite photonic crystal with $2N \times 2N$ square-shaped rods of dielectric constant.

repeated periodically with the period 5Λ is shown in Fig. 5.22. The unit cell is shown as an inset. The structure considered in this diagram is thus a photonic crystal cavity being repeated periodically with the period 5Λ. If the distance between neighbor cavities is sufficiently large, the coupling between neighbor cavities can be ignored. Notice that the number of bands is much larger than in the case of the simple unit cell considered in Fig. 5.19 by roughly a factor 25 due to the larger size of the unit cell. A single discrete and relatively flat band now appears in the bandgap of the photonic crystal. This band represents the cavity mode which is localized to the region of the cavity. The band related to the cavity mode is actually not completely flat. The coupling between neighbor cavities will appear in the band diagram as a small broadening of the defect band such that it will cover a small but finite range of frequencies, and the width of the frequency range depends on the coupling. The larger the distance between cavities the smaller the coupling, and thus the smaller the frequency width of the band.

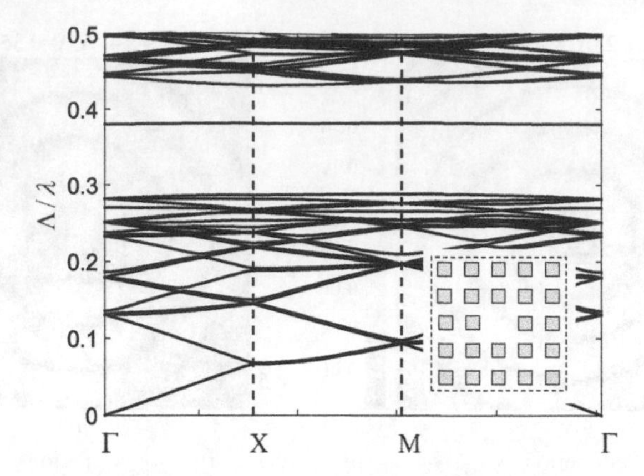

Figure 5.22 Band diagram for a photonic crystal cavity based on the unit cell shown as an inset where an arrangement of $5 \times 5 - 1$ rods are repeated periodically on a square lattice with period 5Λ. The cavity is thus repeated periodically with a period of 5Λ along both x and y.

The Purcell factor for two different array sizes, namely a $3 \times 3 - 1$ and a $5 \times 5 - 1$ geometry as obtained with the GFAIEM, is shown in Fig. 5.23. Notice the peak in

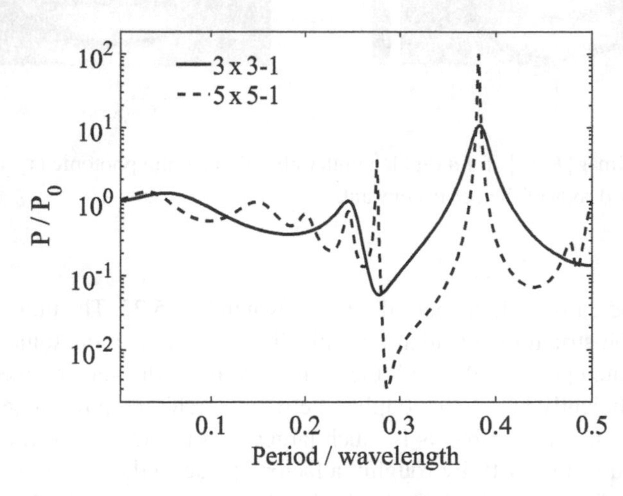

Figure 5.23 Purcell factor for a dipole emitter placed in a cavity in a finite photonic crystal with square-shaped rods of dielectric constant $\varepsilon_2 = 12$ placed on a square lattice with period Λ in a background with dielectric constant $\varepsilon_1 = 1$.

the Purcell factor for a specific $\Lambda/\lambda \approx 0.38$ inside the bandgap. The peak occurs at the frequency of the cavity mode as found in the band diagram in Fig. 5.22. At this frequency the Purcell factor is very large as the dipole excites the cavity mode. Notice also that the peak is both higher and narrower for the $5 \times 5 - 1$ cavity compared with the $3 \times 3 - 1$ cavity. This is because the cavity quality factor increases with the size of the surrounding photonic crystal. With only one layer of rods around the cavity the Purcell factor already reaches approximately a factor of 10, and for two layers of rods it reaches approximately a factor of 100. A Quasi-Normal-Mode theory and a Fourier-Modal-Method were used for studying the Purcell factor (or LDOS enhancement factor) in [133] for a similar cavity placed in the vicinity of a photonic crystal waveguide. The cavity mode and cavity quality factor Q were studied for a photonic crystal microcavity based on air holes in a semiconductor slab in Ref. [134], and the LDOS or Purcell factor was studied for a two-dimensional photonic crystal microcavity with air holes taking into account propagating modes in the third dimension in Ref. [135], and for a photonic crystal slab in Ref. [136].

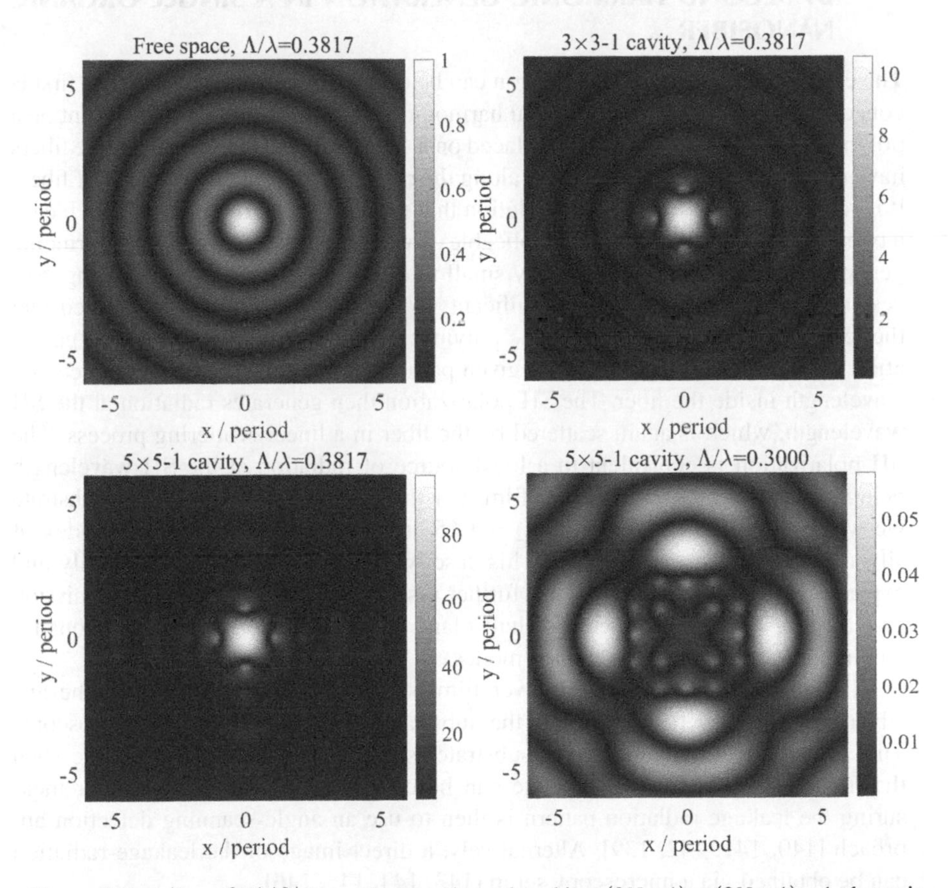

Figure 5.24 $|\mathrm{Imag}\{E(\mathbf{r})\}|$ for a dipole emitter placed in a $(2N+1) \times (2N+1) - 1$ photonic crystal cavity.

Plots of the magnitude of the imaginary part of the field for the cavity are shown in Fig. 5.24 for $3 \times 3 - 1$ and $5 \times 5 - 1$ cavities at the cavity mode frequency $\Lambda/\lambda = 0.3817$, for a $5 \times 5 - 1$ cavity at another frequency of $\Lambda/\lambda = 0.3$, and finally the field is shown for a dipole in free space. In the latter case the imaginary part of the field in the center is unity, implying a Purcell factor of 1 for a dipole in free space. For the two cavities and $\Lambda/\lambda = 0.3817$, the imaginary part in the center is roughly a factor of 10 and 100, respectively, in agreement with Fig. 5.23. The field is highly localized to the cavity region and shows the excitation of the cavity mode. For the frequency $\Lambda/\lambda = 0.3$, the Purcell factor is 0.01 and the dipole emission is thus suppressed by a factor of 100. In this case the dipole is not tuned to the cavity mode.

5.9 EXAMPLE: EXCITATION OF SURFACE PLASMON POLARITONS BY SECOND HARMONIC GENERATION IN A SINGLE ORGANIC NANOFIBER

The example considered in this section can be divided in two steps, where the first is concerned with linear or fundamental harmonic (FH) scattering of light incident on a polymer fiber (CNHP4) [137, 138] placed on a 40-nm silver film on quartz. The fibers have a length on the order of 20 μm along the z-axis. Due to the long length of fibers it is reasonable to assume in a calculation that they are infinitely long, in which case a two-dimensional calculation is applicable. It will be assumed that second-harmonic generation in the fiber is sufficiently small that the influence on the scattering process at the FH wavelength is insignificant. In the second step it is considered that the resulting total FH field inside the polymer fiber causes second-harmonic generation (SHG). This is expressed as a given polarization at the second-harmonic (SH) wavelength inside the fiber. The SH polarization then generates radiation at the SH wavelength, which is again scattered by the fiber in a linear scattering process. The SH polarization is equivalent to a local source of radiation at the SH wavelength being placed directly on the silver film. For the polymer fiber and quartz substrate we will use the refractive indices $n_2 = 1.65$ and $n_{L3} = 1.5$. The refractive index of silver is obtained from Ref. [40]. This case was considered both theoretically and experimentally in Ref. [139]. Here, further results are given including the emission or scattering of light into the upper half-plane and total scattering cross sections including also those related to guided modes.

The purpose of using a thin silver film is that SPPs being excited at the air-silver interface can be detected in the substrate by leakage-radiation microscopy. This requires in practice that the substrate is shaped as, e.g., a half-sphere such that light radiated into the substrate can be coupled out. One approach for measuring the leakage-radiation pattern is then to use an angle-scanning detection approach [140, 141, 142, 139]. Alternatively, a direct image of the leakage-radiation can be obtained via a microscopy setup [143, 144, 145, 146].

The first part concerned with linear scattering of the FH incident plane wave is illustrated in Fig. 5.25. A p-polarized plane wave is normally incident on the polymer fiber. As the incident light is scattered by the fiber, two types of surface plasmon polaritons (SPPs) are excited and are referred to in Fig. 5.25 as SPP1 and SPP2.

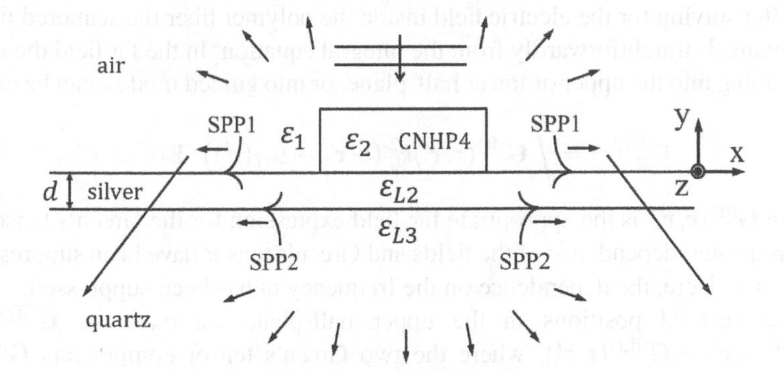

Figure 5.25 Schematic of CNHP4 fiber on a silver film on quartz being illuminated by a normally incident plane wave. Due to scattering, there will be out-of-plane scattering going into the upper and lower half-planes, and SPPs are excited that propagate along the air-silver interface (SPP1) and leak into the substrate. This leakage is observed as a peak in the out-of-plane scattering going into the substrate at a specific angle. In addition another type of non-leaky SPP (SPP2) is excited that propagates along the quartz-silver interface.

SPP1 propagates along the air-silver interface, and due to the presence of the quartz substrate, SPP1 is not a truly bound mode, and it will leak into the substrate at a specific angle determined by conservation of momentum. SPP2 is bound mainly to the quartz-silver interface. The momentum of this SPP does not match any out-of-plane propagating modes, and it is thus truly bound. There will also be out-of-plane scattering going into both the upper and lower half-plane. The leakage of SPP1 into the quartz substrate is observed as a peak at a specific angle in part of the out-of-plane scattering.

We now turn to apply the GFAIEM to the situation in Fig. 5.25. At first we only need the reference field in the upper half-plane of the reference structure (no polymer fiber) is needed. The reference electric field in the upper half-plane including both incident and reflected fields can be expressed as

$$\mathbf{E}_0(\mathbf{r};\omega) = \frac{E_i e^{ik_x x}}{\sqrt{1+\frac{k_x^2}{k_{y1}^2}}}\left\{\left(\hat{\mathbf{x}}+\hat{\mathbf{y}}\frac{k_x}{k_{y1}}\right)e^{-ik_{y1}y} + r_{1,L3}^{(p)}(k_x)\left(-\hat{\mathbf{x}}+\hat{\mathbf{y}}\frac{k_x}{k_{y1}}\right)e^{ik_{y1}y}\right\},$$

$$y > 0. \tag{5.219}$$

The magnitude of the incident field is E_i.

The resulting field in a linear scattering process, when the polymer fiber is present, can be obtained by solving the integral equation

$$\mathbf{E}(\mathbf{r};\omega) = \mathbf{E}_0(\mathbf{r};\omega) + \int \mathbf{G}(\mathbf{r},\mathbf{r}';\omega)\frac{\omega^2}{c^2}\left(\varepsilon(\mathbf{r}') - \varepsilon_{\text{ref}}(\mathbf{r}';\omega)\right)\cdot\mathbf{E}(\mathbf{r}';\omega)d^2r'. \tag{5.220}$$

After solving for the electric field inside the polymer fiber the scattered fields can be obtained straightforwardly from the integral equation. In the far field the scattered field going into the upper or lower half-plane, or into guided modes, can be expressed as

$$\mathbf{E}_{\mathrm{sc}}^{(\mathrm{ff})}(\mathbf{r}) = \int \mathbf{G}^{(\mathrm{ff})}(\mathbf{r},\mathbf{r}')k_0^2\left(\varepsilon(\mathbf{r}') - \varepsilon_{\mathrm{ref}}(\mathbf{r}')\right)\cdot\mathbf{E}(\mathbf{r}')d^2r', \tag{5.221}$$

where $\mathbf{G}^{(\mathrm{ff})}(\mathbf{r},\mathbf{r}')$ is the appropriate far-field expression for the Green's tensor. Here, the frequency-dependence of the fields and Green's tensor have been suppressed, and $k_0 = \omega/c$. Here, the dependence on the frequency ω has been suppressed.

For far-field positions in the upper half-plane we may use $\mathbf{G}^{(\mathrm{ff})}(\mathbf{r},\mathbf{r}') = \mathbf{G}^{(\mathrm{ff},d)}(\mathbf{r},\mathbf{r}') + \mathbf{G}^{(\mathrm{ff},i)}(\mathbf{r},\mathbf{r}')$, where the two Green's tensor components $\mathbf{G}^{(\mathrm{ff},d)}$ and $\mathbf{G}^{(\mathrm{ff},i)}$ are given in Eqs. (5.98) and (5.135). Thus,

$$\mathbf{E}_{\mathrm{sc}}^{(\mathrm{ff})}(r,\theta) = \hat{\theta}\left(A^{(d)} + A^{(i)}\right)\frac{e^{ik_0n_1r}}{\sqrt{r}}, \quad 0<\theta<\pi, \tag{5.222}$$

where

$$A^{(d)} = \hat{\theta}\cdot\left(\frac{1}{4}\sqrt{\frac{2}{\pi k_0 n_1}}e^{i\pi/4}\int e^{-ik_0n_1\hat{\mathbf{r}}\cdot\mathbf{r}'}k_0^2\left(\varepsilon(\mathbf{r}') - \varepsilon_{\mathrm{ref}}(\mathbf{r}')\right)\mathbf{E}(\mathbf{r}')d^2r'\right), \tag{5.223}$$

$$A^{(i)} = (\hat{\mathbf{x}}\sin\theta + \hat{\mathbf{y}}\cos\theta)\frac{1}{4}\sqrt{\frac{2}{\pi k_0 n_1}}e^{i\pi/4}r_{1,L3}^{(p)}(k_0n_1\cos\theta)$$
$$\cdot\left(\int e^{-ik_0n_1\cos\theta x'}e^{ik_0n_1\sin\theta y'}k_0^2\left(\varepsilon(\mathbf{r}') - \varepsilon_{\mathrm{ref}}(\mathbf{r}')\right)\mathbf{E}(\mathbf{r}')d^2r'\right). \tag{5.224}$$

The upward out-of-plane scattering cross section is then obtained as

$$\sigma_{\mathrm{sc},r} = \frac{1}{|E_i|^2}\int_{\theta=0}^{\pi}|\mathbf{E}_{\mathrm{sc}}^{(\mathrm{ff})}(r,\theta)|^2 r d\theta. \tag{5.225}$$

The far field going into the lower half-plane is instead obtained by using $\mathbf{G}^{(\mathrm{ff})}(\mathbf{r},\mathbf{r}') = \mathbf{G}^{(\mathrm{ff},t)}(\mathbf{r},\mathbf{r}')$, where $\mathbf{G}^{(\mathrm{ff},t)}$ is given in Eq. (5.137). Thus

$$\mathbf{E}_{\mathrm{sc}}^{(\mathrm{ff})}(r,\theta) = \hat{\theta}A^{(t)}e^{ik_0n_{L3}r}\sqrt{r}, \quad -\pi<\theta<0, \tag{5.226}$$

where

$$A^{(t)} = \left(\hat{\mathbf{x}}\frac{k_{y1}k_0n_{L3}}{k_0^2\varepsilon_1} + \hat{\mathbf{y}}\cos\theta\frac{\varepsilon_{L3}}{\varepsilon_1}\right)\frac{1}{4}\sqrt{\frac{2}{\pi k_0 n_{L3}}}e^{i\pi/4}t_{1,L3}^{(p)}(k_0n_{L3}\cos\theta)\frac{\varepsilon_1}{\varepsilon_{L3}}\frac{k_{y,L3}}{k_{y,1}}$$
$$\cdot\left(\int e^{-ik_0n_{L3}\cos\theta x'}e^{ik_{y,1}y'}e^{-ik_{y,L3}d}k_0^2\left(\varepsilon(\mathbf{r}') - \varepsilon_{\mathrm{ref}}(\mathbf{r}')\right)\mathbf{E}(\mathbf{r}')d^2r'\right). \tag{5.227}$$

Here $k_{y,i} = \sqrt{k_0^2\varepsilon_i - (k_0n_{L3}\cos\theta)^2}$ with $\mathrm{Imag}\{k_{y,i}\} \geq 0$.

The downward out-of-plane scattering cross section is then obtained as

$$\sigma_{sc,t} = \frac{n_{L3}}{n_1} \frac{1}{|E_i|^2} \int_{\theta=-\pi}^{0} |\mathbf{E}_{sc}^{(ff)}(r,\theta)|^2 r d\theta. \tag{5.228}$$

The differential scattering cross sections are shown versus θ in Fig. 5.26 for a normally incident p-polarized plane wave of wavelength 390 nm, and for a CNHP4 fiber of width 350 nm and height 100 nm. In this case $n_{L2} = 0.05 + i1.9672$ when losses are included, and when losses are ignored $n_{L2} = i1.9672$. The calculation is based on using 40×140 discretization elements that are equivalent to using elements with a size of $\Delta = 2.5 \times 2.5$ nm^2. The cases of scattering going into the quartz substrate

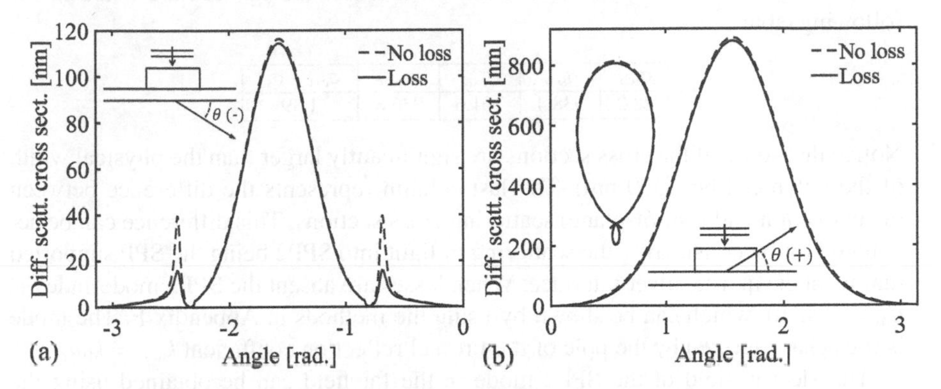

Figure 5.26 Differential scattering cross section for a CNHP4 fiber of width 350 nm and height 100 nm placed on a 40-nm silver film on quartz and being illuminated by a normally incident p-polarized plane wave with wavelength 390 nm. (a) Scattering into the quartz substrate. (b) Scattering into the air superstrate. Solid curves: loss included. Dashed curves: Loss neglected. The inset in (b) is a polar plot of the differential scattering.

and into the air superstrate are considered in Figs. 5.26(a) and (b), respectively. If the differential cross sections are integrated with respect to θ, this leads to the total cross sections in (5.225) and (5.228). Most of the out-of-plane scattering is going into the air superstrate, which is particularly evident from the polar plot of the differential scattering shown as an inset in Fig. 5.26(b).

Notice the two peaks in the differential scattering going into the substrate. These peaks can be identified from phase matching of the SPP1 propagating along the air-silver interface to out-of-plane propagating modes in the substrate. If we assume at first that the silver film would be infinitely thick, then the SPP1 would satisfy the dispersion relation [16]

$$k_{x,SPP1} \approx k_0 \sqrt{\frac{\varepsilon_{L2}}{1+\varepsilon_{L2}}}. \tag{5.229}$$

Now, due to the finite thickness of the silver film, there are out-of-plane propagating modes in the substrate with the same in-plane momentum propagating at an angle α

determined from

$$k_0 n_{L3} \cos\alpha = \text{Real}\{k_{x,SPP1}\}. \tag{5.230}$$

This leads to the angles $\alpha = -0.69$ and -2.45 rad. being in good agreement with the positions of the peaks in Fig. 5.26. Notice that the peaks are higher when loss is ignored as a consequence of no propagation loss of the SPP1 before it leaks into the substrate.

Extinction cross sections for the reflected and transmitted beam can be calculated by inserting the scattered fields for appropriate directions into Eqs. (2.120) and (2.121). These cross sections represent the amount of power removed from the reflected and transmitted beams due to scattering by the polymer fiber. The scattering and extinction cross sections in the case when losses are ignored are listed in the following table

$\sigma_{sc,t}$	$\sigma_{sc,r}$	$\sigma_{ext,t}$	$\sigma_{ext,r}$	$\sigma_{ext} - \sigma_{sc}$
88.2	938.4	101.6	936.8	11.9

Notice that some of the cross sections are significantly larger than the physical width of the polymer fiber (350 nm). The last column represents the difference between the extinction and (out-of-plane) scattering cross sections. This difference can be accounted for by considering the scattering of light into SPP2 being the SPP supported mainly at the quartz-silver interface. When losses are absent the SPP2 mode-index is $n_m = 2.4830$, which can be shown by using the methods in Appendix F. The mode is thus characterized by the pole of the Fresnel reflection coefficient $k_{x,p} = k_0 n_m$.

The electric field of the SPP2 mode in the far field can be obtained using the far-field Green's tensor component related to guided modes (5.140) in the integral (5.221). If we consider $x > x'$ and $y > 0$, the electric far field of the guided mode is given as

$$\mathbf{E}^{(g)}(\mathbf{r}) = \left(-\hat{\mathbf{x}}\frac{k_{y1,p}}{k_1} + \hat{\mathbf{y}}\frac{k_{x,p}}{k_1}\right) e^{ik_{y1,p}y} E_r^{(g)} e^{ik_{x,p}x}, \quad x > x', \quad y > 0, \tag{5.231}$$

where

$$E_r^{(g)} = -\frac{A^{(g)}}{k_{y1,p}} \left(\hat{\mathbf{x}}\frac{k_{y1,p}}{k_1} + \hat{\mathbf{y}}\frac{k_{x,p}}{k_1}\right) \cdot \int e^{-ik_{x,p}x'} e^{ik_{y1,p}y'} k_0^2 \left(\varepsilon(\mathbf{r}') - \varepsilon_{\text{ref}}\right) \mathbf{E}(\mathbf{r}') d^2 r'. \tag{5.232}$$

The corresponding field inside the film ($0 > y > -d_{L2}$), and below the film in the quartz substrate ($y < -d_{L2}$), can be constructed by requiring that the electromagnetics boundary conditions and the wave equation must be satisfied, which leads to

$$\mathbf{E}^{(g)}(\mathbf{r}) = \left\{A\left(-\hat{\mathbf{x}}\frac{k_{y,L2,p}}{k_1} + \hat{\mathbf{y}}\frac{k_{x,p}}{k_1}\right) e^{ik_{y,L2,p}y} + B\left(\hat{\mathbf{x}}\frac{k_{y,L2,p}}{k_1} + \hat{\mathbf{y}}\frac{k_{x,p}}{k_1}\right) e^{-ik_{y,L2,p}y}\right\}$$
$$\times \frac{\varepsilon_1}{\varepsilon_{L2}} E_r^{(g)} e^{ik_{x,p}x}, \quad 0 > y > -d_{L2}, \tag{5.233}$$

$$\mathbf{E}^{(g)}(\mathbf{r}) = C\left(\hat{\mathbf{x}}\frac{k_{y,L3,p}}{k_1} + \hat{\mathbf{y}}\frac{k_{x,p}}{k_1}\right) e^{-ik_{y,L3,p}(y+d_{L2})} \frac{\varepsilon_1}{\varepsilon_{L3}} E_r^{(g)} e^{ik_{x,p}x}, \quad y < -d_{L2} \tag{5.234}$$

where

$$A = \frac{1}{2}\left(1 + \frac{k_{y1,p}}{k_{y,L2,p}}\frac{\varepsilon_{L2}}{\varepsilon_1}\right), \quad B = \frac{1}{2}\left(1 - \frac{k_{y1,p}}{k_{y,L2,p}}\frac{\varepsilon_{L2}}{\varepsilon_1}\right), \tag{5.235}$$

$$C = Ae^{-ik_{y,L2,p}d_{L2}} + Be^{ik_{y,L2,p}d_{L2}}. \tag{5.236}$$

The expression for C was obtained from the continuity of $\varepsilon(\mathbf{r})E_y^{(g)}(\mathbf{r})$ across interfaces. From the continuity of $E_x^{(g)}(\mathbf{r})$ across interfaces it follows equivalently that

$$C = \frac{\varepsilon_{L3}}{\varepsilon_{L2}}\frac{k_{y,L2,p}}{k_{y,L3,p}}\left(-Ae^{-ik_{y,L2,p}d_{L2}} + Be^{ik_{y,L2,p}d_{L2}}\right), \tag{5.237}$$

which may serve as a final check since the same value of C only arises from (5.236) and (5.237) if the pole $k_{x,p}$ has been correctly calculated using, e.g., the methods in Appendix F.

The power in the right-propagating guided mode may now be obtained as

$$P_{\mathrm{r}}^{(g)} = \frac{1}{2}\mathrm{Real}\left\{\int \mathbf{E}^{(g)}(\mathbf{r}) \times \left[\mathbf{H}^{(g)}(\mathbf{r})\right]^* \cdot \hat{\mathbf{x}}dy\right\}. \tag{5.238}$$

The power per unit area of the incident beam is

$$I_i = \frac{1}{2}\left|\mathbf{E}_i \times [\mathbf{H}_i]^*\right| = \frac{1}{2}\frac{k_0 n_1}{\omega\mu_0}|E_i|^2. \tag{5.239}$$

The guided-mode scattering cross section $\sigma_{\mathrm{r}}^{(g)} = P_{\mathrm{r}}^{(g)}/I_i$ can thus be expressed as

$$\sigma_{\mathrm{r}}^{(g)} = \frac{|E^{(g)}|^2|e^{ik_{x,p}x}|^2}{|E_i|^2}\left\{\mathrm{Real}\left\{\frac{k_{x,p}}{k_0 n_1}\right\}\left(\frac{1}{2\mathrm{Imag}\{k_{y1,p}\}} + \frac{|C|^2}{2\mathrm{Imag}\{k_{y,L3,p}\}}\frac{\varepsilon_1}{\varepsilon_{L3}}\right)\right.$$
$$\left. + \mathrm{Real}\left\{\frac{k_{x,p}}{k_0 n_1}\frac{\varepsilon_{L2}^*}{\varepsilon_1}\right\}\left|\frac{\varepsilon_1}{\varepsilon_{L2}}\right|^2 \int_{y=-d}^{0} |Ae^{ik_{y,L2,p}y} + Be^{-ik_{y,L2,p}y}|^2 dy\right\}. \tag{5.240}$$

Since the incident light is normally incident, and the structure is symmetric, the scattering cross section for the left-propagating SPP2 is the same. When using the same parameters as when calculating Fig. 5.26, the scattering cross section for guided modes is obtained as $\sigma_{\mathrm{r}}^{(g)} + \sigma_{\mathrm{l}}^{(g)} = 2\sigma_{\mathrm{r}}^{(g)} = 12.3$. When comparing with the difference between extinction and scattering cross sections, it is seen that the total extinction and total scattering including both out-of-plane modes and guides modes are in good agreement.

We now turn to consider the second part with SHG in the fiber. The corresponding situation is illustrated in Fig. 5.27. The CNHP4 fiber now works as a given source at the SH wavelength, which is equal to half of the FH wavelength. As a model for the SH source polarization distribution, we will assume that the fiber is composed of elongated molecules with the property that SH polarization in the fiber will be in a direction determined by the molecule orientation α, and the magnitude of the polarization is related to the component of the FH field at the same position in the same direction.

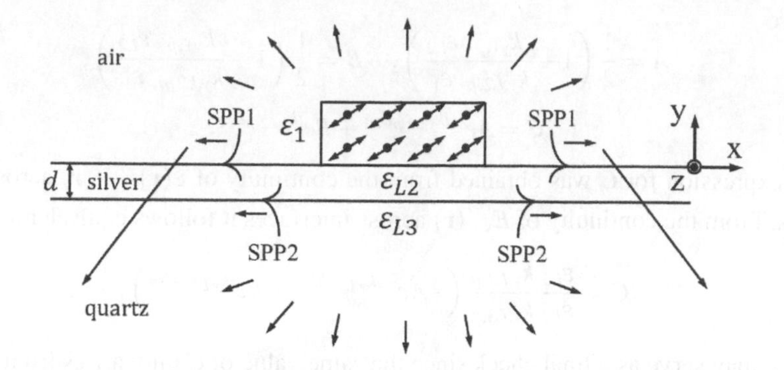

Figure 5.27 Schematic of SHG in a CNHP4 fiber on a silver film on quartz. Radiation from the given SH polarization is further scattered by the fiber leading to excitation of SPP1 and SPP2 modes and to out-of-plane radiation at the SHG wavelength.

Within this model the given SH polarization at the second-harmonic frequency can be expressed as

$$\mathbf{P}(\mathbf{r},2\omega) = \chi^{(2)}\,(\hat{\mathbf{x}}\cos\alpha + \hat{\mathbf{y}}\sin\alpha)\,P(\mathbf{r}), \tag{5.241}$$

where

$$P(\mathbf{r}) = |(\hat{\mathbf{x}}\cos\alpha + \hat{\mathbf{y}}\sin\alpha)\cdot\mathbf{E}(\mathbf{r},\omega)|^2. \tag{5.242}$$

The coefficient $\chi^{(2)}$ governs the strength of the non-linear process. Here, we will simply set this coefficient to unity. In the remainder of this section, k_0 should be replaced with $k_0 = 2\omega/c$ since the SH frequency is twice as large.

The given source polarization itself leads to a second-harmonic field in the reference structure

$$\mathbf{E}_0(\mathbf{r},2\omega) = \int \mathbf{G}(\mathbf{r},\mathbf{r}';2\omega)\frac{(2\omega)^2}{c^2}\cdot\frac{\mathbf{P}(\mathbf{r}';2\omega)}{\varepsilon_0}d^2r'. \tag{5.243}$$

However, a polarization may further be induced in the fiber by this field. Thus, subsequent scattering by the polymer fiber leads to a total field at the second-harmonic frequency given by

$$\mathbf{E}(\mathbf{r};2\omega) = \mathbf{E}_0(\mathbf{r},2\omega) + \int \mathbf{G}(\mathbf{r},\mathbf{r}';2\omega)\frac{(2\omega)^2}{c^2}\left(\varepsilon(\mathbf{r}') - \varepsilon_{\text{ref}}(\mathbf{r}')\right)\cdot\mathbf{E}(\mathbf{r}';2\omega)d^2r'. \tag{5.244}$$

Again, at first the integral equation should be solved for the field inside the fiber, and the integral equation can then be used to subsequently calculate the field at any position outside the fiber. The far field is slightly different from the case of linear scattering since the field $\mathbf{E}_0$ now also contributes to the far field. This field component also originates from sources inside the fiber. The far field may be calculated as follows

$$\mathbf{E}(\mathbf{r},2\omega) = \int \mathbf{G}^{(\mathrm{ff})}(\mathbf{r},\mathbf{r}';2\omega)\frac{(2\omega)^2}{c^2}\cdot\left(\frac{\mathbf{P}(\mathbf{r}';2\omega)}{\varepsilon_0} + (\varepsilon(\mathbf{r}')-\varepsilon_{\mathrm{ref}}(\mathbf{r}'))\,\mathbf{E}(\mathbf{r};2\omega)\right)d^2r'. \tag{5.245}$$

It is also possible in this case to check for energy conservation when losses are absent. In this case the total power radiated from the polymer fiber can be obtained from the near-field as

$$P_{\mathrm{tot}} = -\frac{1}{2}(2\omega\varepsilon_0)\,\mathrm{Real}\left\{\int \mathbf{E}(\mathbf{r};2\omega)\cdot\left[-i\frac{\mathbf{P}(\mathbf{r};2\omega)}{\varepsilon_0}\right]^* d^2r'\right\}, \tag{5.246}$$

where it was used that the polarization is equivalent to a current density $\mathbf{J} = -i2\omega\mathbf{P}$ and then (5.246) follows from (2.41) when there are no absorption losses in the fiber.

The radiated power can also be obtained from the far field. The power radiated into the upper half-plane $P_{\mathrm{OUP},r}$ and lower half-plane $P_{\mathrm{OUP},t}$ are given as

$$P_{\mathrm{OUP},r} = \frac{1}{2}(2\omega\varepsilon_0)\frac{n_1}{k_0}\int_{\theta=0}^{\pi}|\mathbf{E}^{(\mathrm{ff})}(r,\theta;2\omega)|^2 r d\theta, \tag{5.247}$$

$$P_{\mathrm{OUP},t} = \frac{1}{2}(2\omega\varepsilon_0)\frac{n_{L3}}{k_0}\int_{\theta=-\pi}^{0}|\mathbf{E}^{(\mathrm{ff})}(r,\theta;2\omega)|^2 r d\theta. \tag{5.248}$$

The power in the right- and left-propagating SPP2 modes can be expressed as

$$\begin{aligned} P_{\mathrm{u}}^{(g)} &= \frac{1}{2}(2\omega\varepsilon_0)\frac{n_1}{k_0}|E_{\mathrm{u}}^{(g)}|^2|e^{\pm ik_{x,p}x}|^2 \\ &\times\left\{\mathrm{Real}\left\{\frac{k_{x,p}}{k_0 n_1}\right\}\left(\frac{1}{2\mathrm{Imag}\{k_{y1,p}\}} + \frac{|C|^2}{2\mathrm{Imag}\{k_{y,L3,p}\}}\frac{\varepsilon_1}{\varepsilon_{L3}}\right)\right. \\ &\left. +\mathrm{Real}\left\{\frac{k_{x,p}}{k_0 n_1}\frac{\varepsilon_{L2}^*}{\varepsilon_1}\right\}\left|\frac{\varepsilon_1}{\varepsilon_{L2}}\right|^2\int_{y=-d}^{0}|Ae^{ik_{y,L2,p}y} + Be^{-ik_{y,L2,p}y}|^2 dy\right\}, \end{aligned} \tag{5.249}$$

where u $= r$ or l represents either the right- or the left-propagating SPP2 mode, and $\pm$ is plus when u $= r$ and minus when u $= l$. The coefficients $E_{\mathrm{u}}^{(g)}$ are here given as

$$\begin{aligned} E_{\mathrm{u}}^{(g)} &= -\frac{A^{(g)}}{k_{y1,p}}\left(\hat{\mathbf{x}}\frac{k_{y1,p}}{k_1} \pm \hat{\mathbf{y}}\frac{k_{x,p}}{k_1}\right) \\ &\cdot\int e^{+ik_{x,p}x'}e^{ik_{y1,p}y'}k_0^2\left((\varepsilon(\mathbf{r}')-\varepsilon_{\mathrm{ref}})\,\mathbf{E}(\mathbf{r}';2\omega) + \frac{\mathbf{P}(\mathbf{r};2\omega)}{\varepsilon_0}\right)d^2r'. \end{aligned} \tag{5.250}$$

We now consider the case of an FH plane-wave incident field at the wavelength 780 nm. For the calculation at the FH wavelength the silver refractive index $n_{L2} = 0.0335 + i5.4127$ has been used when losses are included, and $n_{L2} = i5.4127$ has been used when losses are neglected. In this case the peaks in scattered light going into the substrate are much sharper compared with the case of the FH wavelength 390 nm. As a result of the FH incident field, there will be an SH polarization in

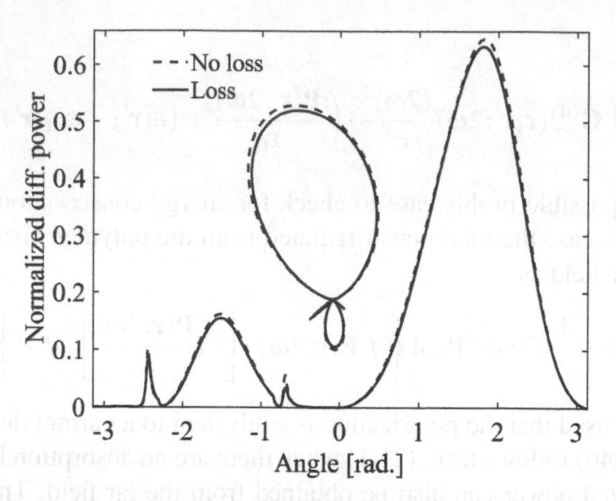

Figure 5.28 Out-of-plane differential radiated power normalized with the total power for SHG in CNHP4 fiber on a silver film on quartz. The CNHP4 molecules are oriented at an angle of $\alpha = 25$ deg. relative to the x-axis, and the SHG wavelength is 390 nm.

the polymer fiber at the wavelength 390 nm, which leads to out-of-plane radiation and excitation of SPP2. The molecule orientation is $\alpha = 25°$. The differential out-of-plane radiated power normalized with the total power P_{tot} is shown in Fig. 5.28. The calculation was carried out using, again, 40×140 square-shaped elements. If the result in Fig. 5.28 is integrated with respect to θ from $-\pi$ to 0, or from 0 to π, the result is $P_{\text{OUP},t}/P_{\text{tot}}$ and $P_{\text{OUP},r}/P_{\text{tot}}$, respectively. The inset shows a polar plot of the out-of-plane radiated power. Clearly, most of the power is radiated into the upper half-plane, and the emission pattern is asymmetric as a consequence of the orientation of the fiber molecules. Also notice again the peaks in the emission going into the substrate. These peaks are again related to excitation of SPP1 and subsequent leakage into the substrate. There is also an asymmetry here for the same reason. Such an asymmetry was observed experimentally in [139].

The normalized upward and downward out-of-plane propagating power, and the power in the SPP2 right- and left-propagating modes, are shown in the following table for the case when losses are neglected:

$P_{\text{OUP},t}/P_{\text{tot}}$	$P_{\text{OUP},r}/P_{\text{tot}}$	$P_r^{(g)}/P_{\text{tot}}$	$P_l^{(g)}/P_{\text{tot}}$
0.1397	0.8529	0.0050	0.0029

Notice that $\left(P_{\text{OUP},t} + P_{\text{OUP},r} + P_r^{(g)} + P_l^{(g)}\right)/P_{\text{tot}} = 1.0005 \approx 1$. The power calculated using either the near field or the far field are thus in good agreement. The excitation of the SPP2 mode is clearly of minor importance. However, it is again clearly necessary to take into account the effect of the SPP2 mode in order to demonstrate conservation of power in the calculation.

5.10 GUIDELINES FOR SOFTWARE IMPLEMENTATION

Structure of GFAIEM program with square-shaped area elements

1. Define the structure and size of area elements, refractive indices, wavelength, polarization, reference field parameters, the region in the *xy*-plane for field calculation, and the number of sampling points along *x* and *y*.
2. Build vectors with center position and dielectric constant for the area elements. Apply the averaging procedure in Sec. 5.2. Plot the structure for a visual check.
3. Construct a vector with the reference field at the positions of the area elements.
4. Construct a table with values of the indirect Green's tensor in case of a layered reference structure.
5. Construct the matrix in (5.52) for *s* polarization or the matrix in (5.96) for *p* polarization. For a layered reference structure the matrices will have the same form except that the Green's function or tensor is different. Solve the matrix equation for the field inside the scatterer.
6. Calculate the near-field. Use the already calculated field at positions inside the scatterer, and apply the integral equation to obtain the field at other positions.
7. Calculate the far-field radiation patterns of scattered or total fields depending on the type of source field. Calculate optical cross sections or emitted power.
8. Plot near-fields and far-field radiation patterns and save the result in a file.

Structure of GFAIEM program with triangle-shaped area elements

1. Define the structure and size of area elements, refractive indices, wavelength, polarization, reference field parameters, the region in the *xy*-plane for field calculation, and the number of sampling points along *x* and *y*.
2. Build a mesh for the region of the scatterer in terms of the matrices $\overline{\overline{P}}$ and $\overline{\overline{T}}$ for the desired size of triangles. This can be done using, e.g., the MATLAB® *pdetool*. Plot the mesh for a visual check.
3. Construct a vector with the incident field at the mesh points (*s* polarization) or sampling points at the center of the triangles (*p* polarization).
4. Construct a vector with the area of all triangles.
5. Construct the matrix in (5.65) for *s* polarization, or the matrix in (5.96) for *p* polarization, and solve for the total field at the mesh points or sampling points. Plot the field.
6. Calculate the field outside the scatterer. One possible approach is to generate a mesh of triangles for the region of interest outside the scatterer, and to calculate the field on the mesh points using the GFAIEM. The field can then be plotted using the MATLAB® command *pdeplot*. Save the result in a file.

5.11 EXERCISES

1. Prove that $\mathbf{G}^{(d)}(\mathbf{r},\mathbf{r}')$ in (5.23) satisfies (5.26) when $\varepsilon_{\text{ref}}(\mathbf{r}) = 1$. One approach is to insert $\mathbf{G}^{(d)}$ in the left-hand side of (5.26) and use the fact that the scalar Green's function satisfies $(\nabla^2+k_0^2)g^{(d)}(\mathbf{r},\mathbf{r}')=-\delta(\mathbf{r}-\mathbf{r}')$.

2. Prove the expression for g_{ii} in (5.49) and verify that $g_{ii}\to 0$ for $\Delta_i\to 0$.

3. Prove (5.69) and (5.74).

4. Derive (5.79).

5. Derive the expression (5.92) for $\mathbf{G}_{ii}$ using the following steps. 1: Apply symmetry considerations to show that $G_{xx,ii}=G_{yy,ii}$ and $G_{xy,ii}=G_{yx,ii}=0$. 2: Show that

$$G_{xx,ii}+G_{yy,ii}=\int_{\Omega_i}\{G_{xx}(\mathbf{r}_i,\mathbf{r}')+G_{yy}(\mathbf{r}_i,\mathbf{r}')\}\,d^2r'$$

$$=\int_{\Omega_i}\left\{2g^{(d)}(\mathbf{r}_i,\mathbf{r}')+\frac{1}{k^2}\nabla'^2g^{(d)}(\mathbf{r}_i,\mathbf{r}')\right\}d^2r'. \tag{5.251}$$

3: Now use $(\nabla'^2+k^2)g^{(d)}(\mathbf{r},\mathbf{r}')=-\delta(\mathbf{r}-\mathbf{r}')$. Finally, apply the expression (5.49).

6. Show that (5.97) leads to (5.98).

7. In this exercise an alternative derivation of the electric field (5.108) inside a circular nanocylinder should be derived in the quasistatic limit. In the quasistatic limit it is assumed that

$$\nabla\times\mathbf{E}(\mathbf{r})\approx\mathbf{0}, \tag{5.252}$$

and thus

$$\mathbf{E}(\mathbf{r})\approx-\nabla\phi, \tag{5.253}$$

where ϕ is the electric potential. Show that in each region, Ω_1 and Ω_2, outside and inside the nanocylinder, respectively, the potential must satisfy

$$\nabla^2\phi(\mathbf{r})=0. \tag{5.254}$$

Here, it is assumed that there are no free charges. Now show that the incident field $\mathbf{E}_0=\hat{\mathbf{x}}E_0$ is given in terms of the potential $\phi_0(\mathbf{r})=-E_0r\cos\theta$.

Consider the following assumptions for the potential in each region

$$\phi(\mathbf{r})=\phi_0(\mathbf{r})+B\frac{x}{r^2},\quad \mathbf{r}\in\Omega_1, \tag{5.255}$$

$$\phi(\mathbf{r})=Ar\cos\theta,\quad \mathbf{r}\in\Omega_2, \tag{5.256}$$

and show that these expressions satisfy (5.254) in each region. Now consider an arbitrary but small cylinder radius a, and apply the electromagnetics boundary conditions across the cylinder surface. Show that these can be satisfied when

$$A=-E_0\frac{2\varepsilon_1}{\varepsilon_1+\varepsilon_2}, \tag{5.257}$$

which is equivalent to the expression (5.108) for the electric field inside the cylinder.

8. Consider the field (5.121) being incident on an interface between media with dielectric constants ε_1 and ε_{L2}. Construct the corresponding reflected and transmitted fields $\mathbf{E}_{r,y}$ and $\mathbf{E}_{t,y}$.

9. Derive (5.140) by using an approach similar to Sec. 4.2.1.

10. Verify (5.181) and (5.187).

11. Verify (5.222) and (5.226) for the OUP scattered far fields.

12. Verify the expression for the guided-mode scattering cross section (5.240).

13. Follow the guidelines in Sec. 5.10 and develop GFAIEM codes (s and p polarization) capable of modeling the near-fields for a plane wave being incident on a circular nanocylinder. Apply both square-shaped discretization elements and triangular-shaped elements. Reproduce the convergence results in Sec. 5.4.

14. Follow the guidelines in Sec. 5.10 and develop GFAIEM codes (s polarization) capable of calculating near and far fields for a point source placed in a photonic crystal or a photonic crystal cavity. Reproduce the results in Sec. 5.8. Experiment further by varying the size of the photonic crystal, the shape of the photonic crystal cavity, and the position of the source. Also consider a plane wave being incident on the photonic crystal instead of a point source.

15. Follow the guidelines in Sec. 5.10 and develop a GFAIEM code (p polarization) capable of calculating far-field radiation patterns for a low-refractive index polymer nanostrip placed on a thin silver film on quartz being illuminated by a plane wave. Reproduce the results in Sec. 5.9.

16. Implement and use the GFAIEM for periodic structures described in Sec. 5.6 for reproducing the results in Sec. 4.3.10 obtained with the GFSIEM.

6 Volume integral equation method for 3D scattering problems

This chapter is concerned with electric-field volume integral equations for 3D scattering problems. The principles are very similar to those considered for the Green's function area integral equation method in Chapter 5 with the main difference that 3D problems are considered instead of 2D problems.

6.1 GREEN'S FUNCTION INTEGRAL EQUATION

The starting point is again the wave equation for the electric field

$$-\nabla \times \nabla \times \mathbf{E}(\mathbf{r}) + k_0^2 \varepsilon(\mathbf{r}) \mathbf{E}(\mathbf{r}) = -i\omega\mu_0 \mathbf{J}(\mathbf{r}), \tag{6.1}$$

where the electric field $\mathbf{E}$ and the current density $\mathbf{J}$ now may vary in all three dimensions, i.e., $\mathbf{r} = \hat{\mathbf{x}}x + \hat{\mathbf{y}}y + \hat{\mathbf{z}}z$. We will consider a geometry that can be described as a scatterer being placed in a reference structure, where the latter has the dielectric function $\varepsilon_{\mathrm{ref}}(\mathbf{r})$, and $\varepsilon(\mathbf{r}) - \varepsilon_{\mathrm{ref}}(\mathbf{r})$ thus only differs from zero for positions inside the scatterer.

In the case where the scatterer is not present the reference field $\mathbf{E}_0$ must be a solution to the equation

$$-\nabla \times \nabla \times \mathbf{E}_0(\mathbf{r}) + k_0^2 \varepsilon_{\mathrm{ref}}(\mathbf{r}) \mathbf{E}_0(\mathbf{r}) = -i\omega\mu_0 \mathbf{J}(\mathbf{r}) \tag{6.2}$$

with appropriate boundary conditions.

If a Green's tensor is known for the reference structure, and it thus satisfies

$$-\nabla \times \nabla \times \mathbf{G}(\mathbf{r},\mathbf{r}') + k_0^2 \varepsilon_{\mathrm{ref}}(\mathbf{r}) \mathbf{G}(\mathbf{r},\mathbf{r}) = -\mathbf{I}\delta(\mathbf{r}-\mathbf{r}'), \tag{6.3}$$

and if the Green's tensor furthermore satisfies appropriate boundary conditions, the reference field can be obtained straightforwardly as

$$\mathbf{E}_0(\mathbf{r}) = i\omega\mu_0 \int \mathbf{G}(\mathbf{r},\mathbf{r}') \cdot \mathbf{J}(\mathbf{r}') d^3 r'. \tag{6.4}$$

If the Green's tensor, for example, satisfies the radiating boundary condition, meaning that $\mathbf{G}(\mathbf{r},\mathbf{r}')$ describes waves that propagate away from the position $\mathbf{r}'$, then the resulting solution for the field (6.4) will also satisfy the radiating boundary condition meaning that the field solution will describe fields that propagate away from the current density.

The Green's tensor is known for a homogeneous medium and for planar layered structures but is not necessarily known for the more complicated structure with dielectric constant $\varepsilon(\mathbf{r})$. In that case we may rewrite the wave equation (6.1) into

$$-\nabla\times\nabla\times\mathbf{E}(\mathbf{r})+k_0^2\varepsilon_{\text{ref}}(\mathbf{r})\mathbf{E}(\mathbf{r})=-k_0^2\left(\varepsilon(\mathbf{r})-\varepsilon_{\text{ref}}(\mathbf{r})\right)\mathbf{E}(\mathbf{r})-i\omega\mu_0\mathbf{J}(\mathbf{r}), \quad (6.5)$$

and by further subtracting Eq. (6.2) it is found that

$$\left(-\nabla\times\nabla\times+k_0^2\varepsilon_{\text{ref}}(\mathbf{r})\right)\left(\mathbf{E}(\mathbf{r})-\mathbf{E}_0(\mathbf{r})\right)=-k_0^2\left(\varepsilon(\mathbf{r})-\varepsilon_{\text{ref}}(\mathbf{r})\right)\mathbf{E}(\mathbf{r}). \quad (6.6)$$

It now follows that the total electric field satisfies the Green's function volume integral equation:

Green's function volume integral equation:

$$\mathbf{E}(\mathbf{r})=\mathbf{E}_0(\mathbf{r})+\int\mathbf{G}(\mathbf{r},\mathbf{r}')k_0^2\cdot\left(\varepsilon(\mathbf{r}')-\varepsilon_{\text{ref}}(\mathbf{r}')\right)\mathbf{E}(\mathbf{r}')d^3r'. \quad (6.7)$$

The only difference compared with the Green's tensor area integral equation (5.31) is that this is a 3D problem.

6.2 SCATTERER IN A HOMOGENEOUS MEDIUM

In this section a homogeneous medium is considered as the reference geometry, and thus $\varepsilon_{\text{ref}}(\mathbf{r})=\varepsilon_{\text{ref}}$ is a constant. The Green's tensor $\mathbf{G}^{(d)}(\mathbf{r},\mathbf{r}')$ for this case, which shall be referred to as the direct Green's tensor similar to Chapter 5, must be a solution to the equation

$$\left(-\nabla\times\nabla\times+k_0^2\varepsilon_{\text{ref}}\right)\mathbf{G}^{(d)}(\mathbf{r},\mathbf{r}')=-\mathbf{I}\delta(\mathbf{r}-\mathbf{r}'). \quad (6.8)$$

Furthermore, the Green's tensor must satisfy the radiating boundary condition. The field $\mathbf{E}$ in (6.7) will thus be in the form of the sum of a given incident field $\mathbf{E}_0$ and a scattered field, where the latter propagates away from the scatterer due to the properties of the Green's tensor.

The Green's tensor that satisfies the radiating boundary condition is given by

$$\mathbf{G}^{(d)}(\mathbf{r},\mathbf{r}')=\left(\mathbf{I}+\frac{1}{k^2}\nabla\nabla\right)g^{(d)}(\mathbf{r},\mathbf{r}'), \quad (6.9)$$

where $k^2=k_0^2\varepsilon_{\text{ref}}$, and the direct scalar Green's function for 3D problems is given by

$$g^{(d)}(\mathbf{r},\mathbf{r}')=\frac{e^{ik|\mathbf{r}-\mathbf{r}'|}}{4\pi|\mathbf{r}-\mathbf{r}'|}. \quad (6.10)$$

The scalar Green's function satisfies

$$\left(\nabla^2+k^2\right)g^{(d)}(\mathbf{r},\mathbf{r}')=-\delta(\mathbf{r}-\mathbf{r}'). \quad (6.11)$$

This can be shown by first using direct calculation to show that

$$(\nabla^2 + k^2)\, g^{(d)}(\mathbf{r},\mathbf{r}') = 0, \quad \mathbf{r} \neq \mathbf{r}'. \tag{6.12}$$

Then we may for simplicity consider $\mathbf{r}' = \mathbf{0}$, in which case the remaining task is to show that

$$\int_{\delta V \to 0} (\nabla^2 + k^2)\, g^{(d)}(\mathbf{r},\mathbf{0}) d^3 r = -1. \tag{6.13}$$

For the volume δV we may consider a spherical volume centered at $\mathbf{r} = \mathbf{0}$. The integral over $k^2 g^{(d)}(\mathbf{r},\mathbf{r}')$ will vanish in the limit $\delta V \to 0$. The remaining integral can be evaluated using Gauss's theorem

$$\int_{\delta V \to 0} \nabla^2 g^{(d)}(\mathbf{r},\mathbf{0}) d^3 r = \oint_{\partial(\delta V)} \hat{\mathbf{r}} \cdot \nabla g^{(d)}(\mathbf{r},\mathbf{0}) d^2 r, \tag{6.14}$$

where the latter integral is over the surface of the spherical volume, $\partial(\delta V)$, and it has been applied that $\hat{\mathbf{r}}$ is the outward normal vector of the surface. In the limit being considered $g(\mathbf{r},\mathbf{0}) \approx 1/4\pi r$, and straightforward evaluation of the integral using that r is constant on the spherical surface now leads to the result -1. It can be shown straightforwardly that the Green's tensor satisfies (6.8) by inserting the expression for the Green's tensor (6.9) into the left-hand side of (6.8) and using (6.11).

The Green's tensor (6.9) can be evaluated analytically, which leads to

$$\mathbf{G}^{(d)}(\mathbf{r},\mathbf{r}') = \left(\mathbf{I} \left[1 + \frac{i}{kR} - \frac{1}{(kR)^2} \right] - \frac{\mathbf{R}\mathbf{R}}{R\,R} \left[1 + \frac{3i}{kR} - \frac{3}{(kR)^2} \right] \right) g^{(d)}(\mathbf{r},\mathbf{r}'), \tag{6.15}$$

where $\mathbf{R} = \mathbf{r} - \mathbf{r}'$ and $R = |\mathbf{R}|$.

A far-field expression for the Green's tensor can be obtained in spherical coordinates r, θ, and ϕ, being related to x, y, and z, as follows

$$x = r \sin\theta \cos\phi, \tag{6.16}$$

$$y = r \sin\theta \sin\phi, \tag{6.17}$$

$$z = r \cos\theta. \tag{6.18}$$

Thus,

$$\mathbf{r} = \hat{\mathbf{r}} r = \hat{\mathbf{x}} r \sin\theta \cos\phi + \hat{\mathbf{y}} r \sin\theta \sin\phi + \hat{\mathbf{z}} r \cos\theta. \tag{6.19}$$

The spherical coordinate unit vectors are defined as

$$\hat{\mathbf{r}} = \mathbf{r}/r, \tag{6.20}$$

$$\hat{\theta} = \hat{\mathbf{x}} \cos\theta \cos\phi + \hat{\mathbf{y}} \cos\theta \sin\phi - \hat{\mathbf{z}} \sin\theta, \tag{6.21}$$

$$\hat{\phi} = -\hat{\mathbf{x}} \sin\phi + \hat{\mathbf{y}} \cos\phi. \tag{6.22}$$

In the limit of large r (far field), and with $\mathbf{r}'$ placed close to the origin in comparison with $\mathbf{r}$, i.e., $R \approx r$, and using that $\mathbf{I} = \hat{\mathbf{r}}\hat{\mathbf{r}} + \hat{\theta}\hat{\theta} + \hat{\phi}\hat{\phi}$, the Green's tensor reduces to

$$\mathbf{G}^{(\mathrm{ff},d)}(\mathbf{r},\mathbf{r}') = (\hat{\theta}\hat{\theta} + \hat{\phi}\hat{\phi}) \frac{e^{ikr}}{4\pi r} e^{-ik\hat{\mathbf{r}}\cdot\mathbf{r}'}. \tag{6.23}$$

Notice that $\mathbf{G}^{(\mathrm{ff},d)}(\mathbf{r},\mathbf{r}') \cdot \mathbf{p}$, with $\mathbf{r}'$ near the origin, corresponds at a far-field position $\mathbf{r}$ to a field that propagates in the direction $\hat{\mathbf{r}}$, and which is polarized in a direction perpendicular to $\hat{\mathbf{r}}$. The field is thus perpendicular to the direction of propagation.

6.2.1 DISCRETIZATION WITH CUBIC VOLUME ELEMENTS

In this subsection a discretization approach is considered where the scatterer is discretized into cubic volume elements of the same size being placed on a cubic lattice. It is assumed that the electric field inside each cube is constant. The center position of cube i is denoted $\mathbf{r}_i$, and the electric field and reference field in cube i are denoted $\mathbf{E}_i \equiv \mathbf{E}(\mathbf{r}_i)$ and $\mathbf{E}_{0,i} \equiv \mathbf{E}_0(\mathbf{r}_i)$, respectively. Similar to the area integral equation method with square-shaped elements, a single dielectric constant tensor may be assigned to each cube. In the case that the actual structure of interest within the region of cube i can be described by two materials separated by an interface with normal vector $\hat{\mathbf{n}}$, the following dielectric constant tensor can be assigned to the element

$$\varepsilon_i = \left(\varepsilon_{\perp,i} - \varepsilon_{\parallel,i}\right)\hat{\mathbf{n}}\hat{\mathbf{n}} + \mathbf{I}\varepsilon_{\parallel,i}, \tag{6.24}$$

where here

$$\varepsilon_{\parallel,i} = \frac{1}{V_i}\int_i \varepsilon(\mathbf{r})d^3r, \tag{6.25}$$

$$\frac{1}{\varepsilon_{\perp,i}} = \frac{1}{V_i}\int_i \frac{1}{\varepsilon(\mathbf{r})}d^3r. \tag{6.26}$$

The integrals are taken over cube i, and V_i is the volume of cube i.

With these assumptions the integral equation (6.7) can be expressed in the discrete form

$$\mathbf{E}_i = \mathbf{E}_{0,i} + \sum_j \mathbf{G}_{i,j}k_0^2\left(\varepsilon_j - \varepsilon_{\text{ref}}\right)\cdot\mathbf{E}_j, \tag{6.27}$$

where the Green's tensor element is

$$\mathbf{G}_{i,j} \equiv \int_j \mathbf{G}^{(d)}(\mathbf{r}_i,\mathbf{r}')d^3r'. \tag{6.28}$$

A simple approximation of $\mathbf{G}_{i,j}$ is given as

$$\mathbf{G}_{i,j} = \begin{cases} \mathbf{G}^{(d)}(\mathbf{r}_i,\mathbf{r}_j)V_j, & i \neq j \\ -\mathbf{I}\frac{1}{3k^2} - \mathbf{I}\frac{1}{k^2}\frac{2}{3}\left[1 - e^{ika}(1 - ika)\right], & i = j \end{cases}, \tag{6.29}$$

where a is the radius of a sphere with the same volume as element i, i.e., $4\pi a^3/3 = V_i$. The case of $i = j$ can be obtained by using that $G_{i,i,xx} = G_{i,i,yy} = G_{i,i,zz}$. It now follows from (6.9) that

$$G_{i,i,xx} = \frac{1}{3}\left(G_{i,i,xx} + G_{i,i,yy} + G_{i,i,zz}\right) = \frac{1}{3}\int_i \left(\frac{1}{k^2}\nabla^2 g^{(d)}(\mathbf{r}_i,\mathbf{r}') + 3g^{(d)}(\mathbf{r}_i,\mathbf{r}')\right)d^3r'. \tag{6.30}$$

By further applying (6.11), and replacing the cubic volume element with a sphere of the same volume, the result for $\mathbf{G}_{i,i}$ in (6.29) is obtained.

Note that

$$\mathbf{G}_{i,i} \to -\mathbf{I}\frac{1}{3k^2} \quad \text{for} \quad V_i \to 0, \tag{6.31}$$

and the self-interaction term thus does not vanish in the limit of vanishing element volume.

The fields may be organized in vectors as

$$\overline{E}_\alpha = \left[E_{\alpha,1} \quad E_{\alpha,2} \quad \ldots \quad E_{\alpha N}\right]^T, \tag{6.32}$$

$$\overline{E}_{0,\alpha} = \left[E_{0,\alpha,1} \quad E_{0,\alpha,2} \quad \ldots \quad E_{0,\alpha N}\right]^T, \tag{6.33}$$

where $\alpha = x$, y, or z. The vectors thus represent the x-, y-, or z-components of the electric field in the elements. In addition, matrices $\overline{\overline{G}}_{\alpha\beta}$ are defined as matrices with the elements $\hat{\alpha} \cdot \mathbf{G}_{i,j} \cdot \hat{\beta}$, and the matrices $\overline{\overline{\varepsilon}}_{\alpha\beta}$ are defined as diagonal matrices with the elements $\hat{\alpha} \cdot \varepsilon_i \cdot \hat{\beta}$.

The integral equation can now be expressed in matrix form as

$$\left(\overline{\overline{I}} - k_0^2 \begin{bmatrix} \overline{\overline{G}}_{xx} & \overline{\overline{G}}_{xy} & \overline{\overline{G}}_{xz} \\ \overline{\overline{G}}_{yx} & \overline{\overline{G}}_{yy} & \overline{\overline{G}}_{yz} \\ \overline{\overline{G}}_{zx} & \overline{\overline{G}}_{zy} & \overline{\overline{G}}_{zz} \end{bmatrix} \left(\begin{bmatrix} \overline{\overline{\varepsilon}}_{xx} & \overline{\overline{\varepsilon}}_{xy} & \overline{\overline{\varepsilon}}_{xz} \\ \overline{\overline{\varepsilon}}_{yx} & \overline{\overline{\varepsilon}}_{yy} & \overline{\overline{\varepsilon}}_{yz} \\ \overline{\overline{\varepsilon}}_{zx} & \overline{\overline{\varepsilon}}_{zy} & \overline{\overline{\varepsilon}}_{zz} \end{bmatrix} - \overline{\overline{I}}\varepsilon_{\text{ref}} \right)\right) \begin{bmatrix} \overline{E}_x \\ \overline{E}_y \\ \overline{E}_z \end{bmatrix} = \begin{bmatrix} \overline{E}_{0,x} \\ \overline{E}_{0,y} \\ \overline{E}_{0,z} \end{bmatrix}. \tag{6.34}$$

This can be readily solved for the field inside the scatterer if the computer has sufficient memory. Consider a 3D geometry where it is desired to use on the order of 100 elements for each direction x, y and z. Then the number of unknowns will be on the order of $3 \cdot 100^3 = 3 \cdot 10^6$. The number of coefficients in the matrix will then be on the order of $(3 \cdot 10^6)^2 = 9 \cdot 10^{12}$, which is a prohibitively large matrix. If the number of unknowns is required to be on that order of magnitude, it is necessary to instead resort to iterative methods combined with the FFT algorithm similar to Sec. 5.7. Thus, the position of each cube will be represented by three indices, i.e.,

$$\mathbf{r}_{i_x,i_y,i_z} = \hat{\mathbf{x}}(x_s + i_x\Delta_x) + \hat{\mathbf{y}}(y_s + i_y\Delta_y) + \hat{\mathbf{z}}(z_s + i_z\Delta_z), \tag{6.35}$$

and the reference field and total field in a cube will be identified by the same three indices, i.e.,

$$\mathbf{E}_{i_x,i_y,i_z} \equiv \mathbf{E}(\mathbf{r}_{i_x,i_y,i_z}), \tag{6.36}$$

$$\mathbf{E}_{0,i_x,i_y,i_z} \equiv \mathbf{E}_0(\mathbf{r}_{i_x,i_y,i_z}). \tag{6.37}$$

Due to the fact that the Green's tensor is in the form $\mathbf{G}(\mathbf{r},\mathbf{r}') = \mathbf{G}(\mathbf{r}-\mathbf{r}')$, the discretized integral equation can be expressed as a three-dimensional discrete convolution

$$\mathbf{E}_{i_x,i_y,i_z} - \sum_{j_x,j_y,j_z} \mathbf{G}_{i_x-j_x,i_y-j_y,i_z-j_z} k_0^2 \left(\varepsilon_{j_x,j_y,j_z} - \varepsilon_{\text{ref}}\right) \cdot \mathbf{E}_{j_x,j_y,j_z} = \mathbf{E}_{0,i_x,i_y,i_z}. \tag{6.38}$$

It is noted that the three-dimensional discrete convolution can be carried out fast using the FFT algorithm, and this can be combined with the conjugate gradient algorithm (Appendix B) or another algorithm such as the Generalized Minimal Residual algorithm [111, 112]. In that case it is not necessary to store any very large matrix, and the computer-memory requirements scale linearly with the number of unknowns, and using several million unknowns is unproblematic.

A few examples of the accuracy of the method are now considered by calculating near-fields and scattering from spherical nano-particles of radius a and dielectric constant ε_2 in a background medium with dielectric constant ε_1. The reference field will be polarized along the z-axis and propagate along the x-axis. Thus,

$$\mathbf{E}_0(\mathbf{r}) = \hat{\mathbf{z}} E_0 e^{ik_0 n_1 x}. \tag{6.39}$$

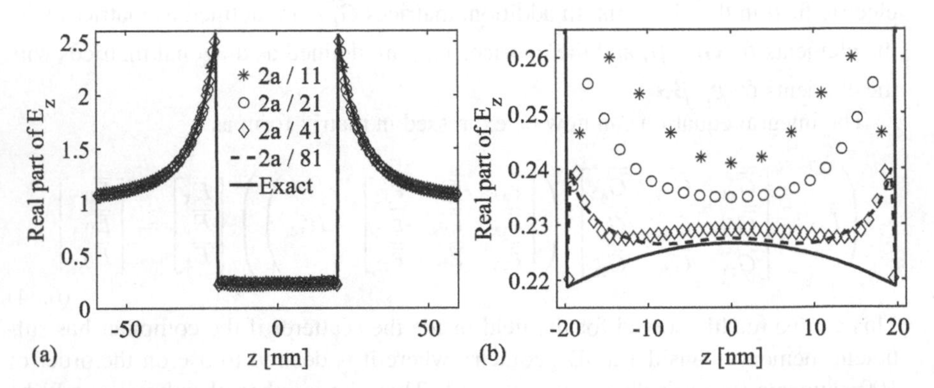

Figure 6.1 A spherical dielectric particle with dielectric constant $\varepsilon_2 = 12$ and radius a = 20 nm that is surrounded by air (ε_1 = 1) is illuminated by a plane wave with wavelength 700 nm propagating along x and being polarized along z. The figure shows the real part of the z-component of the electric field along the z-axis going through the center of the particle calculated with the GFVIEM and cubic discretization elements with different side lengths. (b) Zoom-in on the inside of the particle.

In the first example the near-field is considered for a spherical particle with dielectric constant $\varepsilon_2 = 12$ and radius a = 20 nm surrounded by air ($\varepsilon_1 = 1$) being illuminated by a plane wave with wavelength 700 nm and $E_0 = 1$. The real part of the z-component of the electric field along the z-axis is shown in Fig. 6.1(a) for different side lengths of cubic discretization elements. Also shown for comparison is the exact analytic result obtained using vector spherical harmonics as described in Appendix E. A reasonably good agreement is found already when using cubes of side length 40 nm / 11, especially for positions outside the particle. For positions inside the particle there is a noticeable difference between the numerical and analytical results. A zoom-in on the part of the field inside the spherical particle is shown in Fig. 6.1(b), and here it can be clearly observed how the field obtained with the GFVIEM converges toward the analytic result as the size of cubes decreases. The results were obtained using the conjugate gradient algorithm with the FFT algorithm for fast calculation of matrix-vector products. In the quasi-static limit, which is approached here, the field inside the spherical particle will be constant, and have the value

$$\mathbf{E}_2 \approx \hat{\mathbf{z}} E_0 \frac{3\varepsilon_1}{2\varepsilon_1 + \varepsilon_2} = \hat{\mathbf{z}} E_0 \cdot 0.214. \tag{6.40}$$

This value is in good agreement with the field obtained inside the particle.

We next consider a calculation of the extinction cross section and absorption cross section for lossy spherical particles similar to Ref. [147]. The scattered field $\mathbf{E}_{\text{scat}}(\mathbf{r}) = \mathbf{E}(\mathbf{r}) - \mathbf{E}_0(\mathbf{r})$ is obtained in the far-field limit from the integral equation (6.7) with the far-field Green's tensor (6.23). Thus,

$$\mathbf{E}_{\text{scat}}^{(\text{ff})}(\mathbf{r}) = \frac{e^{ik_0 n_1 r}}{4\pi r} \left(\hat{\theta}\hat{\theta} + \hat{\phi}\hat{\phi}\right) \cdot \int k_0^2 \left(\varepsilon(\mathbf{r}') - \varepsilon_{\text{ref}}\right) e^{-ik_0 n_1 \hat{\mathbf{r}}\cdot\mathbf{r}'} \mathbf{E}(\mathbf{r}') d^3 r'. \tag{6.41}$$

This can be inserted into (2.146) to obtain the scattering cross section.

The extinction cross section can be calculated by inserting the scattered far-field into (2.145) corrected for the polarization and direction of propagation of the reference field. Thus, for the present situation the extinction cross section is

$$\sigma_{\text{ext}} = \frac{-1}{k_0 n_1 |E_0|^2} \text{Real} \left\{ E_0 e^{-i\pi/2} \left(\hat{\mathbf{z}} \cdot \int e^{-ik_0 n_1 x'} k_0^2 \left[\varepsilon(\mathbf{r}') - \varepsilon_1\right] \cdot \mathbf{E}(\mathbf{r}') d^3 r' \right)^* \right\}. \tag{6.42}$$

The corresponding absorption cross section is given as (2.147), which for a tensor dielectric constant can be expressed as

$$\sigma_{\text{abs}} = \frac{k_0/n_1}{|E_0|^2} \int \text{Imag} \left\{ \mathbf{D}(\mathbf{r}) \cdot (\mathbf{E}(\mathbf{r}))^* \right\} d^3 r. \tag{6.43}$$

It is convenient to normalize the extinction and absorption cross sections with the physical cross section πa^2 of the spherical particle, which leads to the so-called extinction and absorption efficiencies.

The extinction and absorption efficiencies calculated with the GFVIEM are shown in Fig. 6.2 for a spherical particle in air with radius a and refractive index $n_2 = 1.7 + i0.1$ versus the normalized particle size $2\pi a/\lambda$. Results are shown for different side lengths of cubic discretization elements along with the exact result (Appendix E). From Fig. 6.2(a) a good agreement is found for small particles, while the numerical results start to deviate from the analytic result for the larger particles with $2\pi a/\lambda \approx 5$. A zoom-in on this region is shown in Fig. 6.2(b) showing convergence with an increasing number of discretization cubes. Note that the same situation was considered in Ref. [147]. Here, however, the number of discretization elements is much higher, and in addition, the results have been obtained using the tensor dielectric constant (6.24), which was not applied in Ref. [147].

A similar calculation of extinction and absorption efficiencies is shown in Fig. 6.3 for a spherical particle in air with radius a and refractive index $n_2 = 0.154 + i4.054$. The refractive index corresponds to gold at the wavelength 700 nm [40]. The calculated extinction efficiency matches quite well with the exact value obtained using Appendix E for the considered range of normalized sizes. The absorption efficiency, however, differs significantly from the exact value.

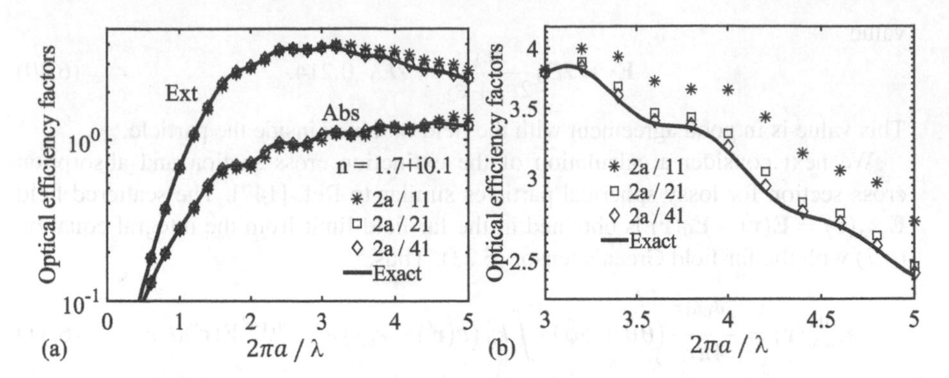

Figure 6.2 (a) Extinction and absorption efficiencies calculated with the GFVIEM for a spherical particle with refractive index $n_2 = 1.7 + i0.1$ and radius a versus normalized particle size $2\pi a/\lambda$. The efficiencies are shown for different side lengths of cubic discretization elements. (b) Zoom-in on the extinction efficiency.

Fig. 6.4 shows calculations of the real part of the z-component of the electric field along the z-axis for a particle with radius a = 20 nm, and for the wavelength 700 nm, for the sizes of cubic elements 41 nm / a and 81 nm /a. The field outside the particle matches well with the exact value [Fig. 6.4(a)]. The field inside the particle oscillates around the exact value but with fairly large oscillations [Fig. 6.4(b)]. The field inside the particle thus poorly represents the exact value. Since the absorption cross section is obtained from the squared magnitude of the field inside the particle, it is not surprising that the calculated absorption cross section deviates significantly from the exact result. The extinction cross section is, on the other hand, obtained from an integral over the field inside the particle, and roughly the average value of the field governs the extinction. The effect of the oscillations in the field are thus less significant when calculating the extinction.

Oscillations in the field inside a cylindrical metal wire were found in Sec. 5.4.2 when using the GFAIEM in the case of p polarization and square-shaped discretization elements. The oscillations seen here for the spherical particle are similar, and may also here be considered as an effect of the stair-cased representation of the particle surface. A solution is to use other shapes of discretization elements that better represent the particle surface. However, when discretizing the integral equation it will then no longer be a discrete convolution, and fast calculation of matrix-vector products can no longer be straightforwardly carried out using the FFT algorithm. Cubic discretization elements appear to work fine for dielectric particles and structures with small imaginary parts of the refractive index. For metals, however, an approach that better represents the particle surface is preferable. A suitable method for such structures is the surface integral equation method presented in Chapter 9.

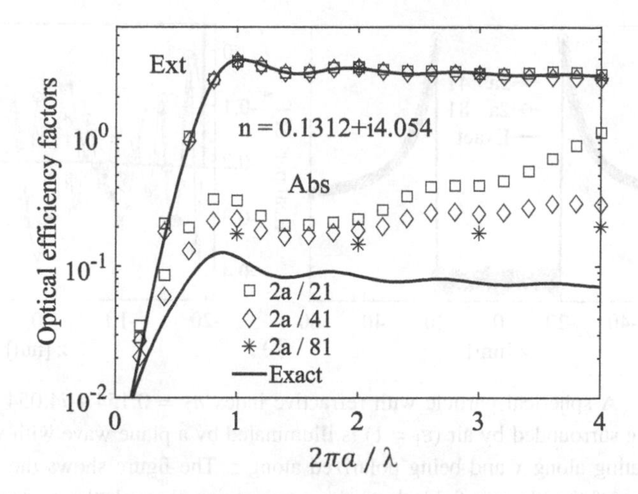

Figure 6.3 Extinction and absorption efficiencies calculated with the GFVIEM for a spherical particle with refractive index $n_2 = 0.154 + i4.054$ and radius a versus normalized particle size $2\pi a/\lambda$. The efficiencies are shown for different side lengths of cubic discretization elements.

6.2.2 DISCRETE DIPOLE APPROXIMATION (DDA)

The Discrete-Dipole Approximation (DDA) [148, 147] is equivalent to the discretized integral equation (6.27). This can be shown by representing each cubic cell with an equivalent dipole. Instead of working with the discretized fields under the assumption that fields are constant within each cubic cell, then in the DDA the unknowns to be calculated are the dipole moments.

The integral equation can be expressed in terms of dipole moments by starting with the general expression for the electric field

$$\mathbf{E}(\mathbf{r}) = \mathbf{E}_0(\mathbf{r}) + i\omega\mu_0 \int \mathbf{G}(\mathbf{r},\mathbf{r}') \cdot \mathbf{J}(\mathbf{r}')d^3r', \tag{6.44}$$

where here $\mathbf{E}_0$ is the incident field, and $\mathbf{J}$ is the induced current density in the scatterer. The current density can be related to the induced polarization density as follows:

$$\mathbf{J}(\mathbf{r}) = -i\omega\mathbf{P}(\mathbf{r}). \tag{6.45}$$

By discretizing (6.44) into cubic volume elements it is now found that the external field in the cubic cell i is given by

$$\mathbf{E}_{\text{ext},i} = \mathbf{E}_{0,i} + \frac{k_0^2}{\varepsilon_0} \sum_{j \neq i} \frac{\mathbf{G}_{i,j}}{V_j} \cdot \mathbf{p}_j, \tag{6.46}$$

where $\mathbf{p}_j = \mathbf{P}_j V_j$ is the dipole moment related to the polarization of cubic cell j. By

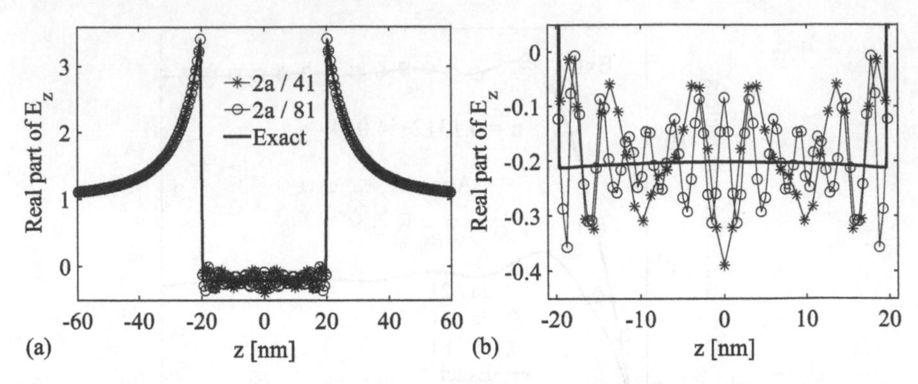

Figure 6.4 A spherical particle with refractive index $n_2 = 0.154 + i4.054$ and radius a = 20 nm being surrounded by air ($\varepsilon_1 = 1$) is illuminated by a plane wave with wavelength 700 nm propagating along x and being polarized along z. The figure shows the real part of the z-component of the electric field along the z-axis going through the center of the particle calculated with the GFVIEM and cubic discretization elements with different side lengths. (b) Zoom-in on the inside of the particle.

comparing with Sec. 6.2.1 the dipole moment can be expressed as

$$\mathbf{p}_j = \varepsilon_0 (\varepsilon_j - \varepsilon_{\text{ref}}) \mathbf{E}_j V_j. \tag{6.47}$$

The external field at element i represents the sum of all contributions to the field in cell i except from cell i itself. By adding this contribution the total field is obtained, i.e.,

$$\mathbf{E}_i = \mathbf{E}_{\text{ext},i} + \frac{k_0^2}{\varepsilon_0} \frac{\mathbf{G}_{i,i}}{V_i} \cdot \mathbf{p}_i. \tag{6.48}$$

The dipole moment of dipole i can now be related to the external field at the dipole location via the polarizability α_i as

$$\mathbf{p}_i = \alpha_i \mathbf{E}_{\text{ext},i}. \tag{6.49}$$

By using Eqs. (6.47)-(6.49) it can be shown that

$$\alpha_i = \left(\mathbf{I} - k_0^2 \mathbf{G}_{i,i} (\varepsilon_i - \varepsilon_{\text{ref}})\right)^{-1} \varepsilon_0 (\varepsilon_i - \varepsilon_{\text{ref}}) V_i. \tag{6.50}$$

In the limit of very small volumes it follows by using (6.31) that

$$\alpha_i \approx \varepsilon_0 \frac{3\varepsilon_{\text{ref}}(\varepsilon_i - \varepsilon_{\text{ref}})}{\varepsilon_i + 2\varepsilon_{\text{ref}}} V_i. \tag{6.51}$$

The discretized integral equation can now be formulated entirely in terms of dipole moments:

$$\mathbf{p}_i = \alpha_i \mathbf{E}_{0,i} + \alpha_i \frac{k_0^2}{\varepsilon_0} \sum_{j \neq i} \frac{\mathbf{G}_{i,j}}{V_j} \cdot \mathbf{p}_j. \tag{6.52}$$

6.3 SCATTERER ON OR NEAR PLANAR SURFACES

In this section the integral equation (6.7) is considered for a scatterer placed on a layered reference structure. The Green's tensor is obtained in cylindrical vector form in Sec. 6.3.1. The advantage of this form is that the Green's tensor can be expressed as integrals over a single in-plane wave number. Far-field expressions for the Green's tensor are derived in Sec. 6.3.2. The conversion between the cylindrical and Cartesian vector form of the Green's tensor is discussed in Sec. 6.3.3. The calculation of optical cross sections is discussed in Sec. 6.3.4, and finally, an example of applying the GFVIEM for modeling of the optics of a scatterer on a nanometer-thin metal film is given in Sec. 6.3.5.

6.3.1 GREEN'S TENSOR FOR LAYERED REFERENCE STRUCTURES

In this section the Green's tensor for layered reference structures will be obtained and expressed in cylindrical coordinates. The approach is that as a first step the plane-wave expansion of the scalar free-space Green's function will be derived and expressed in cylindrical coordinates. The free-space dyadic Green's tensor is then obtained directly from the scalar Green's function. The Green's tensor is then divided into *s*- and *p*-polarized components, and a Fresnel reflection and transmission is constructed for each wave component.

The scalar 3D free-space Green's function can be expressed as the following plane-wave-expansion:

$$g(\mathbf{r},\mathbf{r}')=\frac{e^{ik_1|\mathbf{r}-\mathbf{r}'|}}{4\pi|\mathbf{r}-\mathbf{r}'|}=\frac{i}{2}\frac{1}{(2\pi)^2}\int_{k_x}\int_{k_y}\frac{e^{ik_x(x-x')}e^{ik_y(y-y')}e^{ik_{z,1}|z-z'|}}{k_{z,1}}dk_x dk_y, \quad (6.53)$$

where $k_1=k_0 n_1$, and $k_{z,1}=\sqrt{k_0^2\varepsilon_1-k_x^2-k_y^2}$. This is the Weyl expansion [149]. The scalar Green's function satisfies the equation

$$(\nabla^2+k_0^2\varepsilon_1)\,g(\mathbf{r},\mathbf{r}')=-\delta(\mathbf{r}-\mathbf{r}'), \quad (6.54)$$

and at the same time it satisfies the radiating boundary condition.

The plane-wave expansion (6.53) can be derived by using the mode-expansion method. Consider the following eigenvalue equation

$$(\nabla^2+k_0^2\varepsilon_1)\,E_\lambda(\mathbf{r})=\lambda E_\lambda(\mathbf{r}), \quad (6.55)$$

where λ is the eigenvalue and $E_\lambda(\mathbf{r})$ is the eigenfunction. This can be rewritten as

$$(\nabla^2+k^2)\,E_{\mathbf{k}}(\mathbf{r})=0, \quad (6.56)$$

where $k^2=|\mathbf{k}|^2=k_0^2\varepsilon_1-\lambda_{\mathbf{k}}$.

The eigenfunctions can thus be expressed as

$$E_{\mathbf{k}}(\mathbf{r})=e^{i\mathbf{k}\cdot\mathbf{r}}. \quad (6.57)$$

The eigenfunctions satisfy the orthogonality relation

$$\int \mathbf{E}_{\mathbf{k}}(\mathbf{r}) \left(\mathbf{E}_{\mathbf{k}'}(\mathbf{r})\right)^* d^3r = (2\pi)^3 \delta(\mathbf{k}-\mathbf{k}'). \tag{6.58}$$

The Green's function can now be constructed as follows

$$g(\mathbf{r},\mathbf{r}') = \int \frac{\mathbf{E}_{\mathbf{k}}(\mathbf{r}) \left(\mathbf{E}_{\mathbf{k}}(\mathbf{r}')\right)^*}{(2\pi)^3 \lambda_{\mathbf{k}}} d^3k. \tag{6.59}$$

This is clearly a solution of Eq. (6.54), which can be seen by applying the operator on the left-hand side of (6.54) and using (6.55), and finally by using the following expression for the Dirac delta function:

$$\frac{1}{(2\pi)^3} \int e^{i\mathbf{k}\cdot\mathbf{r}} d^3k = \delta(\mathbf{r}-\mathbf{r}'). \tag{6.60}$$

A small imaginary part is now added to the denominator in (6.59), which is equivalent to adding a homogeneous solution to Eq. (6.54). This leads to

$$g(\mathbf{r},\mathbf{r}') = \int \frac{e^{i\mathbf{k}\cdot(\mathbf{r}-\mathbf{r}')}}{(2\pi)^3 \left(k_z - \left(\sqrt{k_0^2\varepsilon_1 - k_\rho^2 + i\alpha}\right)\right)\left(k_z + \sqrt{k_0^2\varepsilon_1 - k_\rho^2 + i\alpha}\right)} d^3k, \tag{6.61}$$

where $k_\rho^2 = k_x^2 + k_y^2$, and where $\alpha \to 0^+$ is an infinitesimally small positive number. The positiveness of this number causes the resulting Green's function to satisfy the radiating boundary condition.

The integral with respect to k_z along the real axis can now be replaced by an integral along a closed curve as illustrated in Fig. 6.5. The integral must thus be replaced with an integral from $-R$ to R along the real axis, and a semi-circle with radius R going into either the upper or lower half-plane, depending on in which half-plane the magnitude of $\exp(ik_z(z-z'))$ will decrease. As the limit $R \to \infty$ is taken, the integral along the semi-circle will vanish. However, because the integral is now over a closed curve, it can be straightforwardly evaluated. The contribution from the pole in (6.61) being enclosed by the curve now leads to Eq. (6.53).

We now wish to express the Green's function in terms of cylindrical coordinates. The wave numbers and Cartesian coordinates are thus replaced by the following:

$$k_x = k_1 \cos\phi_k, \tag{6.62}$$

$$k_y = k_1 \sin\phi_k, \tag{6.63}$$

$$x - x' = \rho_r \cos\phi_r, \tag{6.64}$$

$$y - y' = \rho_r \sin\phi_r. \tag{6.65}$$

where the subscript r indicates that the cylindrical coordinates ρ_r and angle ϕ_r are defined with respect to relative distances. By inserting these expressions in Eq. (6.53), and applying $\cos(\phi_k - \phi_r) = \cos\phi_r \cos\phi_k + \sin\phi_r \sin\phi_k$, and further using

$$J_0(x) = \frac{1}{2\pi} \int_0^{2\pi} e^{ix\cos\theta} d\theta, \tag{6.66}$$

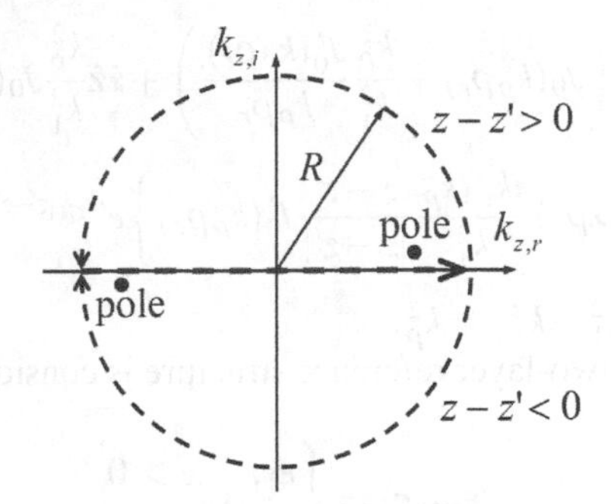

Figure 6.5 Schematic of integration paths when applying residue calculus.

the Green's function in cylindrical coordinates is obtained

$$g(\mathbf{r},\mathbf{r}') = \frac{i}{4\pi}\int_0^\infty \frac{J_0(k_\rho \rho_r)e^{ik_{z,1}|z-z'|}}{k_{z,1}} k_\rho dk_\rho. \tag{6.67}$$

The direct Green's tensor can now be obtained in cylindrical coordinates by using

$$\mathbf{G}^{(d)}(\mathbf{r},\mathbf{r}') = \left(\mathbf{I} + \frac{1}{k_1^2}\nabla\nabla\right) g(\mathbf{r},\mathbf{r}'), \tag{6.68}$$

where $\mathbf{I} = \hat{\rho}_r\hat{\rho}_r + \hat{\phi}_r\hat{\phi}_r + \hat{\mathbf{z}}\hat{\mathbf{z}} = \hat{\mathbf{x}}\hat{\mathbf{x}} + \hat{\mathbf{y}}\hat{\mathbf{y}} + \hat{\mathbf{z}}\hat{\mathbf{z}}$ is the unit dyadic, and the relative cylindrical unit vectors are defined as

$$\hat{\rho}_r = \hat{\mathbf{x}}\cos\phi_r + \hat{\mathbf{y}}\sin\phi_r, \tag{6.69}$$

$$\hat{\phi}_r = -\hat{\mathbf{x}}\sin\phi_r + \hat{\mathbf{y}}\cos\phi_r. \tag{6.70}$$

The corresponding Cartesian unit vectors are then given by

$$\hat{\mathbf{x}} = \hat{\rho}_r\cos\phi_r - \hat{\phi}_r\sin\phi_r, \tag{6.71}$$

$$\hat{\mathbf{y}} = \hat{\rho}_r\sin\phi_r + \hat{\phi}_r\cos\phi_r, \tag{6.72}$$

and the gradient in relative cylindrical coordinates can be expressed as

$$\nabla = \hat{\mathbf{z}}\frac{\partial}{\partial z} + \hat{\rho}_r\frac{\partial}{\partial \rho_r} + \hat{\phi}_r\frac{1}{\rho_r}\frac{\partial}{\partial \phi_r}. \tag{6.73}$$

The expression (6.68) now leads to the direct Green's tensor in cylindrical coordinates

$$\mathbf{G}^{(d)}(\mathbf{r},\mathbf{r}') = \frac{i}{4\pi}\int_0^\infty \left\{\hat{\rho}_r\hat{\rho}_r\left(J_0(k_\rho\rho_r) + \frac{k_\rho^2}{k_1^2}J_0''(k_\rho\rho_r)\right) + \right.$$

$$\hat{\phi}_r\hat{\phi}_r\left(J_0(k_\rho\rho_r)+\frac{k_\rho^2}{k_1^2}\frac{J_0'(k_\rho\rho_r)}{k_\rho\rho_r}\right)+\hat{\mathbf{z}}\hat{\mathbf{z}}\frac{k_\rho^2}{k_1^2}J_0(k_\rho\rho_r)+$$

$$(\hat{\rho}_r\hat{\mathbf{z}}+\hat{\mathbf{z}}\hat{\rho}_r)\frac{ik_{z,1}k_\rho}{k_1^2}\frac{z-z'}{|z-z'|}J_0'(k_\rho\rho_r)\Big\}e^{ik_{z,1}|z-z'|}\frac{k_\rho}{k_{z,1}}dk_\rho. \tag{6.74}$$

It has been used that $k_1^2-k_{z,1}^2=k_\rho^2$.

In the following, a two-layer reference structure is considered with an interface at $z=0$ such that

$$\varepsilon_{\text{ref}}(\mathbf{r})=\begin{cases}\varepsilon_1, & z>0\\ \varepsilon_{L2}, & z<0\end{cases}. \tag{6.75}$$

The Green's tensor for this situation and $z'>0$ is now going to be constructed.

Consider first the following incident field corresponding to $z'>z>0$:

$$\mathbf{E}_{i,z}(\mathbf{r})=\mathbf{G}^{(d)}(\mathbf{r},\mathbf{r}')\cdot\hat{\mathbf{z}}=$$

$$\frac{i}{4\pi}\int_0^\infty\left\{\hat{\mathbf{z}}J_0(k_\rho\rho_r)\frac{k_\rho^2}{k_1^2}-\hat{\rho}_r\frac{ik_{z,1}k_\rho}{k_1^2}J_0'(k_\rho\rho_r)\right\}e^{ik_{z,1}(z'-z)}\frac{k_\rho}{k_{z,1}}dk_\rho,\quad z>0. \tag{6.76}$$

This is essentially a sum of p-polarized waves propagating towards the interface at $z=0$. For each wave, the corresponding reflected and transmitted fields are given by Fresnel reflection. The reflected ($z>0$) and transmitted ($z<0$) fields can thus be expressed as

$$\mathbf{E}_{r,z}(\mathbf{r})=\frac{i}{4\pi}\int_0^\infty r^{(p)}(k_\rho)\left\{\hat{\mathbf{z}}J_0(k_\rho\rho_r)\frac{k_\rho^2}{k_1^2}+\hat{\rho}_r\frac{ik_{z,1}k_\rho}{k_1^2}J_0'(k_\rho\rho_r)\right\}e^{ik_{z,1}(z'+z)}\frac{k_\rho}{k_{z,1}}dk_\rho, \tag{6.77}$$

$$\mathbf{E}_{t,z}(\mathbf{r})=\frac{i}{4\pi}\int_0^\infty t^{(p)}(k_\rho)\frac{\varepsilon_1}{\varepsilon_{L2}}\left\{\hat{\mathbf{z}}J_0(k_\rho\rho_r)\frac{k_\rho^2}{k_1^2}-\hat{\rho}_r\frac{ik_{z,L2}k_\rho}{k_1^2}J_0'(k_\rho\rho_r)\right\}$$

$$\times e^{ik_{z,1}z'}e^{-ik_{z,L2}z}\frac{k_\rho}{k_{z,1}}dk_\rho, \tag{6.78}$$

where

$$r^{(p)}(k_\rho)=\frac{\varepsilon_{L2}k_{z,1}-\varepsilon_1k_{z,L2}}{\varepsilon_{L2}k_{z,1}+\varepsilon_1k_{z,L2}}, \tag{6.79}$$

$$t^{(p)}(k_\rho)=1+r^{(p)}(k_\rho) \tag{6.80}$$

are the Fresnel reflection and transmission coefficients for p-polarized light that have been applied previously for the magnetic field in 2D.

Next, consider the incident field

$$\mathbf{E}_{i,x}(\mathbf{r})=\mathbf{G}^{(d)}(\mathbf{r},\mathbf{r}')\cdot\hat{\mathbf{x}}=\mathbf{E}_{i,x}^{(s)}(\mathbf{r})+\mathbf{E}_{i,x}^{(p)}(\mathbf{r}),\quad z'>z>0, \tag{6.81}$$

where

$$\mathbf{E}_{i,x}^{(s)}(\mathbf{r}) = \frac{i}{4\pi}\int_0^\infty \left(\hat{\phi}_r \sin\phi_r J_0''(k_\rho\rho_r) - \hat{\rho}_r \cos\phi_r \frac{J_0'(k_\rho\rho_r)}{k_\rho\rho_r}\right) e^{ik_{z,1}(z'-z)}\frac{k_\rho}{k_{z,1}}dk_\rho, \tag{6.82}$$

$$\mathbf{E}_{i,x}^{(p)}(\mathbf{r}) = \frac{i}{4\pi}\int_0^\infty \left(\hat{\phi}_r \sin\phi_r \frac{k_{z,1}^2}{k_1^2}\frac{J_0'(k_\rho\rho_r)}{k_\rho\rho_r} - \hat{\rho}_r \cos\phi_r \frac{k_{z,1}^2}{k_1^2} J_0''(k_\rho\rho_r) + \right.$$
$$\left. \hat{\mathbf{z}}\cos\phi_r \frac{-ik_{z,1}k_\rho}{k_1^2} J_0'(k_\rho\rho_r)\right) e^{ik_{z,1}(z'-z)}\frac{k_\rho}{k_{z,1}}dk_\rho, \tag{6.83}$$

are the s- and p-polarized components of the field. Both components (6.82) and (6.83) satisfy the wave equation and are divergence-free. In order to arrange the field into these two components, the following relation has been used

$$J_0''(x) = -J_0(x) - \frac{1}{x}J_0'(x). \tag{6.84}$$

The total field may now be expressed as

$$\mathbf{E}_x(\mathbf{r}) = \begin{cases} \mathbf{E}_{i,x}^{(s)}(\mathbf{r}) + \mathbf{E}_{i,x}^{(p)}(\mathbf{r}) + \mathbf{E}_{r,x}^{(s)}(\mathbf{r}) + \mathbf{E}_{r,x}^{(p)}(\mathbf{r}), & z > 0 \\ \mathbf{E}_{t,x}^{(s)}(\mathbf{r}) + \mathbf{E}_{t,x}^{(p)}(\mathbf{r}), & z < 0 \end{cases}, \tag{6.85}$$

where the reflected and transmitted s- and p-polarized fields may be constructed as

$$\mathbf{E}_{r,x}^{(s)}(\mathbf{r}) = \frac{i}{4\pi}\int_0^\infty \left(\hat{\phi}_r \sin\phi_r J_0''(k_\rho\rho_r) - \hat{\rho}_r \cos\phi_r \frac{J_0'(k_\rho\rho_r)}{k_\rho\rho_r}\right)$$
$$\times r^{(s)}(k_\rho) e^{ik_{z,1}(z+z')}\frac{k_\rho}{k_{z,1}}dk_\rho, \tag{6.86}$$

$$\mathbf{E}_{r,x}^{(p)}(\mathbf{r}) = \frac{i}{4\pi}\int_0^\infty \left(-\hat{\phi}_r \sin\phi_r \frac{k_{z,1}^2}{k_1^2}\frac{J_0'(k_\rho\rho_r)}{k_\rho\rho_r} + \hat{\rho}_r \cos\phi_r \frac{k_{z,1}^2}{k_1^2} J_0''(k_\rho\rho_r) + \right.$$
$$\left. \hat{\mathbf{z}}\cos\phi_r \frac{-ik_{z,1}k_\rho}{k_1^2} J_0'(k_\rho\rho_r)\right) r^{(p)}(k_\rho) e^{ik_{z,1}(z+z')}\frac{k_\rho}{k_{z,1}}dk_\rho, \tag{6.87}$$

$$\mathbf{E}_{t,x}^{(s)}(\mathbf{r}) = \frac{i}{4\pi}\int_0^\infty \left(\hat{\phi}_r \sin\phi_r J_0''(k_\rho\rho_r) - \hat{\rho}_r \cos\phi_r \frac{J_0'(k_\rho\rho_r)}{k_\rho\rho_r}\right)$$
$$\times t^{(s)}(k_\rho) e^{ik_{z,1}z'} e^{-ik_{z,L2}z}\frac{k_\rho}{k_{z,1}}dk_\rho, \tag{6.88}$$

$$\mathbf{E}_{t,x}^{(p)}(\mathbf{r}) = \frac{i}{4\pi}\int_0^\infty \left(\frac{k_{z,L2}}{k_{z,1}}\left(\hat{\phi}_r \sin\phi_r \frac{k_{z,1}^2}{k_1^2}\frac{J_0'(k_\rho\rho_r)}{k_\rho\rho_r} - \hat{\rho}_r \cos\phi_r \frac{k_{z,1}^2}{k_1^2} J_0''(k_\rho\rho_r)\right)\right.$$
$$\left. + \hat{\mathbf{z}}\cos\phi_r \frac{-ik_{z,1}k_\rho}{k_1^2} J_0'(k_\rho\rho_r)\right) t^{(p)}(k_\rho)\frac{\varepsilon_1}{\varepsilon_{L2}} e^{ik_{z,1}z'} e^{-ik_{z,L2}z}\frac{k_\rho}{k_{z,1}}dk_\rho. \tag{6.89}$$

Here, the reflection and transmission coefficients for s polarization are

$$r^{(s)}(k_\rho) = \frac{k_{z,1} - k_{z,L2}}{k_{z,1} + k_{z,L2}}, \tag{6.90}$$

$$t^{(s)}(k_\rho) = 1 + r^{(s)}(k_\rho). \tag{6.91}$$

A similar procedure can be applied when using the incident field

$$\mathbf{E}_{i,y}(\mathbf{r}) = \mathbf{G}^{(d)}(\mathbf{r},\mathbf{r}') \cdot \hat{\mathbf{y}}. \tag{6.92}$$

Furthermore, note that

$$\mathbf{G} = \mathbf{G} \cdot (\hat{\mathbf{x}}\hat{\mathbf{x}} + \hat{\mathbf{y}}\hat{\mathbf{y}} + \hat{\mathbf{z}}\hat{\mathbf{z}}), \tag{6.93}$$

and, for example,

$$\mathbf{G} \cdot (\hat{\mathbf{x}}\hat{\mathbf{x}}) = (\mathbf{G} \cdot \hat{\mathbf{x}}) \left(\hat{\rho}_r \cos\phi_r - \hat{\phi}_r \sin\phi_r \right). \tag{6.94}$$

Thus, by putting everything together we finally arrive at the dyadic Green's tensor in cylindrical coordinates for a two-layer structure

$$\mathbf{G}(\mathbf{r},\mathbf{r}') = \begin{cases} \mathbf{G}^{(d)}(\mathbf{r},\mathbf{r}') + \mathbf{G}^{(i)}(\mathbf{r},\mathbf{r}'), & z > 0,\ z' > 0 \\ \mathbf{G}^{(t)}(\mathbf{r},\mathbf{r}'), & z < 0,\ z' > 0 \end{cases}, \tag{6.95}$$

where

$$\mathbf{G}^{(i)}(\mathbf{r},\mathbf{r}') = \mathbf{E}_{r,x}\hat{\mathbf{x}} + \mathbf{E}_{r,y}\hat{\mathbf{y}} + \mathbf{E}_{r,z}\hat{\mathbf{z}}, \tag{6.96}$$

and

$$\mathbf{G}^{(t)}(\mathbf{r},\mathbf{r}') = \mathbf{E}_{t,x}\hat{\mathbf{x}} + \mathbf{E}_{t,y}\hat{\mathbf{y}} + \mathbf{E}_{t,z}\hat{\mathbf{z}}. \tag{6.97}$$

The indirect Green's tensor is now obtained from (6.96) and is given by

$$\mathbf{G}^{(i)}(\mathbf{r},\mathbf{r}') = \frac{i}{4\pi}\int_0^\infty \Bigg\{ r^{(p)}(k_\rho)\Bigg(\hat{\mathbf{z}}\hat{\mathbf{z}} J_0(k_\rho\rho_r)\frac{k_\rho^2}{k_1^2} + \hat{\phi}_r\hat{\phi}_r \frac{J_0'(k_\rho\rho_r)}{k_\rho\rho_r}\frac{k_{z,1}^2}{k_1^2} +$$
$$\hat{\rho}_r\hat{\rho}_r J_0''(k_\rho\rho_r)\frac{k_{z,1}^2}{k_1^2} - (\hat{\mathbf{z}}\hat{\rho}_r - \hat{\rho}_r\hat{\mathbf{z}})\frac{ik_\rho k_{z,1}}{k_1^2} J_0'(k_\rho\rho_r)\Bigg)$$
$$-r^{(s)}(k_\rho)\left(\hat{\phi}_r\hat{\phi}_r J_0''(k_\rho\rho_r) + \hat{\rho}_r\hat{\rho}_r \frac{J_0'(k_\rho\rho_r)}{k_\rho\rho_r}\right)\Bigg\} e^{ik_{z,1}(z+z')}\frac{k_\rho}{k_{z,1}} dk_\rho, \tag{6.98}$$

and the transmitted part of the Green's tensor is obtained from (6.97) and is given by

$$\mathbf{G}^{(t)}(\mathbf{r},\mathbf{r}') = \frac{i}{4\pi}\int_0^\infty \Bigg\{ t^{(p)}(k_\rho)\frac{\varepsilon_1}{\varepsilon_{L2}}\Bigg(\hat{\mathbf{z}}\hat{\mathbf{z}} J_0(k_\rho\rho_r)\frac{k_\rho^2}{k_1^2} - \hat{\phi}_r\hat{\phi}_r \frac{J_0'(k_\rho\rho_r)}{k_\rho\rho_r}\frac{k_{z,1}k_{z,L2}}{k_1^2} -$$
$$\hat{\rho}_r\hat{\rho}_r J_0''(k_\rho\rho_r)\frac{k_{z,1}k_{z,L2}}{k_1^2} - \left(\hat{\mathbf{z}}\hat{\rho}_r + \hat{\rho}_r\hat{\mathbf{z}}\frac{k_{z,L2}}{k_{z,1}}\right) i\frac{k_\rho k_{z,1}}{k_1^2} J_0'(k_\rho\rho_r)\Bigg)$$

$$-t^{(s)}(k_\rho)\left(\hat{\phi}_r\hat{\phi}_r J_0''(k_\rho\rho_r)+\hat{\rho}_r\hat{\rho}_r\frac{J_0'(k_\rho\rho_r)}{k_\rho\rho_r}\right)\Big\}e^{ik_{z,1}z'}e^{-ik_{z,L2}z}\frac{k_\rho}{k_{z,1}}dk_\rho. \quad (6.99)$$

The indirect part of the Green's tensor has been previously derived by using another type of mode-expansion method in Ref. [150] based on using, from the start, the vectorial modes of the reference structure. The Green's tensor for the layered structure can also be found in the Appendix of Ref. [151].

Now consider a three-layer reference geometry such that the dielectric constant is given by

$$\varepsilon(\mathbf{r})=\begin{cases}\varepsilon_1, & z>0\\ \varepsilon_{L2}, & 0>z>-d\\ \varepsilon_{L3}, & z<-d\end{cases}. \quad (6.100)$$

In that case the expression for the indirect part of the Green's tensor (6.98) will be unchanged except that the reflection coefficients are now those for a three-layer geometry [see Eq. (2.54)]. In the case of $z<-d$, the expression for the transmitted part of the Green's tensor (6.99) will be slightly modified. In that case we must make the replacement

$$t_{1,L2}^{(p)}(k_\rho)\frac{\varepsilon_1}{\varepsilon_{L2}}e^{-ik_{z,L2}z}\rightarrow t_{1,L3}^{(p)}(k_\rho)\frac{\varepsilon_1}{\varepsilon_{L3}}e^{-ik_{z,L3}(z+d)}, \quad (6.101)$$

and $k_{z,L2}$ must be replaced by $k_{z,L3}$. For positions inside the layer ($0>z>-d$) the Green's tensor can be expressed as

$$\mathbf{G}(\mathbf{r},\mathbf{r}')=\frac{i}{4\pi}\int_0^\infty\Big\{A^{(p)}(k_\rho)\left(\hat{\mathbf{z}}\hat{\mathbf{z}}J_0(k_\rho\rho_r)\frac{k_\rho^2}{k_1^2}-\hat{\phi}_r\hat{\phi}_r\frac{J_0'(k_\rho\rho_r)}{k_\rho\rho_r}\frac{k_{z,1}k_{z,L2}}{k_1^2}-\right.$$
$$\left.\hat{\rho}_r\hat{\rho}_r J_0''(k_\rho\rho_r)\frac{k_{z,1}k_{z,L2}}{k_1^2}-\left(\hat{\mathbf{z}}\hat{\rho}_r+\hat{\rho}_r\hat{\mathbf{z}}\frac{k_{z,L2}}{k_{z,1}}\right)i\frac{k_\rho k_{z,1}}{k_1^2}J_0'(k_\rho\rho_r)\right)$$
$$-A^{(s)}(k_\rho)\left(\hat{\phi}_r\hat{\phi}_r J_0''(k_\rho\rho_r)+\hat{\rho}_r\hat{\rho}_r\frac{J_0'(k_\rho\rho_r)}{k_\rho\rho_r}\right)\Big\}e^{ik_{z,1}z'}e^{-ik_{z,L2}z}\frac{k_\rho}{k_{z,1}}dk_\rho+$$
$$\frac{i}{4\pi}\int_0^\infty\Big\{B^{(p)}(k_\rho)\left(\hat{\mathbf{z}}\hat{\mathbf{z}}J_0(k_\rho\rho_r)\frac{k_\rho^2}{k_1^2}+\hat{\phi}_r\hat{\phi}_r\frac{J_0'(k_\rho\rho_r)}{k_\rho\rho_r}\frac{k_{z,1}k_{z,L2}}{k_1^2}+\right.$$
$$\left.\hat{\rho}_r\hat{\rho}_r J_0''(k_\rho\rho_r)\frac{k_{z,1}k_{z,L2}}{k_1^2}-\left(\hat{\mathbf{z}}\hat{\rho}_r-\hat{\rho}_r\hat{\mathbf{z}}\frac{k_{z,L2}}{k_{z,1}}\right)i\frac{k_\rho k_{z,1}}{k_1^2}J_0'(k_\rho\rho_r)\right)$$
$$-B^{(s)}(k_\rho)\left(\hat{\phi}_r\hat{\phi}_r J_0''(k_\rho\rho_r)+\hat{\rho}_r\hat{\rho}_r\frac{J_0'(k_\rho\rho_r)}{k_\rho\rho_r}\right)\Big\}e^{ik_{z,1}z'}e^{ik_{z,L2}(z+2d)}\frac{k_\rho}{k_{z,1}}dk_\rho, \quad (6.102)$$

where

$$A^{(s)}(k_\rho)=\frac{t_{1,L2}^{(s)}(k_\rho)}{1+r_{1,L2}^{(s)}(k_\rho)r_{L2,L3}^{(s)}(k_\rho)e^{2ik_{z,L2}d}}, \quad (6.103)$$

$$B^{(s)}(k_\rho) = \frac{t^{(s)}_{1,L2}(k_\rho) r^{(s)}_{L2,L3}(k_\rho)}{1 + r^{(s)}_{1,L2}(k_\rho) r^{(s)}_{L2,L3}(k_\rho) e^{2ik_{z,L2}d}}, \tag{6.104}$$

$$A^{(p)}(k_\rho) = \frac{\varepsilon_1}{\varepsilon_{L2}} \frac{t^{(p)}_{1,L2}(k_\rho)}{1 + r^{(p)}_{1,L2}(k_\rho) r^{(p)}_{L2,L3}(k_\rho) e^{2ik_{z,L2}d}}, \tag{6.105}$$

$$B^{(p)}(k_\rho) = \frac{\varepsilon_1}{\varepsilon_{L2}} \frac{t^{(p)}_{1,L2}(k_\rho) r^{(p)}_{L2,L3}(k_\rho)}{1 + r^{(p)}_{1,L2}(k_\rho) r^{(p)}_{L2,L3}(k_\rho) e^{2ik_{z,L2}d}}, \tag{6.106}$$

and $r^{(u)}_{i,j}$ and $t^{(u)}_{i,j}$ are the single-interface reflection and transmission coefficients.

6.3.2 FAR-FIELD GREEN'S TENSOR

In order to calculate far-field scattering radiation patterns and scattering cross sections the far-field Green's tensor is needed, and it will be derived in the following starting with the indirect far-field Green's tensor. Consider the following zz-component of the indirect part of the Green's tensor:

$$G^{(i)}_{zz} = \frac{i}{4\pi} \int_0^\infty J_0(k_\rho \rho_r) \frac{k_\rho^2}{k_1^2} r^{(p)}(k_\rho) e^{ik_{z,1}(z+z')} \frac{k_\rho}{k_{z,1}} dk_\rho. \tag{6.107}$$

For large z the integration interval can be immediately reduced to $0 \leq k_\rho \leq k_1$. For $k_\rho > k_1$ the resulting $k_{z,1}$ is purely imaginary and the exponential term $\exp(ik_{z,1}z)$ will make the integrand vanish. Furthermore, the following coordinate transformation may be applied

$$z = r\cos\theta, \quad \rho = r\sin\theta, \quad 0 < \theta \leq \pi/2, \tag{6.108}$$

$$x = \rho\cos\phi, \quad y = \rho\sin\phi, \tag{6.109}$$

$$k_{z,1} = k_1\cos\alpha, \quad k_\rho = k_1\sin\alpha. \tag{6.110}$$

The integral thus reduces to

$$G^{(i,\text{ff})}_{zz} \approx \frac{i}{4\pi} \int_{\alpha=0}^{\pi/2} J_0(k_\rho \rho_r) \frac{k_\rho^2}{k_1^2} r^{(p)}(k_\rho) e^{ik_{z,1}(z+z')} k_\rho d\alpha. \tag{6.111}$$

Note that ρ and ϕ are the usual cylindrical coordinates. In addition, the usual cylindrical-coordinate unit vectors are defined as

$$\hat{\rho} = \hat{\mathbf{x}}\cos\phi + \hat{\mathbf{y}}\sin\phi. \tag{6.112}$$

$$\hat{\phi} = -\hat{\mathbf{x}}\sin\phi + \hat{\mathbf{y}}\cos\phi. \tag{6.113}$$

The Bessel function may, for large arguments, be expressed as

$$J_0(x) \approx \sqrt{\frac{2}{\pi x}}\cos(x - \pi/4), \quad x >> 1. \tag{6.114}$$

Further note that when $\mathbf{r}'$ is near the origin and $\mathbf{r}$ is at a large distance from the origin, then $\rho_r \approx \rho - \mathbf{r}' \cdot \hat{\rho}$, and thus

$$\cos(k_\rho \rho_r - \pi/4) e^{ik_{z,1}z} \approx$$

$$\frac{1}{2}\left(e^{ik_1 r\cos(\alpha-\theta)} e^{-i\pi/4} e^{-ik_\rho \mathbf{r}'\cdot\hat{\rho}} + e^{ik_1 r\cos(\alpha+\theta)} e^{i\pi/4} e^{+ik_\rho \mathbf{r}'\cdot\hat{\rho}}\right). \quad (6.115)$$

The fast varying part of the integrand can now be expressed as

$$e^{ik_1 r\cos(\alpha\mp\theta)} \approx e^{ik_1 r} e^{-ik_1 \frac{1}{2} r(\alpha\mp\theta)^2}. \quad (6.116)$$

Those parts of the integral (6.111) where this term is oscillating fast with α will vanish for large $k_1 r$. The only parts of the integral that will not vanish are those where $\alpha \approx \theta$ and $\alpha + \theta \approx \pi$. The latter can only occur for $\theta = \pi/2$. Let us, for the moment, consider the case $0 \leq \theta < \pi/2$. Then the integral reduces to

$$G_{zz}^{(i,\mathrm{ff})} \approx \frac{ie^{ik_1 r}}{8\pi}\sqrt{\frac{2}{\pi k_1 r\sin^2\theta}} e^{-i\frac{\pi}{4}} r^{(p)}(k_\rho) e^{-ik_\rho \mathbf{r}'\cdot\hat{\rho}} e^{ik_{z,1}z'} \frac{k_\rho^3}{k_1^2}\int_{\alpha=0}^{\pi/2} e^{-i\frac{1}{2}k_1 r(\alpha-\theta)^2} d\alpha, \quad (6.117)$$

where here $k_\rho = k_1 \sin\theta$ and $k_{z,1} = k_1 \cos\theta$. The integral in (6.117) may be extended to an integral with integration limits from $-\infty$ to $+\infty$, since the added integration parts are vanishing due to the fast varying integrand. The integral may now be straightforwardly evaluated using

$$\int_{-\infty}^{+\infty} e^{-ax^2} dx = \sqrt{\pi/a}. \quad (6.118)$$

This leads finally to the far-field Green's tensor component

$$G_{zz}^{(i,\mathrm{ff})} = \frac{e^{ik_1 r}}{4\pi r} e^{-ik_\rho \mathbf{r}'\cdot\hat{\rho}} e^{ik_{z,1}z'} r^{(p)}(k_\rho)\frac{k_\rho^2}{k_1^2}. \quad (6.119)$$

In the case of $\theta = \pi/2$ the same result is obtained. In that case both exponential terms on the right-hand side of (6.115) must be used, but they will only enter with half weight due to the integration limits. The same result (6.119) is thus also obtained for $\theta = \pi/2$.

In the far-field limit, $\hat{\phi}_r \approx \hat{\phi}$ and $\hat{\rho}_r \approx \hat{\rho}$. The $\hat{\phi}_r\hat{\phi}_r$ term related to p polarization in (6.98) and the $\hat{\rho}_r\hat{\rho}_r$ term related to s polarization will vanish in the far field due to the ρ_r in the denominator. These terms vanish faster with distance than $1/r$ and can be ignored in the far field. The far field approximation of other terms in (6.98) can be obtained using the same approach explained above for the zz-component. This leads to the indirect far-field Green's tensor

$$\mathbf{G}^{(i,\mathrm{ff})}(\mathbf{r},\mathbf{r}') = \frac{e^{ik_1 r}}{4\pi r} e^{-ik_\rho \hat{\rho}\cdot\mathbf{r}'} e^{ik_{z,1}z'}\left(r^{(s)}(k_\rho)\hat{\phi}\hat{\phi}\right.$$

$$-r^{(p)}(k_\rho)\hat{\theta}\left(\hat{\mathbf{z}}\frac{k_\rho}{k_1}+\hat{\rho}\frac{k_{z,1}}{k_1}\right)\Bigg), \tag{6.120}$$

where again $k_\rho = k_1 \sin\theta$ and $k_{z,1} = k_1 \cos\theta$.

If it is assumed that ε_{L2} is real and positive, then light can be transmitted and propagate into the substrate to the far field. A far-field approximation to the transmitted Green's tensor (6.99) may then also be obtained. Here, the first step is to reduce the integration limits to go from 0 to $k_{L2} = k_0 n_{L2}$, since for $z << 0$ the integrand will vanish for $k_\rho > k_{L2}$ due to the exponential term. This may then be followed by the coordinate transformation

$$z = r\cos\theta, \quad \rho = r\sin\theta, \quad \pi/2 < \theta \le \pi, \tag{6.121}$$

$$k_\rho = k_{L2}\sin\alpha, \quad k_{z,L2} = k_{L2}\cos\alpha. \tag{6.122}$$

The integral in (6.99) may now be evaluated in the far-field limit using the same principles that were applied to obtain (6.120). When the calculation is extended to a three-layer geometry, this leads to

$$\mathbf{G}^{(t,\mathrm{ff})}(\mathbf{r},\mathbf{r}') = \frac{e^{ik_{L3}r}}{4\pi r}e^{-ik_\rho\hat{\rho}\cdot\mathbf{r}'}e^{ik_{z,1}z'}e^{ik_{z,L3}d}\frac{k_{z,L3}}{k_{z,1}}\Bigg(t^{(s)}(k_\rho)\hat{\phi}\hat{\phi}+$$

$$t^{(p)}_{1,L3}(k_\rho)\frac{\varepsilon_1}{\varepsilon_{L3}}\left[\hat{\mathbf{z}}\hat{\mathbf{z}}\frac{k_\rho^2}{k_1^2}+\hat{\rho}\hat{\rho}\frac{k_{z,1}k_{z,L3}}{k_1^2}+\left(\hat{\mathbf{z}}\hat{\rho}+\hat{\rho}\hat{\mathbf{z}}\frac{k_{z,L3}}{k_{z,1}}\right)\frac{k_{z,1}k_\rho}{k_1^2}\right]\Bigg), \tag{6.123}$$

where here $k_\rho = k_{L3}\sin\theta$ and $k_{z,i} = \sqrt{k_i^2 - k_\rho^2}$ with $k_{L3} = k_0 n_{L3}$.

There is another far-field limit that also needs to be addressed, namely the case where z is small and ρ is large, in which case guided modes can be present. Consider again the case of $z > 0$. In this case the part of the integral (6.98) concerning $k_\rho > k_1$ cannot be ignored if the layered reference structure supports guided modes. Consider the following integral related to the zz-component of the indirect Green's tensor (6.98):

$$I^{(\mathrm{p})} = \frac{i}{4\pi}\int_{k_\rho=k_1}^{\infty} r^{(p)}(k_\rho)J_0(k_\rho\rho_r)\frac{k_\rho^2}{k_1^2}e^{ik_{z,1}(z+z')}\frac{k_\rho}{k_{z,1}}dk_\rho. \tag{6.124}$$

For large ρ the Bessel function can be approximated as in Eq. (6.114), which further leads to

$$J_0(k_\rho\rho_r) \approx \sqrt{\frac{2}{\pi k_\rho \rho}}\frac{1}{2}\left(e^{ik_\rho\rho}e^{-i\pi/4}e^{-ik_\rho\mathbf{r}'\cdot\hat{\rho}} + e^{-ik_\rho\rho}e^{i\pi/4}e^{ik_\rho\mathbf{r}'\cdot\hat{\rho}}\right). \tag{6.125}$$

It is clear that both terms in the parantheses are varying fast with k_ρ as ρ is assumed to be large. Thus, the parts of the integral (6.124) where the rest of the integrand is slowly varying will vanish. If guided modes exist, and the Fresnel reflection coefficient thus has poles, there will be a pole contribution to the integral. Near a pole the Fresnel reflection coefficient (in this case for p polarization) can be expressed as

$$r^{(p)}(k_\rho) \approx \frac{A}{k_\rho - k_{\rho,\mathrm{p}}}. \tag{6.126}$$

The integral with respect to k_ρ along the real axis passing by the pole may now be replaced by an integral over a closed curve as illustrated in Fig. 6.6. Concerning the integral over the term proportional to $\exp(+ik_\rho\rho)$ the integral may be extended with an integral over a semi-circle going into the upper half-plane and surrounding the pole, since the integrand along the semi-circle will vanish and thus make no difference due to the large ρ and k_ρ acquiring a positive imaginary part. Concerning the term proportional to $\exp(-ik_\rho\rho)$, a closed curve going into the lower half plane can be chosen. Also in this case the integral over the semi-circle will vanish. The integrals may now be evaluated using residue calculus. In the latter case the integral thus vanishes since no pole is enclosed by the curve, while the former integral can be evaluated from the pole contribution.

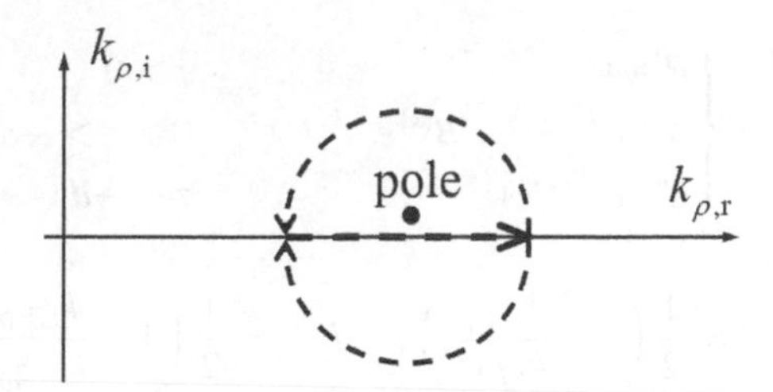

Figure 6.6 Schematic of integration path when evaluating pole contribution.

Residuc calculus now leads to

$$I^{(\text{p})} = -\frac{1}{2}A\frac{k_{\rho,\text{p}}^2}{k_1^2}\sqrt{\frac{2}{\pi k_{\rho,\text{p}}\rho}}\frac{1}{2}e^{-i\pi/4}e^{ik_{z,1,\text{p}}(z+z')}\frac{k_{\rho,\text{p}}}{k_{z,1,\text{p}}}e^{ik_{\rho,\text{p}}\rho}e^{-ik_{\rho,\text{p}}\mathbf{r}'\cdot\hat{\rho}}, \tag{6.127}$$

where $k_{z,1,\text{p}} = \sqrt{k_1^2 - k_{\rho,\text{p}}^2}$. Similar principles can be applied to obtain a far-field expression for other terms in the Green's tensor (6.124). The corresponding indirect Green's tensor related to the pole of the p polarization reflection coefficient can now be expressed as

$$\mathbf{G}^{(i,g,p)}(\mathbf{r},\mathbf{r}') = \frac{A}{4}\sqrt{\frac{2}{\pi k_{\rho,\text{p}}\rho}}e^{-i\pi/4}\frac{k_{\rho,\text{p}}}{k_{z,1,\text{p}}}e^{ik_{\rho,\text{p}}\rho}e^{ik_{z,1,\text{p}}(z+z')}e^{-ik_{\rho,\text{p}}\mathbf{r}'\cdot\hat{\rho}}\times$$

$$\left(-\hat{\mathbf{z}}\hat{\mathbf{z}}\frac{k_{\rho,\text{p}}^2}{k_1^2} + \hat{\rho}\hat{\rho}\frac{k_{z,1,\text{p}}^2}{k_1^2} - (\hat{\mathbf{z}}\hat{\rho} - \hat{\rho}\hat{\mathbf{z}})\frac{k_{\rho,\text{p}}k_{z,1,\text{p}}}{k_1^2}\right), \quad z, z' > 0. \tag{6.128}$$

A similar derivation applies to the s-polarized part of the indirect Green's tensor. In that case the s polarization reflection coefficient near a pole will assume the same form

$$r^{(s)}(k_\rho) \approx \frac{A}{k_\rho - k_{\rho,\text{p}}}. \tag{6.129}$$

The poles $k_{\rho,\mathrm{p}}$ will usually be different from those for p polarization. In terms of the s polarization poles, the indirect Green's tensor will have far-field terms of the form

$$\mathbf{G}^{(i,\mathrm{g},s)}(\mathbf{r},\mathbf{r}') = -\frac{A}{4}\hat{\phi}\hat{\phi}\sqrt{\frac{2}{\pi k_{\rho,\mathrm{p}}\rho}}e^{-i\pi/4}\frac{k_{\rho,\mathrm{p}}}{k_{z,1,\mathrm{p}}}e^{ik_{\rho,\mathrm{p}}\rho}e^{ik_{z,1,\mathrm{p}}(z+z')}e^{-ik_{\rho,\mathrm{p}}\mathbf{r}'\cdot\hat{\rho}}, \quad z,z'>0. \tag{6.130}$$

Considering now all positions z for a three-layer geometry, the latter expression can be extended to

$$\mathbf{G}^{(i,\mathrm{g},s)}(\mathbf{r},\mathbf{r}') = -\frac{A}{4}\hat{\phi}\hat{\phi}\sqrt{\frac{2}{\pi k_{\rho,\mathrm{p}}\rho}}e^{-i\pi/4}\frac{k_{\rho,\mathrm{p}}}{k_{z,1,\mathrm{p}}}e^{ik_{\rho,\mathrm{p}}\rho}e^{ik_{z,1,\mathrm{p}}z'}e^{-ik_{\rho,\mathrm{p}}\mathbf{r}'\cdot\hat{\rho}}$$

$$\times\begin{cases} e^{ik_{z,1,\mathrm{p}}z}, & z>0 \\ A^{(s)}e^{ik_{z,L2,\mathrm{p}}z}+B^{(s)}e^{-ik_{z,L2,\mathrm{p}}z}, & 0>z>-d\,, \\ C^{(s)}e^{-ik_{z,L3,\mathrm{p}}(z+d)}, & z<-d \end{cases} \tag{6.131}$$

where here

$$A^{(s)} = \frac{1}{2}\left(1+\frac{k_{z,1,\mathrm{p}}}{k_{z,L2,\mathrm{p}}}\right), \quad B^{(s)} = \frac{1}{2}\left(1-\frac{k_{z,1,\mathrm{p}}}{k_{z,L2,\mathrm{p}}}\right), \tag{6.132}$$

$$C^{(s)} = A^{(s)}e^{-ik_{z,L2,\mathrm{p}}d}+B^{(s)}e^{+ik_{z,L2,\mathrm{p}}d}. \tag{6.133}$$

The guided-mode far-field Green's tensor for p-polarized guided modes in a three-layer reference structure may, on the other hand, be expressed as

$$\mathbf{G}^{(i,\mathrm{g},p)}(\mathbf{r},\mathbf{r}')\cdot\mathbf{p} =$$

$$\frac{A}{4}\sqrt{\frac{2}{\pi k_{\rho,\mathrm{p}}\rho}}e^{ik_{\rho,\mathrm{p}}\rho}e^{-i\pi/4}\frac{k_{\rho,\mathrm{p}}}{k_{z,1,\mathrm{p}}}\left[\left(\hat{\mathbf{z}}\frac{k_{\rho,\mathrm{p}}}{k_1}+\hat{\rho}\frac{k_{z,1,\mathrm{p}}}{k_1}\right)\cdot\mathbf{p}\right]e^{ik_{z,1,\mathrm{p}}z'}e^{-ik_{\rho,\mathrm{p}}\mathbf{r}'\cdot\hat{\rho}}\times$$

$$\begin{cases} \left(-\hat{\mathbf{z}}\frac{k_{\rho,\mathrm{p}}}{k_1}+\hat{\rho}\frac{k_{z,1,\mathrm{p}}}{k_1}\right)e^{ik_{z,1,\mathrm{p}}z}, & z>0 \\ A^{(p)}\left(-\hat{\mathbf{z}}\frac{k_{\rho,\mathrm{p}}}{k_1}+\hat{\rho}\frac{k_{z,L2,\mathrm{p}}}{k_1}\right)e^{ik_{z,L2,\mathrm{p}}z}+B^{(p)}\left(-\hat{\mathbf{z}}\frac{k_{\rho,\mathrm{p}}}{k_1}-\hat{\rho}\frac{k_{z,L2,\mathrm{p}}}{k_1}\right)e^{-ik_{z,L2,\mathrm{p}}z}, & 0>z>-d\,, \\ C^{(p)}\left(-\hat{\mathbf{z}}\frac{k_{\rho,\mathrm{p}}}{k_1}-\hat{\rho}\frac{k_{z,L3,\mathrm{p}}}{k_1}\right)e^{-ik_{z,L3,\mathrm{p}}(z+d)}, & z<-d \end{cases} \tag{6.134}$$

where here

$$A^{(p)} = \frac{1}{2}\frac{\varepsilon_1}{\varepsilon_{L2}}\left(1+\frac{k_{z,1,\mathrm{p}}}{k_{z,L2,\mathrm{p}}}\frac{\varepsilon_{L2}}{\varepsilon_1}\right), \quad B^{(p)} = \frac{1}{2}\frac{\varepsilon_1}{\varepsilon_{L2}}\left(1-\frac{k_{z,1,\mathrm{p}}}{k_{z,L2,\mathrm{p}}}\frac{\varepsilon_{L2}}{\varepsilon_1}\right), \tag{6.135}$$

$$C^{(p)} = \frac{\varepsilon_{L2}}{\varepsilon_{L3}}\left(A^{(p)}e^{-ik_{z,L2,\mathrm{p}}d}+B^{(p)}e^{ik_{z,L2,\mathrm{p}}d}\right), \tag{6.136}$$

and $\mathbf{p}$ is an arbitrary constant vector (no dependence on ϕ).

In the case of a two-layer reference structure with a metal-dielectric interface supporting SPP waves, the guided-mode far-field Green's tensor may instead be expressed as

$$\mathbf{G}^{(i,g,p)}(\mathbf{r},\mathbf{r}')\cdot\mathbf{p} = \frac{A}{4}\sqrt{\frac{2}{\pi k_{\rho,p}\rho}}e^{ik_{\rho,p}\rho}e^{-i\pi/4}\frac{k_{\rho,p}}{k_{z,1,p}}\left[\left(\hat{\mathbf{z}}\frac{k_{\rho,p}}{k_1}+\hat{\rho}\frac{k_{z,1,p}}{k_1}\right)\cdot\mathbf{p}\right]\times$$

$$e^{ik_{z,1,p}z'}e^{-ik_{\rho,p}\mathbf{r}'\cdot\hat{\rho}}\begin{cases}\left(-\hat{\mathbf{z}}\frac{k_{\rho,p}}{k_1}+\hat{\rho}\frac{k_{z,1,p}}{k_1}\right)e^{ik_{z,1,p}z}, & z>0\\ \frac{\varepsilon_1}{\varepsilon_{L2}}\left(-\hat{\mathbf{z}}\frac{k_{\rho,p}}{k_1}-\hat{\rho}\frac{k_{z,L2,p}}{k_1}\right)e^{-ik_{z,L2,p}z}, & z<0\end{cases}, \tag{6.137}$$

where here $k_{\rho,p} = k_0\sqrt{\varepsilon_1\varepsilon_{L2}/(\varepsilon_1+\varepsilon_{L2})}$.

6.3.3 GREEN'S TENSOR IN CARTESIAN VECTOR FORM

The Green's tensor components (u) derived in Secs. 6.3.1 and 6.3.2 were expressed in the cylindrical vector form

$$\mathbf{G}^{(u)}(\mathbf{r},\mathbf{r}') = \hat{\mathbf{z}}\hat{\mathbf{z}}G^{(u)}_{zz}(\rho_r,z,z') + \hat{\rho}_r\hat{\rho}_r G^{(u)}_{\rho\rho}(\rho_r,z,z') + \hat{\phi}_r\hat{\phi}_r G^{(u)}_{\phi\phi}(\rho_r,z,z')$$
$$+\hat{\rho}_r\hat{\mathbf{z}}G^{(u)}_{\rho z}(\rho_r,z,z') + \hat{\mathbf{z}}\hat{\rho}_r G^{(u)}_{z\rho}(\rho_r,z,z'). \tag{6.138}$$

This can be straightforwardly converted to Cartesian vector form

$$\mathbf{G}^{(u)}(\mathbf{r},\mathbf{r}') = \sum_{\alpha,\beta}\hat{\alpha}\hat{\beta}G^{(u)}_{\alpha\beta}(\rho_r,z,z'), \tag{6.139}$$

where $\alpha,\beta = x,y,z$, by using

$$G^{(u)}_{\alpha\beta}(\rho,z,z') = \hat{\alpha}\cdot\mathbf{G}^{(u)}(\mathbf{r},\mathbf{r}')\cdot\hat{\beta}. \tag{6.140}$$

The Green's tensor in Cartesian vector form is thus related to the cylindrical vector form as follows:

$$G^{(u)}_{xx} = \sin^2\phi_r G^{(u)}_{\phi\phi} + \cos^2\phi_r G^{(u)}_{\rho\rho}, \tag{6.141}$$

$$G^{(u)}_{yy} = \cos^2\phi_r G^{(u)}_{\phi\phi} + \sin^2\phi_r G^{(u)}_{\rho\rho}, \tag{6.142}$$

$$G^{(u)}_{xy} = G^{(u)}_{yx} = \sin\phi_r\cos\phi_r\left(G^{(u)}_{\rho\rho} - G^{(u)}_{\phi\phi}\right), \tag{6.143}$$

$$G^{(u)}_{xz} = \cos\phi_r G^{(u)}_{\rho z}, \tag{6.144}$$

$$G^{(u)}_{yz} = \sin\phi_r G^{(u)}_{\rho z}, \tag{6.145}$$

$$G^{(u)}_{zx} = \cos\phi_r G^{(u)}_{z\rho}, \tag{6.146}$$

$$G^{(u)}_{zy} = \sin\phi_r G^{(u)}_{z\rho}. \tag{6.147}$$

6.3.4 OPTICAL CROSS SECTIONS

The first step in obtaining scattering and extinction cross sections is to calculate the scattered far field. The out-of-plane scattered far field going into the upper half-plane is obtained by using the far-field Green's tensor parts (6.23) and (6.120) in the integral equation (6.7), which leads to

$$\mathbf{E}_{\text{scat}}^{(\text{ff})}(r,\theta,\phi) = \mathbf{E}_{\text{scat}}^{(\text{ff},d)}(r,\theta,\phi) + \mathbf{E}_{\text{scat}}^{(\text{ff},i)}(r,\theta,\phi), \quad z>0, \tag{6.148}$$

where

$$\mathbf{E}_{\text{scat}}^{(\text{ff},d)}(\mathbf{r}) = \frac{e^{ik_1 r}}{4\pi r}\left(\hat{\boldsymbol{\theta}}\hat{\boldsymbol{\theta}} + \hat{\boldsymbol{\phi}}\hat{\boldsymbol{\phi}}\right)\cdot \int k_0^2\left(\varepsilon(\mathbf{r}') - \varepsilon_{\text{ref}}(\mathbf{r}')\right) e^{-ik_1\hat{\mathbf{r}}\cdot\mathbf{r}'}\mathbf{E}(\mathbf{r}')d^3r', \tag{6.149}$$

$$\mathbf{E}_{\text{scat}}^{(\text{ff},i)}(\mathbf{r}) = \frac{e^{ik_1 r}}{4\pi r}\left(-r^{(p)}(k_\rho)\hat{\boldsymbol{\theta}}\left(\hat{\mathbf{z}}\frac{k_\rho}{k_1} + \hat{\boldsymbol{\rho}}\frac{k_{z1}}{k_1}\right) + r^{(s)}(k_\rho)\hat{\boldsymbol{\phi}}\hat{\boldsymbol{\phi}}\right)\cdot$$
$$\int k_0^2\left(\varepsilon(\mathbf{r}') - \varepsilon_{\text{ref}}(\mathbf{r}')\right) e^{-ik_\rho\hat{\boldsymbol{\rho}}\cdot\mathbf{r}'} e^{ik_{z1}z'}\mathbf{E}(\mathbf{r}')d^3r'. \tag{6.150}$$

Here, $k_\rho = k_1 \sin\theta$ and $k_{z1} = k_1\cos\theta$. The extinction and scattering cross sections related to scattering into the upper half-plane are now obtained by inserting the scattered field (6.148) into Eqs. (2.150) and (2.152).

The scattered far field going into the lower half-plane of a three-layer geometry can be expressed as

$$\mathbf{E}_{\text{scat}}^{(\text{ff},t)}(\mathbf{r}) = \frac{e^{ik_{L3}r}}{4\pi r}\int e^{-ik_\rho\hat{\boldsymbol{\rho}}\cdot\mathbf{r}'} e^{ik_{z,1}z'} e^{-ik_{z,L3}d}\frac{k_{z,L3}}{k_{z,1}}$$
$$\left(t^{(s)}(k_\rho)\hat{\boldsymbol{\phi}}\hat{\boldsymbol{\phi}} + t_{1,L3}^{(p)}(k_\rho)\frac{\varepsilon_1}{\varepsilon_{L3}}\left[\hat{\mathbf{z}}\hat{\mathbf{z}}\frac{k_\rho^2}{k_1^2} + \hat{\boldsymbol{\rho}}\hat{\boldsymbol{\rho}}\frac{k_{z,1}k_{z,L3}}{k_1^2} + \left(\hat{\mathbf{z}}\hat{\boldsymbol{\rho}} + \hat{\boldsymbol{\rho}}\hat{\mathbf{z}}\frac{k_{z,L3}}{k_{z,1}}\right)\frac{k_{z,1}k_\rho}{k_1^2}\right]\right)$$
$$\cdot k_0^2\left(\varepsilon(\mathbf{r}') - \varepsilon_{\text{ref}}(\mathbf{r}')\right)\mathbf{E}(\mathbf{r}')d^3r', \tag{6.151}$$

where here $k_\rho = k_{L3}\sin\theta$ and $k_{z,i} = \sqrt{k_i^2 - k_\rho^2}$ with $k_{L3} = k_0 n_{L3}$. The extinction and scattering cross sections related to the lower half-plane are obtained by inserting the scattered field (6.151) into Eqs. (2.151) and (2.153).

For scattering into guided modes we will first consider a two-layer reference geometry where ε_1 is a dielectric, and ε_{L2} is a metal, and it will be assumed that the reference geometry supports surface plasmon polaritons (SPPs). The far-field excitation of SPPs is obtained by inserting the far-field Green's tensor (6.137) in (6.7). When absorption losses are present in the metal it may be preferable to estimate the SPP scattering cross section from the extinction and out-of-plane scatttering cross sections by using

$$\sigma_{\text{SPP}} \approx \sigma_{\text{Ext}} - \sigma_{\text{Scat}}. \tag{6.152}$$

However, for a numerical check of a GFVIEM implementation it is useful to consider a situation where the metal losses are ignored, in which case the SPP field components can be evaluated at a sufficiently large in-plane distance that $\hat{\rho}_r \approx \hat{\rho}$ in (6.137). The SPP scattering cross section is now obtained by inserting the SPP field into (2.159) and then using (2.162). This leads to the following expression:

$$\sigma_{\text{SPP}} = \frac{1}{k_0 n_1 |\mathbf{E}_i|^2} \left\{ \frac{k_{\rho,p}}{2\text{Imag}\{k_{z1,p}\}} + \frac{\varepsilon_1}{\varepsilon_{L2}} \frac{k_{\rho,p}}{2\text{Imag}\{k_{z,L2,p}\}} \right\} \int_{\phi=0}^{2\pi} |f(\phi)|^2 d\phi, \quad (6.153)$$

where

$$f(\phi) = \frac{A}{4}\sqrt{\frac{2}{\pi k_{\rho,p}}} e^{-i\pi/4} \frac{k_{\rho,p}}{k_{z1,p}} \int e^{ik_{z1,p}z'} e^{-ik_{\rho,p}\mathbf{r}'\cdot\hat{\rho}}$$

$$\times \left(\hat{\mathbf{z}}\frac{k_{\rho,p}}{k_1} + \hat{\rho}\frac{k_{z1,p}}{k_1} \right) \cdot k_0^2 \left(\varepsilon(\mathbf{r}') - \varepsilon_{\text{ref}}(\mathbf{r}') \right) \mathbf{E}(\mathbf{r}') d^3 r'. \quad (6.154)$$

Here, $k_{\rho,p} = k_0 \sqrt{\frac{\varepsilon_1 \varepsilon_{L2}}{\varepsilon_1 + \varepsilon_{L2}}}$, and

$$A = \lim_{k_\rho \to k_{\rho,p}} r^{(p)}(k_\rho)\left(k_\rho - k_{\rho,p}\right). \quad (6.155)$$

While (6.153) is specific for a two-layer reference structure, the expression (6.154) applies for any number of layers. For a three-layer geometry, A can be expressed analytically using Eqs. (4.198)-(4.202).

If the reference structure is instead a three-layer geometry, and propagation losses of guided modes are again ignored, then for p-polarized guided modes, a similar expression for the scattering cross section of a guided mode is obtained:

$$\sigma_{\text{g}} = \frac{1}{k_0 n_1 |\mathbf{E}_i|^2} \int_{\phi=0}^{2\pi} |f(\phi)|^2 d\phi$$

$$\times \left\{ \frac{k_{\rho,p}}{2\text{Imag}\{k_{z1,p}\}} + \frac{\varepsilon_{L3}}{\varepsilon_1} |C^{(p)}|^2 \frac{k_{\rho,p}}{2\text{Imag}\{k_{z,L3,p}\}} + \frac{\varepsilon_{L2}}{\varepsilon_1} k_{\rho,\text{p}} I \right\}, \quad (6.156)$$

where

$$I = \int_{z=-d}^{0} \left\{ |A^{(p)}|^2 e^{-2\text{Imag}\{k_{z,L2,\text{p}}\}z} + |B^{(p)}|^2 e^{+2\text{Imag}\{k_{z,L2,\text{p}}\}z} \right.$$

$$\left. + A^{(p)} (B^{(p)})^* e^{i2\text{Real}\{k_{z,L2,\text{p}}\}z} + B^{(p)} (A^{(p)})^* e^{-i2\text{Real}\{k_{z,L2,\text{p}}\}z} \right\} dz, \quad (6.157)$$

and $A^{(p)}$, $B^{(p)}$ and $C^{(p)}$ are defined in (6.135) and (6.136), and the expression (6.154) still applies for $f(\phi)$ except that $k_{\rho,\text{p}}$ is different and A is different.

6.3.5 EXAMPLE: SCATTERING BY A NANOSTRIP ON A THIN METAL FILM

In this section we consider again a polymer nanostrip with refractive index $n_2 = 1.65$ placed on a 40-nm silver film on quartz similar to Sec. 5.9. However, the nanostrip is now a three-dimensional geometry with a finite length, whereas in Sec. 5.9 the

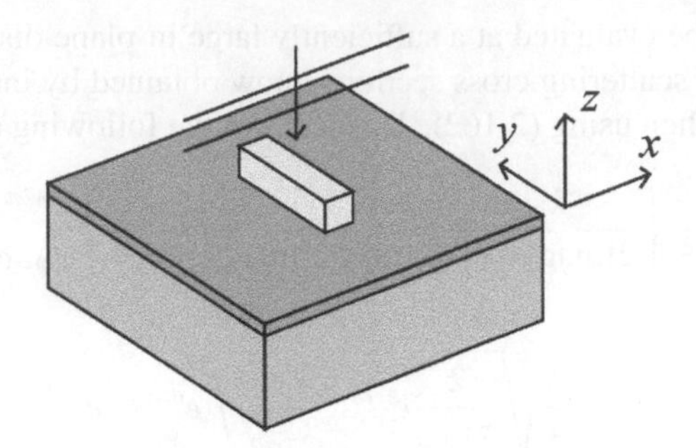

Figure 6.7 Schematic of a nanostrip on a silver-film-on-quartz being illuminated by a normally incident plane wave.

length was infinite. The nanostrip is illuminated by a plane wave propagating along the negative z-axis (normally incident light). The geometry is illustrated in Fig. 6.7. In the following, results will be presented for the three-dimensional out-of-plane scattering radiation patterns and for the extinction and scattering cross sections.

The incident light is a plane wave being polarized with the electric field along the x-axis. The reference electric field for $z > 0$, which is the field that should be used at first in the integral equation (6.7) for finding the field inside the nanostrip, can thus be expressed as

$$\mathbf{E}_0(\mathbf{r}) = \hat{\mathbf{x}}E_0\left(e^{-ik_1 z} + r^{(s)}(0)e^{ik_1 z}\right). \tag{6.158}$$

We will use the approach of discretizing the nanostrip into cubic volume elements, where the field is assumed constant in each element. Then, in order to be able to use a large number of cubic volume elements, the position of the elements, reference field, and total field, will be expressed in terms of three indices similar to Eqs. (6.35)-(6.37). Due to the layered reference structure a contribution from the indirect Green's tensor must be added to (6.38), and the discretized integral equation may now be expressed in the form

$$\mathbf{E}_{i_x,i_y,i_z} - \sum_{j_x,j_y,j_z} \mathbf{G}^{(d)}_{i_x-j_x,i_y-j_y,i_z-j_z} k_0^2\left(\varepsilon_{j_x,j_y,j_z} - \varepsilon_{\text{ref}}\right)\cdot\mathbf{E}_{j_x,j_y,j_z}$$

$$-\sum_{j_x,j_y,j_z} \mathbf{G}^{(i)}_{i_x-j_x,i_y-j_y,i_z+j_z} k_0^2\left(\varepsilon_{j_x,j_y,j_z} - \varepsilon_{\text{ref}}\right)\cdot\mathbf{E}_{j_x,j_y,j_z} = \mathbf{E}_{0,i_x,i_y,i_z}. \tag{6.159}$$

By handling the direct and indirect parts of the Green's tensor separately, it is noted that both sums are discrete convolutions that can be carried out fast using the FFT algorithm. The discrete form of the integral equation (6.159) may thus be solved using, for example, the iterative approach of Appendix B. After solving for the electric field inside the nanostrip, the extinction and scattering cross sections can be calculated using the approach discussed in Sec. 6.3.4.

The first case that will be considered is light with wavelength 390 nm being incident on a nanostrip with width $W = 350$ nm (along x), height $H = 100$ nm (along z), and length $L = 400$ nm (along y). The out-of-plane scattering radiation patterns calculated using $56 \times 64 \times 16 = 57344$ volume elements of size $(6.25\ \text{nm})^3$ are shown in Fig. 6.8. The three-dimensional radiation pattern is shown in Fig. 6.8(b), and cross

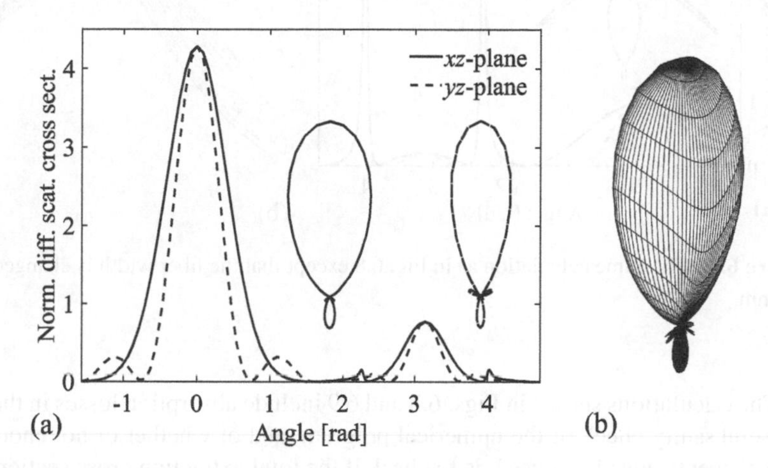

Figure 6.8 Differential out-of-plane scattering for a dielectric nanostrip with refractive index $n_2 = 1.65$ on a 40-nm silver-film-on-quartz being illuminated by a normally incident plane wave with wavelength 390 nm. Nanostrip dimensions: height $H = 100$ nm, length $L = 400$ nm, and width $W = 350$ nm.

sections of the radiation pattern corresponding to the xz-plane and yz-plane, respectively, are shown as insets in Fig. 6.8(a). The radiation pattern is dominated by scattering into the upper half-plane corresponding to angles between $-\pi/2$ and $\pi/2$ in Fig. 6.8(a). The differential out-of-plane scattering has been normalized with the physical cross section of the nanostrip $W \cdot L$. For the xz-plane two small peaks are noticed in the scattering going into the glass substrate similar to Sec. 5.9. These peaks are related to the excitation of surface plasmon polaritons (SPPs) at the air-silver interface leaking into the substrate. Since the SPPs are p-polarized waves, and the field inside the nanostrip will be predominantly polarized along the x-axis, there will be no excitation of SPPs propagating in the yz-plane. Note that the fiber width $W =$ 350 nm is far from optimum. This width is approximately equal to the wavelength of SPPs at an air-silver interface. Thus, for the present situation, radiation into SPPs from different parts of the fiber with half a wavelength between will roughly cancel out, and consequently the SPP-related peaks are small.

Now consider instead a fiber with the width $W = 100$ nm, and L and H unchanged. The radiation pattern for this situation is shown in Fig. 6.9. The radiation pattern is now dominated by radiation going into the lower half-plane at angles corresponding to leakage of SPPs into the quartz substrate.

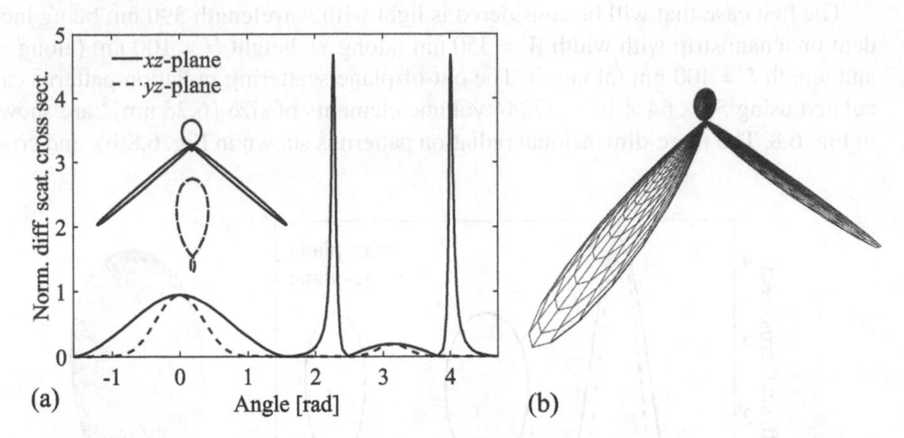

Figure 6.9 The same calculation as in Fig. 6.8 except that the fiber width is changed to W = 100 nm.

The calculations shown in Figs. 6.8 and 6.9 include absorption losses in the metal. A useful sanity check of the numerical program, and of whether or not enough volume elements have been used, is to check if the total extinction cross section equals the total scattering cross section for a situation with losses being neglected. The extinction and scattering cross sections normalized with $W \cdot L$ for the case when absorption loss is neglected is presented in the following table:

W	$\sigma_{sc,t}$	$\sigma_{sc,r}$	σ_g	$\sigma_{ext,t}$	$\sigma_{ext,r}$	$\sigma_{ext} - \sigma_{sc}$
350 nm	0.343	3.220	0.106	0.354	3.314	−0.001
100 nm	0.867	1.208	0.520	0.259	2.332	−0.004

The difference between the sum of normalized extinction and scattering cross sections is 0.001 for W = 350 nm and 0.004 for W = 100 nm. The difference is only this small because the contribution from the guided mode has been taken into account. The guided mode is again, similar to Sec. 5.9, a p-polarized guided mode located primarily at the quartz-silver interface. The SPP that is excited at the air-silver interface appears as scattering into the substrate and is thus included in $\sigma_{sc,t}$.

For comparison the optical cross sections when loss is included in the silver film are as follows:

W	$\sigma_{sc,t}$	$\sigma_{sc,r}$	$\sigma_{ext,t}$	$\sigma_{ext,r}$
350 nm	0.311	3.170	0.343	3.268
100 nm	0.644	1.218	0.242	2.280

The excitation of the SPP at the air-silver interface can be observed from the z-component of the electric field at a small distance above the air-silver interface. The incident field does not have a z-component, and any field along the z-axis must thus be part of the scattered field. The magnitude of the z-component of the electric in the plane z = 110 nm is shown in Fig. 6.10. The field is to a large extent governed by the

Figure 6.10 Magnitude of z-component of electric field in the plane at z = 110 nm for the nanostrip with W = 100 nm, H = 100 nm, and L = 400 nm, being placed on 40-nm-silver-on-quartz and being illuminated by a normally incident plane wave.

excitation of SPP waves propagating mostly along the $\pm x$-axis but spreading out to some extent in the xy-plane.

6.4 PERIODIC SURFACE MICROSTRUCTURES

In this section the GFVIEM is considered for a surface microstructure which is periodic in the xy-plane with period Λ_x along the x-axis, and period Λ_y along the y-axis. Periodic structures have previously been considered with a volume integral equation method in [152] and a surface integral equation method in [153].

The periodic structure can be described by a unit cell which is repeated on a rectangular lattice. The dielectric function of the structure thus satisfies

$$\varepsilon(x+n\Lambda_x, y+m\Lambda_y, z) = \varepsilon(x,y,z), \tag{6.160}$$

for any integers n and m.

For this type of periodic structure the electric field can be expanded in Bloch waves with in-plane Bloch-wave numbers k_x and k_y as

$$\mathbf{E}(\mathbf{r}) = \sum_{k_x,k_y} \mathbf{E}_{k_x,k_y}(\mathbf{r}). \tag{6.161}$$

The Bloch waves in the sum can be expressed in the form

$$\mathbf{E}_{k_x,k_y}(\mathbf{r}) = \mathbf{U}_{k_x,k_y}(\mathbf{r})e^{i(k_x x+k_y y)}, \tag{6.162}$$

where the function $\mathbf{U}_{k_x,k_y}$ is periodic with the same period as the structure, and thus satisfies the periodicity condition

$$\mathbf{U}_{k_x,k_y}(x+n\Lambda_x, y+m\Lambda_y, z) = \mathbf{U}_{k_x,k_y}(x,y,z). \tag{6.163}$$

The Bloch waves on the other hand must satisfy the Bloch boundary condition

$$\mathbf{E}_{k_x,k_y}(x+n\Lambda_x, y+m\Lambda_y, z) = \mathbf{E}_{k_x,k_y}(x,y,z)e^{ik_x\Lambda_x}e^{ik_y\Lambda_y}. \tag{6.164}$$

Now consider a two-layer reference structure with the reference field

$$\mathbf{E}_0(\mathbf{r}) = \begin{cases} \left(\mathbf{E}_{0,i}e^{-ik_{z,1}z} + \mathbf{E}_{0,r}e^{+ik_{z,1}z}\right) e^{ik_x x} e^{ik_y y}, & z > 0 \\ \mathbf{E}_{0,t}e^{-ik_{z,L2}z} e^{ik_x x} e^{ik_y y}, & z < 0 \end{cases} \tag{6.165}$$

then, since the reference field is a Bloch wave with the Bloch-wave numbers k_x and k_y, the corresponding total field due to a periodic surface microstructure will also be a Bloch wave with the same Bloch-wave numbers.

The Bloch-wave electric field can be obtained from a Green's function integral equation being identical in form to (6.7). Here, the Green's function must satisfy the radiating boundary condition along the z-axis and the Bloch boundary condition in the xy-plane for the chosen k_x and k_y. The Green's tensor must thus satisfy the usual condition

$$\left(-\nabla\times\nabla\times + k_0^2\varepsilon_{\text{ref}}(\mathbf{r})\right)\mathbf{G}(\mathbf{r},\mathbf{r}') = -\mathbf{I}\delta(\mathbf{r}-\mathbf{r}'),$$
$$-\frac{\Lambda_x}{2} \le x \le \frac{\Lambda_x}{2}, \quad -\frac{\Lambda_y}{2} \le y \le \frac{\Lambda_y}{2}, \tag{6.166}$$

within one period along with the periodic boundary condition

$$\mathbf{G}(\mathbf{r} + \hat{\mathbf{x}}\Lambda_x + \hat{\mathbf{y}}\Lambda_y, \mathbf{r}') = \mathbf{G}(\mathbf{r},\mathbf{r}')e^{ik_x\Lambda_x}e^{ik_y\Lambda_y}. \tag{6.167}$$

In addition the integral in the integral equation should only be taken over one period of the structure.

The Green's tensor that satisfies the Bloch boundary condition will be constructed in Sec. 6.4.1. Expressions for reflection and transmission of light will be derived in Sec. 6.4.2, and finally an example of applying the GFVIEM for calculating reflection from a glass surface structured with a periodic array of nano-pillars or pyramids is given in Sec. 6.4.3.

6.4.1 GREEN'S TENSOR FOR PERIODIC STRUCTURES

The Green's tensor that satisfies the equation (6.166) and the periodic boundary condition (6.167) will now be constructed. The approach that will be applied is to first construct the scalar direct periodic Greens function. The Greens function will be expressed as a sum of plane waves. The direct periodic Greens tensor is then obtained directly from the scalar Greens function by applying a differential operator. The Greens tensor will then be separated into s- and p-polarized components, and Fresnel reflection and transmission will then be applied to construct reflected waves and transmitted waves for each downward propagating field component in the direct Greens tensor. This will finally lead to the total Greens tensor for a two-layer planar reference structure.

The first step toward the Greens tensor is to calculate the scalar direct Greens function with periodic boundary conditions along x and y. The Greens function is required to satisfy

$$\left(\nabla^2 + k_0^2\varepsilon_1\right) g^{(d)}(\mathbf{r},\mathbf{r}') = -\delta(\mathbf{r}-\mathbf{r}'),$$

$$-\frac{\Lambda_x}{2} \leq x \leq \frac{\Lambda_x}{2}, \; -\frac{\Lambda_y}{2} \leq y \leq \frac{\Lambda_y}{2}, \tag{6.168}$$

along with the radiating boundary condition along the z-axis, and periodic Bloch boundary conditions in the xy-plane. The latter condition can be expressed as

$$g^{(d)}(\mathbf{r}+\hat{\mathbf{x}}\Lambda_x+\hat{\mathbf{y}}\Lambda_y,\mathbf{r}') = g^{(d)}(\mathbf{r},\mathbf{r}')e^{ik_x\Lambda_x}e^{ik_y\Lambda_y}. \tag{6.169}$$

The Greens function can be constructed from the eigenmodes satisfying

$$\left(\nabla^2 + k_0^2\varepsilon_1\right)E_\lambda(\mathbf{r}) = \lambda E_\lambda(\mathbf{r}), \tag{6.170}$$

where here λ is the eigenvalue, and $E_\lambda(\mathbf{r})$ is the corresponding eigenfunction, which must satisfy the Bloch boundary condition.

A complete set of eigenfunctions can be expressed as the plane waves

$$E_{n_x,n_y,k_z}(\mathbf{r}) = e^{i(k_x+n_xG_x)x}e^{i(k_y+n_yG_y)y}e^{ik_zz}, \tag{6.171}$$

where n_x and n_y are integers, $G_x = 2\pi/\Lambda_x$, and $G_y = 2\pi/\Lambda_y$ with the corresponding eigenvalue

$$\lambda_{n_x,n_y,k_z} = k_0^2\varepsilon_1 - (k_x+n_xG_x)^2 - (k_y+n_yG_y)^2 - k_z^2. \tag{6.172}$$

The eigenfunctions satisfy the orthogonality relation

$$\int_z\int_{x=0}^{\Lambda_x}\int_{y=0}^{\Lambda_y} E_{n_x,n_y,k_z}(\mathbf{r})\left(E_{n_x',n_y',k_z'}(\mathbf{r})\right)^* d^3r' = \delta_{n_x,n_x'}\delta_{n_y,n_y'}\delta(k_z-k_z')2\pi\Lambda_x\Lambda_y. \tag{6.173}$$

The Greens function can now be constructed as follows:

$$g^{(d)}(\mathbf{r},\mathbf{r}') = -\sum_{n_x,n_y}\int_{k_z}\frac{E_{n_x,n_y,k_z}(\mathbf{r})\left(E_{n_x,n_y,k_z}(\mathbf{r}')\right)^* dk_z}{2\pi\Lambda_x\Lambda_y\lambda_{n_x,n_y,k_z}}. \tag{6.174}$$

In order to simplify the notation in the following we introduce

$$k_{x,n_x} \equiv k_x + n_xG_x, \tag{6.175}$$

$$k_{y,n_y} \equiv k_y + n_yG_y, \tag{6.176}$$

$$k_{\rho,n_x,n_y} \equiv \sqrt{k_{x,n_x}^2 + k_{y,n_y}^2}, \tag{6.177}$$

$$\hat{\mathbf{k}}_{\rho,n_x,n_y} = \frac{\hat{\mathbf{x}}k_{x,n_x} + \hat{\mathbf{y}}k_{y,n_y}}{k_{\rho,n_x,n_y}}, \tag{6.178}$$

$$\hat{\mathbf{k}}_{\phi,n_x,n_y} = \frac{-\hat{\mathbf{x}}k_{y,n_y} + \hat{\mathbf{y}}k_{x,n_x}}{k_{\rho,n_x,n_y}}, \tag{6.179}$$

$$k_{zi,n_x,n_y} \equiv \sqrt{k_0^2\varepsilon_i - k_{x,n_x}^2 - k_{y,n_y}^2}, \tag{6.180}$$

where $\mathrm{Imag}\left\{k_{zi,n_x,n_y}\right\} \geq 0$.

The Greens function that satisfies the radiating boundary condition is now obtained by replacing the eigenvalue in the denominator of (6.174) by the following

$$\lambda_{n_x,n_y,k_z} \rightarrow -\left(k_z - \left(k_{z1,n_x,n_y} + i\alpha\right)\right)\left(k_z + k_{z1,n_x,n_y} + i\alpha\right), \tag{6.181}$$

and then considering the limit $\alpha \rightarrow 0^+$. This leads to the following Greens function that satisfies the radiating boundary condition along z and the Bloch-wave periodic boundary condition along x and y:

$$g^{(d)}(\mathbf{r},\mathbf{r}') = \frac{i}{\Lambda_x\Lambda_y}\sum_{n_x,n_y}\frac{e^{ik_{x,n_x}(x-x')}e^{ik_{y,n_y}(y-y')}e^{ik_{z1,n_x,n_y}|z-z'|}}{2k_{z1,n_x,n_y}}. \tag{6.182}$$

The corresponding free-space dyadic Greens tensor that satisfies the Bloch boundary condition along x and y, and the radiating boundary condition along z, can now be obtained by using (6.9). This leads straightforwardly to

$$\mathbf{G}^{(d)}(\mathbf{r},\mathbf{r}') = \frac{i}{\Lambda_x\Lambda_y}\sum_{n_x,n_y}\frac{e^{ik_{x,n_x}(x-x')}e^{ik_{y,n_y}(y-y')}e^{ik_{z1,n_x,n_y}|z-z'|}}{2k_{z1,n_x,n_y}}\mathbf{A}_{n_x,n_y,k_z}, \tag{6.183}$$

where

$$\mathbf{A}_{n_x,n_y,k_z} = \left\{\hat{\mathbf{x}}\hat{\mathbf{x}}\left(1-\frac{k_{x,n_x}^2}{k_1^2}\right)+\hat{\mathbf{y}}\hat{\mathbf{y}}\left(1-\frac{k_{y,n_y}^2}{k_1^2}\right)+\hat{\mathbf{z}}\hat{\mathbf{z}}\left(1-\frac{k_{z1,n_x,n_y}^2}{k_1^2}\right)\right.$$
$$+(\hat{\mathbf{x}}\hat{\mathbf{y}}+\hat{\mathbf{y}}\hat{\mathbf{x}})\left(-\frac{k_{x,n_x}k_{y,n_y}}{k_1^2}\right)+(\hat{\mathbf{x}}\hat{\mathbf{z}}+\hat{\mathbf{z}}\hat{\mathbf{x}})\left(-\frac{z-z'}{|z-z'|}\frac{k_{z1,n_x,n_y}k_{x,n_x}}{k_1^2}\right)$$
$$\left.+(\hat{\mathbf{y}}\hat{\mathbf{z}}+\hat{\mathbf{z}}\hat{\mathbf{y}})\left(-\frac{z-z'}{|z-z'|}\frac{k_{z1,n_x,n_y}k_{y,n_y}}{k_1^2}\right)\right\}. \tag{6.184}$$

This can be rearranged into s- and p-polarized components as

$$\mathbf{G}^{(d)}(\mathbf{r},\mathbf{r}') = \mathbf{G}^{(d,p)}(\mathbf{r},\mathbf{r}') + \mathbf{G}^{(d,s)}(\mathbf{r},\mathbf{r}'), \tag{6.185}$$

where

$$\mathbf{G}^{(d,p)}(\mathbf{r},\mathbf{r}') = \frac{i}{\Lambda_x\Lambda_y}\sum_{n_x,n_y}\frac{e^{ik_{x,n_x}(x-x')}e^{ik_{y,n_y}(y-y')}e^{ik_{z1,n_x,n_y}|z-z'|}}{2k_{z1,n_x,n_y}}$$
$$\times\left(\hat{\mathbf{z}}\frac{k_{\rho,n_x,n_y}}{k_1}-\hat{\mathbf{k}}_{\rho,n_x,n_y}\frac{z-z'}{|z-z'|}\frac{k_{z1,n_x,n_y}}{k_1}\right)\left(\hat{\mathbf{z}}\frac{k_{\rho,n_x,n_y}}{k_1}-\hat{\mathbf{k}}_{\rho,n_x,n_y}\frac{z-z'}{|z-z'|}\frac{k_{z1,n_x,n_y}}{k_1}\right), \tag{6.186}$$

and

$$\mathbf{G}^{(d,s)}(\mathbf{r},\mathbf{r}') = \frac{i}{\Lambda_x\Lambda_y}\sum_{n_x,n_y}\frac{e^{ik_{x,n_x}(x-x')}e^{ik_{y,n_y}(y-y')}e^{ik_{z1,n_x,n_y}|z-z'|}}{2k_{z1,n_x,n_y}}\hat{\mathbf{k}}_{\phi,n_x,n_y}\hat{\mathbf{k}}_{\phi,n_x,n_y}. \tag{6.187}$$

The meaning of p and s polarized here means that $\mathbf{G}^{(d,p)}(\mathbf{r},\mathbf{r}')\cdot\mathbf{p}$ for a constant vector $\mathbf{p}$ behaves as a p-polarized wave, and $\mathbf{G}^{(d,s)}(\mathbf{r},\mathbf{r}')\cdot\mathbf{p}$ behaves as a s-polarized wave, since all plane-wave components in the corresponding sum are p- or s-polarized plane waves.

In the case of a two-layer reference structure with dielectric constant

$$\varepsilon_{\text{ref}}(\mathbf{r}) = \begin{cases} \varepsilon_1, & z>0 \\ \varepsilon_{L2}, & z<0 \end{cases}, \tag{6.188}$$

the total Greens tensor can be expressed in terms of the indirect (i) and transmitted (t) Greens tensors as

$$\mathbf{G}(\mathbf{r},\mathbf{r}') = \begin{cases} \mathbf{G}^{(d)}(\mathbf{r},\mathbf{r}') + \mathbf{G}^{(i)}(\mathbf{r},\mathbf{r}'), & z>0, \quad z'>0, \\ \mathbf{G}^{(t)}(\mathbf{r},\mathbf{r}'), & z<0, \quad z'>0 \end{cases}. \tag{6.189}$$

The indirect and transmitted Greens tensors can be constructed from ordinary Fresnel reflection and transmission of each downward propagating plane wave in the expansion of the Greens tensor into plane waves. This leads to

$$\mathbf{G}^{(i)}(\mathbf{r},\mathbf{r}') = \frac{i}{\Lambda_x\Lambda_y}\sum_{n_x,n_y} \frac{e^{ik_{x,n_x}(x-x')}e^{ik_{y,n_y}(y-y')}e^{ik_{z1,n_x,n_y}(z+z')}}{2k_{z1,n_x,n_y}}$$
$$\times\left\{ r^{(s)}_{1,L2}(k_{\rho,n_x,n_y})\hat{\mathbf{k}}_{\phi,n_x,n_y}\hat{\mathbf{k}}_{\phi,n_x,n_y} + r^{(p)}_{1,L2}(k_{\rho,n_x,n_y})\mathbf{k}^{(+)}_{r,n_x,n_y}\mathbf{k}^{(-)}_{r,n_x,n_y}\right\}, \tag{6.190}$$

with

$$\mathbf{k}^{(\pm)}_{r,n_x,n_y} \equiv \left(\hat{\mathbf{z}}\frac{k_{\rho,n_x,n_y}}{k_1} \mp \hat{\mathbf{k}}_{\rho,n_x,n_y}\frac{k_{z1,n_x,n_y}}{k_1}\right), \tag{6.191}$$

and

$$\mathbf{G}^{(t)}(\mathbf{r},\mathbf{r}') = \frac{i}{\Lambda_x\Lambda_y}\sum_{n_x,n_y} \frac{e^{ik_{x,n_x}(x-x')}e^{ik_{y,n_y}(y-y')}e^{ik_{z1,n_x,n_y}z'}e^{-ik_{z,L2,n_x,n_y}z}}{2k_{z1,n_x,n_y}}$$
$$\times\left\{ t^{(s)}_{1,L2}(k_{\rho,n_x,n_y})\hat{\mathbf{k}}_{\phi,n_x,n_y}\hat{\mathbf{k}}_{\phi,n_x,n_y} + \frac{\varepsilon_1}{\varepsilon_{L2}}t^{(p)}_{1,L2}(k_{\rho,n_x,n_y})\mathbf{k}_{t,n_x,n_y}\mathbf{k}^{(-)}_{r,n_x,n_y}\right\}, \tag{6.192}$$

with

$$\mathbf{k}_{t,n_x,n_y} \equiv \hat{\mathbf{z}}\frac{k_{\rho,n_x,n_y}}{k_1} + \hat{\mathbf{k}}_{\rho,n_x,n_y}\frac{k_{z,L2,n_x,n_y}}{k_1}, \tag{6.193}$$

and $r^{(u)}_{1,L2}(k_{\rho,n_x,n_y})$ is the Fresnel reflection coefficient for the two-layer reference structure and u polarization, and $t^{(u)}_{1,L2}(k_{\rho,n_x,n_y}) = 1 + r^{(u)}_{1,L2}(k_{\rho,n_x,n_y})$ is the corresponding transmission coefficient.

Note that in the event that $k_x = k_y = 0$ and $n_x = n_y = 0$, the vectors $\hat{\mathbf{k}}_{\rho,0,0}$, $\hat{\mathbf{k}}_{\phi,0,0}$, $\mathbf{k}^{(\pm)}_{r,0,0}$, and $\mathbf{k}_{t,0,0}$ are not uniquely defined. When taking the limit $k_x \to 0$ and $k_y \to 0$ these vectors will depend on how the limits are taken. The Greens tensor itself will, however, not depend on how the limits are taken. We may thus use $k_y = 0$ and consider the case of $k_x \to 0$, in which case we find $\hat{\mathbf{k}}_{\rho,0,0} = \hat{\mathbf{x}}$, $\hat{\mathbf{k}}_{\phi,0,0} = \hat{\mathbf{y}}$, $\mathbf{k}^{(\pm)}_{r,0,0} = \mp\hat{\mathbf{x}}$, and $\mathbf{k}_{t,0,0} = \hat{\mathbf{x}}n_{L2}/n_1$.

6.4.2 CALCULATING REFLECTION AND TRANSMISSION

It follows directly from the integral equation and the form of the Greens tensor that the reflected field can be expressed as

$$\mathbf{E}_r(\mathbf{r}) = \sum_{n_x,n_y} e^{ik_{x,n_x}x} e^{ik_{y,n_y}y} e^{ik_{z1,n_x,n_y}z} \left(E^{(s)}_{r,n_x,n_y} \hat{\mathbf{k}}_{\phi,n_x,n_y} + E^{(p)}_{r,n_x,n_y} \mathbf{k}^{(+)}_{r,n_x,n_y} \right), \tag{6.194}$$

where the coefficients for s- and p-polarized parts of the reflected field are given by

$$E^{(s)}_{r,n_x,n_y} = \delta_{n_x,0}\delta_{n_y,0}\mathbf{E}_{0,r} \cdot \hat{\mathbf{k}}_{\phi,n_x,n_y} +$$
$$\frac{i}{\Lambda_x\Lambda_y}\frac{1}{2k_{z1,n_x,n_y}} \int \left(e^{-ik_{z1,n_x,n_y}z'} + r^{(s)}_{1,L2}(k_{\rho,n_x,n_y}) e^{ik_{z1,n_x,n_y}z'} \right)$$
$$e^{-ik_{x,n_x}x'} e^{-ik_{y,n_y}y'} k_0^2 \left(\varepsilon(\mathbf{r}') - \varepsilon_{\text{ref}}(\mathbf{r}') \right) \left(\hat{\mathbf{k}}_{\phi,n_x,n_y} \cdot \mathbf{E}(\mathbf{r}') \right) d^3r', \tag{6.195}$$

and

$$E^{(p)}_{r,n_x,n_y} = \delta_{n_x,0}\delta_{n_y,0}\mathbf{E}_{0,r} \cdot \mathbf{k}^{(+)}_{r,n_x,n_y} +$$
$$\frac{i}{\Lambda_x\Lambda_y}\frac{1}{2k_{z1,n_x,n_y}} \int \left(e^{-ik_{z1,n_x,n_y}z'} \mathbf{k}^{(+)}_{r,n_x,n_y} + r^{(p)}_{1,L2}(k_{\rho,n_x,n_y}) e^{ik_{z1,n_x,n_y}z'} \mathbf{k}^{(-)}_{r,n_x,n_y} \right)$$
$$e^{-ik_{x,n_x}x'} e^{-ik_{y,n_y}y'} k_0^2 \left(\varepsilon(\mathbf{r}') - \varepsilon_{\text{ref}}(\mathbf{r}') \right) \cdot \mathbf{E}(\mathbf{r}') d^3r'. \tag{6.196}$$

Similarly, the transmitted field can be expressed as

$$\mathbf{E}_t(\mathbf{r}) = \sum_{n_x,n_y} e^{ik_{x,n_x}x} e^{ik_{y,n_y}y} e^{-ik_{z,L2,n_x,n_y}z} \left(E^{(s)}_{t,n_x,n_y} \hat{\mathbf{k}}_{\phi,n_x,n_y} + E^{(p)}_{t,n_x,n_y} \mathbf{k}_{t,n_x,n_y} \right), \tag{6.197}$$

where

$$E^{(s)}_{t,n_x,n_y} = \delta_{n_x,0}\delta_{n_y,0}\mathbf{E}_{0,t} \cdot \hat{\mathbf{k}}_{\phi,n_x,n_y} +$$
$$\frac{i}{\Lambda_x\Lambda_y}\frac{1}{2k_{z1,n_x,n_y}} \int t^{(s)}_{1,L2}(k_{\rho,n_x,n_y}) e^{ik_{z1,n_x,n_y}z'} e^{-ik_{x,n_x}x'} e^{-ik_{y,n_y}y'}$$
$$k_0^2 \left(\varepsilon(\mathbf{r}') - \varepsilon_{\text{ref}}(\mathbf{r}') \right) \left(\hat{\mathbf{k}}_{\phi,n_x,n_y} \cdot \mathbf{E}(\mathbf{r}') \right) d^3r', \tag{6.198}$$

and

$$E^{(p)}_{t,n_x,n_y} = \delta_{n_x,0}\delta_{n_y,0}\mathbf{E}_{0,t} \cdot \mathbf{k}_{t,n_x,n_y} \frac{\varepsilon_1}{\varepsilon_{L2}} +$$
$$\frac{i}{\Lambda_x\Lambda_y}\frac{1}{2k_{z1,n_x,n_y}} \int \frac{\varepsilon_1}{\varepsilon_{L2}} t^{(p)}_{1,L2}(k_{\rho,n_x,n_y}) e^{ik_{z1,n_x,n_y}z'} \mathbf{k}^{(-)}_{r,n_x,n_y}$$
$$e^{-ik_{x,n_x}x'} e^{-ik_{y,n_y}y'} k_0^2 \left(\varepsilon(\mathbf{r}') - \varepsilon_{\text{ref}}(\mathbf{r}') \right) \cdot \mathbf{E}(\mathbf{r}') d^3r'. \tag{6.199}$$

The reflected power per period is now obtained as

$$P_{\text{r}} = \frac{1}{2}\text{Real}\left\{ \int_{x=0}^{\Lambda_x} \int_{y=0}^{\Lambda_y} \mathbf{E}_r(\mathbf{r}) \times (\mathbf{H}_r(\mathbf{r}))^* \cdot \hat{\mathbf{z}} dx dy \right\}$$

$$= \sum_{n_x,n_y} \frac{\Lambda_x \Lambda_y}{2} \frac{\text{Real}\left\{k_{z1,n_x,n_y}\right\}}{\omega\mu_0} \left(|E^{(s)}_{r,n_x,n_y}|^2 + |E^{(p)}_{r,n_x,n_y}|^2\right). \tag{6.200}$$

The corresponding incident power per period is given as

$$P_{\text{i}} = \frac{\Lambda_x \Lambda_y}{2} \frac{\text{Real}\left\{k_{z1,0,0}\right\}}{\omega\mu_0} |\mathbf{E}_{0,i}|^2. \tag{6.201}$$

The reflectance is thus given by

$$R = \frac{P_{\text{r}}}{P_{\text{i}}} = \sum_{n_x,n_y} R_{n_x,n_y} = \sum_{n_x,n_y} \frac{\text{Real}\left\{k_{z1,n_x,n_y}\right\}}{k_{z1,0,0}} \frac{|E^{(s)}_{r,n_x,n_y}|^2 + |E^{(p)}_{r,n_x,n_y}|^2}{|\mathbf{E}_{0,i}|^2}. \tag{6.202}$$

Here, it has been assumed that the incident field is a propagating wave and that n_1 is purely real and positive.

In a similar way it can be shown that the transmittance is given by

$$T = \sum_{n_x,n_y} T_{n_x,n_y} = \sum_{n_x,n_y} \frac{\text{Real}\left\{k_{z,L2,n_x,n_y}\right\}}{k_{z1,0,0}} \frac{|E^{(s)}_{t,n_x,n_y}|^2 + |E^{(p)}_{t,n_x,n_y} \mathbf{k}_{t,n_x,n_y}|^2}{|\mathbf{E}_{0,i}|^2}. \tag{6.203}$$

Here, it has also been assumed that n_{L2} is purely real and positive. If that is not the case the transmittance may depend on z and the expression must be modified.

6.4.3 EXAMPLE: 2D PERIODIC ANTIREFLECTIVE SURFACE MICROSTRUCTURE

In this section the Greens function volume integral equation (6.7) is applied for calculating the reflectance from a surface with an antireflective periodic microstructure. Two types of surface microstructures will be considered, namely the array of rectangular pillars illustrated in Fig. 6.11(a), and the periodic array of pyramids illustrated in Fig. 6.11(b). These structures are the three-dimensional analogue of the rectangular and triangular groove array considered in Sec. 4.3.10. The period is here set to Λ = 400 nm along the x and y axes. In the specific example shown for the pillar array, the width of pillars is W = 250 nm, and the height is d = 400 nm. For the pyramid array the bottom of the pyramid covers the whole of the period.

The pillars or pyramids are placed on top of a two-layer reference structure with dielectric constant

$$\varepsilon_{\text{ref}}(\mathbf{r}) = \begin{cases} \varepsilon_1, & z > 0, \\ \varepsilon_{L2}, & z < 0 \end{cases}. \tag{6.204}$$

The relevant Greens tensor and method for calculating reflection and transmission are given in detail in Secs. 6.4.1 and 6.4.2.

We will consider the case of a normally incident plane wave ($k_x = k_y = 0$) being polarized along the x-axis, in which case the reference field can be expressed as

$$\mathbf{E}_0(\mathbf{r}) = \begin{cases} \hat{\mathbf{x}}\left(e^{-ik_0 n_1 z} - r^{(p)}_{1,L2}(0) e^{+ik_0 n_1 z}\right), & z > 0 \\ \frac{\varepsilon_1}{\varepsilon_{L2}} t^{(p)}_{1,L2}(0) \hat{\mathbf{x}} \frac{n_{L2}}{n_1} e^{-ik_0 n_{L2} z}, & z < 0 \end{cases}. \tag{6.205}$$

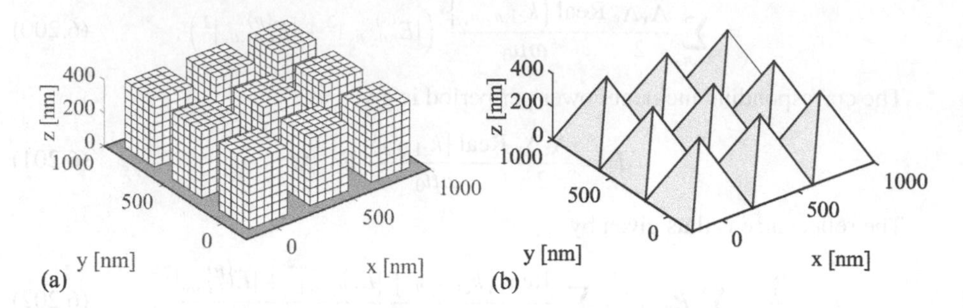

Figure 6.11 Illustration of a periodic array of (a) rectangular pillars, or (b) pyramids, placed on a surface.

For both types of periodically structured surfaces, one period of the structure is discretized with cubic volume elements placed on a cubic grid, and each element is assigned a tensor dielectric constant using the principles of Sec. 6.2.1. The integral equation will here also be expressed in the form (6.159), now with the periodic Greens function, and be solved using iterative approaches involving the FFT algorithm.

The side length of cubic elements is denoted as Δ. The volume of elements is thus $V=\Delta^3$. Now consider discretizing the geometry into N_x steps along x, $N_y=N_x$ steps along y, and N_z steps along z. Thus, for the array of rectangular pillars $\Delta = W/N_x$, and $N_z=\text{ceil}\{H/\Delta\}$, where ceil means that the number is rounded up to an integer value. For the pyramid array, $\Delta=\Lambda/N_x$.

The Greens tensor in (6.159) must be tabulated for the following discrete positions

$$x-x' = j_x\Delta, \quad -(N_x-1)\leq j_x \leq (N_x-1), \tag{6.206}$$

$$y-y' = j_y\Delta, \quad -(N_y-1)\leq j_y \leq (N_y-1), \tag{6.207}$$

and

$$z-z' = j_z\Delta, \quad -(N_z-1)\leq j_z \leq (N_z-1), \tag{6.208}$$

for the direct Greens function, and

$$z+z' = j_z\Delta, \quad 1\leq j_z \leq 2N_z-1, \tag{6.209}$$

for the indirect Greens function.

When $j_z\neq 0$ the sum (6.183) for the direct Greens function can be carried out by considering a sufficient range of n_x and n_y (see (6.183)) such that $|e^{ik_{z1,n_x,n_y}j_z\Delta}|<\delta$ for values outside this range, where δ is a very small number, e.g. $\delta=10^{-8}$. This approach is always valid for the indirect part of the Greens function (6.190) since here $z+z'\geq\Delta$ always. When $j_z=0$ and $j_x\neq 0$ or $j_y\neq 0$, a simple approach for calculating (6.183) is to first carry out the sum using $z-z'=0.2\Delta$ and then using $z-z'=0.1\Delta$. Linear interpolation can then be applied to obtain a good approximation to the value corresponding to having used $z-z'=0$. When $j_x=j_y=j_z=0$ it may be noted that the singular part of the periodic direct Greens function is the same as that of the non-periodic direct Greens function. Thus, a simple approach is to use that

$$\mathbf{G}^{(\mathrm{ns},d)} = \lim_{z-z'\to 0^+} \left(\mathbf{G}^{(d)}(\mathbf{r}-\mathbf{r}') - \mathbf{G}^{(d)}_{\text{non-periodic}}(\mathbf{r}-\mathbf{r}') \right)_{x-x'=y-y'=0} \tag{6.210}$$

is non-singular, where here $\mathbf{G}^{(d)}$ is the periodic direct Greens tensor, and $\mathbf{G}^{(d)}_{\text{non-periodic}}$ is the non-periodic direct Greens tensor given in (6.15).

Thus,

$$\mathbf{G}^{(d)}_{0,0,0} = \int_{x'=x-\Delta/2}^{x'=x+\Delta/2} \int_{y'=y-\Delta/2}^{y'=y+\Delta/2} \int_{z'=z-\Delta/2}^{z'=z+\Delta/2} \mathbf{G}^{(d)}(\mathbf{r}-\mathbf{r}')d^3r'$$
$$\approx \mathbf{G}^{(\mathrm{ns},d)}V - \frac{1}{3k_1^2}\mathbf{I}\left(1+2\left[1-e^{ik_1a}(1-ik_1a)\right]\right), \tag{6.211}$$

where the last term is the integral over one cell of the non-periodic direct Greens tensor from Eq. (6.29) with $4\pi a^3/3 = V$. The term $\mathbf{G}^{(\mathrm{ns},d)}$ can be evaluated at, for example, $z-z' = 0.1\Delta$. In the remaining cases we may simply use

$$\mathbf{G}^{(u)}_{j_x,j_y,j_z} \approx \mathbf{G}^{(u)}(\hat{\mathbf{x}}j_x\Delta + \hat{\mathbf{y}}j_y\Delta + \hat{\mathbf{z}}j_z\Delta)V. \tag{6.212}$$

The approach discussed here is the one that has been used to obtain the results presented in the following for the reflectance and transmittance for the rectangular pillar array and the pyramid array.

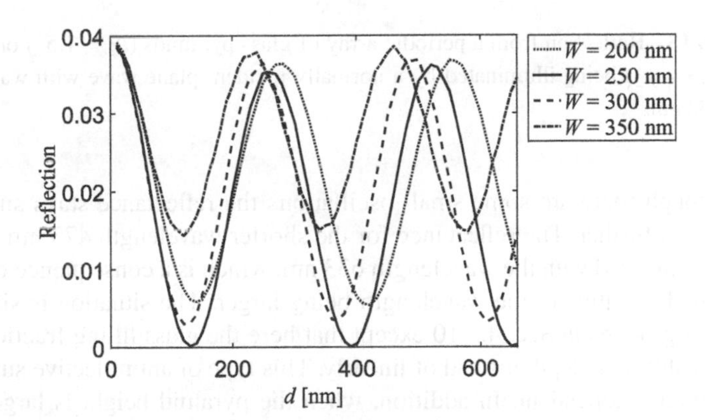

Figure 6.12 Reflection from a periodic array of glass pillars ($\varepsilon_2 = 1.5^2$) on a glass substrate ($\varepsilon_{L2} = \varepsilon_2$) being illuminated by a normally incident plane wave of wavelength 633 nm. The reflectance is shown versus the height of pillars d for different widths of pillars W.

The reflectance for a periodic array of rectangular glass pillars ($\varepsilon_2 = 1.5^2$) on a glass substrate ($\varepsilon_{L2} = 1.5^2$) being illuminated with normally incident light of wavelength 633 nm is shown in Fig. 6.12. The pillar width W = 250 nm, corresponding to a filling factor of $W^2/\Lambda^2 \approx 0.39$, is seen to be optimum since the reflectance can be practically zero for certain pillar heights d. For the pillar width W = 300 nm the minimum reflection has increased to approximately 0.003, and for W = 350 nm it is approximately 0.015. An advantage compared with the array of rectangular grooves in Sec. 4.3.10 is that here the same result is obtained no matter the polarization of incident light (as long as light is normally incident). The depths where the reflectance is minimized will, however, still depend on the wavelength of the incident light.

The reflectance of a periodic array of glass pyramids on a glass substrate is shown in Fig. 6.13 versus the height of pyramids d for the wavelengths 633 nm and 473 nm. For both wavelengths the reflectance is very small already for d = 300 nm,

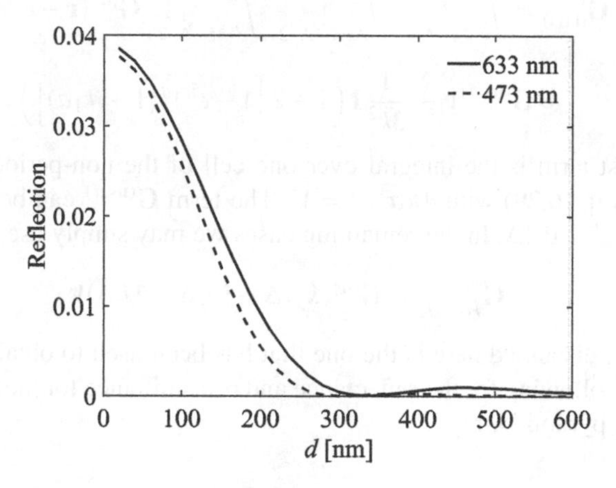

Figure 6.13 Reflection from a periodic array of glass pyramids ($\varepsilon_2 = 1.5^2$) on a glass substrate ($\varepsilon_{L2} = \varepsilon_2$) being illuminated by a normally incident plane wave with wavelength 473 nm or 633 nm.

and although there are some small oscillations the reflectance stays small when d is increased further. The reflectance for the shorter wavelength 473 nm is generally smaller compared with the wavelength 633 nm, which is a consequence of the height of pyramids relative to the wavelength being larger. The situation is similar to the triangular grooves in Sec. 4.3.10 except that here the glass filling fraction increases quadratically with depth instead of linearly. This type of antireflective surface is also polarization independent. In addition, when the pyramid height is large enough, a small reflectance is obtained for a wide range of wavelengths of the incident light, and the anti-reflective effect is now also wavelength insensitive. Note that for the short wavelength of 473 nm, part of the light will be transmitted into higher diffraction orders. This can be avoided by using a smaller period such as, for example, Λ = 300 nm.

6.5 GUIDELINES FOR SOFTWARE IMPLEMENTATION

Structure of GFVIEM program with cubic volume elements

1. Define the reference structure (n_1, n_{L2}, n_{L3}, and d for a three-layer reference structure), scatterer geometry and refractive index (n_2). Define the wavelength and the size of volume elements. Define parameters for calculation of field plots and radiation patterns.
2. Construct three-dimensional arrays with the coordinates of cubic volume elements and the corresponding dielectric constant as in Eqs. (6.35) and (6.38). Construct a three-dimensional array of the direct Greens tensor as in Eq. (6.38) using (6.15) or (6.183). If needed, construct also the indirect Greens tensor as in Eq. (6.159) using either Eq. (6.98) or (6.190) depending on the type of problem.
3. Solve the discretized integral equation using an iterative approach taking advantage of the FFT algorithm for fast calculation of matrix-vector products.
4. By using the calculated field inside the scatterer it is now possible to calculate transmission and reflection for the case of a periodic structure, or to calculate extinction and scattering cross sections, fields, and far-field scattering radiation patterns.

Note that if the Fresnel reflection coefficient has poles, it may be necessary when using (6.98) to extend the integration into the complex plane similar to Sec. 4.2.1. It is also useful for non-periodic structures to first tabulate the Greens tensor elements $G_{zz}^{(i)}$, $G_{z\rho}^{(i)}$, $G_{\rho\rho}^{(i)}$, and $G_{\phi\phi}^{(i)}$ versus ρ_r and $z+z'$. The Cartesian Greens tensor elements depending on $x-x'$, $y-y'$ and $z+z'$ can then be obtained straightforwardly from this table.

6.6 EXERCISES

1. Verify the expression for the direct Greens tensor (6.15) by inserting the scalar direct Greens function (6.10) into (6.9).

2. Verify the expression for $\mathbf{G}_{ii}$ in (6.29) for cubic volume elements.

3. Verify that the field components in (6.82), (6.83), (6.86)-(6.89) that are related to the Greens tensor are divergence-free ($\nabla \cdot \mathbf{E} = 0$).

4. Verify that the electromagnetics boundary conditions across the interface at $z = 0$ in expressions for the Greens tensor for a single-surface geometry leads to the reflection and transmission coefficients in (6.79), (6.80), (6.90), (6.91).

5. Derive (6.103) and (6.104) related to the Greens tensor for positions inside a layer of thickness d in a three-layer geometry by a multiple reflection approach. Hint: The coefficient $A^{(u)}$ for downward propagating waves can be obtained by considering that light must first be transmitted through the first interface. Then inside the layer the light component propagating downward may then originate from a contribution corresponding to 0, 1, 2, ... round-trips in the layer. One round-trip corresponds to two reflections $r_{2,3}^{(u)}$, $r_{2,1}^{(u)}$, and a round-trip propagation phase term $\exp(ik_{z,2}2d)$. This leads to an infinite sum that can be evaluated analytically.

6. Derive the Greens tensor out-of-plane far-field expression (6.123).

7. Verify that the Greens tensor guided-mode far-field expressions (6.131) and (6.134) satisfy the electromagnetics boundary conditions at the interfaces $z = 0$ and $z = -d$.

8. Follow the approach outlined in Sec. 6.5 and develop numerical programs for modeling of either a single scatterer or a periodic array of scatterers placed on a layered reference structure.

7 Volume integral equation method for cylindrically symmetric structures

In this chapter the Green's function volume integral equation is considered for the case of cylindrically symmetric structures. The dielectric function is thus of the form

$$\varepsilon(\mathbf{r}) = \varepsilon(\rho, z). \tag{7.1}$$

The reference structure will be either a homogeneous medium or a layered structure.

It is generally possible to expand the reference electric field and total field in cylindrical harmonics in the following way

$$\mathbf{E}_0(\mathbf{r}) = \sum_n \sum_\alpha \hat{\alpha} E_{0,\alpha}^{(n)}(\rho, z) e^{in\phi}, \tag{7.2}$$

$$\mathbf{E}(\mathbf{r}) = \sum_n \sum_\alpha \hat{\alpha} E_\alpha^{(n)}(\rho, z) e^{in\phi}, \tag{7.3}$$

where $\alpha = \rho, \phi, z$. The expressions (7.2) and (7.3) are not restricted to cylindrically symmetric structures. However, for such structures it is possible to consider one cylindrical harmonic n at a time. The three-dimensional problem is thus reduced to a series of two-dimensional problems, which drastically reduces the computational problem.

It should be noted that $\hat{\rho}$ and $\hat{\phi}$ are ϕ-dependent, and in terms of usual Cartesian coordinates given by $\hat{\rho} = \hat{\mathbf{x}} \cos\phi + \hat{\mathbf{y}} \sin\phi$ and $\hat{\phi} = -\hat{\mathbf{x}} \sin\phi + \hat{\mathbf{y}} \cos\phi$. As we will see in the following sections, the Green's tensor for the reference geometry can be expanded in cylindrical harmonics in the form

$$\mathbf{G}(\mathbf{r}, \mathbf{r}') = \frac{1}{2\pi} \sum_n \left(\sum_{\alpha,\beta} G_{\alpha\beta}^{(n)}(\rho, z; \rho', z') \hat{\alpha} \hat{\beta}' \right) e^{in(\phi - \phi')}, \tag{7.4}$$

where the sums over α and β run over ρ, ϕ and z similar to Eqs. (7.2) and (7.3). The directions of the primed vectors $\hat{\beta}'$ are given by the primed coordinate ϕ', that is $\hat{\rho}' = \hat{\mathbf{x}} \cos\phi' + \hat{\mathbf{y}} \sin\phi'$ and $\hat{\phi}' = -\hat{\mathbf{x}} \sin\phi' + \hat{\mathbf{y}} \cos\phi'$. Note that $\hat{\alpha}\hat{\beta}'$ is a dyad, and there is no dot product between $\hat{\alpha}$ and $\hat{\beta}'$. For each n there are thus 9 elements.

By inserting (7.1)-(7.4) in the Green's function volume integral equation (6.7), and exploiting that $\varepsilon(\mathbf{r})$ is independent of ϕ, it can be seen that the integral equation separates into one equation for each cylindrical harmonic:

Green's function volume integral equation for cylindrical symmetry:

$$E_\alpha^{(n)}(\rho,z) = E_{0,\alpha}^{(n)}(\rho,z) +$$
$$\int \sum_\beta G_{\alpha\beta}^{(n)}(\rho,z;\rho',z') k_0^2 \Delta\varepsilon(\rho',z') E_\beta^{(n)}(\rho',z')\rho' d\rho' dz', \tag{7.5}$$

where $\Delta\varepsilon(\rho,z) = \varepsilon(\rho,z) - \varepsilon_{\text{ref}}(\rho,z)$, and $\alpha, \beta = \rho, \phi, z$.

Similar to the previous chapter, the Green's tensor for a layered structure can be described by a direct, indirect, and transmitted Green's tensor. The rest of this chapter is organized in the following way. In Section 7.1 the cylindrical harmonics of the direct Green's tensor is constructed. The cylindrical harmonics of the indirect and transmitted Green's tensor will be presented in Section 7.2. Far-field expressions of the Green's tensor cylindrical harmonics for out-of-plane propagation and guided modes are obtained in Secs. 7.3 and 7.4, respectively. The calculation of optical cross sections is discussed in Sec. 7.5. A numerical approach for solving the integral equation (7.5) based on ring-shaped elements with a rectangular cross section is discussed in Sec. 7.6. An example of applying the method for a cylindrical scatterer on a dielectric substrate and on a planar waveguide is given in Sec. 7.7. Another example is given in Sec. 7.8 for a microstructured graded-index lens for THz optics.

7.1 EXPANSION OF HOMOGENEOUS-MEDIUM DYADIC GREEN'S TENSOR IN CYLINDRICAL HARMONICS

In this section the cylindrical harmonic direct Green's tensor will be constructed for a medium with refractive index n_1 using cylindrical harmonic vector eigenfunctions. The approach will be similar to C. T. Tai [2] but with some changes. First of all, slightly different expansion functions will be preferred, and secondly, we shall include also longitudinal expansion functions and take into account the contribution from a certain pole when using the residue theorem. The longitudinal expansion functions add an important delta-function contribution.

The direct Green's tensor must satisfy the equation

$$-\nabla\times\nabla\times\mathbf{G}^{(d)}(\mathbf{r},\mathbf{r}') + k_1^2\mathbf{G}^{(d)}(\mathbf{r},\mathbf{r}') = -\mathbf{I}\delta(\mathbf{r}-\mathbf{r}'). \tag{7.6}$$

The Green's tensor can be constructed from eigenfunctions of the left-hand-side operator, which here will be chosen on a cylindrical basis.

7.1.1 EIGENFUNCTIONS

Consider the following eigenvalue problem

$$-\nabla\times\nabla\times\mathbf{E}_\lambda(\mathbf{r}) + k_1^2\mathbf{E}_\lambda(\mathbf{r}) = \lambda\mathbf{E}_\lambda(\mathbf{r}), \tag{7.7}$$

where λ is the eigenvalue, and $\mathbf{E}_\lambda$ is the corresponding vector eigenfunction. This can be rewritten in the form

$$-\nabla\times\nabla\times\mathbf{E}_\lambda(\mathbf{r})+\kappa^2\mathbf{E}_\lambda(\mathbf{r})=0, \tag{7.8}$$

where $\kappa^2=k_1^2-\lambda$. It follows directly from this equation for eigenvalues $\lambda\neq k_1^2$ ($\kappa\neq 0$) that the vector eigenfunction must be transverse or divergence-free ($\nabla\cdot\mathbf{E}_\lambda(\mathbf{r})=0$). On the other hand, when $\kappa=0$, any longitudinal field, which is a field of the form $\mathbf{E}(\mathbf{r})=\nabla f(\mathbf{r})$, will satisfy the equation. Thus, the complete set of eigenfunctions to the left-hand-side operator in Eq. (7.7) also includes longitudinal vector eigenfunctions, and both types of eigenfunctions are needed in the eigenfunction expansion of the dyadic Green's tensor.

A complete set of eigenfunctions of the vector operator in Eq. (7.7) can be constructed from a complete set of eigenfunctions of the following scalar eigenvalue problem

$$(\nabla^2+\kappa^2)f_\lambda(\mathbf{r})=0. \tag{7.9}$$

The eigenfunctions of Eq. (7.9) on a cylindrical basis can be expressed as

$$f_{n,\mu,h}(\mathbf{r})=J_n(\mu\rho)e^{ihz}e^{in\phi}, \tag{7.10}$$

where J_n is the Bessel function of the first kind and order n. Here, n can be any integer, and μ and h are real numbers with $\mu\geq 0$, whereas h can assume positive and negative values. They are related to the eigenvalue in the following way:

$$\mu^2+h^2=\kappa^2=k_1^2-\lambda. \tag{7.11}$$

When $f_{n,\mu,h}(\mathbf{r})$ is a solution to the scalar eigenvalue problem then the following is a transverse solution to the vector eigenvalue problem:

$$\begin{aligned}\mathbf{M}_{n,\mu,h}(\mathbf{r})&=\nabla\times\hat{z}f_{n,\mu,h}(\mathbf{r})\\&=\left(\hat{\rho}\frac{in}{\rho}J_n(\mu\rho)-\hat{\phi}\frac{\partial J_n(\mu\rho)}{\partial\rho}\right)e^{ihz}e^{in\phi}.\end{aligned} \tag{7.12}$$

Another solution to the vector eigenvalue problem is given by

$$\begin{aligned}\mathbf{N}_{n,\mu,h}(\mathbf{r})&=\frac{1}{\kappa}\nabla\times\mathbf{M}_{n,\mu,h}(\mathbf{r})\\&=\frac{1}{\kappa}\left(\hat{\rho}ih\frac{\partial J_n(\mu\rho)}{\partial\rho}-\hat{\phi}\frac{nh}{\rho}J_n(\mu\rho)+\hat{z}\mu^2J_n(\mu\rho)\right)e^{ihz}e^{in\phi}.\end{aligned} \tag{7.13}$$

It can be shown that

$$\mathbf{M}_{n,\mu,h}(\mathbf{r})=\frac{1}{\kappa}\nabla\times\mathbf{N}_{n,\mu,h}(\mathbf{r}), \tag{7.14}$$

and no new vector eigenfunctions can thus be constructed by making additional rotations.

A complete set of longitudinal eigenfunctions to the vector eigenvalue problem can be obtained as

$$\mathbf{L}_{n,\mu,h}(\mathbf{r}) = \nabla f_{n,\mu,h}(\mathbf{r}) = \left(\hat{\rho}\frac{\partial J_n(\mu\rho)}{\partial\rho} + \hat{\phi}\frac{in}{\rho}J_n(\mu\rho) + \hat{z}ihJ_n(\mu\rho)\right)e^{ihz}e^{in\phi}, \tag{7.15}$$

where here $\lambda = k_1^2$ no matter the values of μ and h.

7.1.2 ORTHOGONALITY RELATIONS AND NORMALIZATION

In this subsection the orthogonality and normalization of eigenfunctions will be investigated. The scalar product of two vector functions **A** and **B** will be defined as

$$\langle \mathbf{A}_{n,\mu,h}|\mathbf{B}_{n',\mu',h'}\rangle \equiv \int (\mathbf{A}_{n,\mu,h}(\mathbf{r}))^* \cdot \mathbf{B}_{n',\mu',h'}(\mathbf{r})d^3r$$

$$= \int_{\rho=0}^{\infty}\int_{\phi=0}^{2\pi}\int_{z=-\infty}^{\infty} (\mathbf{A}_{n,\mu,h}(\mathbf{r}))^* \cdot \mathbf{B}_{n',\mu',h'}(\mathbf{r})\rho d\rho d\phi dz. \tag{7.16}$$

The **M**-type and **N**-type vector expansion functions are orthogonal:

$$\langle \mathbf{N}_{n,\mu,h}|\mathbf{M}_{n',\mu',h'}\rangle = 0. \tag{7.17}$$

This can be shown by using

$$\int_{\rho=0}^{\infty} \frac{nh}{\kappa}\left(J_n(\mu'\rho)\frac{\partial J_n(\mu\rho)}{\partial\rho} + J_n(\mu\rho)\frac{\partial J_n(\mu'\rho)}{\partial\rho}\right)d\rho$$

$$= \frac{nh}{\kappa}\left[J_n(\mu'\rho)J_n(\mu\rho)\right]_0^{\infty} = 0. \tag{7.18}$$

For the scalar product of two **M** functions it is found straightforwardly that

$$\langle \mathbf{M}_{n,\mu,h}|\mathbf{M}_{n',\mu',h'}\rangle = (2\pi)^2\delta_{nn'}\delta(h-h')I_M, \tag{7.19}$$

where

$$I_M = \int_{\rho=0}^{\infty}\left[\frac{n^2}{\rho^2}J_n(\mu\rho)J_n(\mu'\rho) + \frac{\partial J_n(\mu\rho)}{\partial\rho}\frac{\partial J_n(\mu'\rho)}{\partial\rho}\right]\rho d\rho. \tag{7.20}$$

The Bessel functions satisfy

$$\frac{n^2}{\rho^2}J_n(\mu\rho) = \mu^2 J_n(\mu\rho) + \frac{1}{\rho}\frac{\partial}{\partial\rho}\left(\rho\frac{\partial J_n(\mu\rho)}{\partial\rho}\right). \tag{7.21}$$

When this is inserted into Eq. (7.20) the result will be a sum of integrals, where one of them except for a constant is

$$\int_0^{\infty} J_n(\mu\rho)J_n(\mu'\rho)\rho d\rho = \frac{1}{\mu}\delta(\mu-\mu'). \tag{7.22}$$

The remaining integral is

$$\int_{\rho=0}^{\infty}\left(J_n(\mu'\rho)\frac{\partial}{\partial\rho}\left(\rho\frac{\partial J_n(\mu\rho)}{\partial\rho}\right)+\frac{\partial J_n(\mu\rho)}{\partial\rho}\frac{\partial J_n(\mu'\rho)}{\partial\rho}\right)\rho d\rho$$

$$=\left[\rho\frac{\partial J_n(\mu\rho)}{\partial\rho}J_n(\mu'\rho)\right]_0^{\infty}, \tag{7.23}$$

which in an operator sense is zero, since inserting a very large ρ for the upper limit in the last bracket will result in a highly oscillatory function, and in the limit of infinite ρ an integration of this expression over any finite interval of μ will vanish. This leads finally to the orthogonality relation

$$\langle\mathbf{M}_{n,\mu,h}|\mathbf{M}_{n',\mu',h'}\rangle=(2\pi)^2\delta_{nn'}\delta(h-h')\mu\delta(\mu-\mu'). \tag{7.24}$$

Similarly, it can be shown that

$$\langle\mathbf{N}_{n,\mu,h}|\mathbf{N}_{n',\mu',h'}\rangle=(2\pi)^2\delta_{nn'}\delta(h-h')\mu\delta(\mu-\mu'), \tag{7.25}$$

$$\langle\mathbf{L}_{n,\mu,h}|\mathbf{L}_{n',\mu',h'}\rangle=(2\pi)^2\delta_{nn'}\delta(h-h')\frac{h^2+\mu^2}{\mu}\delta(\mu-\mu'). \tag{7.26}$$

Finally, it can be shown that $\mathbf{L}$ functions are orthogonal to $\mathbf{M}$ and $\mathbf{N}$ functions:

$$\langle\mathbf{L}_{n,\mu,h}|\mathbf{N}_{n',\mu',h'}\rangle=\langle\mathbf{L}_{n,\mu,h}|\mathbf{M}_{n',\mu',h'}\rangle=0, \tag{7.27}$$

7.1.3 CONSTRUCTING THE DIRECT GREEN'S TENSOR

The direct Green's tensor can now be constructed as

$$\mathbf{G}^{(d)}(\mathbf{r},\mathbf{r}')=\mathbf{G}^T(\mathbf{r},\mathbf{r}')+\mathbf{G}^L(\mathbf{r},\mathbf{r}'), \tag{7.28}$$

where the transverse (T) and longitudinal (L) parts are

$$\mathbf{G}^T(\mathbf{r},\mathbf{r}')=-\sum_{n=-\infty}^{+\infty}\int_{\mu=0}^{\infty}\int_{h=-\infty}^{\infty}\frac{\mathbf{M}_{n,\mu,h}(\mathbf{r})\mathbf{M}^*_{n,\mu,h}(\mathbf{r}')+\mathbf{N}_{n,\mu,h}(\mathbf{r})\mathbf{N}^*_{n,\mu,h}(\mathbf{r}')}{(2\pi)^2\mu(k_1^2-\mu^2-h^2)}d\mu dh, \tag{7.29}$$

and

$$\mathbf{G}^L(\mathbf{r},\mathbf{r}')=-\sum_{n=-\infty}^{+\infty}\int_{\mu=0}^{\infty}\int_{h=-\infty}^{\infty}\frac{\mathbf{L}_{n,\mu,h}(\mathbf{r})\mathbf{L}^*_{n,\mu,h}(\mathbf{r}')}{(2\pi)^2\frac{h^2+\mu^2}{\mu}k_1^2}d\mu dh. \tag{7.30}$$

In order to facilitate the integration with respect to h the following vector functions are defined:

$$\tilde{\mathbf{M}}_{n,\mu}(\mathbf{r})=\left(\hat{\rho}\frac{in}{\rho}J_n(\mu\rho)-\hat{\phi}\frac{\partial J_n(\mu\rho)}{\partial\rho}\right), \tag{7.31}$$

$$\tilde{\mathbf{N}}_{n,\mu,h}(\mathbf{r})=\frac{1}{\kappa}\left(\hat{\rho}ih\frac{\partial J_n(\mu\rho)}{\partial\rho}-\hat{\phi}\frac{nh}{\rho}J_n(\mu\rho)+\hat{z}\mu^2J_n(\mu\rho)\right), \tag{7.32}$$

$$\tilde{\mathbf{L}}_{n,\mu,h}(\mathbf{r}) = \left(\hat{\rho}\frac{\partial J_n(\mu\rho)}{\partial\rho} + \hat{\phi}\frac{in}{\rho}J_n(\mu\rho) + \hat{z}ihJ_n(\mu\rho)\right). \tag{7.33}$$

The longitudinal part of the Green's tensor can now be written as

$$\mathbf{G}^L(\mathbf{r},\mathbf{r}') = -\sum_{n=-\infty}^{+\infty}\int_{\mu=0}^{\infty}\int_{h=-\infty}^{\infty}\frac{\tilde{\mathbf{L}}_{n,\mu,h}(\mathbf{r})\tilde{\mathbf{L}}^*_{n,\mu,h}(\mathbf{r}')e^{ih(z-z')}e^{in(\phi-\phi')}}{(2\pi)^2\frac{h^2+\mu^2}{\mu}k_1^2}d\mu dh. \tag{7.34}$$

Notice that there are poles at $h = \pm i\mu$. Also note that for all matrix elements except the zz-element, the integrand will vanish for sufficiently large values of h. Thus, each of those tensor elements can be expressed as an integral of the type

$$\int_{h=-\infty}^{\infty}\frac{f(h)e^{ih(z-z')}}{h^2+\mu^2}dh = \frac{\pi}{\mu}f\left(\frac{z-z'}{|z-z'|}i\mu\right)e^{-\mu|z-z'|}, \tag{7.35}$$

where the vanishing of the integrand for large absolute values of h is exploited. When this is combined with the exponential decay of $\exp(ih(z-z'))$ when extending the curve of integration into the upper or lower complex plane it is found that the integral along the real axis can be replaced by an integral along a closed curve that includes either the pole $i\mu$ or $-i\mu$ depending on the sign of $z-z'$, and then (7.35) follows from residue calculus.

Let us now consider the zz-component given by

$$\hat{\mathbf{z}}\cdot\mathbf{G}^L(\mathbf{r},\mathbf{r}')\cdot\hat{\mathbf{z}} = -\sum_{n=-\infty}^{+\infty}\int_{\mu=0}^{\infty}\int_{h=-\infty}^{\infty}\frac{h^2J_n(\mu\rho)J_n(\mu\rho')e^{ih(z-z')}e^{in(\phi-\phi')}}{(2\pi)^2\frac{h^2+\mu^2}{\mu}k_1^2}d\mu dh. \tag{7.36}$$

Clearly, the integrand will not vanish with increasing magnitude of h. However, the substitution

$$\frac{h^2}{h^2+\mu^2} = 1 - \frac{\mu^2}{h^2+\mu^2} \tag{7.37}$$

can be used to divide the integral into two integrals. One where residue calculus is again applicable, since the integral is of the form (7.35) such that the integrand vanishes for increasing magnitude of h, and another that can be analytically integrated and shown to be a delta function. The latter term becomes

$$-\sum_{n=-\infty}^{+\infty}\int_{\mu=0}^{\infty}\int_{h=-\infty}^{\infty}\frac{\mu J_n(\mu\rho)J_n(\mu\rho')e^{ih(z-z')}e^{in(\phi-\phi')}}{(2\pi)^2k_1^2}d\mu dh$$

$$= -\frac{1}{2\pi}\frac{1}{k_1^2}\delta(z-z')\frac{\delta(\rho-\rho')}{\rho}\sum_{n=-\infty}^{+\infty}e^{in(\phi-\phi')} = \frac{-1}{k_1^2}\delta(\mathbf{r}-\mathbf{r}'). \tag{7.38}$$

By using these results the longitudinal part of the Green's tensor can be expressed as

$$\mathbf{G}^L(\mathbf{r},\mathbf{r}') =$$

$$\frac{-1}{4\pi}\frac{1}{k_1^2}\sum_{n=-\infty}^{+\infty}\int_{\mu=0}^{\infty}\tilde{\mathbf{L}}_{n,\mu,h}(\mathbf{r})\tilde{\mathbf{L}}^*_{n,\mu,h^*}(\mathbf{r}')e^{-\mu|z-z'|}e^{in(\phi-\phi')}d\mu - \frac{\hat{\mathbf{z}}\hat{\mathbf{z}}}{k_1^2}\delta(\mathbf{r}-\mathbf{r}'), \tag{7.39}$$

where here

$$h \equiv \frac{z-z'}{|z-z'|} i\mu. \tag{7.40}$$

Note that in $\tilde{\mathbf{L}}^*_{n,\mu,h^*}(\mathbf{r}')$ the conjugation of the $\mathbf{L}$ function itself and the additional conjugation of h leads to no conjugation of h (or a double conjugation).

Consider now the $\mathbf{N}_{n,\mu,h}(\mathbf{r})\mathbf{N}^*_{n,\mu,h}(\mathbf{r}')$ term in Eq. (7.29). This product contains a factor

$$\frac{1}{\kappa^2} = \frac{1}{h^2+\mu^2}. \tag{7.41}$$

Thus, as the integration over h is carried out using residue calculus, the poles at $h = \pm i\mu$ must be considered, and a closer examination reveals that this pole contribution is, except for having opposite sign, identical to the term in Eq. (7.39) that was determined via residue calculus. Thus, this pole contribution cancels out with the first term in the longitudinal part of the Green's tensor.

In order to obtain finally a Green's tensor that satisfies appropriate boundary conditions, namely that the resulting Green's tensor must describe fields that propagate away from $\mathbf{r}'$, a small positive imaginary part is added to k_1^2, i.e., $k_1^2 \to k_1^2 + i\varepsilon$, and the limit $\varepsilon \to 0$ is observed. This procedure only adds a homogeneous solution to Eq. (7.6) and does not change the fact that the resulting expression is a particular solution of Eq. (7.6). Thus, there are poles at

$$h = \pm\sqrt{k_1^2 - \mu^2 + i\varepsilon}. \tag{7.42}$$

By using residue calculus it can now be shown that

$$\int_{-\infty}^{+\infty} \frac{f(h)e^{ih(z-z')}}{(h-\sqrt{k_1^2-\mu^2+i\varepsilon})(h+\sqrt{k_1^2-\mu^2+i\varepsilon})} dh = +i\frac{\pi}{k_{z1}} f(h) e^{ik_{z1}|z-z'|}, \tag{7.43}$$

where here

$$k_{z1} \equiv \sqrt{k_1^2 - \mu^2}, \tag{7.44}$$

with $\mathrm{Imag}(k_{z1}) \geq 0$, and also here

$$h \equiv \frac{z-z'}{|z-z'|} k_{z1}. \tag{7.45}$$

This leads finally to the total direct Green's tensor

$$\mathbf{G}^{(d)}(\mathbf{r},\mathbf{r}') = \frac{i}{4\pi} \sum_{n=-\infty}^{+\infty} \int_{\mu=0}^{\infty} \frac{1}{\mu k_{z1}} \left(\tilde{\mathbf{M}}_{n,\mu}(\mathbf{r}) \tilde{\mathbf{M}}^*_{n,\mu}(\mathbf{r}') + \tilde{\mathbf{N}}_{n,\mu,\pm k_{z1}}(\mathbf{r}) \tilde{\mathbf{N}}^*_{n,\mu,\pm k^*_{z1}}(\mathbf{r}') \right)$$

$$\times e^{in(\phi-\phi')} e^{ik_{z1}|z-z'|} d\mu - \hat{\mathbf{z}}\hat{\mathbf{z}} \frac{1}{k_1^2} \frac{\delta(\rho-\rho')}{\rho} \delta(z-z') \frac{1}{2\pi} \sum_n e^{in(\phi-\phi')}. \tag{7.46}$$

Here $+$ is used when $z > z'$ and $-$ is used when $z < z'$. Note that if the small positive imaginary term $i\varepsilon$ is replaced by $-i\varepsilon$ then the resulting expression contains

$\exp(-ik_{z1}|z-z'|)$ instead of $\exp(+ik_{z1}|z-z'|)$, and the radiating boundary condition is then not satisfied for the chosen time convention of time factor $\exp(-i\omega t)$. If we had instead adopted the sign convention of time factor $\exp(i\omega t)$, then it is exactly the Green's tensor with $\exp(-ik_{z1}|z-z'|)$ that fulfills the radiating boundary condition.

Note that in the derivation in [2] the longitudinal part of the Green's tensor was not included. Furthermore, the additional pole contribution from the $\mathbf{N}_{n,\mu,h}(\mathbf{r})\mathbf{N}^*_{n,\mu,h}(\mathbf{r}')$ term was also not included. However, for $\mathbf{r} \neq \mathbf{r}'$ the result agrees with that of Tai [2] after having carried out the integration over h. This is because pole contributions from the longitudinal and transverse Green's function cancel out. The last term in Eq. (7.46) arising from the longitudinal part is highly important when using the volume integral equation.

By comparing with (7.4) it is now straightforward to identify the direct part of the Green's tensor elements $G^{(m)}_{\alpha\beta}$, which we shall refer to as $G^{(d,m)}_{\alpha\beta}$:

$$G^{(d,m)}_{\rho\rho} = \frac{i}{2}\int_{\mu=0}^{\infty}\left(m^2\frac{J_m(\mu\rho)}{\rho}\frac{J_m(\mu\rho')}{\rho'} + \frac{k_{z1}^2}{k_1^2}\frac{\partial J_m(\mu\rho)}{\partial\rho}\frac{\partial J_m(\mu\rho')}{\partial\rho'}\right)\frac{e^{ik_{z1}|z-z'|}}{\mu k_{z1}}d\mu, \tag{7.47}$$

$$G^{(d,m)}_{\rho\phi} = \frac{m}{2}\int_{\mu=0}^{\infty}\left(\frac{J_m(\mu\rho)}{\rho}\frac{\partial J_m(\mu\rho')}{\partial\rho'} + \frac{k_{z1}^2}{k_1^2}\frac{\partial J_m(\mu\rho)}{\partial\rho}\frac{J_m(\mu\rho')}{\rho'}\right)\frac{e^{ik_{z1}|z-z'|}}{\mu k_{z1}}d\mu, \tag{7.48}$$

$$G^{(d,m)}_{\rho z} = -\frac{1}{2}\int_{\mu=0}^{\infty}\frac{\mu}{k_1^2}\frac{\partial J_m(\mu\rho)}{\partial\rho}J_m(\mu\rho')e^{ik_{z1}|z-z'|}d\mu \times \frac{z-z'}{|z-z'|}, \tag{7.49}$$

$$G^{(d,m)}_{\phi\rho} = -\frac{m}{2}\int_{\mu=0}^{\infty}\left(\frac{\partial J_m(\mu\rho)}{\partial\rho}\frac{J_m(\mu\rho')}{\rho'} + \frac{k_{z1}^2}{k_1^2}\frac{J_m(\mu\rho)}{\rho}\frac{\partial J_m(\mu\rho')}{\partial\rho'}\right)\frac{e^{ik_{z1}|z-z'|}}{\mu k_{z1}}d\mu, \tag{7.50}$$

$$G^{(d,m)}_{\phi\phi} = \frac{i}{2}\int_{\mu=0}^{\infty}\left(\frac{\partial J_m(\mu\rho)}{\partial\rho}\frac{\partial J_m(\mu\rho')}{\partial\rho'} + m^2\frac{k_{z1}^2}{k_1^2}\frac{J_m(\mu\rho)}{\rho}\frac{J_m(\mu\rho')}{\rho'}\right)\frac{e^{ik_{z1}|z-z'|}}{\mu k_{z1}}d\mu, \tag{7.51}$$

$$G^{(d,m)}_{\phi z} = -\frac{im}{2}\int_{\mu=0}^{\infty}\frac{\mu}{k_1^2}\frac{J_m(\mu\rho)}{\rho}J_m(\mu\rho')e^{ik_{z1}|z-z'|}d\mu \times \frac{z-z'}{|z-z'|}, \tag{7.52}$$

$$G^{(d,m)}_{z\rho} = \frac{1}{2}\int_{\mu=0}^{\infty}\frac{\mu}{k_1^2}J_m(\mu\rho)\frac{\partial J_m(\mu\rho')}{\partial\rho'}e^{ik_{z1}|z-z'|}d\mu \times \frac{z-z'}{|z-z'|}, \tag{7.53}$$

$$G^{(d,m)}_{z\phi} = -\frac{im}{2}\int_{\mu=0}^{\infty}\frac{\mu}{k_1^2}J_m(\mu\rho)\frac{J_m(\mu\rho')}{\rho'}e^{ik_{z1}|z-z'|}d\mu \times \frac{z-z'}{|z-z'|} \tag{7.54}$$

$$G_{zz}^{(d,m)} = \frac{i}{2}\int_{\mu=0}^{\infty} \frac{\mu^3}{k_1^2 k_{z1}} J_m(\mu\rho)J_m(\mu\rho')e^{ik_{z1}|z-z'|}d\mu - \frac{1}{k_1^2}\delta(z-z')\frac{\delta(\rho-\rho')}{\rho'}. \quad (7.55)$$

In principle the same quantities can also be calculated numerically using another approach, namely using the formula

$$G_{\alpha,\beta}^{(d,m)}(\rho,z;\rho',z') \equiv \int_{\phi=0}^{2\pi} \hat{\alpha}\cdot\mathbf{G}^{(d)}(\mathbf{r},\mathbf{r}')\cdot\hat{\beta}' e^{im(\phi'-\phi)}d\phi', \quad (7.56)$$

where the analytic closed-form expression of the Green's tensor (6.15) is inserted. The latter approach was applied in Refs. [154, 155].

7.2 GREEN'S TENSOR FOR A LAYERED STRUCTURE

Consider now a reference geometry with a layered structure between two homogeneous media such that for $z > 0$ the dielectric constant is $\varepsilon_{\text{ref}} = \varepsilon_1$ and for $z < -d$ the dielectric constant is $\varepsilon_{\text{ref}} = \varepsilon_{LN}$, and in between there can be a layered structure. We will assume here that $z' > 0$.

The dyadic Green's tensor must now satisfy the equation

$$-\nabla\times\nabla\times\mathbf{G}(\mathbf{r},\mathbf{r}') + k_0^2\varepsilon_{\text{ref}}(\mathbf{r})\mathbf{G}(\mathbf{r},\mathbf{r}') = -\mathbf{I}\delta(\mathbf{r}-\mathbf{r}'). \quad (7.57)$$

Similar to the previous chapter the Green's tensor in this case can be obtained by adding an indirect Green's tensor for $z > 0$ and constructing a transmitted Green's tensor for $z < -d$:

$$\mathbf{G}(\mathbf{r},\mathbf{r}') = \begin{cases} \mathbf{G}^{(d)}(\mathbf{r},\mathbf{r}') + \mathbf{G}^{(i)}(\mathbf{r},\mathbf{r}'), & z > 0,\ z' > 0 \\ \mathbf{G}^{(t)}(\mathbf{r},\mathbf{r}'), & z < -d,\ z' > 0 \end{cases}. \quad (7.58)$$

In terms of cylindrical harmonics this can also be expressed as

$$G_{\alpha\beta}^{(m)}(\rho,z;\rho',z') = \begin{cases} G_{\alpha\beta}^{(d,m)}(\rho,z;\rho',z') + G_{\alpha\beta}^{(i,m)}(\rho,z;\rho',z'), & z > 0,\ z' > 0, \\ G_{\alpha\beta}^{(t,m)}(\rho,z;\rho',z'), & z < -d,\ z' > 0 \end{cases}. \quad (7.59)$$

By noting that the direct Green's tensor $\mathbf{G}^{(d)}$ for $z < z'$ describes the excitation of downward propagating waves characterized by different in-plane propagation numbers μ, the appropriate Fresnel reflection and transmission can be constructed for each wave in the decomposition, such that the indirect and transmitted Green's tensors can be written as

$$\mathbf{G}^{(i)}(\mathbf{r},\mathbf{r}') = \frac{i}{4\pi}\sum_{n=-\infty}^{+\infty}\int_{\mu=0}^{\infty}\frac{1}{\mu k_{z1}}\Big(r^{(s)}(\mu)\tilde{\mathbf{M}}_{n,\mu}(\mathbf{r})\tilde{\mathbf{M}}_{n,\mu}^{*}(\mathbf{r}')$$

$$+r^{(p)}(\mu)\tilde{\mathbf{N}}_{n,\mu,+k_{z1}}(\mathbf{r})\tilde{\mathbf{N}}_{n,\mu,-k_{z1}^{*}}^{*}(\mathbf{r}')\Big)e^{in(\phi-\phi')}e^{ik_{z1}(z+z')}d\mu, \quad (7.60)$$

and

$$\mathbf{G}^{(t)}(\mathbf{r},\mathbf{r}') = \frac{i}{4\pi}\sum_{n=-\infty}^{+\infty}\int_{\mu=0}^{\infty}\frac{1}{\mu k_{z1}}\Big(t^{(s)}(\mu)\tilde{\mathbf{M}}_{n,\mu}(\mathbf{r})\tilde{\mathbf{M}}^{*}_{n,\mu}(\mathbf{r}')$$

$$+\frac{k_1}{k_{LN}}t^{(p)}(\mu)\tilde{\mathbf{N}}_{n,\mu,-k_{z,LN}}(\mathbf{r})\tilde{\mathbf{N}}^{*}_{n,\mu,-k^{*}_{z1}}(\mathbf{r}')\Big)e^{in(\phi-\phi')}e^{i(k_{z1}z'-k_{z,LN2}[z+d])}d\mu, \quad (7.61)$$

where $r^{(s)}$, $t^{(s)}$, $r^{(p)}$, and $t^{(p)}$ are the Fresnel reflection and transmission coefficients for a layered structure with thickness d, and for s- and p-polarized light, respectively. In the case of a single interface ($d=0$), the requirement of continuity of tangential electric and magnetic field components across the interface at $z=0$ leads straightforwardly to the usual reflection and transmission coefficients

$$r^{(s)}(\mu) = \frac{k_{z1}-k_{z,L2}}{k_{z1}+k_{z,L2}}, \quad r^{(p)}(\mu) = \frac{\varepsilon_{L2}k_{z1}-\varepsilon_1 k_{z,L2}}{\varepsilon_{L2}k_{z1}+\varepsilon_1 k_{z,L2}}, \quad (7.62)$$

$$t^{(s)}(\mu) = 1+r^{(s)}(\mu), \quad t^{(p)}(\mu) = 1+r^{(p)}(\mu), \quad (7.63)$$

where $k_{z,i} = \sqrt{k_0^2\varepsilon_i-\mu^2}$ with $\mathrm{Imag}\{k_{z,i}\}\geq 0$.

For a numerical implementation, it will be convenient to have expressions for the individual Green's tensor elements for each cylindrical harmonic. By comparing (7.60) and (7.61) with (7.4), the indirect and transmitted parts of the Green's tensor elements $G^{(m)}_{\alpha\beta}$ can be identified. The indirect and transmitted parts shall be referred to as $G^{(i,m)}_{\alpha\beta}$ and $G^{(t,m)}_{\alpha\beta}$, respectively. These components are written down in detail in the following two subsections.

7.2.1 INDIRECT GREEN'S TENSOR: CYLINDRICAL HARMONICS

$$G^{(i,m)}_{\rho\rho} = \frac{i}{2}\int_{\mu=0}^{\infty}\Big(r^{(s)}(\mu)m^2\frac{J_m(\mu\rho)}{\rho}\frac{J_m(\mu\rho')}{\rho'} -$$

$$r^{(p)}(\mu)\frac{k_{z1}^2}{k_1^2}\frac{\partial J_m(\mu\rho)}{\partial\rho}\frac{\partial J_m(\mu\rho')}{\partial\rho'}\Big)\frac{e^{ik_{z1}(z+z')}}{\mu k_{z1}}d\mu, \quad (7.64)$$

$$G^{(i,m)}_{\rho\phi} = \frac{m}{2}\int_{\mu=0}^{\infty}\Big(r^{(s)}(\mu)\frac{J_m(\mu\rho)}{\rho}\frac{\partial J_m(\mu\rho')}{\partial\rho'} -$$

$$r^{(p)}(\mu)\frac{k_{z1}^2}{k_1^2}\frac{\partial J_m(\mu\rho)}{\partial\rho}\frac{J_m(\mu\rho')}{\rho'}\Big)\frac{e^{ik_{z1}(z+z')}}{\mu k_{z1}}d\mu, \quad (7.65)$$

$$G^{(i,m)}_{\rho z} = -\frac{1}{2}\int_{\mu=0}^{\infty}r^{(p)}(\mu)\frac{\mu}{k_1^2}\frac{\partial J_m(\mu\rho)}{\partial\rho}J_m(\mu\rho')e^{ik_{z1}(z+z')}d\mu, \quad (7.66)$$

$$G^{(i,m)}_{\phi\rho} = -\frac{m}{2}\int_{\mu=0}^{\infty}\Big(r^{(s)}(\mu)\frac{\partial J_m(\mu\rho)}{\partial\rho}\frac{J_m(\mu\rho')}{\rho'} -$$

$$r^{(p)}(\mu)\frac{k_{z1}^2}{k_1^2}\frac{J_m(\mu\rho)}{\rho}\frac{\partial J_m(\mu\rho')}{\partial\rho'}\Bigg)\frac{e^{ik_{z1}(z+z')}}{\mu k_{z1}}d\mu, \quad (7.67)$$

$$G_{\phi\phi}^{(i,m)} = \frac{i}{2}\int_{\mu=0}^{\infty}\Bigg(r^{(s)}(\mu)\frac{\partial J_m(\mu\rho)}{\partial\rho}\frac{\partial J_m(\mu\rho')}{\partial\rho'} -$$

$$r^{(p)}(\mu)m^2\frac{k_{z1}^2}{k_1^2}\frac{J_m(\mu\rho)}{\rho}\frac{J_m(\mu\rho')}{\rho'}\Bigg)\frac{e^{ik_{z1}(z+z')}}{\mu k_{z1}}d\mu, \quad (7.68)$$

$$G_{\phi z}^{(i,m)} = -\frac{im}{2}\int_{\mu=0}^{\infty}r^{(p)}(\mu)\frac{\mu}{k_1^2}\frac{J_m(\mu\rho)}{\rho}J_m(\mu\rho')e^{ik_{z1}(z+z')}d\mu, \quad (7.69)$$

$$G_{z\rho}^{(i,m)} = -\frac{1}{2}\int_{\mu=0}^{\infty}r^{(p)}(\mu)\frac{\mu}{k_1^2}J_m(\mu\rho)\frac{\partial J_m(\mu\rho')}{\partial\rho'}e^{ik_{z1}(z+z')}d\mu, \quad (7.70)$$

$$G_{z\phi}^{(i,m)} = \frac{im}{2}\int_{\mu=0}^{\infty}r^{(p)}(\mu)\frac{\mu}{k_1^2}J_m(\mu\rho)\frac{J_m(\mu\rho')}{\rho'}e^{ik_{z1}(z+z')}d\mu, \quad (7.71)$$

$$G_{zz}^{(i,m)} = \frac{i}{2}\int_{\mu=0}^{\infty}r^{(p)}(\mu)\frac{\mu^3}{k_1^2k_{z1}}J_m(\mu\rho)J_m(\mu\rho')e^{ik_{z1}(z+z')}d\mu. \quad (7.72)$$

7.2.2 TRANSMITTED GREEN'S TENSOR: CYLINDRICAL HARMONICS

$$G_{\rho\rho}^{(t,m)} = \frac{i}{2}\int_{\mu=0}^{\infty}\Bigg(t^{(s)}(\mu)m^2\frac{J_m(\mu\rho)}{\rho}\frac{J_m(\mu\rho')}{\rho'}$$

$$+t^{(p)}(\mu)\frac{k_{z1}k_{z,LN}}{k_{LN}^2}\frac{\partial J_m(\mu\rho)}{\partial\rho}\frac{\partial J_m(\mu\rho')}{\partial\rho'}\Bigg)\frac{e^{i(k_{z1}z'-k_{z,LN}[z+d])}}{\mu k_{z1}}d\mu, \quad (7.73)$$

$$G_{\rho\phi}^{(t,m)} = \frac{m}{2}\int_{\mu=0}^{\infty}\Bigg(t^{(s)}(\mu)\frac{J_m(\mu\rho)}{\rho}\frac{\partial J_m(\mu\rho')}{\partial\rho'}+$$

$$t^{(p)}(\mu)\frac{k_{z1}k_{z,LN}}{k_{LN}^2}\frac{\partial J_m(\mu\rho)}{\partial\rho}\frac{J_m(\mu\rho')}{\rho'}\Bigg)\frac{e^{i(k_{z1}z'-k_{z,LN}[z+d])}}{\mu k_{z1}}d\mu, \quad (7.74)$$

$$G_{\rho z}^{(t,m)} = \frac{1}{2}\int_{\mu=0}^{\infty}t^{(p)}(\mu)\frac{\mu}{k_{LN}^2}\frac{k_{z,LN}}{k_{z1}}\frac{\partial J_m(\mu\rho)}{\partial\rho}J_m(\mu\rho')e^{i(k_{z1}z'-k_{z,LN}[z+d])}d\mu, \quad (7.75)$$

$$G_{\phi\rho}^{(t,m)} = -\frac{m}{2}\int_{\mu=0}^{\infty}\Bigg(t^{(s)}(\mu)\frac{\partial J_m(\mu\rho)}{\partial\rho}\frac{J_m(\mu\rho')}{\rho'}+$$

$$t^{(p)}(\mu)\frac{k_{z1}k_{z,LN}}{k_{LN}^2}\frac{J_m(\mu\rho)}{\rho}\frac{\partial J_m(\mu\rho')}{\partial\rho'}\Bigg)\frac{e^{i(k_{z1}z'-k_{z,LN}[z+d])}}{\mu k_{z1}}d\mu, \quad (7.76)$$

$$G_{\phi\phi}^{(t,m)} = \frac{i}{2}\int_{\mu=0}^{\infty}\left(t^{(s)}(\mu)\frac{\partial J_m(\mu\rho)}{\partial\rho}\frac{\partial J_m(\mu\rho')}{\partial\rho'} + \right.$$

$$\left. t^{(p)}(\mu)m^2\frac{k_{z1}k_{z,LN}}{k_{LN}^2}\frac{J_m(\mu\rho)}{\rho}\frac{J_m(\mu\rho')}{\rho'}\right)\frac{e^{i(k_{z1}z'-k_{z,LN}[z+d])}}{\mu k_{z1}}d\mu, \tag{7.77}$$

$$G_{\phi z}^{(t,m)} = \frac{im}{2}\int_{\mu=0}^{\infty} t^{(p)}(\mu)\frac{\mu}{k_{LN}^2}\frac{k_{z,LN}}{k_{z1}}\frac{J_m(\mu\rho)}{\rho}J_m(\mu\rho')e^{i(k_{z1}z'-k_{z,LN}[z+d])}d\mu, \tag{7.78}$$

$$G_{z\rho}^{(t,m)} = -\frac{1}{2}\int_{\mu=0}^{\infty} t^{(p)}(\mu)\frac{\mu}{k_{LN}^2}J_m(\mu\rho)\frac{\partial J_m(\mu\rho')}{\partial\rho'}e^{i(k_{z1}z'-k_{z,LN}[z+d])}d\mu, \tag{7.79}$$

$$G_{z\phi}^{(t,m)} = \frac{im}{2}\int_{\mu=0}^{\infty} t^{(p)}(\mu)\frac{\mu}{k_{LN}^2}J_m(\mu\rho)\frac{J_m(\mu\rho')}{\rho'}e^{i(k_{z1}z'-k_{z,LN}[z+d])}d\mu, \tag{7.80}$$

$$G_{zz}^{(t,m)} = \frac{i}{2}\int_{\mu=0}^{\infty} t^{(p)}(\mu)\frac{\mu^3}{k_{LN}^2 k_{z1}}J_m(\mu\rho)J_m(\mu\rho')e^{i(k_{z1}z'-k_{z,LN}[z+d])}d\mu. \tag{7.81}$$

7.3 OUT-OF-PLANE FAR-FIELD APPROXIMATIONS OF THE CYLINDRICAL GREEN'S TENSOR ELEMENTS

Far-field approximations of the scattered electric and magnetic fields are needed in order to be able to calculate extinction and scattering cross sections. Once the field inside the scattering object has been determined, the far field can be obtained directly from the Green's tensor volume integral equation by inserting the far-field approximation to the Green's tensor.

We will start by finding out-of-plane far-field expressions for the direct Green's tensor elements $G_{\alpha\beta}^{(d,m)}$. We will do this by considering first $G_{\rho\rho}^{(d,m)}$ [see Eq. (7.47)] for a situation where $z > 0$, and $k_1(z-z') >> 1$ ($k_1 = k_0 n_1$). In that case the part of the integral in Eq. (7.47) concerned with $\mu > k_1$ can be ignored due to the factor

$$e^{ik_{z1}|z-z'|} = e^{-\sqrt{\mu^2-k_1^2}|z-z'|} \tag{7.82}$$

becoming vanishingly small. It will be convenient then to make the replacements

$$\mu = k_1 \sin\alpha, \tag{7.83}$$

$$k_{z1} = k_1 \cos\alpha, \tag{7.84}$$

in which case Eq. (7.47) can be approximated by

$$G_{\rho\rho}^{(d,m)} \approx \frac{i}{2}\int_{\alpha=0}^{\pi/2}\left(m^2\frac{J_m(\mu\rho)}{\rho}\frac{J_m(\mu\rho')}{\rho'}+\frac{k_{z1}^2}{k_1^2}\frac{\partial J_m(\mu\rho)}{\partial\rho}\frac{\partial J_m(\mu\rho')}{\partial\rho'}\right)\frac{e^{ik_{z1}|z-z'|}}{\mu}d\alpha. \tag{7.85}$$

For large arguments the following approximations to the Bessel function and its derivative are also useful:

$$J_m(x) \approx \sqrt{\frac{2}{\pi x}}\frac{1}{2}\left(e^{i(x-m\pi/2-\pi/4)}+e^{-i(x-m\pi/2-\pi/4)}\right),\ x>>1, \tag{7.86}$$

$$\frac{\partial J_m(x)}{\partial x} \approx -\sqrt{\frac{2}{\pi x}}\frac{1}{2i}\left(e^{i(x-m\pi/2-\pi/4)}-e^{-i(x-m\pi/2-\pi/4)}\right),\ x>>1. \tag{7.87}$$

The following relations between cylindrical and spherical coordinates should also be recalled:

$$\rho = r\sin\theta, \tag{7.88}$$

$$z = r\cos\theta. \tag{7.89}$$

By also applying the relation

$$\cos(A\pm B) = \cos A\cos B \mp \sin A\sin B, \tag{7.90}$$

part of the integrand can be approximated by

$$J_m(\mu\rho)e^{ik_{z1}z} \approx$$

$$\sqrt{\frac{2}{\pi\mu\rho}}\frac{1}{2}\left(e^{ik_1 r\cos(\alpha-\theta)}e^{-i(m\pi/2+\pi/4)}+e^{ik_1 r\cos(\alpha+\theta)}e^{+i(m\pi/2+\pi/4)}\right), \tag{7.91}$$

$$\frac{\partial J_m(\mu\rho)}{\partial\rho}e^{ik_{z1}z} \approx$$

$$-\sqrt{\frac{2}{\pi\mu\rho}}\frac{\mu}{2i}\left(e^{ik_1 r\cos(\alpha-\theta)}e^{-i(m\pi/2+\pi/4)}-e^{ik_1 r\cos(\alpha+\theta)}e^{+i(m\pi/2+\pi/4)}\right). \tag{7.92}$$

Note that in the far field ($k_1 r >> 1$), a term such as $e^{ik_1 r\cos(\alpha-\theta)}$ will oscillate very fast with α except when $\alpha \approx \theta$. Other factors multiplied onto this term will be slowly varying with α in comparison. Thus, those parts of the integral with fast oscillations can be neglected, and the larger the value of r, the better an approximation it will be to neglect those parts. Thus, in the far field there will only be a contribution to the integral for values of $\alpha \approx \theta$, and in slowly varying terms it is possible to replace α with θ. Note that for $z > 0$, as is considered here, then $0 \leq \theta < \pi/2$, and therefore the other term $e^{ik_1 r\cos(\alpha+\theta)}$ will oscillate fast for all considered α, and this term will thus not contribute to the integral in the far field.

Now, the integral can be evaluated by using

$$\int_{\alpha=\theta-\Delta}^{\theta+\Delta} e^{ik_1 r\cos(\alpha-\theta)} d\alpha \approx \int_{\alpha} e^{ik_1 r} e^{-ik_1 r\frac{1}{2}(\alpha-\theta)^2} d\alpha = e^{ik_1 r}\sqrt{\frac{2\pi}{ik_1 r}}. \tag{7.93}$$

It should be noted that in the far field the term

$$\frac{J_m(\mu\rho)}{\rho} \propto r^{-3/2} \tag{7.94}$$

while

$$\frac{\partial J_m(\mu\rho)}{\partial\rho} \propto r^{-1/2}. \tag{7.95}$$

Terms with a magnitude that decreases faster with r than $1/r$ can be ignored in the far field. In Eq. (7.85) we may therefore discard the term

$$m^2 \frac{J_m(\mu\rho)}{\rho}\frac{J_m(\mu\rho')}{\rho'}. \tag{7.96}$$

These considerations lead finally to the far-field expression

$$G_{\rho\rho}^{(d,\mathrm{ff},m)} = \frac{e^{ik_1 r}}{k_1 r}\frac{i}{2}(-i)^m e^{-ik_{z1}z'}\frac{\cos^2\theta}{\sin\theta}\frac{\partial J_m(k_\rho\rho')}{\partial\rho'}, \tag{7.97}$$

where here $k_{z1} = k_1\cos\theta$, and $k_\rho = k_1\sin\theta$.

Other components of the far-field Green's tensor can be derived using similar considerations. One more example of deriving a far-field Green's tensor component will be given for the transmitted Green's tensor element $G_{\phi\rho}^{(t,m)}$ [See Eq. (7.76)]. Now a situation with $k_{LN}(z+d) << -1$ ($k_{LN} = k_0 n_{LN}$) must be considered. In this case, due to the factor

$$e^{-ik_{z,LN}[z+d])} = e^{-\sqrt{\mu^2-k_{LN}^2}|z+d|}, \tag{7.98}$$

the part of the integral where $\mu > k_{LN}$ can be neglected. Thus, here it is appropriate to make the substitutions

$$\mu = k_{LN}\sin\alpha, \tag{7.99}$$

$$k_{z,LN} = k_{LN}\cos\alpha. \tag{7.100}$$

The second term in the integrand in Eq. (7.76) will behave as

$$\frac{J_m(\mu\rho)}{\rho} \propto r^{-3/2} \tag{7.101}$$

and can thus be immediately discarded. These considerations lead to

$$G_{\phi\rho}^{(t,m)} \approx -\frac{m}{2}\int_{\alpha=0}^{\pi/2} t^{(s)}(\mu)\frac{\partial J_m(\mu\rho)}{\partial\rho}\frac{J_m(\mu\rho')}{\rho'}\frac{e^{i(k_{z1}z'-k_{z,LN}[z+d])}}{\mu k_{z1}} k_{z,LN} d\alpha. \tag{7.102}$$

The fast oscillating part of the integrand is thus

$$\frac{\partial J_m(\mu\rho)}{\partial\rho}e^{-ik_{z,LN}z} \approx$$

$$\sqrt{\frac{2}{\pi\mu\rho}}\frac{-\mu}{2i}\left(e^{-i(m\pi/2+\pi/4)}e^{ik_{LN}r\cos(\alpha-[\pi-\theta])} - e^{i(m\pi/2+\pi/4)}e^{ik_{LN}r\cos(\alpha+[\pi-\theta])}\right), \tag{7.103}$$

and for the values of θ being considered ($\pi/2 < \theta \leq \pi$) the second term in the parantheses will be oscillating fast with α for all considered α, and thus this term can be ignored. The first term can be slowly varying when $\alpha \approx \pi - \theta$. Here, the integral can now be calculated by using

$$\int_\alpha f(\alpha)e^{ik_{LN}r\cos(\alpha-[\pi-\theta])}d\alpha \approx f(\pi-\theta)\sqrt{\frac{2\pi}{ik_{LN}r}}e^{ik_{LN}r}, \tag{7.104}$$

where $f(\alpha)$ is a slowly varying function. These considerations finally lead to the far-field approximation

$$G_{\phi\rho}^{(t,\text{ff},m)} = -\frac{e^{ik_{LN}r}}{k_{LN}r}\frac{m}{2}(-i)^m t^{(s)}(k_\rho)\frac{k_{z,LN}\sin\theta}{k_{z1}}\frac{J_m(k_\rho\rho')}{\rho'}e^{ik_{z1}z'}e^{-ik_{z,LN}d}, \tag{7.105}$$

where here $k_\rho = k_{LN}\sin\theta$, and $k_{z,i} = \sqrt{k_0^2\varepsilon_i - k_\rho^2}$ with Imag$\{k_{z,i}\} \geq 0$. We will consider this as enough examples and will now simply present the results for the far-field homogeneous medium Green's tensor, the far-field reflected Green's tensor, and the far-field transmitted Green's tensor, in the following three subsections.

7.3.1 FAR-FIELD DIRECT GREEN'S TENSOR

$$G_{\rho\rho}^{(d,\text{ff},m)} = \frac{e^{ik_1r}}{k_1r}\frac{i}{2}(-i)^m e^{-ik_1\cos\theta z'}\frac{\cos^2\theta}{\sin\theta}\frac{\partial J_m(k_\rho\rho')}{\partial\rho'}, \tag{7.106}$$

$$G_{\rho\phi}^{(d,\text{ff},m)} = \frac{e^{ik_1r}}{k_1r}\frac{m}{2}(-i)^m e^{-ik_1\cos\theta z'}\frac{\cos^2\theta}{\sin\theta}\frac{J_m(k_\rho\rho')}{\rho'}, \tag{7.107}$$

$$G_{\rho z}^{(d,\text{ff},m)} = \frac{e^{ik_1r}}{k_1r}\frac{-1}{2}(-i)^m e^{-ik_1\cos\theta z'}k_1\cos\theta\sin\theta J_m(k_\rho\rho'), \tag{7.108}$$

$$G_{\phi\rho}^{(d,\text{ff},m)} = \frac{e^{ik_1r}}{k_1r}\frac{-m}{2}(-i)^m e^{-ik_1\cos\theta z'}\frac{1}{\sin\theta}\frac{J_m(k_\rho\rho')}{\rho'}, \tag{7.109}$$

$$G_{\phi\phi}^{(d,\text{ff},m)} = \frac{e^{ik_1r}}{k_1r}\frac{i}{2}(-i)^m e^{-ik_1\cos\theta z'}\frac{1}{\sin\theta}\frac{\partial J_m(k_\rho\rho')}{\partial\rho'}, \tag{7.110}$$

$$G_{\phi z}^{(d,\text{ff},m)} = 0, \tag{7.111}$$

$$G_{z\rho}^{(d,\text{ff},m)} = \frac{e^{ik_1 r}}{k_1 r}\frac{-i}{2}(-i)^m e^{-ik_1\cos\theta z'}\cos\theta\frac{\partial J_m(k_\rho\rho')}{\partial\rho'}, \tag{7.112}$$

$$G_{z\phi}^{(d,\text{ff},m)} = \frac{e^{ik_1 r}}{k_1 r}\frac{-m}{2}(-i)^m e^{-ik_1\cos\theta z'}\cos\theta\frac{J_m(k_\rho\rho')}{\rho'}, \tag{7.113}$$

$$G_{zz}^{(d,\text{ff},m)} = \frac{e^{ik_1 r}}{k_1 r}\frac{1}{2}(-i)^m e^{-ik_1\cos\theta z'}\sin^2\theta k_1 J_m(k_\rho\rho'). \tag{7.114}$$

In this subsection $k_\rho = k_1\sin\theta$.

7.3.2 FAR-FIELD INDIRECT GREEN'S TENSOR

$$G_{\rho\rho}^{(i,\text{ff},m)} = \frac{e^{ik_1 r}}{k_1 r}\frac{i}{2}(-i)^m\left(-r^{(p)}(k_\rho)\right)e^{+ik_{z1}z'}\frac{\cos^2\theta}{\sin\theta}\frac{\partial J_m(k_\rho\rho')}{\partial\rho'}, \tag{7.115}$$

$$G_{\rho\phi}^{(i,\text{ff},m)} = \frac{e^{ik_1 r}}{k_1 r}\frac{m}{2}(-i)^m\left(-r^{(p)}(k_\rho)\right)e^{ik_{z1}z'}\frac{\cos^2\theta}{\sin\theta}\frac{J_m(k_\rho\rho')}{\rho'}, \tag{7.116}$$

$$G_{\rho z}^{(i,\text{ff},m)} = \frac{e^{ik_1 r}}{k_1 r}\frac{-1}{2}(-i)^m r^{(p)}(k_\rho)e^{ik_{z1}z'}\sin\theta J_m(k_\rho\rho')k_{z1}, \tag{7.117}$$

$$G_{\phi\rho}^{(i,\text{ff},m)} = \frac{e^{ik_1 r}}{k_1 r}\frac{-m}{2}(-i)^m r^{(s)}(k_\rho)e^{ik_{z1}z'}\frac{1}{\sin\theta}\frac{J_m(k_\rho\rho')}{\rho'}, \tag{7.118}$$

$$G_{\phi\phi}^{(i,\text{ff},m)} = \frac{e^{ik_1 r}}{k_1 r}\frac{i}{2}(-i)^m r^{(s)}(k_\rho)e^{ik_{z1}z'}\frac{1}{\sin\theta}\frac{\partial J_m(k_\rho\rho')}{\partial\rho'}, \tag{7.119}$$

$$G_{\phi z}^{(i,\text{ff},m)} = 0, \tag{7.120}$$

$$G_{z\rho}^{(i,\text{ff},m)} = \frac{e^{ik_1 r}}{k_1 r}\frac{-i}{2}(-i)^m\left(-r^{(p)}(k_\rho)\right)e^{ik_{z1}z'}\cos\theta\frac{\partial J_m(k_\rho\rho')}{\partial\rho'}, \tag{7.121}$$

$$G_{z\phi}^{(i,\text{ff},m)} = \frac{e^{ik_1 r}}{k_1 r}\frac{-m}{2}(-i)^m\left(-r^{(p)}(k_\rho)\right)e^{ik_{z1}z'}\cos\theta\frac{J_m(k_\rho\rho')}{\rho'}, \tag{7.122}$$

$$G_{zz}^{(i,\text{ff},m)} = \frac{e^{ik_1 r}}{k_1 r}\frac{1}{2}(-i)^m r^{(p)}(k_\rho)e^{ik_{z1}z'}\sin^2\theta k_1 J_m(k_\rho\rho'). \tag{7.123}$$

In this subsection it also applies that $k_\rho = k_1\sin\theta$ and $k_{z1} = \sqrt{k_1^2 - k_\rho^2}$.

7.3.3 FAR-FIELD TRANSMITTED GREEN'S TENSOR

$$G_{\rho\rho}^{(t,\text{ff},m)} = \frac{e^{ik_{LN}r}}{k_{LN}r}\frac{i}{2}(-i)^m t^{(p)}(k_\rho)e^{ik_{z1}z'}e^{-ik_{z,LN}d}\frac{\cos^2\theta}{\sin\theta}\frac{\partial J_m(k_\rho\rho')}{\partial\rho'}, \tag{7.124}$$

$$G_{\rho\phi}^{(t,\text{ff},m)} = \frac{e^{ik_{LN}r}}{k_{LN}r}\frac{m}{2}(-i)^m t^{(p)}(k_\rho)e^{ik_{z1}z'}e^{-ik_{z,LN}d}\frac{\cos^2\theta}{\sin\theta}\frac{J_m(k_\rho\rho')}{\rho'}, \tag{7.125}$$

$$G_{\rho z}^{(t,\text{ff},m)} = \frac{e^{ik_{LN}r}}{k_{LN}r}\frac{1}{2}(-i)^m t^{(p)}(k_\rho)e^{ik_{z1}z'}e^{-ik_{z,LN}d}\sin\theta\frac{k_{z,LN}^2}{k_{z1}}J_m(k_\rho\rho'), \tag{7.126}$$

$$G_{\phi\rho}^{(t,\text{ff},m)} = \frac{e^{ik_{LN}r}}{k_{LN}r}\frac{-m}{2}(-i)^m t^{(s)}(k_\rho)e^{ik_{z1}z'}e^{-ik_{z,LN}d}\frac{k_{z,LN}}{k_{z1}}\frac{1}{\sin\theta}\frac{J_m(k_\rho\rho')}{\rho'}, \tag{7.127}$$

$$G_{\phi\phi}^{(t,\text{ff},m)} = \frac{e^{ik_{LN}r}}{k_{LN}r}\frac{i}{2}(-i)^m t^{(s)}(k_\rho)e^{ik_{z1}z'}e^{-ik_{z,LN}d}\frac{1}{\sin\theta}\frac{k_{z,LN}}{k_{z1}}\frac{\partial J_m(k_\rho\rho')}{\partial\rho'}, \tag{7.128}$$

$$G_{\phi z}^{(t,\text{ff},m)} = 0, \tag{7.129}$$

$$G_{z\rho}^{(t,\text{ff},m)} = \frac{e^{ik_{LN}r}}{k_{LN}r}\frac{i}{2}(-i)^m t^{(p)}(k_\rho)e^{ik_{z1}z'}e^{-ik_{z,LN}d}\frac{k_{z,LN}}{k_{LN}}\frac{\partial J_m(k_\rho\rho')}{\partial\rho'}, \tag{7.130}$$

$$G_{z\phi}^{(t,\text{ff},m)} = \frac{e^{ik_{LN}r}}{k_{LN}r}\frac{m}{2}(-i)^m t^{(p)}(k_\rho)e^{ik_{z1}z'}e^{-ik_{z,LN}d}\frac{k_{z,LN}}{k_{LN}}\frac{J_m(k_\rho\rho')}{\rho'}, \tag{7.131}$$

$$G_{zz}^{(t,\text{ff},m)} = \frac{e^{ik_{LN}r}}{k_{LN}r}\frac{1}{2}(-i)^m t^{(p)}(k_\rho)e^{ik_{z1}z'}e^{-ik_{z,LN}d}\frac{k_{z,LN}}{k_{z1}}\frac{k_\rho^2}{k_{LN}}J_m(k_\rho\rho'). \tag{7.132}$$

In this subsection it applies that $k_\rho = k_{LN}\sin\theta$, $k_{z1} = \sqrt{k_1^2 - k_\rho^2}$, and $k_{z,LN} = \sqrt{k_{LN}^2 - k_\rho^2}$.

7.4 GUIDED-MODE FAR-FIELD APPROXIMATIONS OF THE CYLINDRICAL GREEN'S TENSOR ELEMENTS

For positions z being close to the layered reference structure, and for large values of ρ corresponding to large in-plane distances to the scatterer, there will be a contribution to the far-field Green's tensor being related to guided modes if any exist. In the following it is assumed that $z, z' > 0$.

Consider the expression for $G_{\rho\rho}^{(i,m)}$ in (7.64) as an example. The first term in the integrand is ignored in the far field due to the term $J_m(\mu\rho)/\rho$ decreasing fast with ρ. In the second term it is noted that near a pole the Fresnel reflection coefficient can be expressed in the form given in Eq. (6.126). In addition it can be applied that

$$\frac{\partial J_m(\mu\rho)}{\partial\rho} \approx -\sqrt{\frac{2}{\pi\mu\rho}}\frac{\mu}{2i}\left(e^{i\mu\rho}(-i)^m e^{-i\pi/4} - e^{-i\mu\rho}(+i)^m e^{+i\pi/4}\right). \tag{7.133}$$

The following integral

$$\frac{i}{2}\int_{\mu=k_0 n_1}^{\infty}\left(-r^{(p)}(\mu)\frac{k_{z1}^2}{k_1^2}\frac{\partial J_m(\mu\rho)}{\partial\rho}\frac{\partial J_m(\mu\rho')}{\partial\rho'}\right)\frac{e^{ik_{z1}(z+z')}}{\mu k_{z1}}d\mu \tag{7.134}$$

can now be calculated using residue calculus similar to Sec. 6.3.2. This leads to

$$G_{\rho\rho}^{(g,\mathrm{ff},m)}(\rho,z,\rho',z') = f_\rho^{(p,m)}(\rho',z')\frac{e^{ik_{\rho,\mathrm{p}}\rho}}{\sqrt{\rho}}e^{ik_{z1,\mathrm{p}}z}\left(-\frac{k_{z1,\mathrm{p}}}{k_{\rho,\mathrm{p}}}\right), \tag{7.135}$$

where

$$f_\rho^{(p,m)}(\rho',z') = Ai\frac{\pi}{2}(-i)^m e^{-i\pi/4}\frac{\partial J_m(k_{\rho,\mathrm{p}}\rho')}{\partial(k_{\rho,\mathrm{p}}\rho')}e^{ik_{z1,\mathrm{p}}z'}\sqrt{\frac{2}{\pi k_{\rho,\mathrm{p}}}}\left(-\frac{k_{\rho,\mathrm{p}}^2}{k_1^2}\right), \tag{7.136}$$

and the parameters $k_{\rho,\mathrm{p}}$, $k_{z1,\mathrm{p}}$, and A have the same meaning as in Sec. 6.3.2. By writing $f_\rho^{(p,m)}$ instead of $f_\rho^{(m)}$ it is emphasized that the far field is p polarized, and the A coefficient is related to the pole of $r^{(p)}$. Using the same approach it can be shown that

$$G_{z\rho}^{(g,\mathrm{ff},m)}(\rho,z,\rho',z') = f_\rho^{(p,m)}(\rho',z')\frac{e^{ik_{\rho,\mathrm{p}}\rho}}{\sqrt{\rho}}e^{ik_{z1,\mathrm{p}}z}. \tag{7.137}$$

Note that as a check it can be seen that

$$\nabla\cdot\left(-\hat{\rho}\frac{k_{z1,\mathrm{p}}}{k_{\rho,\mathrm{p}}}+\hat{\mathbf{z}}\right)e^{ik_{\rho,\mathrm{p}}\rho}e^{ik_{z1,\mathrm{p}}z}\approx 0$$

for large ρ.

Other guided-mode far-field Green's tensor components can be calculated in the same way. The other p-polarized terms are

$$G_{\rho z}^{(g,\mathrm{ff},m)}(\rho,z,\rho',z') = f_z^{(p,m)}(\rho',z')\frac{e^{ik_{\rho,\mathrm{p}}\rho}}{\sqrt{\rho}}e^{ik_{z1,\mathrm{p}}z}\left(-\frac{k_{z1,\mathrm{p}}}{k_{\rho,\mathrm{p}}}\right), \tag{7.138}$$

$$G_{\phi z}^{(g,\mathrm{ff},m)}(\rho,z,\rho',z') = 0, \tag{7.139}$$

$$G_{zz}^{(g,\mathrm{ff},m)}(\rho,z,\rho',z') = f_z^{(p,m)}(\rho',z')\frac{e^{ik_{\rho,\mathrm{p}}\rho}}{\sqrt{\rho}}e^{ik_{z1,\mathrm{p}}z}, \tag{7.140}$$

where

$$f_z^{(p,m)}(\rho',z') = A\frac{\pi}{2}(-i)^m e^{-i\pi/4}\frac{k_{\rho,\mathrm{p}}}{k_{z1,\mathrm{p}}}J_m(k_{\rho,\mathrm{p}}\rho')e^{ik_{z1,\mathrm{p}}z'}\sqrt{\frac{2}{\pi k_{\rho,\mathrm{p}}}}\left(-\frac{k_{\rho,\mathrm{p}}^2}{k_1^2}\right), \tag{7.141}$$

and

$$G_{\rho\phi}^{(g,\mathrm{ff},m)}(\rho,z,\rho',z') = f_\phi^{(p,m)}(\rho',z')\frac{e^{ik_{\rho,\mathrm{p}}\rho}}{\sqrt{\rho}}e^{ik_{z1,\mathrm{p}}z}\frac{k_{\rho,\mathrm{p}}k_{z1,\mathrm{p}}}{k_1^2}\left(-\frac{k_{z1,\mathrm{p}}}{k_{\rho,\mathrm{p}}}\right), \tag{7.142}$$

$$G_{z\phi}^{(g,\mathrm{ff},m)}(\rho,z,\rho',z') = f_{\phi}^{(p,m)}(\rho',z')\frac{e^{ik_{\rho,\mathrm{p}}\rho}}{\sqrt{\rho}}e^{ik_{z1,\mathrm{p}}z}. \tag{7.143}$$

where

$$f_{\phi}^{(p,m)}(\rho',z') = Am\frac{\pi}{2}(-i)^m e^{-i\pi/4}\frac{J_m(k_{\rho,\mathrm{p}}\rho')}{k_{\rho,\mathrm{p}}\rho'}e^{ik_{z1,\mathrm{p}}z'}\sqrt{\frac{2}{\pi k_{\rho,\mathrm{p}}}}\left(-\frac{k_{\rho,\mathrm{p}}^2}{k_1^2}\right). \tag{7.144}$$

The guided-mode far-field Green's tensor components for an s-polarized guided mode are

$$G_{\phi\rho}^{(g,\mathrm{ff},m)}(\rho,z,\rho',z') = f_{\rho}^{(s,m)}(\rho',z')\frac{e^{ik_{\rho,\mathrm{p}}\rho}}{\sqrt{\rho}}e^{ik_{z1,\mathrm{p}}z}, \tag{7.145}$$

$$G_{\phi\phi}^{(g,\mathrm{ff},m)}(\rho,z,\rho',z') = f_{\phi}^{(s,m)}(\rho',z')\frac{e^{ik_{\rho,\mathrm{p}}\rho}}{\sqrt{\rho}}e^{ik_{z1,\mathrm{p}}z}, \tag{7.146}$$

where

$$f_{\rho}^{(s,m)}(\rho',z') = Am\frac{\pi}{2}(-i)^m e^{-i\pi/4}\frac{J_m(k_{\rho,\mathrm{p}}\rho')}{k_{\rho,\mathrm{p}}\rho'}e^{ik_{z1,\mathrm{p}}z'}\sqrt{\frac{2}{\pi k_{\rho,\mathrm{p}}}}\left(\frac{k_{\rho,\mathrm{p}}}{k_{z1,\mathrm{p}}}\right), \tag{7.147}$$

$$f_{\phi}^{(s,m)}(\rho',z') = A\frac{\pi}{2}(-i)^m e^{-i\pi/4}\frac{\partial J_m(k_{\rho,\mathrm{p}}\rho')}{\partial(k_{\rho,\mathrm{p}}\rho')}e^{ik_{z1,\mathrm{p}}z'}\sqrt{\frac{2}{\pi k_{\rho,\mathrm{p}}}}\left(-i\frac{k_{\rho,\mathrm{p}}}{k_{z1,\mathrm{p}}}\right), \tag{7.148}$$

and here the coefficient A is related to the pole of $r^{(s)}$.

By applying the electromagnetics boundary conditions the p-polarized guided-mode far field for a three-layer reference structure can now be constructed as

$$\mathbf{E}^{(g,\mathrm{ff},p)}(\rho,\phi,z) = \frac{e^{ik_{\rho,\mathrm{p}}\rho}}{\sqrt{\rho}}f^{(p)}(\phi)\times$$

$$\begin{cases} (\hat{\mathbf{z}}-\hat{\rho}\frac{k_{z1,\mathrm{p}}}{k_{\rho,\mathrm{p}}})e^{ik_{z1,\mathrm{p}}z}, & z>0 \\ A^{(p)}(\hat{\mathbf{z}}-\hat{\rho}\frac{k_{z,L2,\mathrm{p}}}{k_{\rho,\mathrm{p}}})e^{ik_{z,L2,\mathrm{p}}z}+B^{(p)}(\hat{\mathbf{z}}+\hat{\rho}\frac{k_{z,L2,\mathrm{p}}}{k_{\rho,\mathrm{p}}})e^{-ik_{z,L2,\mathrm{p}}z}, & 0>z>-d\,, \\ C^{(p)}(\hat{\mathbf{z}}+\hat{\rho}\frac{k_{z,L3,\mathrm{p}}}{k_{\rho,\mathrm{p}}})e^{-ik_{z,L3,\mathrm{p}}(z+d)}, & z<-d \end{cases} \tag{7.149}$$

where

$$f^{(p)}(\phi) = \sum_m f^{(p,m)}e^{im\phi}, \tag{7.150}$$

with

$$f^{(p,m)} = \int\left\{f_{\rho}^{(p,m)}(\rho',z')E_{\rho}^{(m)}(\rho',z')+f_{\phi}^{(p,m)}(\rho',z')E_{\phi}^{(m)}(\rho',z')+\right.$$

$$\left.f_{z}^{(p,m)}(\rho',z')E_{z}^{(m)}(\rho',z')\right\}k_0^2\Delta\varepsilon(\rho',z')\rho'd\rho'dz'. \tag{7.151}$$

The coefficients $A^{(p)}$, $B^{(p)}$ and $C^{(p)}$ are given in (6.135) and (6.136).

The s-polarized guided-mode far field for a three-layer structure can be constructed as

$$\mathbf{E}^{(g,\text{ff},s)}(\rho,\phi,z)=\frac{e^{ik_{\rho,\text{p}}\rho}}{\sqrt{\rho}}f^{(s)}(\phi)\hat{\phi}\begin{cases}e^{ik_{z1,\text{p}}z}, & z>0\\ A^{(s)}e^{ik_{z,L2,\text{p}}z}+B^{(s)}e^{-ik_{z,L2,\text{p}}z}, & 0>z>-d\,,\\ C^{(s)}e^{-ik_{z,L3,\text{p}}(z+d)}, & z<-d\end{cases}\tag{7.152}$$

where

$$f^{(s)}(\phi)=\sum_m f^{(s,m)}e^{im\phi},\tag{7.153}$$

with

$$f^{(s,m)}=\int\Big\{f_\rho^{(s,m)}(\rho',z')E_\rho^{(m)}(\rho',z')+$$
$$f_\phi^{(s,m)}(\rho',z')E_\phi^{(m)}(\rho',z')\Big\}k_0^2\Delta\varepsilon(\rho',z')\rho'd\rho'dz'.\tag{7.154}$$

The coefficients $A^{(s)}$, $B^{(s)}$ and $C^{(s)}$ are given in (6.132) and (6.133).

7.5 OPTICAL CROSS SECTIONS

The extinction, scattering, and absorption cross sections can be calculated by using the expressions (2.150)-(2.154). In this section these expressions will be formulated in a form suitable for a cylindrical harmonic expansion of the fields.

By inserting the field expression (7.3) into (2.154), and carrying out the integration with respect to ϕ, the absorption cross section becomes

$$\sigma_{\text{abs}}=\frac{2\pi k_0}{n_1|\mathbf{E}_{0,i}|^2}\sum_n\sum_\alpha\int\text{Imag}\{\varepsilon(\rho,z)\}|E_\alpha^{(n)}(\rho,z)|^2\rho d\rho dz.\tag{7.155}$$

By first finding the scattered far field as

$$\mathbf{E}_{\text{sc}}^{(\text{ff})}(r,\theta,\phi)=\sum_n\sum_\alpha\hat{\alpha}E_{\text{sc},\alpha}^{(\text{ff},n)}(r,\theta)e^{in\phi},\tag{7.156}$$

where

$$E_{\text{sc},\alpha}^{(\text{ff},n)}(r,\theta)=\sum_\beta\int G_{\alpha\beta}^{(\text{ff},n)}(r,\theta,\rho',z')k_0^2\Delta\varepsilon(\rho',z')E_\beta^{(n)}(\rho',z')\rho'd\rho'dz',\tag{7.157}$$

and then inserting this expression into (2.152) and (2.153), the out-of-plane scattering cross sections may be expressed as

$$\sigma_{\text{sc},r}=\frac{2\pi}{|\mathbf{E}_0|^2}\sum_n\sum_\alpha\int_{\theta=0}^{\pi/2}|E_{\text{sc},\alpha}^{(\text{ff},n)}(r,\theta)|^2r^2\sin\theta d\theta,\tag{7.158}$$

$$\sigma_{\text{sc},t}=\frac{n_{LN}}{n_1}\frac{2\pi}{|\mathbf{E}_0|^2}\sum_n\sum_\alpha\int_{\pi/2}^{\pi}|E_{\text{sc},\alpha}^{(\text{ff},n)}(r,\theta)|^2r^2\sin\theta d\theta.\tag{7.159}$$

When the incident light is a normally incident plane wave being polarized along the x-axis, the extinction cross section for reflection and transmission can be obtained by inserting the field component

$$E_{\mathrm{sc},x}^{(\mathrm{ff})}(r, \theta = \theta_0, \phi) = \sum_n E_{\mathrm{sc},\rho}^{(\mathrm{ff},n)}(r, \theta = \theta_0) \tag{7.160}$$

with $\theta_0 = 0$ or π into Eqs. (2.150) and (2.151). Here, it has been used that when $\theta = 0$ or π, the angle ϕ can be set to any value.

Guided-mode scattering cross sections for p-polarized modes can be obtained by inserting (7.149) and the corresponding magnetic field in (2.159). If it can be assumed that there are no absorption losses, then this results in the following expression for the scattering cross section:

$$\sigma_g^{(p)} = \frac{k_0 n_1}{k_{\rho,\mathrm{p}}} \frac{2\pi}{|\mathbf{E}_{0,i}|^2} \left\{ \frac{1}{2\mathrm{Imag}\{k_{z1,\mathrm{p}}\}} + \frac{\varepsilon_{L3}}{\varepsilon_1} \frac{|C^{(p)}|^2}{2\mathrm{Imag}\{k_{z,L3,\mathrm{p}}\}} + \frac{\varepsilon_{L2}}{\varepsilon_1} I^{(p)} \right\} \sum_m |f^{(p,m)}|^2, \tag{7.161}$$

where $I^{(p)}$ is given in (6.157).

In a similar way the s-polarized guided mode far field (7.152) can be used to find the cross section for scattering into s-polarized guided modes:

$$\sigma_g^{(s)} = \frac{k_{\rho,\mathrm{p}}}{k_0 n_1} \frac{2\pi}{|\mathbf{E}_{0,i}|^2} \left\{ \frac{1}{2\mathrm{Imag}\{k_{z1,\mathrm{p}}\}} + \frac{|C^{(s)}|^2}{2\mathrm{Imag}\{k_{z,L3,\mathrm{p}}\}} + I^{(s)} \right\} \sum_m |f^{(s,m)}|^2, \tag{7.162}$$

where

$$I^{(s)} = \int_{z=-d}^{0} |A^{(s)} e^{ik_{z,L2,\mathrm{p}} z} + B^{(s)} e^{-ik_{z,L2,\mathrm{p}} z}|^2 dz. \tag{7.163}$$

7.6 NUMERICAL APPROACH: RING ELEMENTS WITH RECTANGULAR CROSS SECTION

In the following a numerical approach is described where the cylindrically symmetric structure is discretized into ring-shaped elements with a rectangular cross section, and one cylindrical harmonic n will be considered at a time. The field in a given ring i will be assumed to be of the form

$$\mathbf{E}_i(\mathbf{r}) = \left(\hat{\rho} E_{\rho,i}^{(n)} + \hat{\phi} E_{\phi,i}^{(n)} + \hat{\mathbf{z}} E_{z,i}^{(n)} \right) e^{in\phi}, \tag{7.164}$$

with $E_{\alpha,i}^{(n)}$ assumed constant. The field components $E_\alpha^{(n)}$ entering in the integral equation (7.5) are thus assumed constant in rectangular-shaped regions in the ρz-plane.

A ring j with rectangular cross section is defined by a lower and upper value of ρ referred to as ρ_j^- and ρ_j^+, and a lower and upper value of z referred to as z_j^- and z_j^+, respectively. The field values belonging to ring j can be collected in a vector defined by

$$\overline{E}_j^{(n)} = \begin{bmatrix} E_{\rho,j}^{(n)} \\ E_{\phi,j}^{(n)} \\ E_{z,j}^{(n)} \end{bmatrix}. \tag{7.165}$$

At first it is sufficient to only discretize the regions where $\Delta\varepsilon \neq 0$ and calculate the fields $\overline{E}_j^{(n)}$ for those regions. The field may then be obtained elsewhere from the integral equation (7.5).

Matrix elements that govern the coupling between elements i and j can be defined in the following way:

$$M_{\alpha\beta,ij}^{(n)} = \int_{\rho'=\rho_j^-}^{\rho_j^+} \int_{z=z_j^-}^{z_j^+} G_{\alpha\beta}^{(n)}(\rho_i, z_i; \rho', z')\rho' d\rho' dz'. \tag{7.166}$$

The matrix elements can be further divided into a part $M_{\alpha\beta,ij}^{(d,n)}$ involving the direct Green's tensor, and a part $M_{\alpha\beta,ij}^{(i,n)}$ involving the indirect Green's tensor, with $M_{\alpha\beta,ij}^{(n)} = M_{\alpha\beta,ij}^{(d,n)} + M_{\alpha\beta,ij}^{(i,n)}$. The matrix elements can be collected in a 3×3 matrix defined as

$$\overline{\overline{M}}_{ij}^{(n)} = \begin{bmatrix} M_{\rho\rho,ij}^{(n)} & M_{\rho\phi,ij}^{(n)} & M_{\rho z,ij}^{(n)} \\ M_{\phi\rho,ij}^{(n)} & M_{\phi\phi,ij}^{(n)} & M_{\phi z,ij}^{(n)} \\ M_{z\rho,ij}^{(n)} & M_{z\phi,ij}^{(n)} & M_{zz,ij}^{(n)} \end{bmatrix}. \tag{7.167}$$

The dielectric constant of ring j is denoted ε_j. It is in principle possible to assign a tensor dielectric constant to each ring. However, for simplicity we consider a single value of the dielectric constant, and define the following matrix

$$\overline{\overline{\Delta\varepsilon}} = \begin{bmatrix} \overline{\overline{I}}_3(\varepsilon_1 - \varepsilon_{\text{ref},1}) & \overline{\overline{0}}_3 & \cdots & \overline{\overline{0}}_3 \\ \overline{\overline{0}}_3 & \overline{\overline{I}}_3(\varepsilon_2 - \varepsilon_{\text{ref},2}) & \cdots & \overline{\overline{0}}_3 \\ \vdots & \vdots & \ddots & \vdots \\ \overline{\overline{0}}_3 & \overline{\overline{0}}_3 & \cdots & \overline{\overline{I}}_3(\varepsilon_N - \varepsilon_{\text{ref},N}) \end{bmatrix}, \tag{7.168}$$

where $\overline{\overline{I}}_3$ is a 3×3 unit matrix, and $\overline{\overline{0}}_3$ is a 3×3 zero matrix.

The integral equation (7.5) can now be formulated in discrete form as:

$$\left(\overline{\overline{I}} - \begin{bmatrix} \overline{\overline{M}}_{11} & \overline{\overline{M}}_{12} & \cdots & \overline{\overline{M}}_{1N} \\ \overline{\overline{M}}_{21} & \overline{\overline{M}}_{22} & \cdots & \overline{\overline{M}}_{2N} \\ \vdots & \vdots & \ddots & \vdots \\ \overline{\overline{M}}_{N1} & \overline{\overline{M}}_{N2} & \cdots & \overline{\overline{M}}_{NN} \end{bmatrix} \overline{\overline{\Delta\varepsilon}} k_0^2 \right) \begin{bmatrix} \overline{E}_1^{(n)} \\ \overline{E}_2^{(n)} \\ \vdots \\ \overline{E}_N^{(n)} \end{bmatrix} = \begin{bmatrix} \overline{E}_{0,1}^{(n)} \\ \overline{E}_{0,2}^{(n)} \\ \vdots \\ \overline{E}_{0,N}^{(n)} \end{bmatrix}. \tag{7.169}$$

Note that in Eq. (7.166) the term $G_{\alpha\beta}^{(n)}$ itself requires an integration over μ as can be seen from the expressions (7.47)-(7.55) and (7.64)-(7.72). It is convenient, however, when calculating $M_{\alpha\beta,ij}^{(d,n)}$ that the integration with respect to z' can be carried out analytically as follows:

$$Z_{ij}(\mu) = \int_{z_j^-}^{z_j^+} e^{ik_{z1}|z-z'|}dz' = \begin{cases} \frac{-1}{ik_{z1}}\left(e^{ik_{z1}(z_i-z_j^+)} - e^{ik_{z1}(z_i-z_j^-)}\right), & z_i > z_j^+, z_j^- \\ \frac{1}{ik_{z1}}\left(e^{ik_{z1}|z_i-z_j^+|} - e^{ik_{z1}|z_i-z_j^-|}\right), & z_i < z_j^+, z_j^- \\ \frac{2}{ik_{z1}}\left(e^{ik_{z1}(z_j^+-z_j)} - 1\right), & z_i = z_j \end{cases}. \tag{7.170}$$

When calculating $M_{\alpha\beta,ij}^{(i,n)}$ the integration with respect to z' can also be carried out analytically using

$$\tilde{Z}_{ij}(\mu) = \int_{z_j^-}^{z_j^+} e^{ik_{z1}(z+z')}dz' = \frac{1}{ik_{z1}}e^{ik_{z1}z}\left(e^{ik_{z1}z_j^+} - e^{ik_{z1}z_j^-}\right). \tag{7.171}$$

In addition, for $n = \pm 1$ some integrals with respect to ρ' can also be calculated analytically:

$$R_j^{(1)}(\mu) = \int_{\rho_j^-}^{\rho_j^+} J_1(\mu\rho')d\rho' = \frac{-1}{\mu}\left(J_0(\mu\rho_j^+) - J_0(\mu\rho_j^-)\right), \tag{7.172}$$

$$R_j^{(2)}(\mu) = \int_{\rho_j^-}^{\rho_j^+} \frac{\partial J_1(\mu\rho')}{\partial\rho'}\rho' d\rho' =$$

$$\left(J_1(\mu\rho_j^+)\rho_j^+ - J_1(\mu\rho_j^-)\rho_j^- + \frac{1}{\mu}J_0(\mu\rho_j^+) - \frac{1}{\mu}J_0(\mu\rho_j^-)\right), \tag{7.173}$$

$$R_j^{(3)}(\mu) = \int_{\rho_j^-}^{\rho_j^+} J_1(\mu\rho')\rho' d\rho' =$$

$$-\frac{1}{\mu}J_0(\mu\rho_j^+)\rho_j^+ + \frac{1}{\mu}J_0(\mu\rho_j^-)\rho_j^- + \frac{1}{\mu}\int_{\rho_j^-}^{\rho_j^+} J_0(\mu\rho')d\rho'. \tag{7.174}$$

The latter expression $R_j^{(3)}$ cannot be reduced further in terms of any simple analytic functions and must be carried out numerically. The integral in (7.174) can, for example, be handled as a sum of 5-10 terms. Note that Z_{ij} is a function of μ since $k_{z1} = \sqrt{k_1^2 - \mu^2}$.

7.7 EXAMPLE: NANOCYLINDER ON A LAYERED STRUCTURE

In this section the cylindrical GFVIEM is applied to calculate scattering by a single dielectric cylinder placed on a two- or three-layer reference geometry. The dielectric cylinder is illuminated by a normally incident plane wave (Fig. 7.1(a)). A schematic of the structure in the ρz-plane is shown in Fig. 7.1(b).

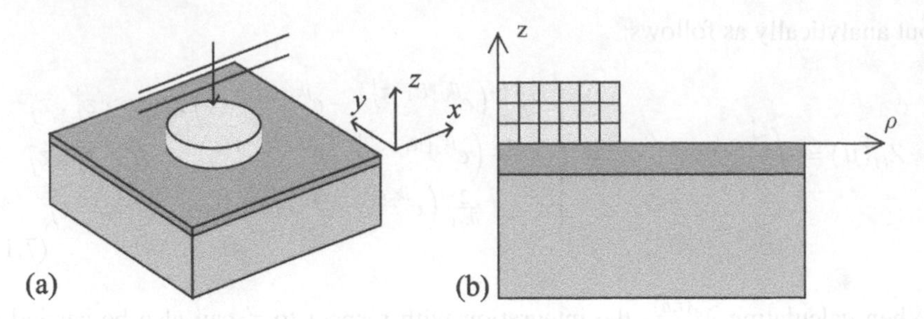

Figure 7.1 (a) Schematic of a nanocylinder placed on a three-layer reference geometry. (b) Schematic of the structure in the ρz-plane.

The reference field for $z > 0$ including also the reflected wave is given by

$$\mathbf{E}_0(\mathbf{r}) = \hat{\mathbf{x}} E_0 \left(e^{-ik_1 z} + r^{(s)}(0) e^{ik_1 z} \right), \quad z > 0. \tag{7.175}$$

The angular momentum components of the incident field $E_{0,\alpha}^{(n)}(\rho, z)$ can be generally obtained using the procedure

$$E_{0,\alpha}^{(n)}(\rho, z) = \frac{1}{2\pi} \int_{\phi=0}^{2\pi} \mathbf{E}_0(\mathbf{r}) \cdot \hat{\alpha} e^{-in\phi} d\phi. \tag{7.176}$$

This leads to

$$E_{0,\rho}^{(\pm 1)}(\rho, z) = \frac{1}{2} E_0 \left(e^{-ik_1 z} + r^{(s)}(0) e^{ik_1 z} \right), \quad z > 0, \tag{7.177}$$

$$E_{0,\phi}^{(\pm 1)}(\rho, z) = \pm \frac{i}{2} E_0 \left(e^{-ik_1 z} + r^{(s)}(0) e^{ik_1 z} \right), \quad z > 0, \tag{7.178}$$

$$E_{0,z}^{(\pm 1)}(\rho, z) = 0. \tag{7.179}$$

All other $E_{0,\alpha}^{(n)}$ are equal to zero. Thus, for the chosen incident field it is sufficient to consider the angular momenta $n = \pm 1$. Furthermore, from symmetry, the result for $n = -1$ can be obtained from the result for $n = +1$, and it is thus sufficient to solve the integral equation (7.5) for a single angular momentum. The three-dimensional problem is thereby effectively reduced to a single two-dimensional problem.

7.7.1 CYLINDRICAL SCATTERER ON A DIELECTRIC SUBSTRATE

The first example is a cylindrical amorphous-silicon scatterer with refractive index 3.918, height 50 nm, and radius 50 nm, which is placed on a quartz substrate ($n_{L2} = n_{L3} = 1.5$), and which is being illuminated by a normally incident plane wave polarized along the x-axis with wavelength 800 nm. The scatterer is discretized in the ρz-plane into 20×20 square-shaped elements. The calculated differential scattering

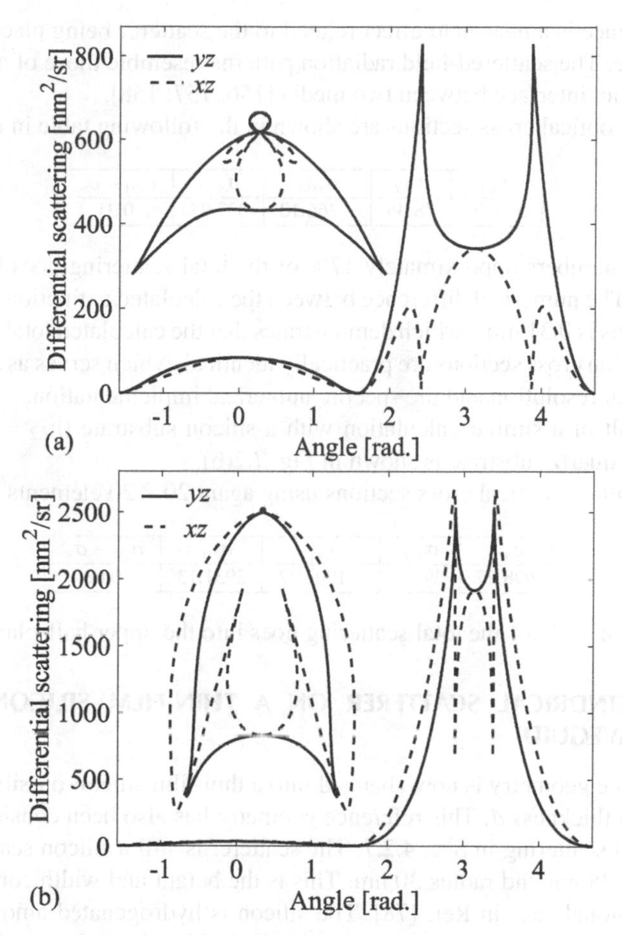

Figure 7.2 (a) Differential scattering cross section for a cylinder with refractive index $n_2 = 3.918$, radius 50 nm, and height 50 nm, being placed on a quartz substrate ($n_{L2} = n_{L3} = 1.5$), and being illuminated by a normally incident x-polarized plane wave. Inset: polar plot of differential scattering cross section. (b) Similar calculation for a silicon substrate.

cross section for the yz- and xz-planes is shown in Fig. 7.2(a). The angles from $-\pi/2$ to $+\pi/2$ correspond to light propagating into the upper half-plane, and angles from $\pi/2$ to $3\pi/2$ correspond to light propagating into the lower half-plane. It is noticed that most of the scattered light goes into the substrate. The differential cross section for the xz-plane is quite similar to the result shown in Fig. 4.33 for two-dimensional scattering of p-polarized light by a nanostrip on a quartz substrate. Here, the light scattered into the xz-plane is also p-polarized. The scattering for the yz-plane is, on the other hand, s polarized, and the differential scattering is quite different. In both cases scattering into the lower half-plane at angles above the critical angle for an air-quartz interface is a near-field effect related to the scatterer being placed directly on the substrate. The scattered-field radiation patterns resemble those of a single dipole placed near an interface between two media [156, 157, 158].

The total optical cross sections are shown in the following table in units of nm^2:

$\sigma_{sc,t}$	$\sigma_{sc,r}$	$\sigma_{ext,t}$	$\sigma_{ext,r}$	$\sigma_{ext} - \sigma_{sc}$
393.20	78.36	−366.10	837.97	0.31

From these numbers approximately 17% of the total scattering goes into the upper half-plane. The numerical difference between the calculated extinction and scattering cross sections is 0.31 nm^2, which demonstrates that the calculated total scattering and total extinction cross sections are practically identical, which serves as a useful check of the chosen resolution and the specific numerical implementation.

The result of a similar calculation with a silicon substrate ($n_{L2} = n_{L3} = 3.918$) instead of a quartz substrate is shown in Fig. 7.2(b).

The calculated optical cross sections using again 20×20 elements are:

$\sigma_{sc,t}$	$\sigma_{sc,r}$	$\sigma_{ext,t}$	$\sigma_{ext,r}$	$\sigma_{ext} - \sigma_{sc}$
998.27	29.95	−1902.22	2931.13	0.68

Thus, now only 3% of the total scattering goes into the upper half-plane.

7.7.2 CYLINDRICAL SCATTERER ON A THIN-FILM SILICON-ON-SILVER WAVEGUIDE

The reference geometry is now changed into a thin-film silicon-on-silver waveguide with silicon thickness d. This reference geometry has also been considered for two-dimensional scattering in Sec. 4.2.7. The scatterer is still a silicon scatterer but now with height 25 nm and radius 30 nm. This is the height and width considered in the two-dimensional case in Ref. [78]. The silicon is hydrogenated amorphous silicon (aSi:H) with refractive index n_{Si} as given in Ref. [78]. The refractive index of silver n_{Ag} is as given in Ref. [40]. Thus, the refractive indices are: $n_1 = 1$, $n_2 = n_{L2} = n_{Si}$, and $n_{L3} = n_{Ag}$.

The first waveguide thickness to be considered is $d = 50$ nm. Optical cross-section spectra for this case calculated using 12×10 elements are shown in Fig. 7.3. Absorption losses are here ignored by neglecting the imaginary part of n_{Si} and the real part of n_{Ag}. In this case the scattering into guided modes is perfectly well defined. Guided modes only exist for s polarization for wavelengths λ below approximately 1000 nm. The scattering cross section for s-polarized guided modes is thus only non-zero when $\lambda < 1000$ nm. The peak in the extinction at wavelengths around 1000 nm is thus close to this cut-off wavelength. Two-dimensional calculations for p polarization and a silicon strip on the same waveguide also lead to an extinction peak at a wavelength near 1000 nm [78]. Contrary to the two-dimensional case considered in Sec. 4.2.7, both s- and p-polarized guided modes are here excited at the same time for $\lambda < 1000$ nm. For wavelengths beyond 1000 nm, the only guided mode that light

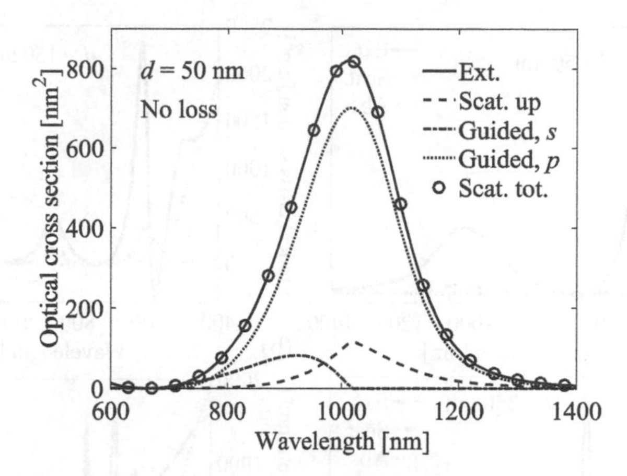

Figure 7.3 Optical cross sections for a cylindrical silicon scatterer with height 25 nm and radius 30 nm placed on a thin-film silicon-on-silver waveguide with silicon layer thickness 50 nm. Absorption losses are ignored.

can be coupled into is the p-polarized surface plasmon polariton. For the considered wavelengths the out-of-plane scattering (Scat. up. or $\sigma_{sc,r}$) is relatively small compared with scattering into guided modes.

Notice that the extinction is in good agreement with the sum of out-of-plane scattering and scattering into s- and p-polarized guided modes (Scat. tot.), which can be used as a check. Consider the wavelength 800 nm. The refractive indices in this case are $n_1 = 1$, $n_2 = n_{L2} = 3.918$, and $n_{L3} = i5.5616$. With these parameters the reference geometry supports one s-polarized and one p-polarized guided mode, which can also be seen from Figs. 4.48 and 4.49. The calculated cross sections for scattering into the upper half-plane, and into the s- and p-polarized guided modes, and the extinction cross section, are shown in the following table for different numbers of discretization elements $N_\rho \times N_z$:

$N_\rho \times N_z$	$\sigma_{sc,r}$	$\sigma_{g(s)}$	$\sigma_g^{(p)}$	$\sigma_{ext,r}$	$\sigma_{ext} - \sigma_{sc}$
6×5	1.93	33.04	49.17	84.31	0.17
12×10	2.12	36.09	54.28	92.54	0.05
24×20	2.19	37.26	56.13	95.59	0.01

The total scattering and the extinction are in good agreement, and more so with increasing numbers of elements.

Optical cross-section spectra for four different waveguide thicknesses d = 50 nm, 150 nm, 250 nm, and 500 nm, are shown in Fig. 7.4 for a broader wavelength range and with absorption losses included. In this case, for the wavelengths with small absorption in the materials ($\lambda > 700$ nm) the excitation of waveguide modes can be estimated as the difference between extinction and out-of-plane scattering. Notice

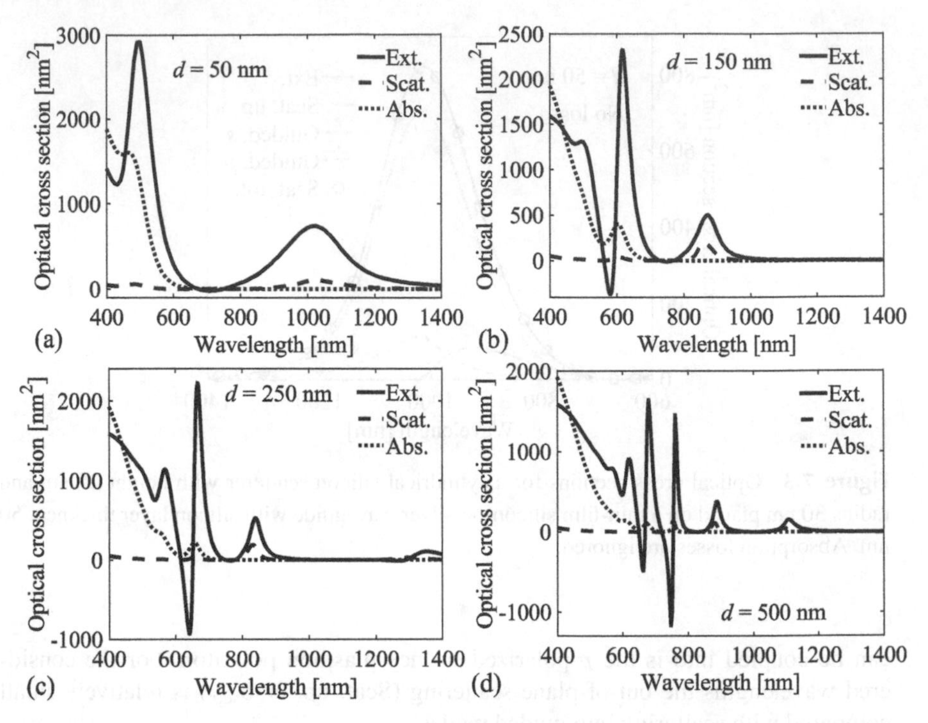

Figure 7.4 Optical cross sections for a cylindrical silicon scatterer with height 25 nm and radius 30 nm placed on a thin-film silicon-on-silver waveguide with silicon layer thickness d. (a) d = 50 nm, (b) d = 150 nm, (c) d = 250 nm, and (d) d = 500 nm.

that in some cases the extinction is negative. This can occur when losses are significant and the waveguide thickness is optimum for antireflection by interference effects. By adding the nanocylinder the thickness is locally changed into one without antireflection, which has the effect that the reflected beam power is increased and the extinction becomes negative. Also notice that with increasing waveguide thickness the number of peaks in spectra increases similar to Sec. 4.2.7.

7.8 EXAMPLE: MICROSTRUCTURED GRADIENT-INDEX LENS

The GRIN lens that has previously been considered as a two-dimensional geometry in Sec. 4.2.8 is in this section re-examined as a three-dimensional geometry. A series of silicon rings of height 100 μm and varying width are placed on a silicon layer of thickness $d = 400$ μm. The geometry is shown to scale in Fig. 7.5. The right half of the structure in Sec. 4.2.8 and the structure here in the ρz-plane shown in Fig. 7.5(b) are identical.

The source is a dipole polarized along x placed just below the silicon layer. The corresponding reference field is discussed in Sec. 7.8.1. The calculation of total emit-

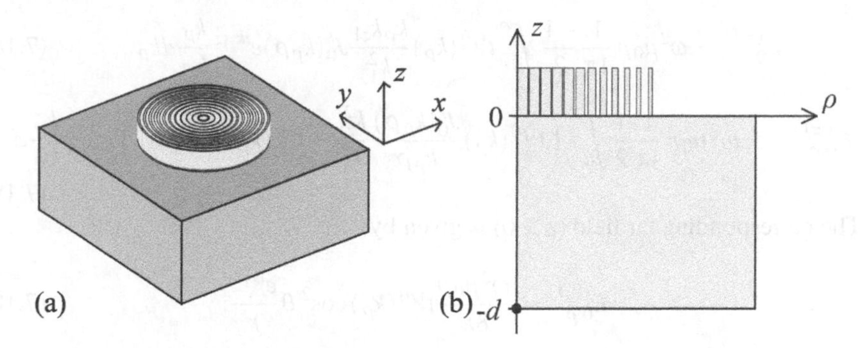

Figure 7.5 (a) Schematic of silicon GRIN lens placed on a silicon film. (b) Schematic of the structure in the ρz-plane.

ted power, the power emitted into the upper and lower half-planes, and the power going into guided modes, is discussed in Sec. 7.8.2. Finally, the differential and total emitted power is presented in Sec. 7.8.3 for a dipole emitting at 1 THz and 0.5 THz.

7.8.1 DIPOLE REFERENCE FIELD

The reference field that will be considered is the field generated by a point-dipole with dipole moment $\mathbf{p} = \hat{\mathbf{x}}p$ placed just below the silicon film on the axis of symmetry, i.e., at $\mathbf{r}_0 = -\hat{\mathbf{z}}d$:

$$\mathbf{E}_0(\mathbf{r}) = \omega^2 \mu_0 \mathbf{G}(\mathbf{r}, \mathbf{r}_0) \cdot \hat{\mathbf{p}}. \tag{7.180}$$

The case of a dipole embedded in a dielectric slab has been considered in [159]. For $z > 0$ the field can be obtained by slightly modifying the Green's tensor in (6.99) to take into account that here the source point is below the film, and the observation point is above the film, whereas (6.99) considers the reverse situation. In addition, since the dipole is placed on the axis of symmetry it applies that $\rho_r = \rho$, $\hat{\rho}_r = \hat{\rho}$, and $\hat{\phi}_r = \hat{\phi}$ (see (6.99)). Thus,

$$\mathbf{E}_0(\mathbf{r}) = \omega^2 \mu_0 p \frac{i}{4\pi} \int_0^\infty \Bigg\{ t^{(p)}(k_\rho) \Bigg(\hat{\phi} \sin\phi \frac{J_0'(k_\rho \rho)}{k_\rho \rho} \frac{k_{z,1}^2}{k_1^2} - \hat{\rho} \cos\phi J_0''(k_\rho \rho) \frac{k_{z,1}^2}{k_1^2} + \hat{\mathbf{z}} \cos\phi \frac{i k_\rho k_{z,1}}{k_1^2} J_0'(k_\rho \rho) \Bigg) - t^{(s)}(k_\rho) \Bigg(-\hat{\phi} \sin\phi J_0''(k_\rho \rho) + \hat{\rho} \cos\phi \frac{J_0'(k_\rho \rho)}{k_\rho \rho} \Bigg) \Bigg\} e^{i k_{z,1} z} \frac{k_\rho}{k_{z,1}} dk_\rho, \quad z > 0. \tag{7.181}$$

The angular momentum components of the reference field ($z > 0$) are thus given by:

$$E_{0,\rho}^{(\pm 1)} = \omega^2 \mu_0 p \frac{i}{4\pi} \frac{-1}{2} \int_0^\infty \left(t^{(p)}(k_\rho) J_0''(k_\rho \rho) \frac{k_{z1}^2}{k_1^2} + t^{(s)}(k_\rho) \frac{J_0'(k_\rho \rho)}{k_\rho \rho} \right) e^{i k_{z1} z} \frac{k_\rho}{k_{z1}} dk_\rho, \tag{7.182}$$

$$E_{0,z}^{(\pm 1)} = \omega^2 \mu_0 p \frac{1}{4\pi} \frac{-1}{2} \int_0^\infty t^{(p)}(k_\rho) \frac{k_\rho k_{z1}}{k_1^2} J_0'(k_\rho \rho) e^{ik_{z1}z} \frac{k_\rho}{k_{z1}} dk_\rho, \tag{7.183}$$

$$E_{0,\phi}^{(\pm 1)} = \pm \omega^2 \mu_0 p \frac{1}{4\pi} \frac{1}{2} \int_0^\infty \left(t^{(p)}(k_\rho) \frac{J_0'(k_\rho \rho)}{k_\rho \rho} \frac{k_{z1}^2}{k_1^2} + t^{(s)}(k_\rho) J_0''(k_\rho \rho) \right) e^{ik_{z1}z} \frac{k_\rho}{k_{z1}} dk_\rho. \tag{7.184}$$

The corresponding far field ($z > 0$) is given by:

$$E_{0,\rho}^{(\mathrm{ff},1)} = \frac{\omega^2 \mu_0 p}{8\pi} t^{(p)}(k_\rho) \cos^2\theta \frac{e^{ik_1 r}}{r}, \tag{7.185}$$

$$E_{0,z}^{(\mathrm{ff},1)} = -\frac{\omega^2 \mu_0 p}{8\pi} t^{(p)}(k_\rho) \cos\theta \sin\theta \frac{e^{ik_1 r}}{r}, \tag{7.186}$$

$$E_{0,\phi}^{(\mathrm{ff},1)} = i \frac{\omega^2 \mu_0 p}{8\pi} t^{(s)}(k_\rho) \frac{e^{ik_1 r}}{r}, \tag{7.187}$$

where here $k_\rho = k_1 \sin\theta$.

For $z < -d$ the reference field is given by

$$\mathbf{E}_0(\mathbf{r}) = \mathbf{E}_0^{(d)}(\mathbf{r}) + \mathbf{E}_0^{(i)}(\mathbf{r}), \tag{7.188}$$

where the direct contribution is related to the direct Green's tensor (6.74) and is given by

$$\mathbf{E}_0^{(d)}(\mathbf{r}) = \omega^2 \mu_0 p \frac{i}{4\pi} \int_0^\infty \left\{ -\hat{\boldsymbol{\phi}} \sin\phi \left(J_0(k_\rho \rho) + \frac{k_\rho^2}{k_1^2} \frac{J_0'(k_\rho \rho)}{k_\rho \rho} \right) \right.$$
$$+ \hat{\boldsymbol{\rho}} \cos\phi \left(J_0(k_\rho \rho) + \frac{k_\rho^2}{k_1^2} J_0''(k_\rho \rho) \right)$$
$$\left. - \hat{\mathbf{z}} \cos\phi \frac{ik_{z1} k_\rho}{k_1^2} J_0'(k_\rho \rho) \right\} e^{-ik_{z,1}(z+d)} \frac{k_\rho}{k_{z,1}} dk_\rho, \quad z < -d. \tag{7.189}$$

The corresponding angular momentum components of the direct contribution can be simplified into

$$E_{0,z}^{(d,1)}(\rho, z) = -\frac{1}{2} \omega^2 \mu_0 p \frac{(z+d)\rho}{R^2} \left(1 + \frac{3i}{k_1 R} - \frac{3}{(k_1 R)^2} \right) \frac{e^{ik_1 R}}{4\pi R}, \tag{7.190}$$

$$E_{0,\rho}^{(d,1)}(\rho, z) = \frac{1}{2} \omega^2 \mu_0 p \left[\left(1 + \frac{i}{k_1 R} - \frac{1}{(k_1 R)^2} \right) - \frac{\rho^2}{R^2} \left(1 + \frac{3i}{k_1 R} - \frac{3}{(k_1 R)^2} \right) \right] \frac{e^{ik_1 R}}{4\pi R}, \tag{7.191}$$

$$E_{0,\phi}^{(d,1)}(\rho, z) = \frac{i}{2} \omega^2 \mu_0 p \left(1 + \frac{i}{k_1 R} - \frac{1}{(k_1 R)^2} \right) \frac{e^{ik_1 R}}{4\pi R}, \tag{7.192}$$

where $R = \sqrt{(z+d)^2 + \rho^2}$.

The indirect contribution is related to (6.98) and is given by

$$\mathbf{E}_0^{(i)}(\mathbf{r}) = \omega^2\mu_0 p\frac{i}{4\pi}\int_0^\infty \Bigg\{ r^{(p)}(k_\rho)\Bigg(-\hat{\phi}\sin\phi\frac{J_0'(k_\rho\rho)}{k_\rho\rho}\frac{k_{z,1}^2}{k_1^2} +$$

$$\hat{\rho}\cos\phi J_0''(k_\rho\rho)\frac{k_{z,1}^2}{k_1^2} + \hat{\mathbf{z}}\cos\phi\frac{ik_\rho k_{z,1}}{k_1^2}J_0'(k_\rho\rho)\Bigg)$$

$$-r^{(s)}(k_\rho)\Bigg(-\hat{\phi}\sin\phi J_0''(k_\rho\rho) + \hat{\rho}\cos\phi\frac{J_0'(k_\rho\rho)}{k_\rho\rho}\Bigg)\Bigg\} e^{-ik_{z,1}(z+d)}\frac{k_\rho}{k_{z,1}}dk_\rho,\quad z<-d. \tag{7.193}$$

The corresponding far-field angular momentum components ($z<-d$) are given by

$$E_{0,\rho}^{(\mathrm{ff},1)} = \frac{\omega^2\mu_0 p}{8\pi}\left(1 - r^{(p)}(k_\rho)\right)\cos^2\theta e^{-ik_{z,1}d}\frac{e^{ik_1 r}}{r}, \tag{7.194}$$

$$E_{0,z}^{(\mathrm{ff},1)} = \frac{\omega^2\mu_0 p}{8\pi}\left(-1 + r^{(p)}(k_\rho)\right)\cos\theta\sin\theta e^{-ik_{z,1}d}\frac{e^{ik_1 r}}{r}, \tag{7.195}$$

$$E_{0,\phi}^{(\mathrm{ff},1)} = i\frac{\omega^2\mu_0 p}{8\pi}\left(1 + r^{(s)}(k_\rho)\right)e^{-ik_{z1}d}\frac{e^{ik_1 r}}{r}, \tag{7.196}$$

where here, again $k_\rho = k_1\sin\theta$, and $k_{z1} = \sqrt{k_0^2\varepsilon_1 - k_\rho^2}$.

7.8.2 CALCULATION OF EMITTED POWER

An expression for the total power emitted by the dipole, assuming the dipole moment $\mathbf{p}$ is purely real, can be obtained by using (2.41). Thus,

$$P_{\mathrm{tot}} = \frac{\omega p}{2}\mathrm{Imag}\{\mathbf{E}(\mathbf{r}_0)\cdot\hat{\mathbf{x}}\}, \tag{7.197}$$

where $\mathbf{E}(\mathbf{r}) = \mathbf{E}_0(\mathbf{r}) + \mathbf{E}_{\mathrm{sc}}(\mathbf{r})$.

By using (7.189) and (7.193) it can be straightforwardly shown that

$$\mathrm{Imag}\{\mathbf{E}_0^{(d)}(\mathbf{r}_0)\cdot\hat{\mathbf{x}}\} = \frac{\omega^2\mu_0 p k_1}{6\pi}, \tag{7.198}$$

$$\mathrm{Imag}\{\mathbf{E}_0^{(i)}(\mathbf{r}_0)\cdot\hat{\mathbf{x}}\} = \frac{\omega^2\mu_0 p}{8\pi}\mathrm{Imag}\left\{ i\int_0^\infty \left(-r^{(p)}(k_\rho)\frac{k_{z1}^2}{k_1^2} + r^{(s)}(k_\rho)\right)\frac{k_\rho}{k_{z1}}dk_\rho\right\}. \tag{7.199}$$

Note that when all refractive indices are real-valued and positive, the upper integration limit in (7.199) can be set to $k_0 n_{\max}$, where $n_{\max}$ is the largest of the refractive indices. When the reflection coefficients have poles this can be handled by changing the integration path along the real axis into a path going into the complex plane similar to what has been described several times in the book for the calculation of, for example, the indirect Green's tensor.

For the specific situation, the contribution from the scattered field can be calculated using, for example,

$$\mathbf{E}_{sc}(\mathbf{r}_0)\cdot\hat{\mathbf{x}} = 2\sum_{\alpha}\int G_{\rho\alpha}^{(t,1)}(0,-d,\rho',z')E_{\alpha}^{(1)}(\rho',z')$$

$$\times k_0^2\left(\varepsilon(\rho',z') - \varepsilon_{\text{ref}}(\rho',z')\right)\rho' d\rho' dz'. \tag{7.200}$$

In the limit $\rho \to 0$ being considered here, the expressions for $G_{\rho\alpha}^{(t,1)}$ given in Sec. 7.2.2 can be greatly simplified.

The total radiation into the upper and lower half-plane is given by

$$P_{\text{OUP}} = 2\pi\frac{1}{2}\frac{k_1}{\omega\mu_0}\sum_n\sum_\alpha\int_{\theta=0}^{\pi}|E_\alpha^{(\text{ff},n)}(r,\theta)|r^2\sin\theta d\theta. \tag{7.201}$$

For the upper half-plane ($0 \leq \theta < \pi/2$) the field components $E_\alpha^{(\text{ff},n)}$ consist of the reference field (7.185)-(7.187) and a contribution from the scattered field. For the lower half-plane ($\pi/2 < \theta \leq \pi$) the reference far field is given in (7.194)-(7.196).

The power scattered into the guided modes of the waveguide film can now be obtained as

$$P_g = P_{\text{tot}} - P_{\text{OUP}}. \tag{7.202}$$

Note that if $0 < n_{L2} < n_1 = n_{L3}$, then there are no guided modes in the reference structure, and $P_{\text{tot}} = P_{\text{OUP}}$, which can be used as a check.

7.8.3 EMISSION PATTERNS AND EMITTED POWERS

The emission patterns for the xz- and yz-planes for a dipole emitting at 1 THz, and being polarized along x and placed below the GRIN lens geometry, are shown in Fig. 7.6(a,b). The differential emission has been normalized with the total emission calculated from the local field at the dipole position. The angles from $-\pi/2$ to $\pi/2$ correspond to emission into the upper half-plane, while the angles from $\pi/2$ to $3\pi/2$ correspond to emission into the lower half-plane. The emission into the upper half-plane is significantly stronger than the emission into the lower half-plane. The radiation in the xz-plane is p polarized. Here, the emission is mainly going in the forward direction (along z being equivalent to the angle 0) except for some quite small side lobes. The emission in the yz-plane is s polarized, and here for the emission into the upper half-plane, the peak of the side lobes is slightly higher than the radiation along the z-axis.

The normalized total emitted power P_{tot} calculated from the local field at the dipole position, the power emitted into the upper half-plane P_{up}, the power emitted into the lower half-plane P_{down}, and the power coupled into guided modes $P_g = P_{\text{tot}} - P_{\text{up}} - P_{\text{down}}$, are shown in the following table as calculated with different numbers of discretization elements:

$N_\rho \times N_z$	P_{tot}	P_{up}	P_{down}	P_g
29×10	3.308	0.697	0.370	2.241
58×20	3.329	0.701	0.358	2.228

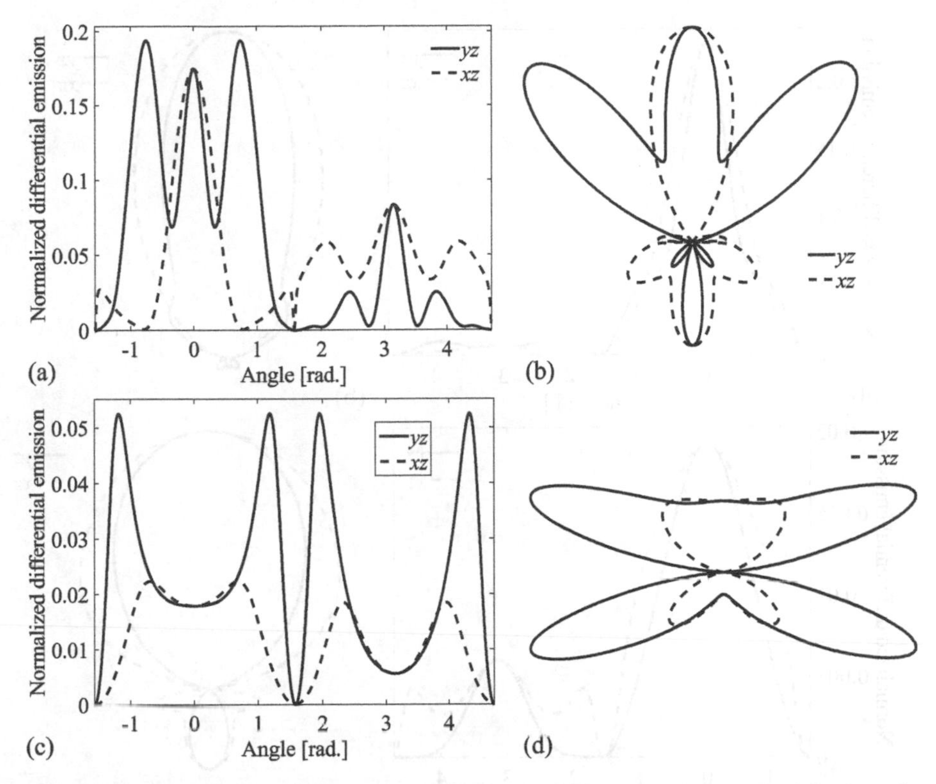

Figure 7.6 (a) Emission pattern for the xz-plane and yz-plane for an x-polarized dipole emitting at 1 THz placed just below the geometry in Fig. 7.5 with a GRIN lens on a 400 μm silicon film. (b) Polar plot of the emission pattern in (a). (c,d) Emission pattern and polar plot when the GRIN lens is absent and the dipole is placed just below a bare 400-μm silicon film.

The powers have been normalized with the total power emitted from the same dipole placed in free space. The total emitted power is seen to be enhanced by a factor 3.3 compared with a dipole in free space, which is due to the Purcell effect discussed in Sec. 5.8. Roughly 67% of the total emitted power is coupled into guided modes, while 21% is coupled into the upper half-plane.

For comparison, the emission patterns are shown in Fig. 7.6(c,d) for the case where the GRIN lens is absent, and the dipole is placed just below a bare silicon film of thickness 400 μm. The corresponding normalized powers are:

P_{tot}	P_{up}	P_{down}	P_{g}
3.572	0.288	0.249	3.035

Here, 85% of the light is coupled into guided modes, while only 8% is coupled into the upper half-plane. The GRIN lens does, therefore, significantly improve the fraction of emission into the upper half-plane.

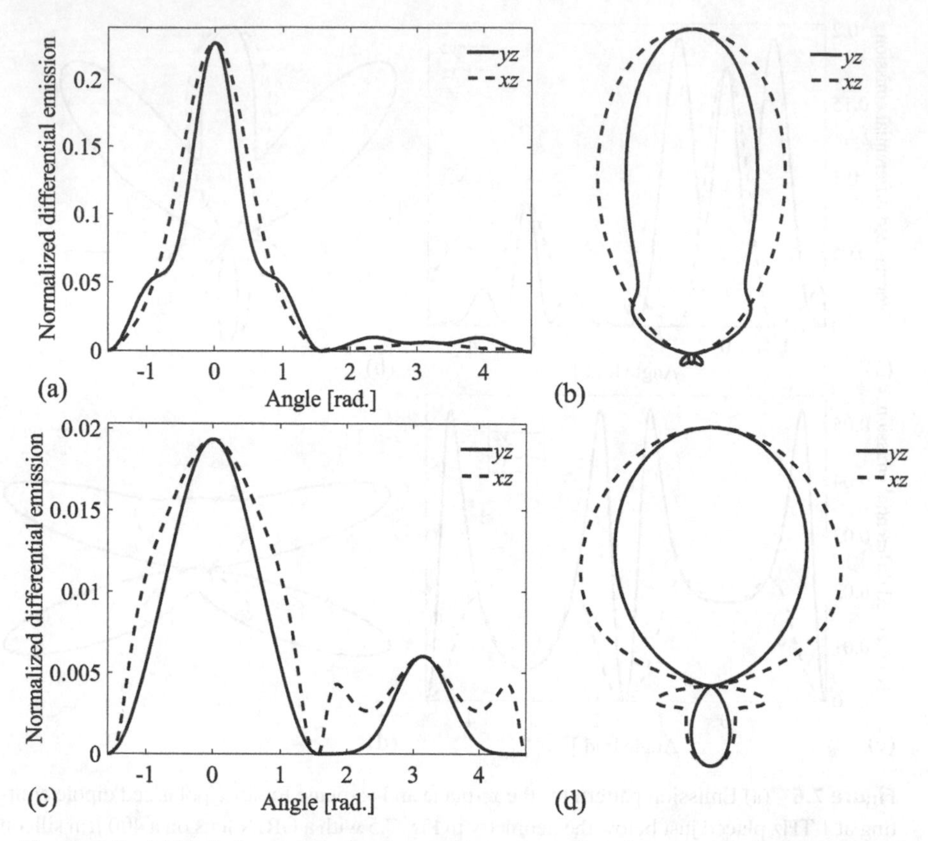

Figure 7.7 (a) Emission pattern for the xz-plane and yz-plane for an x-polarized dipole emitting at 0.5 THz placed just below the geometry in Fig. 7.5 with a GRIN lens on a 400-μm silicon film. (b) Polar plot of the emission pattern in (a). (c,d) Emission pattern and polar plot when the GRIN lens is absent.

One more example is given in Fig. 7.7(a,b) with the same GRIN lens for the frequency 0.5 THz, and in Fig. 7.7(c,d) at 0.5 THz when the GRIN lens is absent. Here, the normalized powers with and without the GRIN lens are ($N_\rho \times N_z = 58 \times 20$):

Lens / no lens	P_{tot}	P_{up}	P_{down}	P_{g}
Lens	2.767	0.562	0.045	2.160
No lens	3.303	0.103	0.027	3.173

7.9 GUIDELINES FOR SOFTWARE IMPLEMENTATION

Structure of GFVIEM program for cylindrical symmetry and ring elements

1. Define the reference structure (n_1, n_{L2}, n_{L3}, and d for a three-layer reference structure), scatterer geometry, and refractive index (n_2). Define the wavelength and the cylindrical harmonic n to be considered. For a calculation of scattering into guided modes also the mode indices of s- and p-polarized guided modes must be obtained.
2. Divide the scatterer into ring elements. Construct arrays with the sampling points (ρ_i, z_i) within elements in the ρz-plane. For a simple geometry, such as a single cylindrical scatterer, an option is to use the same height and width of all elements in the ρz-plane. For a more complex geometry, such as the GRIN lens, it is favorable to choose the width of elements to match the ring widths. This choice can also be stored in arrays.
3. Calculate the reference field at the sampling points.
4. Calculate the matrix elements $M^{(d,n)}_{\alpha\beta,ij}$ and $M^{(i,n)}_{\alpha\beta,ij}$ and construct the matrix equation (7.169). When calculating the matrix elements, the expressions (7.170)-(7.174) can be applied to handle part of the integration analytically. It is thus sufficient to integrate with respect to one parameter μ. Matrix elements that are identical should preferably only be calculated once. For the examples in this chapter, $M^{(d,n)}_{\alpha\beta,ij} = M^{(d,n)}_{\alpha\beta,i'j'}$ when $\rho_i = \rho_{i'}$, $\rho_j = \rho_{j'}$ and $z_i - z_j = z_{i'} - z_{j'}$, while $M^{(i,n)}_{\alpha\beta,ij} = M^{(i,n)}_{\alpha\beta,i'j'}$ when instead $z_i + z_j = z_{i'} + z_{j'}$.
5. Solve the matrix equation.
6. Calculate the scattered cylindrical far-field components. (a) If a plane wave was used as the incident wave, then use these to calculate the optical cross sections and differential scattering cross sections. (b) If a dipole source was used, then add the reference far field to the scattered far field, and calculate the differential emission patterns and total out-of-plane emitted power. Also calculate the total emitted power from the near-field. The difference can now be used to estimate the coupling into guided modes.
7. The procedure can be inserted in a loop where the wavelength is varied to obtain spectra of optical cross sections.

Note that poles in Fresnel reflection and transmission coefficients can be handled when calculating the indirect and transmitted Green's tensor, and when calculating the dipole reference field, by extending the integration into the complex plane. Also note that when $z_i = z_j$, the direct use of $Z_{ij}(\mu)$ (see (7.170)) in (7.166) leads to a difficult integral with respect to μ, since the integrand will not eventually decrease exponentially with increasing μ. An option then is to replace $Z_{ij}(\mu)$ by

$$\tilde{Z}_{ij}(\mu, z) = \frac{2}{ik_{z1}} \left(e^{ik_{z1}(z_j^+ - z_j)} - e^{ik_{z1}(z - z_j)} \right) \tag{7.203}$$

and calculate the corresponding matrix element $\tilde{M}^{(d,n)}_{\alpha\beta,ij}(z)$ that now depends on z. By calculating the matrix element for a few values of z in the range from z_j to $z_j^\dagger$ it is now possible to apply interpolation to find $M_{\alpha\beta,ij}^{(d,n)} = \lim_{z\to z_j} \tilde{M}_{\alpha\beta,ij}^{(d,n)}(z)$.

7.10 EXERCISES

1. Verify the orthogonality relation (7.25).

2. Verify the expressions for the far-field direct Green's tensor cylindrical harmonics in Sec. 7.3.1.

3. Show that the far-field expression for the p-polarized guided mode (7.149) is approximately divergence-free ($\nabla \cdot \mathbf{E}^{(g,\mathrm{ff},p)} \approx 0$).

4. Insert the expression (6.74) into (7.56) and calculate analytically the cylindrical Green's tensor component $G_{\rho\rho}^{(d,1)}(\rho, z, \rho' = 0, z')$. Verify that this agrees with (7.47).

5. Verify the integral relations (7.172)-(7.174).

6. Verify the expressions related to the power emitted from a dipole in a layered reference structure (7.198) and (7.199).

7. Verify the expressions for the direct component of the dipole reference field (7.190)-(7.192). Apply the analytic expression for the direct Green's tensor (6.15) and use the procedure given in Eqs. (7.176).

8. Follow the approach outlined in Sec. 7.9 and develop numerical programs for modeling of a cylindrical scatterer on a layered reference structure.

8 Surface integral equation method for the quasi-static limit

In this chapter a Green's function surface integral equation method (GFSIEM) will be presented that can be applied for calculating the absorption cross section of a metal nanoparticle in the quasi-static limit. In this limit it is assumed that the nanoparticle is much smaller than the wavelength in the surrounding medium, and smaller than the penetration depth for electromagnetic radiation into the nanoparticle material. For a nanoparticle with a characteristic size a it is thus required that

$$a << \lambda/n_1, \tag{8.1}$$

where n_1 is the refractive index of the surrounding medium, and

$$a < \lambda/(2\pi \text{Imag}\{n_2\}), \tag{8.2}$$

where n_2 is the refractive index of the nanoparticle.

In the vicinity and inside the nanoparticle, the electric field can be approximated in this limit by

$$\mathbf{E} \approx -\nabla\phi, \tag{8.3}$$

where ϕ is the scalar electric potential. It is thus possible to work with a scalar function instead of vector fields, which greatly simplifies calculations. A GFSIEM for larger particles is considered in Chapter 9. Note that the fields and potentials are not static but have a time variation $\exp(-i\omega t)$.

8.1 GREEN'S FUNCTION INTEGRAL EQUATIONS

Assuming that there are no free charges, the electric field must satisfy

$$\nabla \cdot (\varepsilon(\mathbf{r})\mathbf{E}(\mathbf{r})) = 0. \tag{8.4}$$

We will consider a nano-particle with dielectric constant ε_2 in a region Ω_2. Outside the nanoparticle the dielectric constant is ε_1, and this region is referred to as Ω_1. In each of the regions Ω_1 and Ω_2, the dielectric function is thus independent of position, and for each of these regions (8.4) simplifies to the Laplace equation

$$\nabla^2\phi = 0, \quad \mathbf{r} \in \Omega_1 \quad \text{or} \quad \mathbf{r} \in \Omega_2. \tag{8.5}$$

We now introduce the Green's function

$$g(\mathbf{r},\mathbf{r}') = \frac{1}{4\pi|\mathbf{r}-\mathbf{r}'|} \tag{8.6}$$

that satisfies

$$\nabla^2 g(\mathbf{r},\mathbf{r}') = -\delta(\mathbf{r}-\mathbf{r}'). \tag{8.7}$$

In addition, the Green's function satisfies the boundary condition that it vanishes with increasing distance between $\mathbf{r}$ and $\mathbf{r}'$. This is the equivalent of the radiating boundary condition for the quasi-static case.

The potential in each of the regions Ω_1 and Ω_2 shall now be referred to as ϕ_1 and ϕ_2, respectively. It can be shown that the potential in each region can be obtained from the potential and its normal derivative at the inner and outer boundaries S_2 and S_1 of the nanoparticle from the following surface integral equations

$$\phi_2(\mathbf{r}) = \oint_{S_2} \{\hat{\mathbf{n}}'\cdot\nabla'\phi_2(\mathbf{r}')g(\mathbf{r},\mathbf{r}') - \hat{\mathbf{n}}'\cdot\nabla' g(\mathbf{r},\mathbf{r}')\phi_2(\mathbf{r}')\}\, d^2r', \quad \mathbf{r}\in\Omega_2, \tag{8.8}$$

$$\phi_1(\mathbf{r}) = \phi_0(\mathbf{r}) - \oint_{S_1} \{\hat{\mathbf{n}}'\cdot\nabla'\phi_1(\mathbf{r}')g(\mathbf{r},\mathbf{r}') - \hat{\mathbf{n}}'\cdot\nabla' g(\mathbf{r},\mathbf{r}')\phi_1(\mathbf{r}')\}\, d^2r', \quad \mathbf{r}\in\Omega_1, \tag{8.9}$$

where $\hat{\mathbf{n}}$ is the surface normal vector pointing out of the nanoparticle, and ϕ_0 represents the potential that corresponds to the applied electric field. In the case of an applied field being polarized along the z-axis, the applied potential will, for example, be

$$\phi_0(\mathbf{r}) = -E_0 z, \tag{8.10}$$

corresponding to the applied field

$$\mathbf{E}_0 = -\nabla\phi_0 = \hat{\mathbf{z}}E_0. \tag{8.11}$$

For positions $\mathbf{r}$ placed on just opposite sides of the particle surface (on S_1 and S_2) the potential must satisfy the boundary conditions

$$\phi_1 = \phi_2, \quad \text{at surface} \tag{8.12}$$

$$\varepsilon_1\hat{\mathbf{n}}\cdot\nabla\phi_1 = \varepsilon_2\hat{\mathbf{n}}\cdot\nabla\phi_2, \quad \text{at surface}, \tag{8.13}$$

where (8.13) follows from Eq. (8.4).

For positions at the nanoparticle surface it is convenient to introduce the shorthand notation

$$\psi_u = \hat{\mathbf{n}}\cdot\nabla\phi_u. \tag{8.14}$$

Due to the boundary condition (8.12) it is no longer necessary to use a subscript to indicate whether the potential is in region Ω_1 or Ω_2. Self-consistent equations for the surface potential and its normal derivative $\phi(\mathbf{s})$ and $\psi_u(\mathbf{s})$ can be obtained by letting the position $\mathbf{r}$ approach the nanoparticle surface infinitesimally from either side. Thus

$$\phi(\mathbf{s}) + \lim_{\mathbf{r}\to\mathbf{s}_1}\oint_{S_1} \{g(\mathbf{r},\mathbf{s}')\psi_1(\mathbf{s}') - \hat{\mathbf{n}}'\cdot\nabla' g(\mathbf{r},\mathbf{s}')\phi(\mathbf{s}')\}\, d^2s' = \phi_0(\mathbf{s}), \tag{8.15}$$

$$\phi(\mathbf{s}) - \lim_{\mathbf{r}\to\mathbf{s}_2}\oint_{S_2} \left\{g(\mathbf{r},\mathbf{s}')\frac{\varepsilon_1}{\varepsilon_2}\psi_1(\mathbf{s}') - \hat{\mathbf{n}}'\cdot\nabla' g(\mathbf{r},\mathbf{s}')\phi(\mathbf{s}')\right\} d^2s' = 0. \tag{8.16}$$

In these equations $\mathbf{s}_1$ and $\mathbf{s}_2$ refer to positions on S_1 and S_2, respectively, being on just opposite sides of the position $\mathbf{s}$ on the particle surface.

In the limit of $\mathbf{r} \to \mathbf{s}$ from one or the other side it can then be applied that

$$\hat{\mathbf{n}}' \cdot \nabla' g(\mathbf{r}, \mathbf{s}') d^2 s' = \frac{-1}{4\pi |\mathbf{r} - \mathbf{s}'|^2} \frac{\mathbf{s}' - \mathbf{r}}{|\mathbf{s}' - \mathbf{r}|} \cdot \hat{\mathbf{n}}' d^2 s' = \pm \frac{d\Omega}{4\pi}. \tag{8.17}$$

Here $d\Omega$ is the solid angle corresponding to the area $d^2 s'$ as seen from the position $\mathbf{r}$. The signs $+$ or $-$ must be applied depending on whether $\mathbf{r} \in \Omega_1$ or $\mathbf{r} \in \Omega_2$. In the case of a smooth surface, then in the limit $\mathbf{r} \to \mathbf{s}$ the integral over the infinitesimal part of the surface right next to $\mathbf{r}$ will correspond to a total solid angle of 2π. Regarding the other term in the integral equation, it can be noted that the singularity of $g(\mathbf{r}, \mathbf{s}')$ is much weaker, and the contribution to the integral from the infinitesimal part of the surface right next to $\mathbf{r}$ in the limit $\mathbf{r} \to \mathbf{s}$ will vanish.

For a smooth surface these considerations lead to the integral equations

$$\frac{1}{2}\phi(\mathbf{s}) + P \oint_S \left\{ \psi(\mathbf{s}') g(\mathbf{s}, \mathbf{s}') - \hat{\mathbf{n}}' \cdot \nabla' g(\mathbf{s}, \mathbf{s}') \phi(\mathbf{s}') \right\} d^2 s' = \phi_0(\mathbf{s}), \tag{8.18}$$

$$\frac{1}{2}\phi(\mathbf{s}) - P \oint_S \left\{ \frac{\varepsilon_1}{\varepsilon_2} \psi(\mathbf{s}') g(\mathbf{s}, \mathbf{s}') - \hat{\mathbf{n}}' \cdot \nabla' g(\mathbf{s}, \mathbf{s}') \phi(\mathbf{s}') \right\} d^2 s' = 0. \tag{8.19}$$

Here, P refers to the principal value of the integrals meaning that the singularity point $\mathbf{s}' = \mathbf{s}$ is excluded from the integration. In addition, now $\psi \equiv \psi_1$.

The power being absorbed in the nanoparticle can be obtained as

$$P_{\text{abs}} = \frac{1}{2} \text{Real} \left\{ \int_{\Omega_2} (\mathbf{E}(\mathbf{r}))^* \cdot \mathbf{J}(\mathbf{r}) d^3 r \right\}. \tag{8.20}$$

By using Ohms law $\mathbf{J} = \sigma \mathbf{E}$ and $\mathbf{E} = -\nabla \phi$, we find

$$P_{\text{abs}} = \frac{1}{2} \text{Real} \left\{ \int_{\Omega_2} \sigma (\nabla \phi(\mathbf{r}))^* \cdot \nabla \phi(\mathbf{r}) d^3 r \right\}. \tag{8.21}$$

Since $\nabla^2 \phi = 0$ the following applies:

$$\nabla \cdot (\phi \nabla \phi^*) = \nabla \phi \cdot \nabla \phi^* \tag{8.22}$$

By further using Gauss's law the expression (8.21) can now be reduced to

$$P_{\text{abs}} = \frac{1}{2} \sigma \left(\frac{\varepsilon_1}{\varepsilon_2} \right)^* \oint \phi \psi^* d^2 r, \tag{8.23}$$

which can readily be calculated once the integral equations have been solved for ϕ and ψ on the particle surface.

The conductivity is given by

$$\sigma = \omega \varepsilon_0 \varepsilon_2'', \tag{8.24}$$

where the dielectric constant of the particle is divided into its real and imaginary parts as

$$\varepsilon_2 = \varepsilon_2' + i\varepsilon_2''. \tag{8.25}$$

Thus, $\varepsilon_2' = \text{Real}\{\varepsilon_2\}$ and $\varepsilon_2'' = \text{Imag}\{\varepsilon_2\}$.

8.2 NUMERICAL APPROACH: PULSE EXPANSION

We are now going to consider a simple approach for solving the integral equations. The nanoparticle surface may be discretized into N area elements, where element i has the area Δ_i, center position $\mathbf{s}_i$, and normal vector at the center position $\hat{\mathbf{n}}_i$. In a simple approximation we may assume that the applied potential ϕ_0, the potential ϕ, and its normal derivative ψ, can be assumed constant over each element, and these will be evaluated at the center position of the element. Thus, $\phi_{0,i} \equiv \phi_0(\mathbf{s}_i)$, $\phi_i \equiv \phi(\mathbf{s}_i)$, and $\psi_i \equiv \psi_1(\mathbf{s}_i)$. This is the pulse expansion approach.

The integral equations can thus be expressed in discrete form as

$$\frac{1}{2}\phi_i + \sum_j \{A_{i,j}\psi_j - B_{i,j}\phi_j\} = \phi_{0,i}, \tag{8.26}$$

$$\frac{1}{2}\phi_i - \sum_j \left\{\frac{\varepsilon_1}{\varepsilon_2}A_{i,j}\psi_j - B_{i,j}\phi_j\right\} = 0, \tag{8.27}$$

where the matrix elements are given by

$$A_{i,j} = P\int_j g(\mathbf{s}_i, \mathbf{s}')d^2s', \tag{8.28}$$

$$B_{i,j} = P\int_j \hat{\mathbf{n}}' \cdot \nabla' g(\mathbf{s}_i, \mathbf{s}')d^2s'. \tag{8.29}$$

In a crude approximation the integral expressions for the matrix elements can be replaced by

$$A_{i,j} \approx \begin{cases} g(\mathbf{s}_i, \mathbf{s}_j)\Delta_j, & i \neq j \\ \sqrt{\Delta_j/4\pi}, & i = j \end{cases}, \tag{8.30}$$

$$B_{i,j} \approx \begin{cases} \frac{-(\mathbf{s}_j - \mathbf{s}_i)\cdot\hat{\mathbf{n}}_j}{4\pi|\mathbf{s}_i - \mathbf{s}_j|^3}\Delta_j, & i \neq j \\ 0, & i = j \end{cases}. \tag{8.31}$$

The expression for $A_{i,i}$ was obtained by replacing the integral over the element i by an integral over a flat circular disk of the same total area.

The integral equations can also be expressed in matrix form

$$\left(\frac{1}{2}\overline{\overline{I}} - \overline{\overline{B}}\right)\overline{\phi} + \overline{\overline{A}}\overline{\psi} = \overline{\phi}_0, \tag{8.32}$$

$$\left(\frac{1}{2}\overline{\overline{I}} + \overline{\overline{B}}\right)\overline{\phi} - \frac{\varepsilon_1}{\varepsilon_2}\overline{\overline{A}}\overline{\psi} = \overline{0}, \tag{8.33}$$

where the matrices $\overline{\overline{A}}$ and $\overline{\overline{B}}$ are defined by the matrix elements $A_{i,j}$ and $B_{i,j}$, and

$$\overline{\phi}_0 = \begin{bmatrix}\phi_{0,1} & \phi_{0,2} & \ldots & \phi_{0,N}\end{bmatrix}^T, \tag{8.34}$$

$$\overline{\phi} = \begin{bmatrix}\phi_1 & \phi_2 & \ldots & \phi_N\end{bmatrix}^T, \tag{8.35}$$

$$\overline{\psi} = \left[\psi_1 \; \psi_2 \; \ldots \; \psi_N \right]^T. \tag{8.36}$$

The matrix equations can, for example, be solved in two steps by first finding the potential from

$$\overline{\phi} = \left[\frac{1}{2}\left(1 + \frac{\varepsilon_1}{\varepsilon_2}\right)\overline{\overline{I}} + \left(1 - \frac{\varepsilon_1}{\varepsilon_2}\right)\overline{\overline{B}} \right]^{-1} \frac{\varepsilon_1}{\varepsilon_2}\overline{\phi}_0, \tag{8.37}$$

and next the normal derivative of the potential can be obtained from

$$\overline{\psi} = \overline{\overline{A}}^{-1}\frac{\varepsilon_2}{\varepsilon_1}\left(\frac{1}{2}\overline{\overline{I}} + \overline{\overline{B}}\right)\overline{\phi}. \tag{8.38}$$

At this point the potential and its normal derivative at the scatterer surface have been obtained, and it is possible to obtain the potential at any position inside and outside the nanoparticle by using the integral equations (8.8) and (8.9).

In terms of the discrete values for the potential and its normal derivative the absorption can be expressed as

$$P_{\text{abs}} = \frac{1}{2}\sigma \text{Real}\left\{ \left(\frac{\varepsilon_1}{\varepsilon_2}\right)^* \sum_{i=1}^{N} \phi_i \psi_i{}^* \Delta_i \right\}. \tag{8.39}$$

It is suggested to take the real part in the latter expression as a small imaginary part may result due to numerical reasons. By taking the imaginary part instead, it can be checked if this part is small as it is supposed to be. If it is not, it may be necessary to use a higher number of elements N.

Finally, if the absorbed power is normalized by the magnitude of the Poynting vector of the incident plane wave, the absorption cross section is obtained as

$$\sigma_{\text{abs}} = \frac{k_0 \varepsilon_2'' \text{Real}\left\{ \left(\frac{\varepsilon_1}{\varepsilon_2}\right)^* \sum_{i=1}^{N} \phi_i \psi_i{}^* \Delta_i \right\}}{n_1 |E_0|^2}. \tag{8.40}$$

In order to test the above method we may consider the absorption cross section for a spherical gold nanoparticle with radius R in water. For this geometry an analytic expression exists for the absorption cross section given by (see for example [29])

$$\sigma_{\text{abs, sphere}} = \frac{24\pi^2 R^3 n_1^3}{\lambda} \frac{\varepsilon_2''}{\left(\varepsilon_2' + 2\varepsilon_1\right)^2 + \varepsilon_2''^2}. \tag{8.41}$$

The surface of the spherical particle may, for example, be discretized by using the discrete spherical coordinate angles

$$\phi = \frac{2\pi}{N_\phi}(j_\phi - 0.5), \quad 1 \le j_\phi \le N_\phi, \tag{8.42}$$

$$\theta = \frac{\pi}{N_\theta}(j_\theta - 0.5), \quad 1 \le j_\theta \le N_\theta. \tag{8.43}$$

For each of the $N = N_\phi N_\theta$ angle combinations, the corresponding element positions and normal vectors are given by

$$\mathbf{r} = R\left(\hat{\mathbf{z}}\cos\theta + \hat{\mathbf{x}}\sin\theta\cos\phi + \hat{\mathbf{y}}\sin\theta\sin\phi\right), \tag{8.44}$$

$$\hat{\mathbf{n}} = \mathbf{r}/R, \tag{8.45}$$

where R is the radius of the sphere, and the area of the element is given by

$$\Delta = \int R^2 \sin\theta d\theta d\phi = R^2 \frac{2\pi}{N_\phi}\left(\cos(\theta - \pi/2N_\theta) - \cos(\theta + \pi/2N_\theta)\right). \tag{8.46}$$

The results obtained using different numbers of discretization elements $N = N_\phi N_\theta$ with $N_\phi = 2N_\theta$ are shown in Fig. 8.1. The refractive index of water was set to $n_1 = 1.33$, and n_2 for gold or silver has been obtained from Ref. [40]. The particle radius has been set to 5 nm. The numerical results are compared to the exact analytic result (8.41). The case $N_s = 1$ represents the case of using Eqs. (8.30) and (8.31), and $N_s = 121$ represents carrying out the integrals in (8.28) and (8.29) by numerical subsampling of the integration area into N_s subelements. In the case of gold (Fig. 8.1(a))

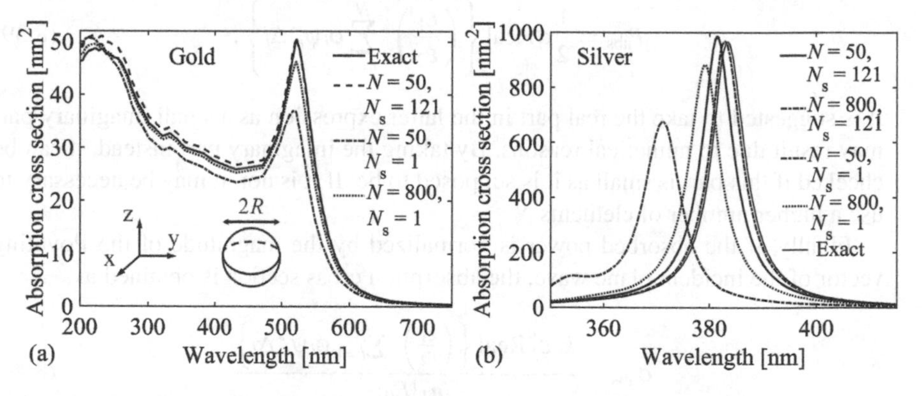

Figure 8.1 Numerical results for the absorption cross section of a spherical (a) gold and (b) silver particle of radius 5 nm in water versus the number of discretization elements $N = N_\phi N_\theta$ with $N_\phi = 2N_\theta$. Also shown is the exact analytic result.

using $N = 50$ and $N_s = 1$ already leads to a result that captures the shape of the exact curve and the position of the resonance. Using instead $N = 800$ and $N_s = 1$ leads to a result being fairly close to the exact result. However, when using again $N = 50$ but $N_s = 121$, the result is actually better compared with $N = 800$ and $N_s = 1$. It is thus possible to use much fewer elements if the integrals are carried out to a higher precision. In the case of silver (Fig. 8.1(b)), the absorption resonance is sharper and it becomes more important to improve the precision of matrix elements. Here $N = 50$ and $N_s = 1$ leads to a resonance peak which is clearly displaced from the peak in the exact result. Again $N = 50$ and $N_s = 121$ is better compared with $N = 800$ and $N_s = 1$. If the position of the resonance should be accurate within 1 nm, then here it is sufficient to use $N = 800$ and $N_s = 121$.

8.3 FINITE-ELEMENT-APPROACH: LINEAR EXPANSION

In this section an approach is considered where the particle surface is approximated by dividing it into triangles and using linear expansion functions. Each triangle k on the surface has three corners $\mathbf{r}_k^{(i)}$, $i = 1,2,3$, and the position on the triangle can be parameterized by

$$\mathbf{r}_k(u,v) = \sum_{i=1}^{3} \mathbf{r}_k^{(i)} f^{(i,\mathrm{L})}(u,v), \quad 0 \le u \le 1, \quad 0 \le v \le 1-u, \tag{8.47}$$

where

$$f^{(1,\mathrm{L})}(u,v) = 1-u-v, \tag{8.48}$$

$$f^{(2,\mathrm{L})}(u,v) = u, \tag{8.49}$$

$$f^{(3,\mathrm{L})}(u,v) = v. \tag{8.50}$$

A linear expansion of ϕ and ψ on the triangle can be expressed as

$$\phi(\mathbf{r}_k(u,v)) = \sum_{i=1}^{3} \phi_k^{(i)} f^{(i,\mathrm{L})}(u,v), \tag{8.51}$$

$$\psi(\mathbf{r}_k(u,v)) = \sum_{i=1}^{3} \psi_k^{(i)} f^{(i,\mathrm{L})}(u,v). \tag{8.52}$$

Here, the coefficients $\phi_k^{(i)}$ and $\psi_k^{(i)}$ thus represent the value of ϕ and ψ at the corners of the triangle k. A similar approach with linear expansion and triangular elements has been applied in Chapter 5 for the area integral equation method.

The functions ϕ and ψ on the entire particle surface can be expressed as

$$\phi(\mathbf{s}) = \sum_k \sum_{i=1}^{3} \phi_k^{(i)} f_k^{(i,\mathrm{L})}(\mathbf{s}), \tag{8.53}$$

$$\psi(\mathbf{s}) = \sum_k \sum_{i=1}^{3} \psi_k^{(i)} f_k^{(i,\mathrm{L})}(\mathbf{s}), \tag{8.54}$$

where here $\mathbf{s}$ is a position on the particle surface, and

$$f_k^{(i)}(\mathbf{s}) = \begin{cases} f^{(i,\mathrm{L})}(u,v), & \mathbf{s} \in k, \ \mathbf{s} = \mathbf{r}_k(u,v) \\ 0, & \text{otherwise} \end{cases}. \tag{8.55}$$

In practice we will not need to construct the functions $f_k^{(i)}(\mathbf{s})$ though.

One approach for approximating a surface with triangles is to use a standard mesh generator for meshing a 2D region and then mapping this onto a 3D surface. Consider first a rectangle in the (t_1,t_2)-plane within the region $0 \le t_1 \le 2\pi R$ and $0 \le t_2 \le$

H. The rectangular region can be meshed and then subsequently mapped onto the surface of a cylinder with height H and radius R using, e.g., the rule

$$\mathbf{r}(t_1,t_2) = \hat{\mathbf{r}}_0 + \hat{\mathbf{z}}t_2 + \hat{\mathbf{x}}R\cos\left(\frac{t_1}{R}+\phi_0\right) + \hat{\mathbf{y}}R\sin\left(\frac{t_1}{R}+\phi_0\right). \tag{8.56}$$

This procedure is illustrated in Fig. 8.2 for a cylinder with height 15 and radius 10. Another orientation of the cylinder or another geometry can be obtained by rotating

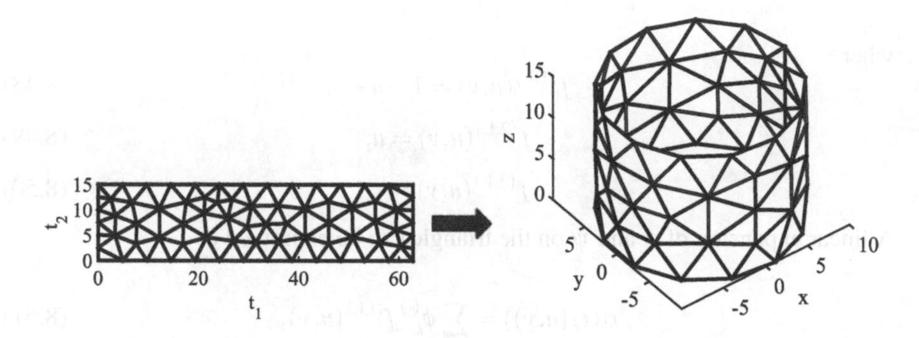

Figure 8.2 Illustration of mapping of a triangulated rectangle onto a cylinder surface.

the cylinder by an angle α around an axis with direction $\hat{\mathbf{n}}_c$ going through a position $\mathbf{r}_c$ by using

$$\mathbf{r} \to \mathbf{r}_c + \hat{\mathbf{n}}_c\hat{\mathbf{n}}_c \cdot (\mathbf{r}-\mathbf{r}_c) + \cos\alpha\tilde{\mathbf{r}} + \sin\alpha\mathbf{n}_c \times \tilde{\mathbf{r}}, \tag{8.57}$$

where

$$\tilde{\mathbf{r}} = (\mathbf{r}-\mathbf{r}_c) - \hat{\mathbf{n}}_c\hat{\mathbf{n}}_c \cdot (\mathbf{r}-\mathbf{r}_c). \tag{8.58}$$

The surface of a quarter of a half-sphere can be mapped to two dimensions by, for example, expressing a position on the surface as

$$\mathbf{r} = \mathbf{r}_c + \hat{\mathbf{z}}R\cos\left(\frac{t_1}{R}\right) + R\sin\left(\frac{t_1}{R}\right)\hat{\rho}(t_1,t_2), \tag{8.59}$$

where

$$\hat{\rho}(t_1,t_2) = \left(\hat{\mathbf{x}}\cos\left(\frac{t_2}{R\sin\left(\frac{t_1}{R}\right)}+\alpha_0\right) + \hat{\mathbf{y}}\sin\left(\frac{t_2}{R\sin\left(\frac{t_1}{R}\right)}+\alpha_0\right)\right). \tag{8.60}$$

Here, t_1/R and $t_2/(R\sin(t_1/R))$ are equivalent to spherical coordinate angles. However, t_1 and t_2 are physical distances along the spherical surface. For a quarter half-sphere the distances are restricted to

$$t_1 \in \left[0; R\frac{\pi}{2}\right], \tag{8.61}$$

$$t_2 \in \left[-\frac{\pi}{4} R \sin\left(\frac{t_1}{R}\right); \frac{\pi}{4} R \sin\left(\frac{t_1}{R}\right)\right]. \tag{8.62}$$

An example of the mesh of the two-dimensional region defined by (8.61) and (8.62), and the mapping of the mesh onto the corresponding quarter of a half-sphere is shown in Fig. 8.3 for a half-sphere with radius 10. Many simple geometries can be con-

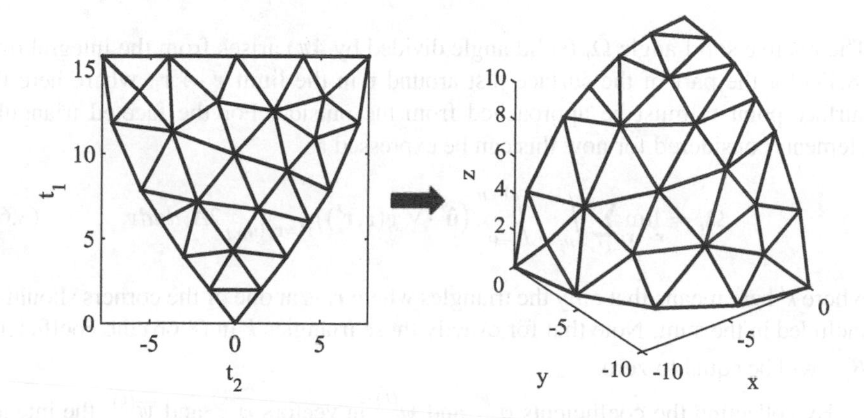

Figure 8.3 Illustration of mapping of a triangulated rectangle onto a cylinder surface.

structed by joining together parts of spheres, cylinders, and flat geometries.

The mesh of the geometry can be described by a point matrix $\overline{\overline{P}}$ and a connectivity matrix $\overline{\overline{T}}$, where each column in $\overline{\overline{P}}$ represents the coordinates of a point, and each column of $\overline{\overline{T}}$ identifies three points that are connected in a triangle. Additional information may be added to a column of $\overline{\overline{T}}$ that describes the normal vector of the element and the type of element, e.g. if it is a flat element, part of a cylinder, or if it is part of a sphere.

We shall now refer to the potential, its normal derivative, and the applied potential, at point i of the mesh, as ϕ_i, ψ_i, and $\phi_{0,i}$, respectively. By using the expressions (8.53) and (8.54) in the integral equations, and letting $\mathbf{r}$ approach one of the mesh points i with position $\mathbf{r}_i$, we arrive at the discretized integral equations

$$(1-\tilde{\Omega}_i)\phi_i + \sum_k \sum_{j=1}^{3} \left\{ A_{i,j}^{(k)} \psi_j^{(k)} - B_{i,j}^{(k)} \phi_j^{(k)} \right\} = \phi_{0,i}, \tag{8.63}$$

$$\tilde{\Omega}_i \phi_i - \sum_k \sum_{j=1}^{3} \left\{ \frac{\varepsilon_1}{\varepsilon_2} A_{i,j}^{(k)} \psi_j^{(k)} - B_{i,j}^{(k)} \phi_j^{(k)} \right\} = 0, \tag{8.64}$$

with the matrix elements

$$A_{i,j}^{(k)} = P \int_{u=0}^{1} \int_{v=0}^{1-u} g(\mathbf{r}_i, \mathbf{r}_k(u,v)) f^{(j,\mathrm{L})}(u,v) 2A_k du dv, \tag{8.65}$$

$$B_{i,j}^{(k)} = P\int_{u=0}^{1}\int_{v=0}^{1-u}\left(\hat{\mathbf{n}}'\cdot\nabla' g(\mathbf{r}_i,\mathbf{r}')\right)_{\mathbf{r}'=\mathbf{r}_k(u,v)} f^{(j,\mathrm{L})}(u,v) 2A_k du dv. \tag{8.66}$$

Here, the area of triangle k is given by

$$A_k = \frac{1}{2}\left|\frac{\partial \mathbf{r}_k(u,v)}{\partial u}\times\frac{\partial \mathbf{r}_k(u,v)}{\partial v}\right|. \tag{8.67}$$

The relative solid angle $\tilde{\Omega}_i$ (solid angle divided by 4π) arises from the integral over (8.17) for the part of the surface just around $\mathbf{r}$ in the limit $\mathbf{r}\rightarrow\mathbf{r}_i$, where here the surface point $\mathbf{r}_i$ must be approached from the outside. For the faceted triangular elements considered for now this can be expressed as

$$\tilde{\Omega}_i = \lim_{\mathbf{r}\rightarrow\mathbf{r}_i}\sum_{k'}\int_{u=0}^{1}\int_{v=0}^{1-u}\left(\hat{\mathbf{n}}'\cdot\nabla' g(\mathbf{r},\mathbf{r}')\right)_{\mathbf{r}'=\mathbf{r}_k(u,v)} 2A_k du dv, \tag{8.68}$$

where k' here means that only the triangles where $\mathbf{r}_i$ is at one of the corners should be included in the sum. Note that for exactly those triangles k in (8.66) the coefficients $B_{i,j}^{(k)}$ will be equal to zero.

By collecting the coefficients $\phi_j^{(k)}$ and $\psi_j^{(k)}$ in vectors $\overline{\phi}^{(k)}$ and $\overline{\psi}^{(k)}$, the integral equations can be written in matrix form

$$\left(\overline{\overline{I}}-\overline{\overline{\Omega}}\right)\overline{\phi}+\sum_k\left\{\overline{\overline{A}}^{(k)}\overline{\psi}^{(k)}-\overline{\overline{B}}^{(k)}\overline{\phi}^{(k)}\right\}=\overline{\phi}_0, \tag{8.69}$$

$$\overline{\overline{\Omega}}\,\overline{\phi}-\sum_k\left\{\frac{\varepsilon_1}{\varepsilon_2}\overline{\overline{A}}^{(k)}\overline{\psi}^{(k)}-\overline{\overline{B}}^{(k)}\overline{\phi}^{(k)}\right\}=\overline{0}, \tag{8.70}$$

where $\overline{\overline{\Omega}}$ is a diagonal matrix with $\tilde{\Omega}_i$ as the diagonal elements.

By further collecting ϕ_i and ψ_i in vectors $\overline{\phi}$ and $\overline{\psi}$, and noting that the element points are equivalent to mesh points via a relation of the form

$$\overline{\phi}^{(k)}=\overline{\overline{P}}_k\overline{\phi}, \tag{8.71}$$

$$\overline{\psi}^{(k)}=\overline{\overline{P}}_k\overline{\psi}, \tag{8.72}$$

we obtain the final matrix equations

$$\left(\overline{\overline{I}}-\overline{\overline{\Omega}}-\overline{\overline{B}}\right)\overline{\phi}+\overline{\overline{A}}\,\overline{\psi}=\overline{\phi}_0, \tag{8.73}$$

$$\left(\overline{\overline{\Omega}}+\overline{\overline{B}}\right)\overline{\phi}-\frac{\varepsilon_1}{\varepsilon_2}\overline{\overline{A}}\,\overline{\psi}=\overline{0}, \tag{8.74}$$

that are similar to (8.32) and (8.33) except that $\overline{\phi}$ and $\overline{\psi}$ have a different meaning, and the matrices $\overline{\overline{A}}$ and $\overline{\overline{B}}$ are now given by

$$\overline{\overline{A}}=\sum_k\overline{\overline{A}}^{(k)}\overline{\overline{P}}_k, \tag{8.75}$$

$$\overline{\overline{B}} = \sum_k \overline{\overline{B}}^{(k)} \overline{\overline{P}}_k. \tag{8.76}$$

The absorbed power can now be calculated using

$$P_{\text{abs}} = \frac{1}{2}\text{Real}\left\{ \sigma \frac{\varepsilon_1}{\varepsilon_2^*} \sum_k \overline{\phi}_k^T \begin{bmatrix} 2 & 1 & 1 \\ 1 & 2 & 1 \\ 1 & 1 & 2 \end{bmatrix} \overline{\psi}_k^* \frac{1}{12} A_k \right\}. \tag{8.77}$$

An example of the performance of the GFSIEM with triangular elements and linear expansion functions is shown in Fig. 8.4 for a spherical silver particle with radius 5 nm. Note that it is highly important to accurately calculate the solid angles

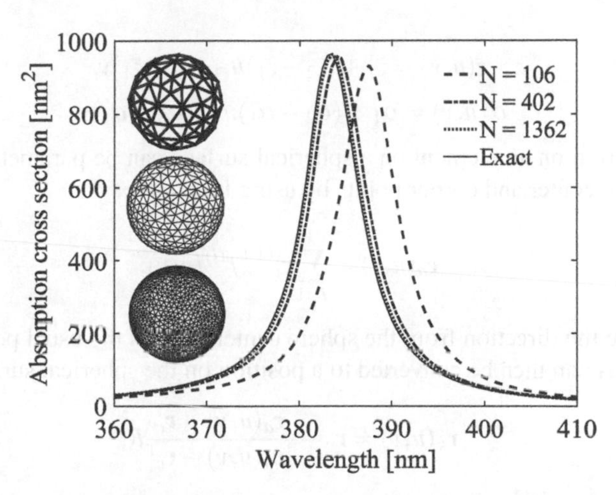

Figure 8.4 Calculation of absorption cross section of a spherical silver particle with radius 5 nm obtained by the GFSIEM. The particle surface is approximated by triangles where N is the number of mesh points. A linear expansion of ϕ and ψ is applied on each triangle.

$\tilde{\Omega}_i$ due to the faceted nature of the surface.

Another possible approach is to use the exact surface rather than the faceted approximation considered above. We will still describe an element from its three corner points as above, and use the same point and connectivity information. However, for each element we will now add information about the type of element, namely if it is a flat element, if it is part of a cylinder surface, or if it is part of a spherical surface. In the first case it is sufficient to also add information about the normal vector to the surface. Regarding the cylinder, additional information is required about the axis of rotation given by a point on the axis $\mathbf{r}_a$ and a unit vector in the direction of the cylinder axis $\hat{\mathbf{n}}_a$. In the case of a spherical surface, information about the center of the sphere $\mathbf{r}_c$ must be added. In the latter cases, information about the direction of the normal vector must also be added.

The position on an element k being part of a cylinder can be parameterized in terms of the three corner points $\mathbf{r}_k^{(i)}$ and the information about the cylinder axis by first defining

$$\tilde{z}_i = \hat{\mathbf{n}}_a \cdot (\mathbf{r}_k^{(i)} - \mathbf{r}_a), \quad i = 1,2,3, \tag{8.78}$$

$$\tilde{\mathbf{r}}_i = (\mathbf{r}_k^{(i)} - \mathbf{r}_a) - \tilde{z}_i \hat{\mathbf{n}}_a, \quad i = 1,2,3, \tag{8.79}$$

and the angles

$$\alpha_i = \sin^{-1}\left(\frac{\hat{\mathbf{n}}_a \cdot (\tilde{\mathbf{r}}_1 \times \tilde{\mathbf{r}}_i)}{|\tilde{\mathbf{r}}_1|^2}\right), \quad i = 1,2,3. \tag{8.80}$$

The position within the element can now be expressed as

$$\mathbf{r}_k(u,v) = \mathbf{r}_a + \hat{\mathbf{n}}_a \tilde{z}(u,v) + \tilde{\mathbf{r}}_1 \cos\alpha(u,v) + \hat{\mathbf{n}}_a \times \tilde{\mathbf{r}}_1 \sin\alpha(u,v), \tag{8.81}$$

where

$$\tilde{z}(u,v) = \tilde{z}_1 + (\tilde{z}_2 - \tilde{z}_1)u + (\tilde{z}_3 - \tilde{z}_1)v, \tag{8.82}$$

$$\alpha(u,v) = \alpha_1 + (\alpha_2 - \alpha_1)u + (\alpha_3 - \alpha_1)v. \tag{8.83}$$

The position on an element on a spherical surface can be parameterized in terms of the sphere center and corner points by using first

$$\mathbf{r}_a(u,v) = \sum_{i=1}^{3} \mathbf{r}_k^{(i,\mathrm{L})} f^{(i)}(u,v), \tag{8.84}$$

to determine the direction from the sphere center. This is the usual position on a flat triangle. This can then be converted to a position on the spherical surface by using

$$\mathbf{r}_k(u,v) = \mathbf{r}_c + \frac{\mathbf{r}_a(u,v) - \mathbf{r}_c}{|\mathbf{r}_a(u,v) - \mathbf{r}_c|} R, \tag{8.85}$$

where $R = |\mathbf{r}_k^{(i)} - \mathbf{r}_c|$ is the radius of the sphere.

The form of the discretized integral equations (8.63) and (8.64) will be unchanged. Here though, if the actual surface does not have any sharp corners, and is thus locally flat at all sampling points, then $\tilde{\Omega}_i = 1/2$. In addition, when carrying out the integrals (8.65) and (8.66) the area parameter A_k given in (8.67) may now be a function of u and v, where this is not the case for a flat triangle. This also means that the expression (8.77) no longer applies for the absorbed power. Instead the absorption can be calculated by numerical integration.

As an example of the performance when using elements that follow the exact surface, the spherical silver particle with radius 5 nm in water is considered once more in Fig. 8.5. The absorption cross section calculated for the faceted particle with $N = 106$ mesh points (also shown in Fig. 8.4) is compared with the corresponding calculation using the same mesh and the exact surface of the particle. In both cases a linear expansion of ϕ and ψ is applied on the elements. The case of using elements that follow the exact surface significantly improves the convergence toward the exact analytic result. Also shown is a result obtained when using $N = 98$ mesh points and a quadratic polynomial expansion of ϕ and ψ on elements being explained in the following section. This approach leads to even better convergence using fewer mesh points.

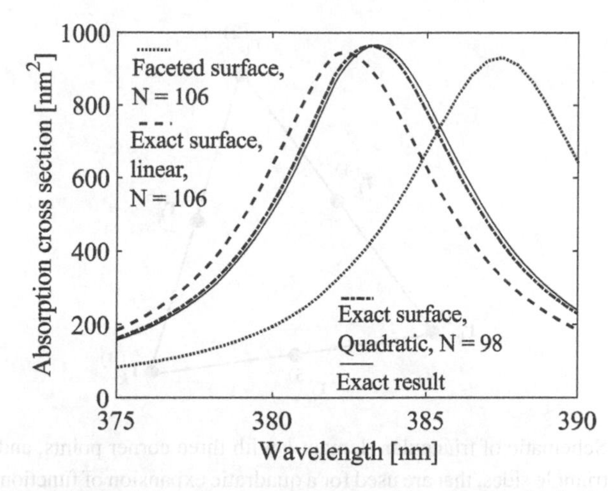

Figure 8.5 Calculations of the absorption cross section of a spherical silver particle with radius 5 nm obtained by the GFSIEM. Two calculations are shown with a linear polynomial expansion of ϕ and ψ with either a faceted surface or the exact surface, and N = 106 mesh points. Also shown is a calculation with a quadratic polynomial expansion and N = 98 mesh points.

8.4 FINITE-ELEMENT-APPROACH: QUADRATIC EXPANSION

In this section we will use a quadratic polynomial expansion of ϕ and ψ on elements k. There are now 6 points $\mathbf{r}_k^{(i)}$, $i = 1,2,\ldots,6$ associated with the element as illustrated in Fig. 8.6. Thus, each column of the connectivity matrix $\overline{\overline{T}}$ must now contain the labels of 6 points that belong to the element instead of 3 points. The points $\mathbf{r}_k^{(1)}$, $\mathbf{r}_k^{(2)}$ and $\mathbf{r}_k^{(3)}$ are, as before, the corners of the triangle, and the additional points $\mathbf{r}_k^{(4)}$, $\mathbf{r}_k^{(5)}$ and $\mathbf{r}_k^{(6)}$ are placed on the lines between corner points with the ordering shown in Fig. 8.6.

A position on the element can now be parameterized as

$$\mathbf{r}_k(u,v) = \sum_{i=1}^{6} \mathbf{r}_k^{(i)} f^{(i,Q)}(u,v), \quad 0 \le u \le 1, \quad 0 \le v \le 1-u, \tag{8.86}$$

where the quadratic expansion functions are given by [160]:

$$f^{(1,Q)}(u,v) = (1-u-v)(1-2u-2v), \tag{8.87}$$

$$f^{(2,Q)}(u,v) = (2u-1)u, \tag{8.88}$$

$$f^{(3,Q)}(u,v) = (2v-1)v, \tag{8.89}$$

$$f^{(4,Q)}(u,v) = 4u(1-u-v), \tag{8.90}$$

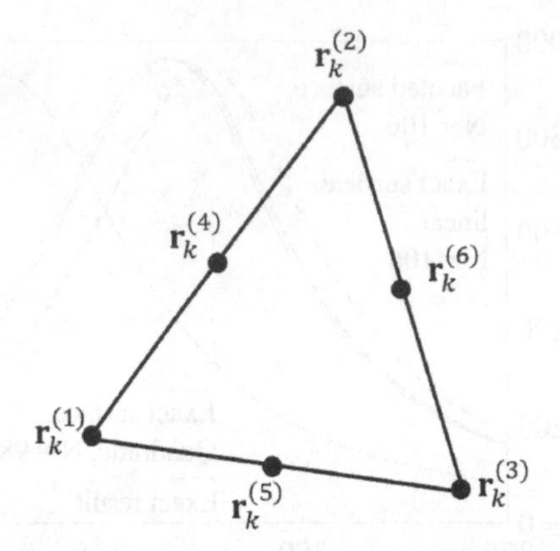

Figure 8.6 Schematic of triangular element k with three corner points, and three additional points on the triangle sides, that are used for a quadratic expansion of functions on the element.

$$f^{(5,Q)}(u,v) = 4v(1-u-v), \tag{8.91}$$

$$f^{(6,Q)}(u,v) = 4uv. \tag{8.92}$$

Note that (8.86) reduces to the linear expansion (8.47) when $\mathbf{r}_k^{(4)}$, $\mathbf{r}_k^{(5)}$ and $\mathbf{r}_k^{(6)}$ are placed midway between the corner points of the triangle. The points do not need to be placed in that way and can be chosen in a way such that (8.86) can be used to describe a curved surface. However, the approach explained in Sec. 8.3 for describing the exact surface profile of curved surfaces also applies here, and that is the method which will be used in the following.

A quadratic polynomial expansion of ϕ and ψ on element k is now given by

$$\phi(\mathbf{r}_k(u,v)) = \sum_{i=1}^{6} \phi_k^{(i)} f^{(i,Q)}(u,v), \tag{8.93}$$

$$\psi(\mathbf{r}_k(u,v)) = \sum_{i=1}^{6} \psi_k^{(i)} f^{(i,Q)}(u,v), \tag{8.94}$$

where the coefficients $\phi_k^{(i)}$ and $\psi_k^{(i)}$ represent the value of ϕ and ψ at the six positions $\mathbf{r}_k^{(i)}$ on the triangle.

By inserting the quadratic expansion of ϕ and ψ in the integral equations, and taking the limit of $\mathbf{r} \to \mathbf{r}_i$, where $\mathbf{r}_i$ is a mesh point, the following discretized integral equations are obtained for a particle with a smooth surface

$$\frac{1}{2}\phi_i + \sum_k \sum_{j=1}^{6} \left\{ A_{i,j}^{(k)} \psi_j^{(k)} - B_{i,j}^{(k)} \phi_j^{(k)} \right\} = \phi_{0,i}, \tag{8.95}$$

$$\frac{1}{2}\phi_i - \sum_k \sum_{j=1}^{6} \left\{ \frac{\varepsilon_1}{\varepsilon_2} A_{i,j}^{(k)} \psi_j^{(k)} - B_{i,j}^{(k)} \phi_j^{(k)} \right\} = 0, \tag{8.96}$$

where the matrix elements are now defined in terms of the quadratic expansion functions

$$A_{i,j}^{(k)} = P \int_{u=0}^{1} \int_{v=0}^{1-u} g(\mathbf{r}_i, \mathbf{r}_k(u,v)) f^{(j,Q)}(u,v) 2A_k(u,v) du dv, \tag{8.97}$$

$$B_{i,j}^{(k)} = P \int_{u=0}^{1} \int_{v=0}^{1-u} (\hat{\mathbf{n}}' \cdot \nabla' g(\mathbf{r}_i, \mathbf{r}'))_{\mathbf{r}'=\mathbf{r}_k(u,v)} f^{(j,Q)}(u,v) 2A_k(u,v) du dv. \tag{8.98}$$

This can be expressed in matrix form similar to (8.69)-(8.76), except that $\overline{\overline{A}}^{(k)}$, $\overline{\overline{B}}^{(k)}$, and $\overline{\overline{P}}_k$, are now $N \times 6$ and $6 \times N$ matrices instead of $N \times 3$ and $3 \times N$ matrices.

8.5 EXAMPLES OF ABSORPTION CROSS SECTIONS OF 3D SILVER NANOPARTICLES

We are now going to consider a few examples of calculations of the absorption cross section for different shapes of nanoparticles. An example for a silver nanoparticle which consists of two half spheres separated by a cylinder is shown in Fig. 8.7. The half spheres have radius R = 5 nm, and the cylinder has the length d and the same radius R. The cylinder is surrounded by water ($n_1 = 1.33$). In Fig. 8.7(a) the applied electric field is in the xz direction ($\mathbf{E}_0 = E_0 (\hat{\mathbf{x}} + \hat{\mathbf{z}}) / \sqrt{2}$) and d is varied. Two resonance peaks are visible, and the separation between the peaks increases with d. The case of a fixed d = 6 nm is considered in Fig. 8.7(b). Here, the direction of the applied field is varied such that the field is either along the x, z, or the xz direction. Different resonances are related to fields along the cylinder axis and fields perpendicular to the axis. This is seen here from the fact that in the x and z directions of the field, only one resonance peak is visible. In the case of the xz direction, both resonances appear in the spectrum.

A final example is presented in Fig. 8.8 for a silver particle with three distinct symmetry axes. The particle has the lengths 14 nm, 10 nm, and 12 nm, along the x, y and z directions, respectively. Corners and edges are rounded by a radius of 5 nm. This is equivalent to the particle in Fig. 8.7 with d = 2 nm being extended by 4 nm along the x-axis. It can be seen from Fig. 8.8(a) that different resonances are excited when the electric field is oriented along each of the three symmetry axes x, y, and z. From Fig. 8.8(b) it can be seen that two resonances are excited when the electric field is oriented in the xy or xz directions, and three resonances are excited when the electric field is oriented along the xyz direction.

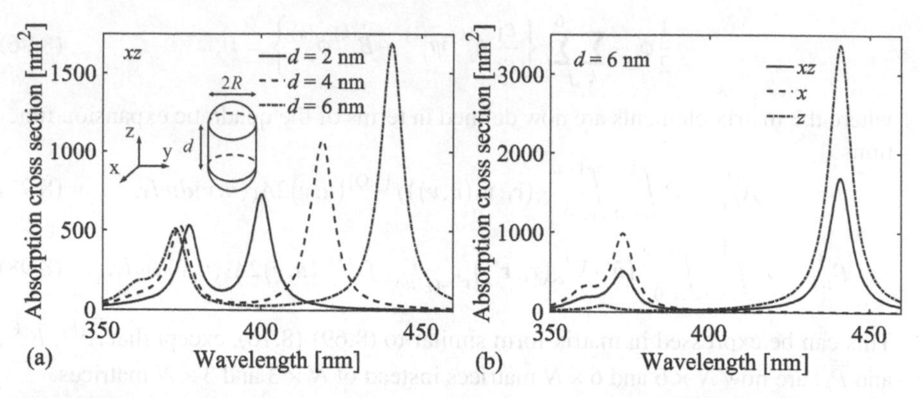

Figure 8.7 Numerical results for the absorption cross section of a silver particle consisting of two half-spheres of radius $R = 5$ nm displaced by a silver cylinder of length d and also radius $R = 5$ nm. The particle is surrounded by water. (a) $d = 2, 4$, and 6 nm. Applied electric field (xz): $\mathbf{E}_0 = E_0(\hat{\mathbf{x}} + \hat{\mathbf{z}})/\sqrt{2}$. Inset: Schematic of particle. (b) $d = 6$ nm and different applied fields (xz): $\mathbf{E}_0 = E_0(\hat{\mathbf{x}} + \hat{\mathbf{z}})/\sqrt{2}$; ($z$): $\mathbf{E}_0 = E_0\hat{\mathbf{z}}$; ($x$): $\mathbf{E}_0 = E_0\hat{\mathbf{x}}$.

8.6 GUIDELINES FOR SOFTWARE IMPLEMENTATION

Structure of GFSIEM program with pulse expansion

1. Define the particle surface, and divide it into N elements.
2. Construct vectors that contains the coordinates, areas, and normal vectors of elements.
3. Construct the $\overline{\overline{A}}$ and $\overline{\overline{B}}$ matrices using (8.28) and (8.29) or approximations to these expressions depending on the required precision.
4. Construct a vector with the reference field at elements $\overline{\phi}_0$, and solve for $\overline{\phi}$ and $\overline{\psi}$ using the steps described in Sec. 8.2.
5. Use the result to calculate the absorption cross section.

Structure of GFSIEM program with linear or quadratic expansion on triangular elements

1. Define the particle surface, and divide it into triangular elements described by the point matrix $\overline{\overline{P}}$ and $\overline{\overline{T}}$. Add for each element information about element type and normal vector.
2. Construct the $\overline{\overline{A}}$ and $\overline{\overline{B}}$ matrices by following the steps in Sec. 8.3 or Sec. 8.4.
3. Construct a vector with the reference field at mesh points $\overline{\phi}_0$, and solve for $\overline{\phi}$ and $\overline{\psi}$.
4. Use the result to calculate the absorption cross section.

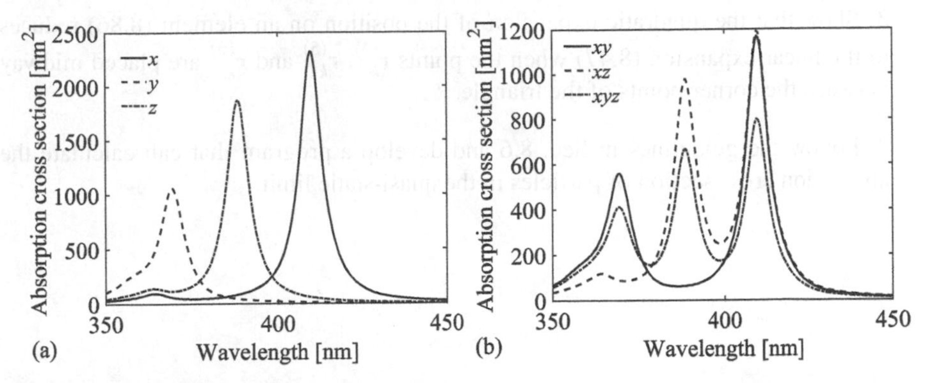

Figure 8.8 Numerical results for the absorption cross section of a silver particle with lengths 14 nm along x, 10 nm along y and 12 nm along z. The particle is surrounded by water. Corners and edges are rounded by a radius of 5 nm. (a) Spectra for the electric field being oriented along the x, y, and z directions. (b) Spectra for the electric field along the xy, xz, and xyz directions.

8.7 EXERCISES

1. In this exercise the analytic expression for the absorption cross section of a spherical particle of radius R in (8.41) will be obtained. Use the applied potential $\phi_0(\mathbf{r}) = -E_0 z$. Assume that the potential can be expressed as

$$\phi(\mathbf{r}) = A_1 z + B_1 \frac{z}{r^3}, \quad r > R, \tag{8.99}$$

$$\phi(\mathbf{r}) = A_2 z + B_2 \frac{z}{r^3}, \quad r < R. \tag{8.100}$$

Apply spherical coordinates ($z = r\cos\theta$). First use appropriate boundary conditions for $r \to 0$ and $r \to \infty$ to determine the values of A_1 and B_2. Afterward the coefficients A_2 and B_1 can be determined from the electromagnetics boundary conditions across the particle surface. Obtain the field inside the particle from (8.100). Express the absorbed power in the particle in terms of the field inside the particle, and normalize with the intensity of an incident plane wave having the electric field magnitude E_0.

2. Consider again the spherical particle from exercise **1** but for the case when the applied field is absent. In this case it is assumed that $A_1 = B_2 = 0$. Determine the ratio of $\varepsilon_1/\varepsilon_2$ where it is still possible with the expressions (8.99) and (8.100) to solve the equations resulting from the electromagnetics boundary condition.

Consider the discretized integral equations (8.32) and (8.33) in the case when the applied potential is absent ($\phi_0 = 0$). Convert the equations into an eigenvalue problem in terms of the eigenvalue $\lambda = \varepsilon_1/\varepsilon_2$. When sufficient numerical accuracy is used, the same value $\varepsilon_1/\varepsilon_2$ just found analytically can be obtained as an eigenvalue.

3. Show that the quadratic expansion of the position on an element (8.86) reduces to the linear expansion (8.47) when the points $\mathbf{r}_k^{(4)}$, $\mathbf{r}_k^{(5)}$ and $\mathbf{r}_k^{(6)}$ are placed midway between the corner points of the triangle.

4. Follow the guidelines in Sec. 8.6 and develop a program that can calculate the absorption cross section of particles in the quasi-static limit.

9 Surface integral equation method for 3D scattering problems

This chapter is concerned with the Green's function surface integral equation method (GFSIEM) for three-dimensional scattering problems. Compared with Chapter 8, retardation will now be taken fully into account. The same problems can in principle be considered as with the Green's function volume integral equation method (GFVIEM) considered in Chapter 6. With the GFVIEM it was necessary for large problems to, for example, use cubic discretization elements placed on a cubic lattice in order to obtain a discretized integral equation in the form of a discrete convolution, since this allowed the use of memory-efficient iterative methods and fast calculation of matrix-vector products by using the FFT algorithm. However, cubic elements give a stair-cased representation of the surface of a scatterer, which may lead to calculated fields inside especially metallic scatterers with large deviations from the correct value. Such structures require a better representation of the scatterer surface, and this is achieved when using the GFSIEM. The discretization of a surface instead of a volume also means discretization in two dimensions rather than three dimensions, with a resulting dramatic decrease in the number of coefficients in the resulting matrix equation. For structures with cylindrical symmetry it was found in Chapter 7 that the GFVIEM could be reduced to a series of problems with discretization in two dimensions. Here, the GFSIEM further allows us to reduce such problems to a series of problems with discretization in only one dimension.

The chapter is organized as follows. The surface integral equations will be reviewed for a scatterer in a homogeneous medium in Sec. 9.1. The calculation of optical cross sections is briefly discussed in Sec. 9.2. A node-based numerical approach for solving the integral equations for structures with a general shape is presented in Sec. 9.3. A numerical approach that takes advantage of cylindrical symmetry is presented in Sec. 9.4. Finally, examples of using the GFSIEM is presented for cylindrical metal-nano-disc resonators in Sec. 9.5.

9.1 SURFACE INTEGRAL EQUATIONS

In this section the electric-field Green's function integral equations are reviewed for a three-dimensional scatterer with dielectric constant ε_2 surrounded by a homogeneous medium with dielectric constant ε_1. A detailed derivation of the surface integral equations for a three-dimensional scatterer in free space has been given, for example, in [161]. The integral equations have also been accounted for in, for example, Refs. [23, 22]. An alternative is to instead work with the vector potential [162].

The scattering situation is illustrated in Fig. 9.1. The respective regions inside and outside the scatterer are denoted Ω_2 and Ω_1, and the scatterer is illuminated by an incident field $\mathbf{E}_{\mathrm{inc}}$. It should be imagined that the scatterer is now a three-dimensional object with a surface in three dimensions.

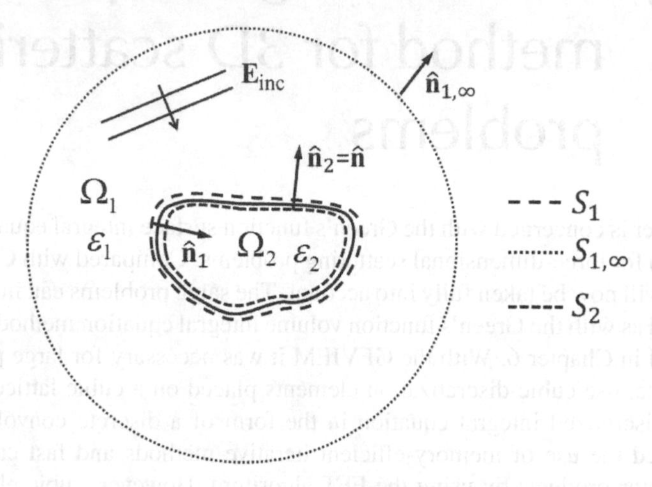

Figure 9.1 Schematic of a nanoparticle with dielectric constant ε_2 surrounded by a homogeneous dielectric with dielectric constant ε_1. The regions inside and outside the nanoparticle are referred to as Ω_2 and Ω_1, respectively. S_1 and S_2 refer to boundaries placed, respectively, just outside and just inside the nanoparticle surface, and $S_{1,\infty}$ is a spherical surface at a very large distance from the scatterer. $\hat{\mathbf{n}}_i$ is the normal vector pointing out of region Ω_i.

In both of the regions Ω_1 and Ω_2 the dielectric function is assumed constant, and the overall structure is piecewise homogeneous. Surface integral equations for a homogeneous region Ω_i can be obtained from first principles by starting from the vector wave equation for the electric field in that region, and the equation for the Green's tensor of a homogeneous medium with the same dielectric constant

$$-\nabla\times\nabla\times\mathbf{E}(\mathbf{r})+k_0^2\varepsilon_i\mathbf{E}(\mathbf{r})=-i\omega\mu_0\mathbf{J}(\mathbf{r}),\quad \mathbf{r}\in\Omega_i, \tag{9.1}$$

and

$$-\nabla\times\nabla\times\mathbf{G}_i(\mathbf{r},\mathbf{r}')+k_0^2\varepsilon_i\mathbf{G}(\mathbf{r},\mathbf{r}')=-\mathbf{I}\delta(\mathbf{r}-\mathbf{r}'). \tag{9.2}$$

By multiplying (9.1) from the right with $\mathbf{G}_i(\mathbf{r},\mathbf{r}')$, multiplying (9.2) from the left with $\mathbf{E}(\mathbf{r})$, and subtracting one resulting equation from the other, the following is obtained

$$(-\nabla\times\nabla\times\mathbf{E}(\mathbf{r}))\cdot\mathbf{G}_i(\mathbf{r},\mathbf{r}')+\mathbf{E}(\mathbf{r})\cdot(\nabla\times\nabla\times\mathbf{G}_i(\mathbf{r},\mathbf{r}'))=$$
$$-i\omega\mu_0\mathbf{J}(\mathbf{r})\cdot\mathbf{G}_i(\mathbf{r},\mathbf{r}')+\mathbf{E}(\mathbf{r})\delta(\mathbf{r}-\mathbf{r}'). \tag{9.3}$$

Now the left- and right-hand sides are integrated over region Ω_i. For the right-hand side (*RHS*) this leads to

$$RHS=-\mathbf{E}_{\mathrm{inc},i}(\mathbf{r}')+\mathbf{E}(\mathbf{r}'),\quad \mathbf{r}'\in\Omega_i, \tag{9.4}$$

where the incident field in region Ω_i is defined as

$$\mathbf{E}_{\mathrm{inc},i}(\mathbf{r}') = i\omega\mu_0 \int_{\Omega_i} \mathbf{G}_i(\mathbf{r}',\mathbf{r}) \cdot \mathbf{J}(\mathbf{r}) d^3r, \tag{9.5}$$

where reciprocity was used:

$$\mathbf{J}(\mathbf{r}) \cdot \mathbf{G}_i(\mathbf{r},\mathbf{r}') = \mathbf{G}_i(\mathbf{r}',\mathbf{r}) \cdot \mathbf{J}(\mathbf{r}). \tag{9.6}$$

It is straightforward to see that (9.6) is satisfied for the free-space Green's tensor (6.15).

By using the following vector formula

$$\nabla \cdot (\mathbf{A} \times \mathbf{B}) = (\nabla \times \mathbf{A}) \cdot \mathbf{B} - \mathbf{A} \cdot (\nabla \times \mathbf{B}), \tag{9.7}$$

where $\mathbf{A}$ is a vector, and $\mathbf{B}$ is a vector or a dyad, the integral of the left-hand side of (9.3) over Ω_i can be written as

$$LHS = -\int_{\Omega_i} \nabla \cdot \left(\{\nabla \times \mathbf{E}(\mathbf{r})\} \times \mathbf{G}_i(\mathbf{r},\mathbf{r}') + \mathbf{E}(\mathbf{r}) \times \{\nabla \times \mathbf{G}_i(\mathbf{r},\mathbf{r}')\} \right) d^3r. \tag{9.8}$$

By applying Gauss's law, this may be converted to a surface integral over the surface of Ω_i being here referred to as $\partial\Omega_i$:

$$LHS = -\oint_{\partial\Omega_i} \hat{\mathbf{n}}_i \cdot \left(\{\nabla \times \mathbf{E}(\mathbf{r})\} \times \mathbf{G}_i(\mathbf{r},\mathbf{r}') + \mathbf{E}(\mathbf{r}) \times \{\nabla \times \mathbf{G}_i(\mathbf{r},\mathbf{r}')\} \right) d^2r, \tag{9.9}$$

where $\hat{\mathbf{n}}_i$ is the surface normal vector pointing out of Ω_i.

We may now further use the vector formulas

$$\mathbf{A} \cdot (\mathbf{B} \times \mathbf{C}) = -\mathbf{B} \cdot (\mathbf{A} \times \mathbf{C}) = (\mathbf{A} \times \mathbf{B}) \cdot \mathbf{C}, \tag{9.10}$$

where $\mathbf{A}$ and $\mathbf{B}$ are vectors, and $\mathbf{C}$ is either a vector or a dyad. This leads to

$$LHS = -\oint_{\partial\Omega_i} \left(\{\hat{\mathbf{n}}_i \times (\nabla \times \mathbf{E}(\mathbf{r}))\} \cdot \mathbf{G}_i(\mathbf{r},\mathbf{r}') + \{\hat{\mathbf{n}}_i \times \mathbf{E}(\mathbf{r})\} \cdot \nabla \times \mathbf{G}_i(\mathbf{r},\mathbf{r}') \right) d^2r. \tag{9.11}$$

It is noted that $\nabla \times \mathbf{E}(\mathbf{r}) = i\omega\mu_0 \mathbf{H}(\mathbf{r})$.

The equivalent electric and magnetic surface current densities on the surface of Ω_i are now defined as

$$\mathbf{J}_{\mathrm{s},i}(\mathbf{r}) \equiv -\hat{\mathbf{n}}_i \times \mathbf{H}(\mathbf{r}), \quad \mathbf{r} \text{ on } \partial\Omega_i, \tag{9.12}$$

$$\mathbf{M}_{\mathrm{s},i}(\mathbf{r}) \equiv \hat{\mathbf{n}}_i \times \mathbf{E}(\mathbf{r}), \quad \mathbf{r} \text{ on } \partial\Omega_i. \tag{9.13}$$

In addition, reciprocity as in (9.6) can be applied once more, and it can be used that

$$\mathbf{M}_{\mathrm{s},i}(\mathbf{r}) \cdot \left[\nabla \times \mathbf{G}_i(\mathbf{r},\mathbf{r}')\right] = -\left[\nabla \times \mathbf{G}_i(\mathbf{r},\mathbf{r}')\right] \cdot \mathbf{M}_{\mathrm{s},i}(\mathbf{r}). \tag{9.14}$$

Finally, $\mathbf{r}$ and $\mathbf{r}'$ should be interchanged, and here it can be used for the homogeneous-medium Green's tensor that

$$\nabla' \times \mathbf{G}_i(\mathbf{r}',\mathbf{r}) = -\nabla \times \mathbf{G}_i(\mathbf{r},\mathbf{r}'). \tag{9.15}$$

These considerations lead to the integral equation

$$\mathbf{E}(\mathbf{r}) = \mathbf{E}_{\text{inc},i}(\mathbf{r}) + i\omega\mu_0 \oint_{\partial\Omega_i} \mathbf{G}(\mathbf{r},\mathbf{r}')\cdot\mathbf{J}_{\text{s},i}(\mathbf{r}')d^2r' - \oint_{\partial\Omega_i} \nabla\times\mathbf{G}_i(\mathbf{r},\mathbf{r}')\cdot\mathbf{M}_{\text{s},i}(\mathbf{r}')d^2r'. \tag{9.16}$$

By considering a position $\mathbf{r}$ on $\partial\Omega_i$ and applying $\hat{\mathbf{n}}_i\times$ to both the left- and right-hand sides of (9.16), and applying (9.13), the equation is cast in terms of only $\mathbf{J}$, $\mathbf{M}$ and $\mathbf{E}_{\text{inc},i}$.

Now consider a scatterer with dielectric constant ε_2 in region Ω_2 being surrounded by a homogeneous medium region Ω_1 with dielectric constant ε_1 as illustrated in Fig. 9.1. When using the Green's tensor in Ω_1 that satisfies the radiating boundary condition, it can be shown that the integrals over the outer surface placed at infinity vanish. This can be seen by using a spherical outer surface with surface normal vector $\hat{\mathbf{r}}'$, by using the expression for the far-field Green's tensor (6.23), and the relation between the electric and magnetic far fields

$$i\omega\mu_0\mathbf{H}^{(\text{ff})}(\mathbf{r}) = ik_1\hat{\mathbf{r}}\times\mathbf{E}^{(\text{ff})}(\mathbf{r}). \tag{9.17}$$

The two integrals over the outer surface will then have integrands that are identical except for having opposite sign, and the two integrals over the outer surface thus cancel each other. In the right-hand side of (9.16) it is thus sufficient to only consider the integral over the scatterer surface. In addition there will only be an incident field in Ω_1 and no such field in Ω_2. Thus $\mathbf{E}_{\text{inc},2} = \mathbf{0}$. In the following, the incident field in region 1 will therefore be denoted simply $\mathbf{E}_{\text{inc}}$.

For a dielectric scatterer or a metal scatterer, but not a perfect electric (or magnetic) conductor, the tangential electric field and tangential magnetic field must be continuous across the scatterer surface. It is thus convenient to redefine the electric and magnetic surface currents (9.12) and (9.13) in terms of only the outward surface normal vector of the scatterer, since then the electric and magnetic currents on both sides of the scatterer surface will be identical. The resulting electric-field integral equations are summarized in the following text box:

Green's function surface integral equations for a scatterer with dielectric constant ε_2 in Ω_2 embedded in a homogeneous region Ω_1 with dielectric constant ε_1:

The electric field at a position $\mathbf{r} \in \Omega_1$ outside a scatterer in a homogeneous medium is given from the integral over the outer surface S_1 of the scatterer

$$\mathbf{E}(\mathbf{r}) = \mathbf{E}_{\text{inc}}(\mathbf{r}) + i\omega\mu_0 \oint_{S_1} \mathbf{G}_1(\mathbf{r},\mathbf{r}') \cdot \mathbf{J}_s(\mathbf{r}') d^2r' - \oint_{S_1} \nabla \times \mathbf{G}_1(\mathbf{r},\mathbf{r}') \cdot \mathbf{M}_s(\mathbf{r}') d^2r', \tag{9.18}$$

and the electric field at a position $\mathbf{r} \in \Omega_2$ inside the scatterer is given from the integral over the inner surface S_2 of the scatterer

$$\mathbf{E}(\mathbf{r}) = -i\omega\mu_0 \oint_{S_2} \mathbf{G}_2(\mathbf{r},\mathbf{r}') \cdot \mathbf{J}_s(\mathbf{r}') d^2r' + \oint_{S_2} \nabla \times \mathbf{G}_2(\mathbf{r},\mathbf{r}') \cdot \mathbf{M}_s(\mathbf{r}') d^2r'. \tag{9.19}$$

The electric and magnetic surface currents are given by

$$\mathbf{J}_s(\mathbf{r}) = \hat{\mathbf{n}} \times \mathbf{H}(\mathbf{r}), \quad \mathbf{r} \text{ on scatterer surface}, \tag{9.20}$$

$$\mathbf{M}_s(\mathbf{r}) = -\hat{\mathbf{n}} \times \mathbf{E}(\mathbf{r}), \quad \mathbf{r} \text{ on scatterer surface}, \tag{9.21}$$

where $\hat{\mathbf{n}}$ is the surface normal vector pointing into Ω_1.

The surface currents can be obtained self-consistently from (9.18) and (9.19) by letting the position $\mathbf{r}$ approach the scatterer surface from either side, which leads to the electric field integral equations

$$-\mathbf{M}_s(\mathbf{r}) = \hat{\mathbf{n}} \times \mathbf{E}_{\text{inc}}(\mathbf{r}) + i\omega\mu_0 \oint_{S_1} [\hat{\mathbf{n}} \times \mathbf{G}_1(\mathbf{r},\mathbf{r}')] \cdot \mathbf{J}_s(\mathbf{r}') d^2r'$$
$$- \oint_{S_1} [\hat{\mathbf{n}} \times (\nabla \times \mathbf{G}_1(\mathbf{r},\mathbf{r}'))] \cdot \mathbf{M}_s(\mathbf{r}') d^2r',$$
$$\mathbf{r} \text{ infinitesimally outside scatterer surface}, \tag{9.22}$$

$$-\mathbf{M}_s(\mathbf{r}) = -i\omega\mu_0 \oint_{S_2} [\hat{\mathbf{n}} \times \mathbf{G}_2(\mathbf{r},\mathbf{r}')] \cdot \mathbf{J}_s(\mathbf{r}') d^2r' +$$
$$\oint_{S_2} [\hat{\mathbf{n}} \times (\nabla \times \mathbf{G}_2(\mathbf{r},\mathbf{r}'))] \cdot \mathbf{M}_s(\mathbf{r}') d^2r',$$
$$\mathbf{r} \text{ infinitesimally inside scatterer surface}. \tag{9.23}$$

It can be noted that there are two vector unknowns $\mathbf{J}_s$ and $\mathbf{M}_s$ and two vector equations, which is sufficient to obtain the surface currents. It is also possible to formulate similar integral equations for the magnetic field instead [23, 22, 161] that can also be applied to calculate the same surface currents. Sometimes the electric-field integral equations (EFIEs) and the magnetic field integral equations (MFIEs) are combined to create mixed equations, which can improve stability. In this book only the EFIE is considered.

For a perfect electric conductor (PEC) it is assumed that both **H** and **E** are zero just inside the scatterer surface. At the surface of a PEC the tangential component of **E** vanishes, and thus also $\mathbf{M}_s$ vanishes. In that case it is sufficient to consider only one vector equation (9.22) with one vector unknown, $\mathbf{J}_s$. In this case the magnetic field will jump across the surface, and $\hat{\mathbf{n}} \times (\mathbf{H}_1 - \mathbf{H}_2) = \hat{\mathbf{n}} \times \mathbf{H}_1 = \mathbf{J}_s$ will be the physical surface current.

As the position **r** approaches a position **s** on, for example, the surface S_1, the singularity of the Green's tensor must be dealt with. The integral involving $\hat{\mathbf{n}} \times (\nabla \times \mathbf{G}_1(\mathbf{r},\mathbf{r}'))$ may be split into two parts, one integral over an infinitesimal area δS_1 in the immediate vicinity of **r** and an integral over the rest of S_1. For the homogeneous-medium Green's tensor it can be shown that

$$\left[\hat{\mathbf{n}} \times (\nabla \times \mathbf{G}_1(\mathbf{r},\mathbf{r}'))\right] \cdot \mathbf{M}_s(\mathbf{r}') =$$

$$\left(ik_1 - \frac{1}{R}\right) \frac{e^{ik_1 R}}{4\pi R} \left[\frac{\mathbf{R}}{R}(\hat{\mathbf{n}}(\mathbf{r}) \cdot \mathbf{M}_s(\mathbf{r}')) - \mathbf{M}_s(\mathbf{r}')\left(\hat{\mathbf{n}}(\mathbf{r}) \cdot \frac{\mathbf{R}}{R}\right)\right], \tag{9.24}$$

where $\mathbf{R} = \mathbf{r} - \mathbf{r}'$ and $R = |\mathbf{R}|$.

The term involving $\hat{\mathbf{n}}(\mathbf{s}) \cdot \mathbf{M}_s(\mathbf{r}')$ will vanish for $\mathbf{r}' \to \mathbf{s}$ for a smooth surface, since the surface currents and the normal vector will be orthogonal in this limit. This term is thus unproblematic. For the other term involving $\left(\hat{\mathbf{n}}(\mathbf{s}) \cdot \frac{\mathbf{R}}{R}\right)$ it may be noted that for a flat surface, and $\mathbf{r}'$ at a finite distance from **r**, also $\hat{\mathbf{n}}$ and $\mathbf{R}/R$ will be orthogonal. However, as $\mathbf{r}' \to \mathbf{s}$ it is found that

$$\left(\hat{\mathbf{n}}(\mathbf{r}) \cdot \frac{\mathbf{R}}{R^3}\right) d^2r' \approx \frac{d\Omega}{4\pi}, \tag{9.25}$$

where $d\Omega$ is the solid angle seen from **r** that corresponds to the area d^2r'. If the position **r** approaches the scatterer surface from the other side of the surface, as will be the case in the term $\hat{\mathbf{n}} \times (\nabla \times \mathbf{G}_2(\mathbf{r},\mathbf{r}'))$, the sign will be opposite.

Thus, for a smooth surface

$$-\oint_{S_1} \left[\hat{\mathbf{n}} \times (\nabla \times \mathbf{G}_1(\mathbf{r},\mathbf{r}'))\right]_{\mathbf{r}\to\mathbf{s}} \cdot \mathbf{M}_s(\mathbf{r}') d^2r' = \tag{9.26}$$

$$-\frac{1}{2}\mathbf{M}_s(\mathbf{s}) - P\int_{S_1} \left[\hat{\mathbf{n}} \times (\nabla \times \mathbf{G}_1(\mathbf{r},\mathbf{r}'))\right]_{\mathbf{r}=\mathbf{s}} \cdot \mathbf{M}_s(\mathbf{r}') d^2r', \tag{9.27}$$

and

$$+\oint_{S_2} \left[\hat{\mathbf{n}} \times (\nabla \times \mathbf{G}_2(\mathbf{r},\mathbf{r}'))\right]_{\mathbf{r}\to\mathbf{s}} \cdot \mathbf{M}_s(\mathbf{r}') d^2r' = \tag{9.28}$$

$$-\frac{1}{2}\mathbf{M}_s(\mathbf{s}) + P\int_{S_2} \left[\hat{\mathbf{n}} \times (\nabla \times \mathbf{G}_2(\mathbf{r},\mathbf{r}'))\right]_{\mathbf{r}=\mathbf{s}} \cdot \mathbf{M}_s(\mathbf{r}') d^2r', \tag{9.29}$$

where P means that the infinitesimal region just around **s** should be excluded from the integral.

The integral involving $\hat{\mathbf{n}} \times \mathbf{G}_1(\mathbf{r},\mathbf{r}')$ is in principle unproblematic. For the homogeneous-medium Green's tensor

$$\hat{\mathbf{n}} \times \mathbf{G}_1(\mathbf{r},\mathbf{r}') \cdot \mathbf{J}_s(\mathbf{r}') = \hat{\mathbf{n}} \times \left[\left(\mathbf{I} + \frac{1}{k_1^2} \nabla\nabla \right) \frac{e^{ik_1|\mathbf{r}-\mathbf{r}'|}}{4\pi|\mathbf{r}-\mathbf{r}'|} \right] \cdot \mathbf{J}_s(\mathbf{r}'). \tag{9.30}$$

Now consider a flat circular surface with radius a lying in the xy-plane with center position $\mathbf{s} = \mathbf{0}$. The position $\mathbf{r}$ is placed just above $\mathbf{s}$ in Ω_1, i.e., $\mathbf{r} = \mathbf{s} + \hat{\mathbf{z}}\delta$, where $\delta \to 0$. The integral over this surface will be carried out in only the xy-plane. If we assume that the disk is sufficiently small that $\mathbf{J}_s(\mathbf{r}') \approx \mathbf{J}_s(\mathbf{s})$ over the disk, then the integral can be carried out analytically, i.e.,

$$\int_{\text{disc}} \hat{\mathbf{n}} \times \mathbf{G}_1(\mathbf{s},\mathbf{r}') \cdot \mathbf{J}_s(\mathbf{s}) d^2r' \approx$$

$$\frac{1}{k_1^2} \hat{\mathbf{n}} \times \int_{\rho'=0}^{a} \int_{\phi'=0}^{2\pi} \left(\frac{-1}{4\pi R^3} \mathbf{I} - \frac{\mathbf{RR}}{R^2} \frac{-3}{4\pi R^3} \right) \rho' d\rho' d\phi' \cdot \mathbf{J}_s(\mathbf{s}) =$$

$$\frac{1}{k_1^2} \int_{\rho'=0}^{a} \left(\frac{-1}{2R^3} + \frac{3}{4} \frac{\rho'^2}{R^2} \frac{1}{R^3} \right) \rho' d\rho' \hat{\mathbf{n}} \times \mathbf{J}_s(\mathbf{s}) = -\frac{1}{4k_1^2 a} \hat{\mathbf{n}} \times \mathbf{J}_s(\mathbf{s}). \tag{9.31}$$

Here $\rho' = |\mathbf{r}' - \mathbf{s}|$, $\mathbf{R} = \mathbf{r} - \mathbf{r}'$, and $R = |\mathbf{R}|$. The integral will not depend on whether $\mathbf{r}$ is on one or the other side of the disc. For very small disc radii, the integral will scale as $1/a$, and thus the smaller the disc the larger the value of the integral. The integral involving $\hat{\mathbf{n}} \times \mathbf{G}_1(\mathbf{r},\mathbf{r}')$ can be carried out straightforwardly when a position $\mathbf{r}$ is chosen that is a finite but very small distance outside the scatterer surface.

9.2 CALCULATING OPTICAL CROSS SECTIONS

The normal component of the Poynting vector $\mathbf{S}$ at the surface of the scatterer can be directly calculated using the electric and magnetic surface currents:

$$\hat{\mathbf{n}} (\hat{\mathbf{n}} \cdot \mathbf{S}(\mathbf{r})) = \frac{1}{2} \text{Real} \left\{ (\mathbf{J}(\mathbf{r}))^* \times \mathbf{M}(\mathbf{r}) \right\}. \tag{9.32}$$

The power absorbed in the scatterer is thus given by minus the outward Poynting vector flux as follows

$$P_{\text{abs}} = -\oint \hat{\mathbf{n}} \cdot \frac{1}{2} \text{Real} \left\{ (\mathbf{J}(\mathbf{r}))^* \times \mathbf{M}(\mathbf{r}) \right\} d^2r. \tag{9.33}$$

In the case of the incident light being a plane wave with electric field $\mathbf{E}_{\text{inc}}$, the absorption cross section is given by

$$\sigma_{\text{abs}} = \frac{P_{\text{abs}}}{\frac{1}{2} n_1 |\mathbf{E}_{\text{inc}}|^2 \sqrt{\varepsilon_0/\mu_0}}. \tag{9.34}$$

The scattering and extinction cross sections can be obtained by first finding the scattered far field by using the integral equation (9.18) with the far-field Green's tensor (6.23), and by subsequently using the far fields in Eqs. (2.145) and (2.146).

9.3 NUMERICAL APPROACH: GENERAL STRUCTURE

In this section a numerical approach is considered where the surface of the scatterer is divided into a number of triangle-like elements similar to the finite-element approach used in Chapter 8. The scatterer surface is thus described by a point matrix with a set of points on the surface, and a connectivity matrix describing which points are connected in triangles. The matrices may also contain additional information regarding the normal and tangential unit vectors and the type of element, etc.

The point matrix $\overline{\overline{P}}$ may have the form

$$\overline{\overline{P}} = \begin{bmatrix} x_1 & x_2 & \dots & x_N \\ y_1 & y_2 & \dots & y_N \\ z_1 & z_2 & \dots & z_N \\ n_{x,1} & n_{x,2} & \dots & n_{x,N} \\ n_{y,1} & n_{y,2} & \dots & n_{y,N} \\ n_{z,1} & n_{z,2} & \dots & n_{z,N} \\ t^{(1)}_{x,1} & t^{(1)}_{x,2} & \dots & t^{(1)}_{x,N} \\ t^{(1)}_{y,1} & t^{(1)}_{y,2} & \dots & t^{(1)}_{y,N} \\ t^{(1)}_{z,1} & t^{(1)}_{z,2} & \dots & t^{(1)}_{z,N} \\ t^{(2)}_{x,1} & t^{(2)}_{x,2} & \dots & t^{(2)}_{x,N} \\ t^{(2)}_{y,1} & t^{(2)}_{y,2} & \dots & t^{(2)}_{y,N} \\ t^{(2)}_{z,1} & t^{(2)}_{z,2} & \dots & t^{(2)}_{z,N} \end{bmatrix}. \tag{9.35}$$

Each column represents a node point on the surface of the scatterer, where the first three coefficients of column i denoted x_i, y_i, and z_i are the coordinates of the point. The next three coefficients $n_{x,i}$, $n_{y,i}$, and $n_{z,i}$ are the components of the surface normal unit vector at the point. The coefficients $t^{(a)}_{x,i}$, $t^{(a)}_{y,i}$, and $t^{(a)}_{z,i}$ are the components of a surface tangential unit vector at the point where $a = 1$ or 2 is used for distinguishing between two different tangential unit vectors. Such vectors were not needed in Chapter 8. There are many ways of choosing the tangential unit vectors but they should always span the same tangent plane. The tangential unit vectors will here be chosen orthogonal. In terms of the tangent unit vectors

$$\hat{\mathbf{t}}^{(a)}_i = \hat{\mathbf{x}} t^{(a)}_{x,i} + \hat{\mathbf{y}} t^{(a)}_{y,i} + \hat{\mathbf{z}} t^{(a)}_{z,i}, \tag{9.36}$$

the electric and magnetic surface currents at point i can be expressed as

$$\hat{\mathbf{J}}_i = \hat{\mathbf{t}}^{(1)}_i J^{(1)}_i + \hat{\mathbf{t}}^{(2)}_i J^{(2)}_i, \tag{9.37}$$

$$\hat{\mathbf{M}}_i = \hat{\mathbf{t}}^{(1)}_i M^{(1)}_i + \hat{\mathbf{t}}^{(2)}_i M^{(2)}_i. \tag{9.38}$$

The coefficients $J^{(a)}_i$ and $M^{(a)}_i$, $a = 1$ or 2, are global values describing the currents at the node points. These can be conveniently organized in vectors

$$\overline{J}^{(a)} = \left[J^{(a)}_1 \quad J^{(a)}_2 \quad \dots \quad J^{(a)}_N \right]^T, \tag{9.39}$$

$$\overline{M}^{(a)} = \left[M_1^{(a)} \quad M_2^{(a)} \quad \dots \quad M_N^{(a)} \right]^T. \tag{9.40}$$

It may also be useful to express $\hat{\mathbf{n}} \times \mathbf{E}_{\text{inc}}$ at node i as

$$\hat{\mathbf{n}}_i \times \hat{\mathbf{E}}_{\text{inc}}(\mathbf{r}_i) = \hat{\mathbf{t}}_i^{(1)} E_{\text{inc},i}^{(1)} + \hat{\mathbf{t}}_i^{(2)} E_{\text{inc},i}^{(2)}, \tag{9.41}$$

and to organize the global reference field coefficients in vectors as

$$\overline{E}_{\text{inc}}^{(a)} = \left[E_{\text{inc},1}^{(a)} \quad E_{\text{inc},2}^{(a)} \quad \dots \quad E_{\text{inc},N}^{(a)} \right]^T. \tag{9.42}$$

The connectivity matrix $\overline{\overline{T}}$ has the form

$$\overline{\overline{T}} = \begin{bmatrix} \dots & i_k^{(1)} & \dots \\ \dots & i_k^{(2)} & \dots \\ \dots & i_k^{(3)} & \dots \\ \dots & \text{element type} & \dots \\ \dots & \text{additional data} & \dots \end{bmatrix}. \tag{9.43}$$

The column k represents a triangular-like element, where the first three coefficients $i_k^{(1)}$, $i_k^{(2)}$, and $i_k^{(3)}$ identify the corner points of the element. The coordinates of the corners can be obtained from columns $i_k^{(u)}$ of the point matrix. The fourth coefficient in column k can be used to specify the type of element, e.g. if it is a flat triangle, if it is part of a spherical surface, or part of a cylindrical surface. Additional data can be added to the column for the description of curved elements, e.g. in the case of a spherical surface, the sphere center and radius can be added, etc. This is similar to the discussion in Chapter 8. In a node-based approach it is important that the surface is smooth and has no sharp corners, especially at the node points, since otherwise the surface currents are not well defined at the node points.

Similar to the discussion in Chapter 8 the position on an element k may be parameterized by two parameters u and v and given by an expression in the form $\mathbf{r} = \mathbf{r}_k(u,v)$. Two surface tangential unit vectors on the element may now be defined as

$$\hat{\mathbf{t}}^{(k,1)}(u,v) = \frac{\partial \mathbf{r}_k(u,v)}{\partial u} \Big/ \left| \frac{\partial \mathbf{r}_k(u,v)}{\partial u} \right|, \tag{9.44}$$

$$\hat{\mathbf{t}}^{(k,2)}(u,v) = \frac{\partial \mathbf{r}_k(u,v)}{\partial v} \Big/ \left| \frac{\partial \mathbf{r}_k(u,v)}{\partial v} \right|. \tag{9.45}$$

Note that these two tangential unit vectors are not necessarily orthogonal, and in the case of a curved element, their direction will change with the position on the element.

A linear expansion of the electric and magnetic surface currents on element k can be expressed in terms of these vectors as

$$\mathbf{J}(\mathbf{r}_k(u,v)) = \sum_{j=1}^{3} \left(\hat{\mathbf{t}}^{(k,1)}(u,v) J_k^{(j,1)} + \hat{\mathbf{t}}^{(k,2)}(u,v) J_k^{(j,2)} \right) f^{(j,\mathrm{L})}(u,v), \tag{9.46}$$

$$\mathbf{M}(\mathbf{r}_k(u,v)) = \sum_{j=1}^{3} \left(\hat{\mathbf{t}}^{(k,1)}(u,v) M_k^{(j,1)} + \hat{\mathbf{t}}^{(k,2)}(u,v) M_k^{(j,2)} \right) f^{(j,\mathrm{L})}(u,v), \tag{9.47}$$

where the linear expansion functions $f^{(j,L)}(u,v)$ are defined in (8.48)-(8.50).

The coefficients $J_k^{(j,a)}$ and $M_k^{(j,a)}$ are local variables for the surface currents, that can be organized in vectors as

$$\overline{J}_k^{(a)} = \left[J_k^{(1,a)} \quad J_k^{(2,a)} \quad J_k^{(3,a)} \right]^T, \tag{9.48}$$

$$\overline{M}_k^{(a)} = \left[M_k^{(1,a)} \quad M_k^{(2,a)} \quad M_k^{(3,a)} \right]^T. \tag{9.49}$$

Due to the definition of the linear expansion functions $f^{(j,\mathrm{L})}(u,v)$, the total current at node $i_k^{(1)}$ is given as $\mathbf{J}(\mathbf{r}_k(u=0,v=0))$, the total current at node $i_k^{(2)}$ is given as $\mathbf{J}(\mathbf{r}_k(u=1,v=0))$, and the total current at node $i_k^{(3)}$ is given as $\mathbf{J}(\mathbf{r}_k(u=0,v=1))$. The currents at nodes can also be expressed in terms of the global variables (see (9.37) and (9.38)), and the local and global variables are thus related by expressions of the form

$$\begin{bmatrix} \overline{J}_k^{(1)} \\ \overline{J}_k^{(2)} \end{bmatrix} = \overline{\overline{P}}_k \begin{bmatrix} \overline{J}^{(1)} \\ \overline{J}^{(2)} \end{bmatrix}, \tag{9.50}$$

$$\begin{bmatrix} \overline{M}_k^{(1)} \\ \overline{M}_k^{(2)} \end{bmatrix} = \overline{\overline{P}}_k \begin{bmatrix} \overline{M}^{(1)} \\ \overline{M}^{(2)} \end{bmatrix}. \tag{9.51}$$

A matrix equation related to the position i can be obtained from the integral equation (9.22) by using $\mathbf{r} = \mathbf{r}_i$, by taking the dot product from the left with the vector $\hat{\mathbf{t}}_i^{(a)}$, and by dividing the surface integrals into a sum of surface integrals over each surface element with the expansions of currents (9.46) and (9.47). Thus,

$$E_{\mathrm{inc},i}^{(a)} = -\frac{1}{2} M_i^{(a)} - \sum_k \left[\overline{\overline{K}}_{1,a,1,i,k}^{(J)} \quad \overline{\overline{K}}_{1,a,2,i,k}^{(J)} \right] \begin{bmatrix} \overline{J}_k^{(1)} \\ \overline{J}_k^{(2)} \end{bmatrix}$$

$$+ \sum_k \left[\overline{\overline{K}}_{1,a,1,i,k}^{(M)} \quad \overline{\overline{K}}_{1,a,2,i,k}^{(M)} \right] \begin{bmatrix} \overline{M}_k^{(1)} \\ \overline{M}_k^{(2)} \end{bmatrix}. \tag{9.52}$$

A similar matrix equation can be constructed from the integral equation (9.23). The matrices $\overline{\overline{K}}_{c,a,b,i,k}^{(U)}$ are row vectors with 3 elements ($j = 1,2,3$) governing the coupling between element k and the global point i, and are given by

$$\left[\overline{\overline{K}}_{c,a,b,i,k}^{(J)} \right]_j =$$

$$i\omega\mu_0 \int_{u=0}^{1} \int_{v=0}^{1-u} \hat{\mathbf{t}}_i^{(a)} \cdot \hat{\mathbf{n}}_i \times \mathbf{G}_c(\mathbf{r}_i, \mathbf{r}_k(u,v)) \cdot \hat{\mathbf{t}}^{(k,b)}(u,v) f^{(j,L)}(u,v) 2A_k(u,v) du dv, \tag{9.53}$$

$$\left[\overline{\overline{K}}^{(M)}_{c,a,b,i,k}\right]_j =$$

$$\int_{u=0}^{1}\int_{v=0}^{1-u} \hat{\mathbf{t}}_i^{(a)} \cdot \hat{\mathbf{n}}_i \times [\nabla \times \mathbf{G}_c(\mathbf{r}, \mathbf{r}_k(u,v))]_{\mathbf{r}=\mathbf{r}_i} \cdot \hat{\mathbf{t}}^{(k,b)}(u,v) f^{(j,L)}(u,v) 2A_k(u,v)dudv. \tag{9.54}$$

The area function $A_k(u,v)$ is defined in (8.67).

By using (9.50) and (9.51) the grand matrix equation can now be formulated as

$$\begin{bmatrix} \overline{E}^{(1)}_{\text{inc}} \\ \overline{E}^{(2)}_{\text{inc}} \\ \overline{0} \\ \overline{0} \end{bmatrix} = \begin{bmatrix} -\overline{\overline{K}}^{(J)}_1 & \left(\overline{\overline{K}}^{(M)}_1 - \frac{1}{2}\overline{\overline{I}}\right) \\ \overline{\overline{K}}^{(J)}_2 & \left(-\overline{\overline{K}}^{(M)}_2 - \frac{1}{2}\overline{\overline{I}}\right) \end{bmatrix} \begin{bmatrix} \overline{J}^{(1)} \\ \overline{J}^{(2)} \\ \overline{M}^{(1)} \\ \overline{M}^{(2)} \end{bmatrix}, \tag{9.55}$$

where the involved matrices are given by

$$\overline{\overline{K}}^{(U)}_c = \sum_i \overline{\overline{P}}'_i \sum_k \begin{bmatrix} \overline{\overline{K}}^{(U)}_{c,1,1,i,k} & \overline{\overline{K}}^{(U)}_{c,1,2,i,k} \\ \overline{\overline{K}}^{(U)}_{c,2,1,i,k} & \overline{\overline{K}}^{(U)}_{c,2,2,i,k} \end{bmatrix} \overline{\overline{P}}_k. \tag{9.56}$$

The matrix $\overline{\overline{P}}'_i$ is given by

$$\overline{\overline{P}}'_i = \begin{bmatrix} \overline{P}''_i & \overline{0} \\ \overline{0} & \overline{P}''_i \end{bmatrix}, \tag{9.57}$$

where $\overline{P}''_i$ is a column vector of length N with all zeros except for a 1 at position i. The matrices $\overline{\overline{K}}^{(U)}_c$ are of size $2N \times 2N$.

The matrix equation (9.55) can now be solved for the global values of the currents at the node points. The current on each surface element k may then be obtained by using (9.50) and (9.51) to obtain the local variables for the element, and then the current on the element is given by (9.46) and (9.47). These currents may now be applied in the integral equations (9.18) and (9.19) to calculate the electric field at any position in the near and far field. The surface currents can be directly applied to calculate the absorption in the scatterer by using (9.33), and the far fields can be used to calculate the extinction and scattering cross sections.

9.4 NUMERICAL APPROACH: CYLINDRICALLY SYMMETRIC STRUCTURE

The surface of a structure with cylindrical symmetry can be parameterized via two coordinates t and ϕ, where t represents the distance along the surface in the ρz-plane from a starting point, and ϕ is the azimuthal angle. The surface position will be of the form

$$\mathbf{s}(t,\phi) = \hat{\mathbf{z}} s_z(t) + \hat{\rho} s_\rho(t), \tag{9.58}$$

where the ϕ-dependence enters via

$$\hat{\rho} = \hat{\mathbf{x}}\cos\phi + \hat{\mathbf{y}}\sin\phi. \tag{9.59}$$

For each angular momentum, the dependence on ϕ will be removed from the problem, and effectively it is sufficient to discretize the structure surface in only one dimension, namely along the surface in the ρz-plane. The position on the surface in the ρz-plane determined via the distance parameter t is illustrated in Fig. 9.2.

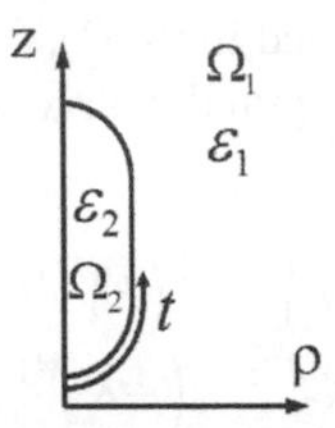

Figure 9.2 Schematic of a nanoparticle with cylindrical symmetry in the ρz-plane. In the region of the nanoparticle Ω_2 the dielectric constant is ε_2, and in the region Ω_1 outside the nanoparticle the dielectric constant is ε_1. The position along the surface is parameterized by the parameter t representing the distance along the surface from a starting position.

The electric and magnetic surface currents at the position $\mathbf{s}(t,\phi)$ may have a component in the direction $\hat{\mathbf{t}}$ along the surface in the ρz-plane, and a component along the azimuth $\hat{\phi}$, where these directions are defined as

$$\hat{\mathbf{t}}(t,\phi) = \frac{\partial \mathbf{s}(t,\phi)}{\partial t} / \left| \frac{\partial \mathbf{s}(t,\phi)}{\partial t} \right| \tag{9.60}$$

$$\hat{\phi}(\phi) = -\hat{\mathbf{x}} \sin\phi + \hat{\mathbf{y}} \cos\phi. \tag{9.61}$$

The electric and magnetic surface currents can be expanded in cylindrical harmonics as

$$\mathbf{J}(t,\phi) = \sum_n \left(\hat{\mathbf{t}}(t,\phi) J_t^{(n)}(t) + \hat{\phi}(\phi) J_\phi^{(n)}(t) \right) e^{in\phi}, \tag{9.62}$$

$$\mathbf{M}(t,\phi) = \sum_n \left(\hat{\mathbf{t}}(t,\phi) M_t^{(n)}(t) + \hat{\phi}(\phi) M_\phi^{(n)}(t) \right) e^{in\phi}. \tag{9.63}$$

With an appropriate choice of direction of increasing t the surface normal vector pointing into Ω_1 can be obtained as

$$\hat{\mathbf{n}}(t,\phi) = \hat{\phi}(\phi) \times \hat{\mathbf{t}}(t,\phi). \tag{9.64}$$

The incident electric field on the scatterer surface may likewise be expanded as

$$\mathbf{E}_{\text{inc}}(t,\phi) = \sum_n \left(\hat{\mathbf{t}}(t,\phi) E_{\text{inc},t}^{(n)}(t) + \hat{\phi}(\phi) E_{\text{inc},\phi}^{(n)}(t) + \hat{\mathbf{n}}(t,\phi) E_{\text{inc},n}^{(n)}(t) \right) e^{in\phi}. \tag{9.65}$$

The incident field enters in the integral equations as

$$\hat{\mathbf{n}}(t,\phi) \times \mathbf{E}_{\text{inc}}(t,\phi) = \sum_n \left(-\hat{\phi}(\phi) E_{\text{inc},t}^{(n)}(t) + \hat{\mathbf{t}}(t,\phi) E_{\text{inc},\phi}^{(n)}(t) \right) e^{in\phi}, \tag{9.66}$$

and the component of the incident field in the direction of the normal vector is thus eliminated.

The primed surface position coordinate in the integral equations (9.22) and (9.23) may also be parameterized in two coordinates t' and ϕ':

$$\mathbf{r}' = \hat{\mathbf{z}} s_z(t') + \hat{\rho}' s_\rho(t'), \tag{9.67}$$

where

$$\hat{\rho}' = \hat{\mathbf{x}}\cos\phi' + \hat{\mathbf{y}}\sin\phi'. \tag{9.68}$$

Integral equations for only one of the cylindrical harmonics n can be obtained by multiplying both the left- and the right-hand side of (9.22) and (9.23) by $\exp(-in\phi)/2\pi$ and integrating with respect to ϕ from 0 to 2π. This leads to

$$\begin{bmatrix} E^{(n)}_{\mathrm{inc},\phi}(t) \\ -E^{(n)}_{\mathrm{inc},t}(t) \end{bmatrix} = \frac{1}{2}\begin{bmatrix} -M^{(n)}_t(t) \\ -M^{(n)}_\phi(t) \end{bmatrix} - \int_{t'=0}^{L} \begin{bmatrix} K^{(J,n)}_{1,tt}(t,t') & K^{(J,n)}_{1,t\phi}(t,t') \\ K^{(J,n)}_{1,\phi t}(t,t') & K^{(J,n)}_{1,\phi\phi}(t,t') \end{bmatrix} \begin{bmatrix} J^{(n)}_t(t') \\ J^{(n)}_\phi(t') \end{bmatrix} dt'$$

$$+P\int_{t'=0}^{L} \begin{bmatrix} K^{(M,n)}_{1,tt}(t,t') & K^{(M,n)}_{1,t\phi}(t,t') \\ K^{(M,n)}_{1,\phi t}(t,t') & K^{(M,n)}_{1,\phi\phi}(t,t') \end{bmatrix} \begin{bmatrix} M^{(n)}_t(t') \\ M^{(n)}_\phi(t') \end{bmatrix} dt', \tag{9.69}$$

$$\begin{bmatrix} 0 \\ 0 \end{bmatrix} = \frac{1}{2}\begin{bmatrix} -M^{(n)}_t(t) \\ -M^{(n)}_\phi(t) \end{bmatrix} + \int_{t'=0}^{L} \begin{bmatrix} K^{(J,n)}_{2,tt}(t,t') & K^{(J,n)}_{2,t\phi}(t,t') \\ K^{(J,n)}_{2,\phi t}(t,t') & K^{(J,n)}_{2,\phi\phi}(t,t') \end{bmatrix} \begin{bmatrix} J^{(n)}_t(t') \\ J^{(n)}_\phi(t') \end{bmatrix} dt'$$

$$-P\int_{t'=0}^{L} \begin{bmatrix} K^{(M,n)}_{2,tt}(t,t') & K^{(M,n)}_{2,t\phi}(t,t') \\ K^{(M,n)}_{2,\phi t}(t,t') & K^{(M,n)}_{2,\phi\phi}(t,t') \end{bmatrix} \begin{bmatrix} M^{(n)}_t(t') \\ M^{(n)}_\phi(t') \end{bmatrix} dt'. \tag{9.70}$$

Here, P means that the position $t' = t$ is excluded from the integral.

The integration Kernels that are functions of t and t' are given by

$$K^{(J,n)}_{u,\alpha\beta}(t,t') =$$

$$i\omega\mu_0 \int_{\phi'=0}^{2\pi} \hat{\alpha}(t,\phi)\cdot\left[\hat{\mathbf{n}}\times\mathbf{G}_u(\mathbf{s}(t,\phi),\mathbf{s}(t',\phi'))\right]\cdot\hat{\beta}(t',\phi')s_\rho(t')e^{in(\phi'-\phi)}d\phi', \tag{9.71}$$

$$K^{(M,n)}_{u,\alpha\beta}(t,t') =$$

$$\int_{\phi'=0}^{2\pi} \hat{\alpha}(t,\phi)\cdot\left[\hat{\mathbf{n}}\times\left[\nabla\times\mathbf{G}_u(\mathbf{s}(t,\phi),\mathbf{s}(t',\phi'))\right]\right]\cdot\hat{\beta}(t',\phi')s_\rho(t')e^{in(\phi'-\phi)}d\phi', \tag{9.72}$$

and the unit vectors $\hat{\alpha}$ and $\hat{\beta}$ represent either $\hat{\mathbf{t}}$ or $\hat{\phi}$. Note that the result will be independent of the specific choice of the value of ϕ. When calculating $K^{(J,n)}_{u,\alpha\beta}(t,t')$, the singularity can be handled by using a position $\mathbf{s}(t,\phi)$ placed slightly outside or inside the surface.

The integral equations can now be solved by discretizing in a way which is similar to the 2D GFSIEM in Chapter 4. One slight difference though is that the boundary of interest in the ρz-plane may or may not be a closed curve. In the schematic in Fig. 9.2 the boundary is not a closed curve, while if a torus is considered instead, the boundary in the ρz-plane will be a closed curve.

The surface of the scatterer may be divided into N_k sections, where the start and end of section k is identified by the positions $t_k^{(s)}$ and $t_k^{(e)}$. Within each section the electric and magnetic surface currents can be expressed as a polynomial expansion:

$$J_\alpha^{(n)}(t) = \sum_{k=1}^{N_k}\sum_{v=0}^{m} f_k^{(m,v)}(t) J_{\alpha,k}^{(v,n)}, \tag{9.73}$$

$$M_\alpha^{(n)}(t) = \sum_{k=1}^{N_k}\sum_{v=0}^{m} f_k^{(m,v)}(t) M_{\alpha,k}^{(v,n)}, \tag{9.74}$$

where $\alpha = t$ or ϕ, and

$$f_k^{(m,v)}(t) = f^{(m,v)}\left(\frac{t - t_k^{(s)}}{t_k^{(e)} - t_k^{(s)}}\right) \tag{9.75}$$

are the polynomial expansion functions of order m as given in (4.33). The sampling positions within section k are referred to as $t_k^{(m,v)}$, $v = 0, 1, \ldots, m$, and $t_k^{(m,0)} = t_k^{(s)}$ and $t_k^{(m,m)} = t_k^{(e)}$.

The coefficients in the expansions (9.73) and (9.74) can be identified as the values of the currents at specific positions. In order to handle the possibility of both open and closed curves in the ρz-plane it is convenient to define global values of surface currents and fields $J_{\alpha,i}^{(n)}$, $M_{\alpha,i}^{(n)}$, and $E_{\text{inc},\alpha,i}^{(n)}$ at specific sampling points i, and to organize these in vectors:

$$\overline{J}_\alpha^{(n)} = \begin{bmatrix} J_{\alpha,1}^{(n)} & J_{\alpha,2}^{(n)} & \cdots & J_{\alpha,N}^{(n)} \end{bmatrix}^T, \tag{9.76}$$

$$\overline{M}_\alpha^{(n)} = \begin{bmatrix} M_{\alpha,1}^{(n)} & M_{\alpha,2}^{(n)} & \ldots & M_{\alpha,N}^{(n)} \end{bmatrix}^T, \tag{9.77}$$

$$\overline{E}_{\text{inc},\alpha}^{(n)} = \begin{bmatrix} E_{\text{inc},\alpha,1}^{(n)} & E_{\text{inc},\alpha,2}^{(n)} & \cdots & E_{\text{inc},\alpha,N}^{(n)} \end{bmatrix}^T. \tag{9.78}$$

The coefficients related to a specific section may like-wise be organized in vectors

$$\overline{J}_{\alpha,k}^{(n)} = \begin{bmatrix} J_{\alpha,k}^{(0,n)} & J_{\alpha,k}^{(1,n)} & \ldots & J_{\alpha,k}^{(m,n)} \end{bmatrix}^T, \tag{9.79}$$

$$\overline{M}_{\alpha,k}^{(n)} = \begin{bmatrix} M_{\alpha,k}^{(0,n)} & M_{\alpha,k}^{(1,n)} & \ldots & M_{\alpha,k}^{(m,n)} \end{bmatrix}^T, \tag{9.80}$$

$$\overline{E}_{\text{inc},\alpha,k}^{(n)} = \begin{bmatrix} E_{\text{inc},\alpha,k}^{(0,n)} & E_{\text{inc},\alpha,k}^{(1,n)} & \cdots & E_{\text{inc},\alpha,k}^{(m,n)} \end{bmatrix}^T. \tag{9.81}$$

These coefficients are related to the global values via relations of the form

$$\overline{J}_{\alpha,k}^{(n)} = \overline{\overline{P}}_k \overline{J}_\alpha, \quad \overline{M}_{\alpha,k}^{(n)} = \overline{\overline{P}}_k \overline{M}_\alpha, \quad \overline{E}_{\text{inc},\alpha,k}^{(n)} = \overline{\overline{P}}_k \overline{E}_{\text{inc},\alpha}, \tag{9.82}$$

where $\overline{\overline{P}}_k$ is a matrix that relates the local and global coefficients. This relation can be easily constructed from a matrix $\overline{\overline{T}}$ where column k identifies the numbers of the global positions related to section k. In addition it may be useful to add to column k information about the type of element, e.g. if it is part of a circle or if it is a straight line, and information about the direction of the outward normal vector. It may also be useful to construct a point matrix for the global positions $\overline{\overline{P}}$ where column i contains the coordinates of sampling point i. The normal vector at point i in the ρz-plane may also be added to column i.

The equation (9.69) may now be formulated in matrix form as

$$\begin{bmatrix} \overline{E}_{\text{inc},\phi}^{(n)} \\ -\overline{E}_{\text{inc},t}^{(n)} \end{bmatrix} = -\frac{1}{2}\begin{bmatrix} \overline{M}_t^{(n)} \\ \overline{M}_\phi^{(n)} \end{bmatrix} - \sum_{i=1}^{N} \overline{\overline{P}}_i' \sum_{k=1}^{N_k} \begin{bmatrix} \overline{\overline{K}}_{1,tt,i,k}^{(J,n)} & \overline{\overline{K}}_{1,t\phi,i,k}^{(J,n)} \\ \overline{\overline{K}}_{1,\phi t,i,k}^{(J,n)} & \overline{\overline{K}}_{1,\phi\phi,i,k}^{(J,n)} \end{bmatrix} \begin{bmatrix} \overline{J}_{t,k}^{(n)} \\ \overline{J}_{\phi,k}^{(n)} \end{bmatrix}$$

$$+\sum_{i=1}^{N} \overline{\overline{P}}_i' \sum_{k=1}^{N_k} \begin{bmatrix} \overline{\overline{K}}_{1,tt,i,k}^{(M,n)} & \overline{\overline{K}}_{1,t\phi,i,k}^{(M,n)} \\ \overline{\overline{K}}_{1,\phi t,i,k}^{(M,n)} & \overline{\overline{K}}_{1,\phi\phi,i,k}^{(M,n)} \end{bmatrix} \begin{bmatrix} \overline{M}_{t,k}^{(n)} \\ \overline{M}_{\phi,k}^{(n)} \end{bmatrix}. \tag{9.83}$$

Here, the matrices $\overline{\overline{K}}_{u,\alpha\beta,i,k}^{(U,n)}$ govern the coupling between electric and magnetic surface currents on element (or section) k and the global position i.

The matrices $\overline{\overline{K}}_{u,\alpha\beta,i,k}^{(U,n)}$ have a single row with $m+1$ values ($\nu = 0, 1, \ldots, m$) given by

$$\left[\overline{\overline{K}}_{u,\alpha\beta,i,k}^{(U,n)}\right]_\nu = \int_{t'} K_{u,\alpha\beta}^{(U,n)}(t_i, t') f_k^{(m,\nu)}(t')dt', \tag{9.84}$$

with $U = J$ or M, and α, $\beta = t$ or ϕ. A similar matrix equation can be formulated for (9.70). The matrix $\overline{\overline{P}}_i'$ is similar to (9.57).

By using (9.82) the matrix problem can now be formulated

$$\begin{bmatrix} \overline{E}_{\text{inc},\phi}^{(n)} \\ -\overline{E}_{\text{inc},t}^{(n)} \\ \overline{0} \\ \overline{0} \end{bmatrix} = \begin{bmatrix} -\overline{\overline{K}}_{1,tt}^{(J,n)} & -\overline{\overline{K}}_{1,t\phi}^{(J,n)} & \left(\overline{\overline{K}}_{1,tt}^{(M,n)} - \frac{1}{2}\overline{\overline{I}}\right) & \overline{\overline{K}}_{1,t\phi}^{(M,n)} \\ -\overline{\overline{K}}_{1,\phi t}^{(J,n)} & -\overline{\overline{K}}_{1,\phi\phi}^{(J,n)} & \overline{\overline{K}}_{1,\phi t}^{(M,n)} & \left(\overline{\overline{K}}_{1,\phi\phi}^{(M,n)} - \frac{1}{2}\overline{\overline{I}}\right) \\ \overline{\overline{K}}_{2,tt}^{(J,n)} & \overline{\overline{K}}_{2,t\phi}^{(J,n)} & \left(-\overline{\overline{K}}_{2,tt}^{(M,n)} - \frac{1}{2}\overline{\overline{I}}\right) & -\overline{\overline{K}}_{2,t\phi}^{(M,n)} \\ \overline{\overline{K}}_{2,\phi t}^{(J,n)} & \overline{\overline{K}}_{2,\phi\phi}^{(J,n)} & -\overline{\overline{K}}_{2,\phi t}^{(M,n)} & \left(-\overline{\overline{K}}_{2,\phi\phi}^{(M,n)} - \frac{1}{2}\overline{\overline{I}}\right) \end{bmatrix} \begin{bmatrix} \overline{J}_t^{(n)} \\ \overline{J}_\phi^{(n)} \\ \overline{M}_t^{(n)} \\ \overline{M}_\phi^{(n)} \end{bmatrix}, \tag{9.85}$$

where the matrices are given by

$$\overline{\overline{K}}_{u,\alpha\beta}^{(U,n)} = \sum_i \sum_k \overline{\overline{P}}_i'' \overline{\overline{K}}_{u,\alpha\beta,i,k}^{(U,n)} \overline{\overline{P}}_k. \tag{9.86}$$

Note that these are $N \times N$ matrices, and $\overline{\overline{P}}_i''$ is a column vector with N elements that are all zeros except for a 1 at position i. The sum can be carried out by adding, for each i and k, the $m+1$ elements of $\overline{\overline{K}}_{u,\alpha\beta,i,k}^{(U,n)}$ to $m+1$ positions in a $N \times N$ matrix.

The expression for the power absorbed in the scatterer (9.33) simplifies in the case of cylindrical symmetry to

$$P_{\text{abs}} = -2\pi \sum_n \int_t \frac{1}{2}\text{Real}\left\{ \left(J_\phi^{(n)}(t)\right)^* M_t^{(n)}(t) - \left(J_t^{(n)}(t)\right)^* M_\phi^{(n)}(t) \right\} s_\rho(t)dt, \tag{9.87}$$

and the absorption cross section is then again calculated using (9.34).

The cylindrical harmonics of the scattered far field can be obtained by using the far-field Green's tensor (6.23) in (9.18). Thus

$$\mathbf{E}_{\text{scat}}^{(\text{ff},n)}(\mathbf{r}) = \sum_{\alpha=\theta,\phi} \hat{\alpha} E_{\text{scat},\alpha}^{(\text{ff},n)}(\rho,z) e^{in\phi} =$$

$$i\omega\mu_0 \left(\hat{\theta}\hat{\theta} + \hat{\phi}(\phi)\hat{\phi}(\phi)\right) \frac{e^{ik_1 r}}{4\pi r}$$

$$\cdot \int_{t'} \int_{\phi'} \left\{ \hat{\mathbf{t}}(t',\phi') J_t^{(n)}(t') + \hat{\phi}(\phi') J_\phi^{(n)}(t') \right\} e^{in\phi'} e^{-ik_1 \hat{\mathbf{r}}\cdot\mathbf{s}(t',\phi')} s_\rho(t') dt' d\phi'$$

$$-ik_1 \left(\hat{\phi}(\phi)\hat{\theta} - \hat{\theta}\hat{\phi}(\phi)\right) \frac{e^{ik_1 r}}{4\pi r}$$

$$\cdot \int_{t'} \int_{\phi'} \left\{ \hat{\mathbf{t}}(t',\phi') M_t^{(n)}(t') + \hat{\phi}(\phi') M_\phi^{(n)}(t') \right\} e^{in\phi'} e^{-ik_1 \hat{\mathbf{r}}\cdot\mathbf{s}(t',\phi')} s_\rho(t') dt' d\phi', \quad (9.88)$$

where here $\hat{\alpha}$, $\hat{\theta} = -\hat{\mathbf{z}}\sin\theta + \hat{\rho}\cos\theta$ and $\hat{\mathbf{r}}$ are spherical coordinate unit vectors defined with respect to the un-primed coordinates θ and ϕ. The scattered far field can then be used to calculate the scattering and extinction cross sections similar to the description in Sec. 7.5.

In order to illustrate the convergence of the method, the example of a spherical gold particle previously considered with the GFVIEM in Fig. 6.4 is revisited. The particle has radius 20 nm and is illuminated by a plane wave with wavelength 700 nm. When using the GFVIEM and cubic volume elements, the calculated field inside that nanoparticle had large oscillations around the exact value. The corresponding calculation with the GFSIEM exploiting cylindrical symmetry is shown in Fig. 9.3 for different numbers of sampling points $N = 9$, 11 and 13 in a scheme with 2nd-order polynomial expansion functions ($m = 2$). Also shown is the exact result. It is clear that already $N = 9$, corresponding to dividing the surface in the ρz-plane into 4 sections, gives a result that is very close to the exact values. Only when making a zoom-in on the calculated field (Fig. 9.3(b)) is it possible to notice any difference between the result obtained with the GFSIEM and the exact result. As N increases, the calculated result converges toward the exact result, and even on the scale in Fig. 9.3(b), the calculated result will not be distinguishable from the exact values when using, e.g., $N = 21$. The optical cross sections for this case using different N are shown in the following table and compared with the analytical result:

N	σ_{abs}	σ_{scat}	σ_{ext}	$\sigma_{\text{ext}} - \sigma_{\text{abs}} - \sigma_{\text{scat}}$
11	16.175	5.330	21.107	−0.398
21	15.629	5.326	20.910	−0.045
41	15.551	5.326	20.879	−0.0002
exact	15.532	5.326	20.859	0

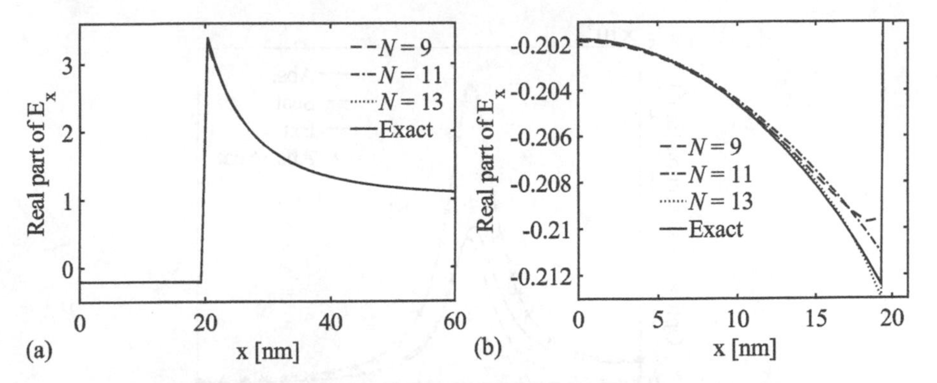

Figure 9.3 Calculation of electric field in a spherical gold nanoparticle with radius 20 nm using the cylindrical GFSIEM with different numbers of sampling points N and a 2nd-order polynomial field expansion. The particle is illuminated by a plane wave with wavelength 700 nm propagating along the z-axis and being polarized along the x-axis. (a) The real part of E_x along the x-axis. $x = 0$ corresponds to the center of the particle. (b) Zoom-in on the field in (a).

9.5 EXAMPLE: METAL NANO-DISC RESONATORS

In this section, examples of metal disc resonators with cylindrical symmetry are considered. The first example is a single gold disc in quartz ($n_1 = 1.5$) with radius 33 nm and thickness 20 nm. The nano-disc is illuminated by a normally incident plane wave propagating along the axis of symmetry $-z$ and which is polarized along x. In the ρz-plane the corners are rounded with the corner radius 3 nm. The calculation of optical cross-section spectra for the three-dimensional gold nano-disc is shown in Fig. 9.4(a). The calculation was made using the approach described in Sec. 9.4 with a 2nd-order polynomial expansion of the currents, and using for the boundary in the ρz-plane, 4 elements for each of the top and bottom straight sides, 2 elements for each of the two corners, and 2 elements for the vertical straight side, which is equivalent to 29 sampling points. The spectra show a resonance peak in the optical cross-section spectra around the wavelength 650 nm. The sum of absorption and scattering spectra are in excellent agreement with the extinction spectrum, which serves as a useful check of the calculation. The magnitude of the electric field in the xz-plane (normalized relative to the incident field) is shown in Fig. 9.4(b). The field magnitude clearly peaks in the center of the nanostrip. The highest field values are found just outside the ends of the cylinder $x = \pm 33$ nm, since the normal component of the field jumps across interfaces by a factor $\varepsilon_2/\varepsilon_1$. The field is similar to the fundamental resonance field of a single nanostrip considered with two-dimensional calculations in Fig. 4.17. The corresponding field in the yz-plane is shown in Fig. 9.4(c). Here the field is everywhere parallel to the structure interfaces, and the field thus does not jump.

As a second example the configuration of two gold nano-discs placed in quartz is considered. The gold discs are identical to those considered in Fig. 9.4, and they are

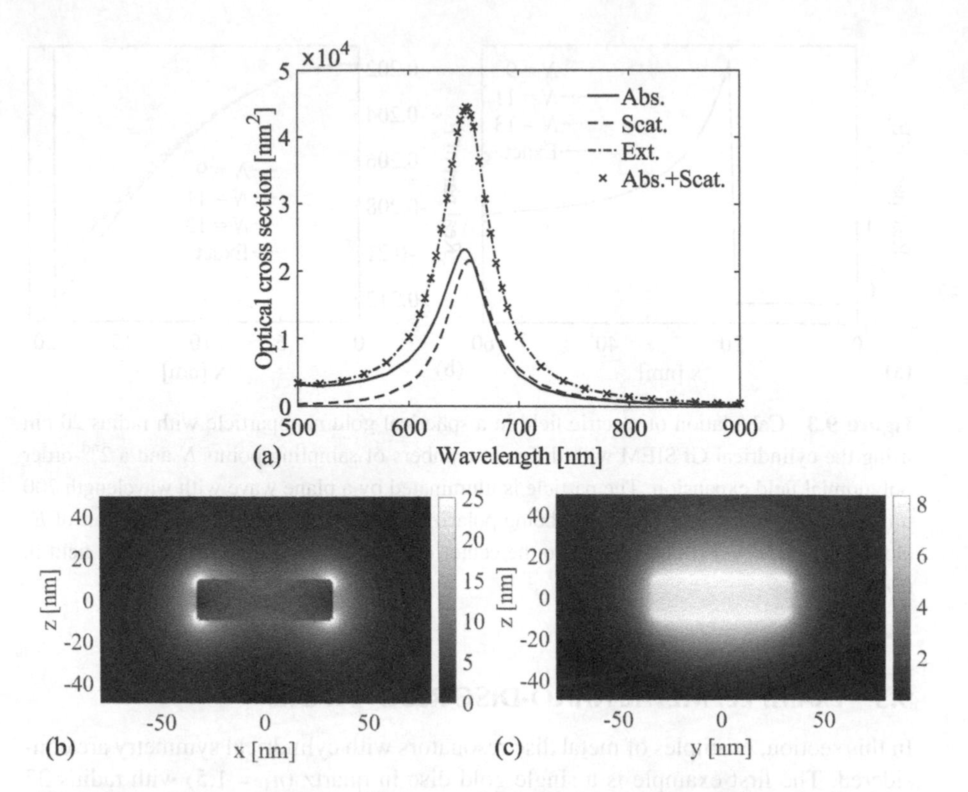

Figure 9.4 (a) Optical cross-section spectra for a cylindrical gold nano-disc of radius 33 nm and thickness 20 nm in quartz (n_1 = 1.5) being illuminated by a normally incident plane wave propagating along the symmetry axis ($-z$). (b) Magnitude of electric field in the xz-plane for the wavelength 650 nm. (c) Magnitude of the electric field in the yz-plane for the wavelength 650 nm.

discretized in the same wave. The number of sampling points is thus doubled. The gold discs are vertically separated (along z) by a 20-nm gap. The diameter, thickness, and gap are thus chosen as identical to the strip width, thickness, and gap for a configuration of two gold nanostrips in quartz considered with two-dimensional calculations in Sec. 4.1.8. The calculation of optical cross-section spectra for the two gold nano-discs is shown in Fig. 9.5(a). A calculation with twice as many elements at 9 different wavelengths showed no difference on the scale of the figure.

A resonance peak is seen in the absorption and extinction cross section at wavelengths near 700 nm. The peak is only just barely visible as a shoulder in the scattering cross section. In the case of two nanostrips instead (see Fig. 4.21) a peak in the optical cross sections is also found at the wavelength 700 nm. In that case a peak was also found in the scattering cross section although it is not very pronounced. In both cases the peak is related to the excitation of a gap-SPP resonance. The magnitude of

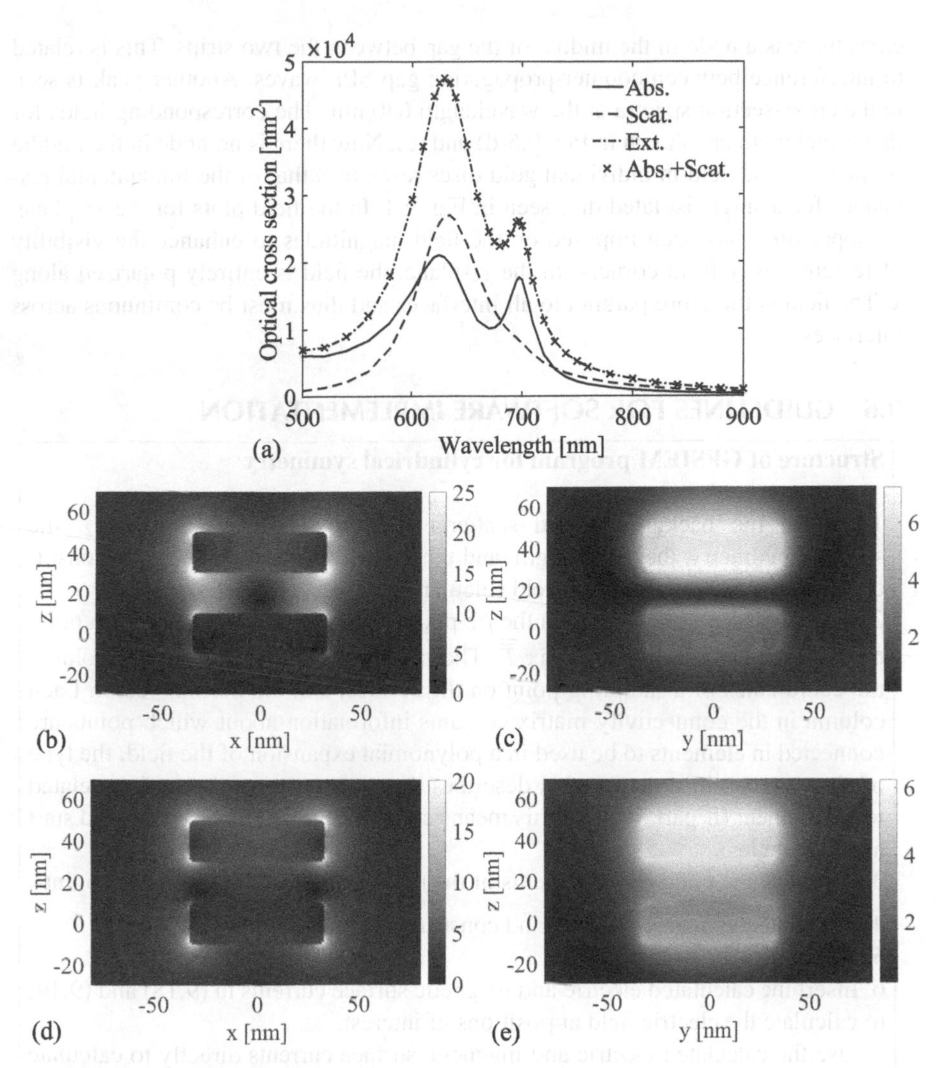

Figure 9.5 (a) Optical cross-section spectra for two cylindrical gold discs of radius 33 nm and thickness 20 nm in quartz being separated by 20 nm along the *z*-axis. The discs are illuminated by a normally incident plane wave propagating along the symmetry axis. (b) Magnitude of electric field in the *xz*-plane for the wavelength 700 nm. (c) Field in *yz*-plane for the wavelength 700 nm. (d) Field in *xz*-plane for the wavelength 630 nm. (e) Field in *yz*-plane for the wavelength 630 nm.

the electric field for the wavelength 700 nm is shown in Fig. 9.5(b) and (c) for the *xz*-plane and the *yz*-plane, respectively. The field in the *xz*-plane looks quite similar to the corresponding two-dimensional calculation for the nanostrip in Fig. 4.21. In both cases there is a node in the middle of the gap between the two strips. This is related to interference between counter-propagating gap-SPP waves. Another peak is seen in the cross-section spectra at the wavelength 630 nm. The corresponding fields for that wavelength are shown in Fig. 9.5(d) and (e). Now there is no node in the middle of the gap. The field in individual gold discs resembles that of the fundamental resonance for a single isolated disc seen in Fig. 9.4. In the field plots for the xz-plane, an upper limit has been imposed on the field magnitudes to enhance the visibility of features away from corners. In the yz-plane, the field is entirely polarized along x. The field is therefore parallel to all interfaces and thus must be continuous across interfaces.

9.6 GUIDELINES FOR SOFTWARE IMPLEMENTATION

Structure of GFSIEM program for cylindrical symmetry

1. Define the background and scatterer refractive indices n_1 and n_2, the scatterer geometry, the wavelength, and the cylindrical harmonic n to be considered. Specify parameters for a field calculation.
2. Divide the scatterer surface in the ρz-plane into N_k sections. Construct a point matrix $\overline{\overline{P}}$ and connectivity matrix $\overline{\overline{T}}$. The point matrix contains in each column the coordinates of a sampling point on the surface, and the normal vector. Each column in the connectivity matrix contains information about which points are connected in elements to be used in a polynomial expansion of the field, the type of element, and information that describes the exact curve of the surface related to the element (if part of circle this means center position, circle radius, and start and end angles).
3. Calculate the $\hat{\mathbf{t}}$ and $\hat{\phi}$ components of the reference field at the sampling points.
4. Calculate the matrices $\overline{\overline{K}}_{u,\alpha\beta}^{(U,n)}$ and construct the matrix equation (9.85).
5. Solve the matrix equation.
6. Insert the calculated electric and magnetic surface currents in (9.18) and (9.19) to calculate the electric field at positions of interest.
7. Use the calculated electric and magnetic surface currents directly to calculate the absorption cross section.
8. Use again (9.18) and (9.19) with the far-field Green's tensor to calculate the cylindrical harmonics of the scattered far-field for a range of angles θ. Use the far field harmonics to calculate both the scattering and extinction cross sections, and the differential scattering cross section.

When calculating $\overline{\overline{K}}_{u,\alpha\beta,i,k}^{(J,n)}$ for elements k, where the corresponding surface crosses the position i, a limiting procedure can be applied where the position i is either placed slightly inside or outside the surface. Such positions can be obtained from the actual surface position by a small displacement in the direction of the normal vector. From a few different displacements, the result corresponding to an infinitesimal displacement can be obtained via interpolation. When calculating $\overline{\overline{K}}_{u,\alpha\beta,i,k}^{(M,n)}$ for a similar situation, then instead of using an interpolation procedure, a very small part of the integration range for t' can be simply left out, since the integrand is well behaved.

9.7 EXERCISES

1. Verify that the integral over the outer surface of Ω_1 at infinity in (9.16) vanishes when the fields and the Green's tensor satisfy the radiating boundary condition.

2. Consider the vector wave equation for the magnetic field in a homogeneous region

$$-\nabla\times\nabla\times\mathbf{H}(\mathbf{r})+k_0^2\varepsilon_i\mathbf{H}(\mathbf{r})=-\nabla\times\mathbf{J}(\mathbf{r}),\quad \mathbf{r}\in\Omega_i, \tag{9.89}$$

and follow a procedure similar to Sec. 9.1. Show that instead of (9.16) the procedure now leads to the magnetic field integral equation

$$\mathbf{H}(\mathbf{r})=\mathbf{H}_{\text{inc},i}(\mathbf{r})-i\omega\varepsilon_0\varepsilon_i\oint_{\partial\Omega_i}\mathbf{G}(\mathbf{r},\mathbf{r}')\cdot\mathbf{M}_{s,i}(\mathbf{r}')d^2r'+\oint_{\partial\Omega_i}\nabla\times\mathbf{G}_i(\mathbf{r},\mathbf{r}')\cdot\mathbf{J}_{s,i}(\mathbf{r}')d^2r'. \tag{9.90}$$

Verify this expression directly from (9.16) and Maxwell's equations.

3. Consider the integral equation (9.19) for a spherical particle in the quasi-static limit. Assume that the incident field is polarized along z and propagating along x. The electric field inside the particle is approximately given by (6.40). Use this to obtain the magnetic current **M**. Estimate the magnetic field by using (2.17) and use this to obtain the electric current **J**. Now show that integrals in the right-hand side of (9.19) again lead to the result (6.40). The important lesson here is that with the surface integral equation, the magnetic field cannot be ignored, and the contribution from electric currents is significant. With the volume integral equation on the other hand, the magnetic field can be neglected in the quasi-static limit.

4. Consider the matrix equation (9.55) in the limit of a perfect electric conductor. Explain how the matrix equation simplifies in that case.

5. Follow the approach outlined in Sec. 9.6 and develop numerical programs for calculating fields and optical cross sections in a scattering situation with a nanoparticle using either a general approach or the restriction to cylindrical symmetry.

Residue theorem

Residue theorem. Let $f(z)$ be analytic inside a simple closed path C and on C, except for finitely many singular points $z_1, z_2, \cdots, z_n$ inside C. Then the integral of $f(z)$ taken counterclockwise around C equals $2\pi i$ times the sum of residues of $f(z)$ at $z_1, z_2, \cdots, z_n$:

$$\oint_C f(z)dz = 2\pi i \sum_{j=1}^{n} [\mathrm{Res} f(z)]_{z=z_j}.$$

Residue at a simple pole

$$[\mathrm{Res} f(z)]_{z=z_0} = \lim_{z \to z_0} (z - z_0) f(z).$$

For a simple pole the above limit will converge to a finite non-zero number as $z \to z_0$. It is also possible to find the residue at poles of higher order but that will not be needed in this text.

Analyticity

A complex function $f(z)$ is said to be *analytic* in a domain D if $f(z)$ is defined and differentiable at all points of D. The function $f(z)$ is said to be *analytic at a point* $z = z_0$ in D if $f(z)$ is analytic in a neighborhood of z_0.

B Conjugate gradient algorithm

The conjugate gradient algorithm presented in the following is from [147, 163].

Conjugate gradient algorithm:
The algorithm finds solutions to the equation

$$\hat{C}y = x, \tag{B.1}$$

where $\hat{C}$ is an operator, and y and x represent functions, vectors or matrices depending on the formulation of the problem, and $\hat{C}y$ is a new function, vector or matrix. The first step is to choose an inital guess y_0 and set

$$z \equiv \hat{C}^\dagger x, \;\; g_0 = z - \hat{C}^\dagger \hat{C} y_0, \tag{B.2}$$

$$p_0 = g_0, \;\; w_0 = \hat{C} y_0, \;\; v_0 = \hat{C} p_0, \tag{B.3}$$

where $\hat{C}^\dagger$ is the Hermitian conjugate of the operator $\hat{C}$. The estimate y_i is now iteratively improved ($i = 0, 1, 2,...$) by repeating the following steps:

$$\alpha_i = \frac{\langle g_i | g_i \rangle}{\langle v_i | v_i \rangle}, \;\; y_{i+1} = y_i + \alpha_i p_i, \tag{B.4}$$

$$w_{i+1} = w_i + \alpha_i v_i, \;\; g_{i+1} = z - \hat{C}^\dagger w_{i+1}, \tag{B.5}$$

$$\beta_i = \frac{\langle g_{i+1} | g_{i+1} \rangle}{\langle g_i | g_i \rangle}, \;\; p_{i+1} = g_{i+1} + \beta_i p_i, \tag{B.6}$$

$$v_{i+1} = \hat{C} g_{i+1} + \beta_i v_i. \tag{B.7}$$

The above expressions for v_i and w_i are equivalent to

$$v_i \equiv \hat{C} p_i, \;\; w_i \equiv \hat{C} y_i. \tag{B.8}$$

These expressions can be used for e.g. every 10 iterations to avoid rounding-off errors. When y_i is close to the actual solution y, then w_i and x must be nearly the same. A measure of the error in y_i is

$$\text{error} = 1 - \frac{\langle w_i | x \rangle \langle x | w_i \rangle}{\langle w_i | w_i \rangle \langle x | x \rangle}. \tag{B.9}$$

Iterations are continued until *error* is below a threshold.

The notation $\langle \ldots | \ldots \rangle$ is the same as is conventionally used in quantum mechanics. If, e.g. v represents a vector $\overline{v}$, as will be the case if the algorithm is applied to the GFIEM in Chapter 3, then $\langle v|v\rangle$ must be interpreted as

$$\langle v|v\rangle \rightarrow \sum\{(\overline{v}^*).*\overline{v}\}. \tag{B.10}$$

Here $(\overline{v}^*).*\overline{v}$ represents element-wise multiplication of the conjugate of $\overline{v}$ and $\overline{v}$ itself. Afterward the elements of the resulting vector are summed. If v represents a matrix $\overline{\overline{v}}$ as in Sec. 5.7 then $\langle v|v\rangle$ must similarly be interpreted as

$$\langle v|v\rangle \rightarrow \sum\left\{\left(\overline{\overline{v}}^*\right).*\overline{\overline{v}}\right\}. \tag{B.11}$$

This is again element-wise multiplication of two matrices and then summation of all elements in the resulting matrix. This principle can be straightforwardly extended to three-dimensional arrays as considered in Chapter 6.

C Bessel functions

Bessel functions are the radial component of solutions to the 2D Helmholtz equation when requiring that solutions should be a product of a function of r, and a function of θ, respectively, where $x = r\cos\theta$ and $y = r\sin\theta$. Bessel functions of the first kind and order m are solutions to

$$(\nabla^2 + k^2)\, J_m(kr)e^{im\theta} = 0, \tag{C.1}$$

and the Bessel function itself satisfies

$$\left(\frac{d^2}{d(kr)^2} + \frac{1}{kr}\frac{d}{d(kr)} + 1 - \frac{m^2}{(kr)^2}\right) J_m(kr) = 0, \tag{C.2}$$

and can be expressed or defined as

$$J_m(x) = \sum_{k=0}^{\infty} \frac{(-1)^k (x/2)^{m+2k}}{k!(m+k)!} = \frac{1}{\pi}\int_0^{\pi} \cos(m\theta - x\sin\theta)d\theta, \quad m = 0,1,2,\ldots. \tag{C.3}$$

For small arguments

$$J_m(x) \propto x^m, \quad m = 0,1,2,\ldots, \quad |x| << 1, \tag{C.4}$$

and the Bessel function of the first kind thus does not have a singularity.

Bessel functions of the second kind and order m, Y_m, are also solutions to Eq. (C.2), except at $r = 0$, and can be expressed as

$$Y_m(x) = \frac{2}{\pi}\{\ln(x/2) + \gamma\}\, J_m(x) - \frac{1}{\pi}\sum_{k=0}^{m-1} \frac{(m-k-1)!}{k!}\left(\frac{x}{2}\right)^{2k-m}$$
$$- \frac{1}{\pi}\sum_{k=0}^{\infty} (-1)^k \{\Phi(k) + \Phi(m+k)\} \frac{(x/2)^{m+2k}}{k!(m+k)!}, \quad m = 0,1,2,\ldots, \tag{C.5}$$

where $\gamma = 0.5772156649015329\ldots$ is Euler's constant, and

$$\Phi(p) = 1 + \frac{1}{2} + \frac{1}{3} + \ldots + \frac{1}{p}, \quad \Phi(0) = 0. \tag{C.6}$$

The Bessel function of the second kind and order 0 thus has a logarithmic singularity, and those of higher order have stronger singularities.

For negative orders, the Bessel functions can be obtained from those with positive orders m from

$$J_{-m}(x) = (-1)^m J_m(x) \quad \text{and} \quad Y_{-m}(x) = (-1)^m Y_m(x). \tag{C.7}$$

Bessel functions of the first and second kind have the asymptotic expansions

$$J_m(x) \approx \sqrt{\frac{2}{\pi x}}\cos\left(x - \frac{m\pi}{2} - \frac{\pi}{4}\right), \quad x >> 1, \tag{C.8}$$

$$Y_m(x) \approx \sqrt{\frac{2}{\pi x}} \sin\left(x - \frac{m\pi}{2} - \frac{\pi}{4}\right), \quad x >> 1. \tag{C.9}$$

The Hankel functions of the 1st and 2nd kind and order m, are defined as

$$H_m^{(1)}(x) \equiv J_m(x) + iY_m(x), \quad \text{and} \quad H_m^{(2)}(x) \equiv J_m(x) - iY_m(x). \tag{C.10}$$

From the asymptotic expressions for J_m and Y_m it follows that the Hankel function of the 1st kind behaves like an out-going cylindrical wave:

$$H_m^{(1)}(kr) \approx \sqrt{\frac{2}{\pi kr}} \exp\left(i\left(kr - \frac{m\pi}{2} - \frac{\pi}{4}\right)\right), \quad |kr| >> 1, \tag{C.11}$$

and the Hankel function of the 2nd kind behaves like an in-going cylindrical wave.

Some useful recurrence formulas:

$$J_{m+1}(x) = \frac{2m}{x} J_m(x) - J_{m-1}(x), \tag{C.12}$$

$$J'_m(x) \equiv \frac{dJ_m(x)}{dx} = \frac{1}{2}\left\{J_{m-1}(x) - J_{m+1}(x)\right\}, \tag{C.13}$$

$$xJ'_m(x) = xJ_{m-1}(x) - mJ_m(x), \tag{C.14}$$

$$xJ'_m(x) = mJ_m(x) - xJ_{m+1}(x). \tag{C.15}$$

The Bessel functions of the second kind, $Y_m(x)$, satisfy identical recurrence formulas.

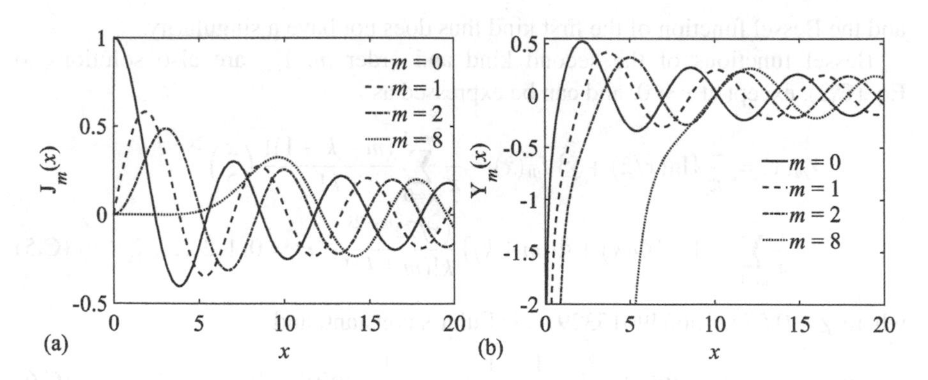

Figure C.1 Examples of Bessel functions of the (a) first kind and (b) of the second kind, and different order m.

A few examples of Bessel functions of the first and second kind are shown in Fig. C.1. For the Bessel functions of the first kind it should be noticed that as the order increases, the argument x needs to also be increasingly larger before the value of $J_m(x)$ deviates substantially from zero. For both types of Bessel functions the asymptotic behavior described in Eqs. (C.8) and (C.9) is noticeable for large x.

The above and many more results and details on Bessel functions can be found in, for example, Refs. [164, 165].

D Analytic scattering from a circular cylinder

In this appendix the two-dimensional scattering problem of a plane wave being incident on a circular cylinder will be considered analytically using cylindrical mode-expansion theory. The method provides exact results for the fields and optical cross sections that can be used for comparison with numerical calculations based on Green's function integral equation methods.

The cylinder has radius a, dielectric constant ε_2, and it is placed in a reference medium with dielectric constant ε_1. The cylinder is placed with its center at the origin of the coordinate system, and the dielectric function is thus given by

$$\varepsilon(\mathbf{r}) = \begin{cases} \varepsilon_1, & r > a \\ \varepsilon_2, & r < a \end{cases}, \tag{D.1}$$

where $\mathbf{r} = \hat{\mathbf{x}} r \cos\theta + \hat{\mathbf{y}} r \sin\theta$. We will also later need the vectors $\hat{\mathbf{r}} = \mathbf{r}/r$ and $\hat{\theta} = -\hat{\mathbf{x}} \sin\theta + \hat{\mathbf{y}} \cos\theta$.

We first consider the case of illuminating the cylinder with a p-polarized plane wave propagating along the x-axis. The magnetic field is thus polarized along the z-axis, and the incident magnetic field can be expanded in cylindrical harmonics as follows [36]:

$$H_0(\mathbf{r}) = e^{ik_0 n_1 x} = \sum_{n=-\infty}^{+\infty} i^n J_n(k_0 n_1 r) e^{-in\theta}, \tag{D.2}$$

where J_n is the Bessel function of the first kind and order n.

The magnetic field inside the cylinder can be expressed as

$$H_2(\mathbf{r}) = \sum_{n=-\infty}^{+\infty} B_n i^n J_n(k_0 n_2 r) e^{-in\theta}, \quad r < a. \tag{D.3}$$

Each term in this expansion satisfies the scalar wave equation for waves in a medium with dielectric constant ε_2. Outside the cylinder the total magnetic field is the sum of the incident field and a scattered field that propagates away from the cylinder given by

$$H_{1,\text{scat}}(\mathbf{r}) = \sum_{n=-\infty}^{+\infty} A_n i^n H_n^{(1)}(k_0 n_1 r) e^{-in\theta}, \quad r > a. \tag{D.4}$$

The scattered field has been expanded in Hankel functions of the first kind as these are equivalent to outward propagating waves, and this field component thus satisfies the radiating boundary condition.

The requirement of continuity of H at $r = a$ leads to

$$J_n(k_0 n_1 a) + A_n H_n^{(1)}(k_0 n_1 a) = B_n J_n(k_0 n_2 a), \tag{D.5}$$

while the requirement of continuity of the tangential component of the electric field, being equivalent to the requirement of continuity of $(\hat{\mathbf{n}}\cdot\nabla H)/\varepsilon = (\partial H/\partial r)/\varepsilon$, leads to

$$\frac{1}{\varepsilon_1}\left(J_n'(k_0n_1a)+A_nH_n^{(1)'}(k_0n_1a)\right)k_0n_1 = \frac{1}{\varepsilon_2}B_nJ_n'(k_0n_2a)k_0n_2. \tag{D.6}$$

Here $J_n'(x) = \partial J_n(x)/\partial x$ and $H_n^{(1)'}(x) = \partial H_n^{(1)}(x)/\partial x$.

By solving the equations (D.5) and (D.6) we now find the coefficients A_n governing the scattered field:

$$A_n = \frac{n_2J_n'(k_0n_1a)J_n(k_0n_2a)-n_1J_n(k_0n_1a)J_n'(k_0n_2a)}{n_1H_n^{(1)}(k_0n_1a)J_n'(k_0n_2a)-n_2H_n^{(1)'}(k_0n_1a)J_n(k_0n_2a)}. \tag{D.7}$$

The coefficients B_n governing the field inside the cylinder can subsequently be obtained from (D.5). In the far field the Hankel functions can be replaced by their approximation valid for large arguments and we thus find

$$H_{\text{scat}}^{(\text{ff})} = \sqrt{\frac{2}{\pi k_0n_1r}}e^{ik_0n_1r}e^{-i\pi/4}\sum_n A_ne^{-in\theta}. \tag{D.8}$$

When this is used in the expressions for the optical cross sections presented in Sec. 2.3.1 these can finally be written

$$\sigma_{\text{ext}} = \frac{-4}{k_0n_1}\text{Re}\left\{\sum_n A_n\right\}, \tag{D.9}$$

$$\sigma_{\text{scat}} = \frac{4}{k_0n_1}\sum_n |A_n|^2, \tag{D.10}$$

$$\sigma_{\text{abs}} = -2\pi a\text{Imag}\left\{\sum_n\left(\left(J_n(k_0n_1a)+A_nH_n^{(1)}(k_0n_1a)\right)^*\left(J_n'(k_0n_1a)+A_nH_n^{(1)'}(k_0n_1a)\right)\right)\right\}. \tag{D.11}$$

The corresponding electric field for the same situation can be obtained from

$$\mathbf{E}(\mathbf{r}) = \frac{i}{\omega\varepsilon_0\varepsilon(\mathbf{r})}\nabla\times\hat{\mathbf{z}}H(\mathbf{r}). \tag{D.12}$$

The electric field can be further normalized by multiplying with the factor $\sqrt{\varepsilon_0\varepsilon_1/\mu_0}$, which leads to the incident electric field

$$\mathbf{E}_0(\mathbf{r}) = \hat{\mathbf{y}}e^{ik_0n_1x}, \tag{D.13}$$

the scattered electric field

$$\mathbf{E}_{1,\text{scat}}(\mathbf{r}) = i\sum_n A_ni^ne^{-in\theta}\left(-\hat{\theta}H_n^{(1)'}(k_0n_1r)+\hat{\mathbf{r}}H_n^{(1)}(k_0n_1r)\frac{-in}{k_0n_1r}\right),\quad r>a, \tag{D.14}$$

and the electric field inside the cylinder

$$\mathbf{E}_2(\mathbf{r}) = i\frac{n_2}{n_1}\sum_n B_n i^n e^{-in\theta}\left(-\hat{\boldsymbol{\theta}}J_n'(k_0 n_2 r) + \hat{\mathbf{r}}J_n(k_0 n_2 r)\frac{-in}{k_0 n_2 r}\right), \quad r < a. \tag{D.15}$$

We now turn to the case of s polarization where the electric field is polarized along the z-axis. The incident electric field, the scattered field, and the field inside the particle, are given by the same expressions as for p polarization, that is by Eqs. (D.2)-(D.4), except that H should be replaced by E. The electric field is now continuous across the surface of the cylinder, and the boundary condition (D.5) also applies to s polarization. However, the tangential component of the magnetic field is now continuous across the cylinder surface, and the other boundary condition (D.6) must be replaced by

$$\left(J_n'(k_0 n_1 a) + A_n H_n^{(1)\prime}(k_0 n_1 a)\right) n_1 = B_n J_n'(k_0 n_2 a) n_2. \tag{D.16}$$

By solving the equations (D.6) and (D.16) we find the coefficients A_n for s polarization as

$$A_n = -\frac{n_1 J_n'(k_0 n_1 a) J_n(k_0 n_2 a) - n_2 J_n(k_0 n_1 a) J_n'(k_0 n_2 a)}{n_2 H_n^{(1)}(k_0 n_1 a) J_n'(k_0 n_2 a) - n_1 H_n^{(1)\prime}(k_0 n_1 a) J_n(k_0 n_2 a)}, \tag{D.17}$$

and B_n can subsequently be obtained from (D.6).

The electric far field is also still given by (D.8) except that $H_{\text{scat}}^{(\text{ff})}$ should be replaced by $E_{\text{scat}}^{(\text{ff})}$. The expressions for the optical cross sections (D.9)-(D.11) also apply for s polarization.

E Analytic scattering from a spherical particle

In this appendix the scattering of a plane wave by a spherical particle is considered. The plane wave is propagating in a medium with dielectric constant ε_1, and the particle has radius a, dielectric constant ε_2, and its center is placed in the origin. Spherical coordinates (r, θ, ϕ) will be used in the following.

The dielectric function of the considered geometry is given by

$$\varepsilon(\mathbf{r}) = \begin{cases} \varepsilon_1, & r > a \\ \varepsilon_2, & r < a \end{cases}. \tag{E.1}$$

The incident electric field will be chosen to propagate along the z-axis and to be polarized along the x-axis. The incident field can be expressed using spherical wave functions as [36, 166]

$$\mathbf{E}_0(\mathbf{r}) = \hat{\mathbf{x}} E_0 e^{ik_1 z} = E_0 \sum_{n=1}^{\infty} i^n \frac{2n+1}{n(n+1)} \left(\mathbf{m}_{n,1}^{(1)}(\mathbf{r}) - i\mathbf{n}_{n,1}^{(1)}(\mathbf{r}) \right), \tag{E.2}$$

where $k_i = k_0 n_i = k_0 \sqrt{\varepsilon_i}$, and the spherical wave functions are defined as

$$\mathbf{m}_{n,i}^{(1)}(\mathbf{r}) = \hat{\theta} \frac{\cos\phi}{\sin\theta} P_n^1(\cos\theta) j_n(k_i r) - \hat{\phi} \sin\phi \frac{dP_n^1(\cos\theta)}{d\theta} j_n(k_i r), \tag{E.3}$$

$$\mathbf{n}_{n,i}^{(1)}(\mathbf{r}) = \hat{\mathbf{r}} n(n+1) \cos\phi P_n^1(\cos\theta) \frac{j_n(k_i r)}{k_i r} +$$

$$\hat{\theta} \cos\phi \frac{dP_n^1(\cos\theta)}{d\theta} \frac{[k_i r j_n(k_i r)]'}{k_i r} - \hat{\phi} \frac{\sin\phi}{\sin\theta} P_n^1(\cos\theta) \frac{[k_i r j_n(k_i r)]'}{k_i r}. \tag{E.4}$$

Here, P_n^m is the associated Legendre function of the first kind given by

$$P_n^m(x) = \left(1 - x^2\right)^{m/2} \frac{d^m}{dx^m} P_n(x), \tag{E.5}$$

with the Legendre polynomial

$$P_n(x) = \frac{1}{2^n n!} \frac{d^n}{dx^n} \left(x^2 - 1\right)^n, \tag{E.6}$$

and

$$j_n(x) = \sqrt{\frac{\pi}{2x}} J_{n+1/2}(x), \tag{E.7}$$

is the spherical Bessel function, where J_n is the Bessel function of the first kind. In addition, the following short-hand notation has been applied:

$$[xj_n(x)]' \equiv \frac{d}{dx}[xj_n(x)]. \tag{E.8}$$

We will also define the functions $\mathbf{m}_{n,i}^{(2)}(\mathbf{r})$ and $\mathbf{n}_{n,i}^{(2)}(\mathbf{r})$ as being identical to the expressions for $\mathbf{m}_{n,i}^{(1)}(\mathbf{r})$ and $\mathbf{n}_{n,i}^{(1)}(\mathbf{r})$, except that the spherical Bessel function should be replaced by the spherical Hankel function

$$h_n(x) = \sqrt{\frac{\pi}{2x}}\left(J_{n+1/2}(x) + iY_{n+1/2}(x)\right) = \sqrt{\frac{\pi}{2x}} H_{n+1/2}^{(1)}(x). \tag{E.9}$$

Here, Y_n is the Bessel function of the second kind, and $H_n^{(1)}$ is the Hankel function of the first kind. Note that $h_n(k_1 r)$ behaves as an outgoing spherical wave when $k_1 r >> 1$.

The resulting field inside the spherical particle can be expressed as

$$\mathbf{E}(\mathbf{r}) = E_0 \sum_{n=1}^{\infty} i^n \frac{2n+1}{n(n+1)} \left(c_n \mathbf{m}_{n,2}^{(1)}(\mathbf{r}) - i d_n \mathbf{n}_{n,2}^{(1)}(\mathbf{r})\right), \quad r < a, \tag{E.10}$$

and outside the particle the resulting field is

$$\mathbf{E}(\mathbf{r}) = \mathbf{E}_0(\mathbf{r}) + \mathbf{E}_{\text{scat}}(\mathbf{r}), \quad r > a, \tag{E.11}$$

where the scattered field can be expressed as

$$\mathbf{E}_{\text{scat}}(\mathbf{r}) = E_0 \sum_{n=1}^{\infty} i^n \frac{2n+1}{n(n+1)} \left(a_n \mathbf{m}_{n,1}^{(2)}(\mathbf{r}) - i b_n \mathbf{n}_{n,1}^{(2)}(\mathbf{r})\right). \tag{E.12}$$

The expansion coefficients can be obtained from the requirement of continuity of the tangential component of the electric and magnetic fields across the particle surface, and are given by

$$a_n = -\frac{j_n(k_2 a)\,[k_1 a j_n(k_1 a)]' - j_n(k_1 a)\,[k_2 a j_n(k_2 a)]'}{j_n(k_2 a)\,[k_1 a h_n(k_1 a)]' - h_n(k_1 a)\,[k_2 a j_n(k_2 a)]'}, \tag{E.13}$$

$$b_n = -\frac{\varepsilon_1 j_n(k_1 a)\,[k_2 a j_n(k_2 a)]' - \varepsilon_2 j_n(k_2 a)\,[k_1 a j_n(k_1 a)]'}{\varepsilon_1 h_n(k_1 a)\,[k_2 a j_n(k_2 a)]' - \varepsilon_2 j_n(k_2 a)\,[k_1 a h_n(k_1 a)]'}, \tag{E.14}$$

$$c_n = \frac{j_n(k_1 a)\,[k_1 a h_n(k_1 a)]' - h_n(k_1 a)\,[k_1 a j_n(k_1 a)]'}{j_n(k_2 a)\,[k_1 a h_n(k_1 a)]' - h_n(k_1 a)\,[k_2 a j_n(k_2 a)]'}, \tag{E.15}$$

$$d_n = \frac{j_n(k_1 a)\,[k_1 a h_n(k_1 a)]' - h_n(k_1 a)\,[k_1 a j_n(k_1 a)]'}{\frac{n_2}{n_1} j_n(k_2 a)\,[k_1 a h_n(k_1 a)]' - \frac{n_1}{n_2} h_n(k_1 a)\,[k_2 a j_n(k_2 a)]'}. \tag{E.16}$$

The extinction and scattering cross sections can be expressed in terms of the expansion coefficients governing the field outside the scatterer as

$$\sigma_{\text{ext}} = -\frac{2\pi}{k_1^2}\sum_{n=1}^{\infty}(2n+1)\text{Real}\{a_n + b_n\}, \tag{E.17}$$

$$\sigma_{\text{scat}} = \frac{2\pi}{k_1^2}\sum_{n=1}^{\infty}(2n+1)\left\{|a_n|^2 + |b_n|^2\right\}. \tag{E.18}$$

The absorption cross section is the difference

$$\sigma_{\text{abs}} = \sigma_{\text{ext}} - \sigma_{\text{scat}}. \tag{E.19}$$

Further details and derivations can be found in, for example, Refs. [167, 36, 35, 166].

F Calculating guided modes of planar waveguides

In this appendix we will consider methods for calculating the guided modes of the planar waveguides introduced in Sec. 2.2.2. The modes will be bound to the waveguide along the y-axis and propagate along the x-axis. For s polarization the electric field (z-component) of the guided modes will have the form:

$$E(x,y) = E(y)e^{ik_0 n_{\rm m} x}. \tag{F.1}$$

A similar expression applies for p polarization but with electric field E replaced by magnetic field H. The x-dependent part of this field has the same form as a plane wave propagating along the x-axis in a medium with refractive index $n_{\rm m}$, and we thus refer to $n_{\rm m}$ as the mode index. The mode index is a convenient parameter since its value can be compared with the refractive indices of the different media of the planar waveguide structure.

We will first consider a three-layer symmetric dielectric waveguide, and search for guided modes with polarization u = p or s. It was found in Sec. 2.2 that the guided-mode wave numbers of three-layer waveguides can be found as solutions to the equation

$$f(n_{\rm m}) = 1 + r^{(u)}_{1,2}(k_x) r^{(u)}_{2,3}(k_x) e^{2ik_{y2}d} = 0, \tag{F.2}$$

where $k_x = k_0 n_{\rm m}$, and $k_{y2} = \sqrt{k_0^2 \varepsilon_2 - k_x^2}$ is thus also a function of $n_{\rm m}$.

For a symmetric three-layer dielectric waveguide with real and positive refractive indices satisfying

$$n_2 > n_1 = n_3 \tag{F.3}$$

the mode index of guides modes must be in the interval

$$n_1 < n_{\rm m} < n_2. \tag{F.4}$$

The requirement $n_{\rm m} > n_1$ ensures that the mode is bound, meaning that the field will decrease exponentially into media 1 and 3 away from the layer 2 since $k_{y1} = k_{y3}$ will be purely imaginary. Instead of solving Eq. (F.2) it is convenient to instead solve the equivalent equation

$$f(n_{\rm m}) = e^{-ik_{y2}d}\left(r^{(u)}_{1,2}(k_x)\right)^{-1} + r^{(u)}_{2,3}(k_x) e^{ik_{y2}d} = 0, \tag{F.5}$$

since for a mode index in the considered range, and the fact that $r^{(u)}_{1,2}(k_x) = -r^{(u)}_{2,3}(k_x) = e^{i\phi}$ for a symmetric waveguide, then $f(n_{\rm m})$ in Eq. (F.5) will be purely imaginary. As an example, the function $f(n_{\rm m})$ is shown in Fig. F.1(a) for a dielectric

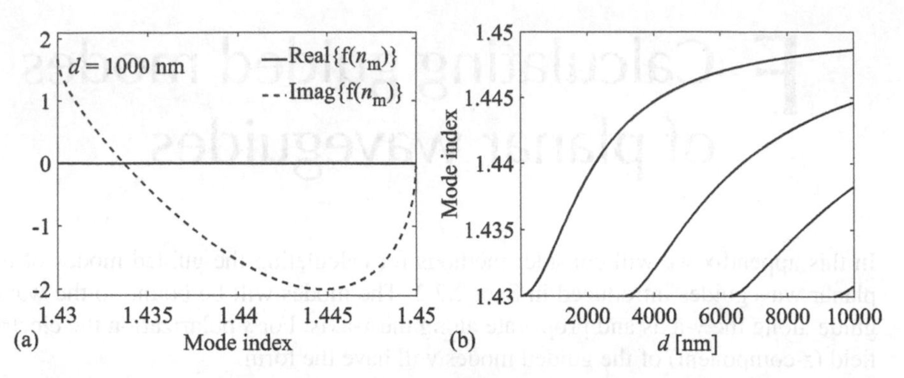

Figure F.1 (a) Real and imaginary part of $f(n_m)$ for a dielectric waveguide with $n_1 = n_3 = 1.43$, $n_2 = 1.45$, waveguide thickness $d = 1000$ nm, p polarization, and wavelength $\lambda = 1500$ nm. (b) Mode-indices as af function of the waveguide thickness d.

waveguide with $n_1 = n_3 = 1.43$, $n_2 = 1.45$, waveguide thickness $d = 1000$ nm, p polarization, and wavelength $\lambda = 1500$ nm. Since the real part of f is identically zero we only need to find n_m such that $\mathrm{Imag}\{f(n_\mathrm{m})\} = 0$. For the considered interval, the square roots in $k_{yi} = \sqrt{k_0^2\varepsilon_i - k_0^2 n_\mathrm{m}^2}$ are well behaved, and the function f is a smooth continuous function.

The simplest method to find the modes now is to use the method of *bisection*, where the interval from n_1 to n_2 is divided into a suitable number of points n_m, and we then calculate $\mathrm{Imag}\{f(n_\mathrm{m})\}$ for each point. If a single solution to Eq. (F.2) exists between two neighbor points, e.g. $n_{\mathrm{m}1} = 1.432$ and $n_{\mathrm{m}2} = 1.435$ in Fig. F.1, the function f must change sign or be zero. Thus, if

$$\mathrm{Imag}\{f(n_{\mathrm{m}1})\}\,\mathrm{Imag}\{f(n_{\mathrm{m}2})\} \leq 0, \tag{F.6}$$

then a solution exists in the interval from $n_{\mathrm{m}1}$ to $n_{\mathrm{m}2}$. The procedure now is to divide the interval in the middle using $n_{\mathrm{m}3} = (n_{\mathrm{m}1} + n_{\mathrm{m}2})/2$, and again observe if f changes sign between $n_{\mathrm{m}1}$ and $n_{\mathrm{m}3}$, or between $n_{\mathrm{m}3}$ and $n_{\mathrm{m}2}$. In the first case we may then set $n_{\mathrm{m}2} = n_{\mathrm{m}3}$, and in the latter case we set $n_{\mathrm{m}1} = n_{\mathrm{m}3}$. The procedure is repeated until the difference $n_{\mathrm{m}2} - n_{\mathrm{m}1}$ is smaller than a threshold that depends on the required resolution. Finally, we may set $n_\mathrm{m} = (n_{\mathrm{m}1} + n_{\mathrm{m}2})/2$. This method is not particularly fast but it is very robust, and for the cases considered in this book it is more than fast enough. The solutions for the mode index as a function of d obtained using *bisection* are shown in Fig. F.1(b). It can be seen that for symmetric waveguides, a mode exists even for very small d. In addition the mode index of a guided mode tends to n_2 as d increases. It can also be seen that as d increases, the waveguide supports an increasing number of modes.

In the case of an asymmetric waveguide it is not straightforward to construct f such that either the real or imaginary part is always zero. The case of an air-silicon-quartz waveguide ($n_1 = 1$, $n_2 = 3.5$, and $n_3 = 1.45$) is considered in Fig. F.2 for

s polarization and wavelength 1500 nm. The function f as formulated in Eq. (F.2) is shown in Fig. F.2(a). The solution to Eq. (F.2) is indicated with an arrow. In the example, the real part does not change sign as $n_{\rm m}$ crosses the solution to $f(n_{\rm m}) = 0$. The real part is also not zero on either side. It is thus not sufficient to look for sign changes in both the real and imaginary part of f to determine if a solution exists in a small interval. Instead we can consider the phase of f which jumps across the solution point as seen in Fig. F.2(b). We may thus still use the method of bisection to search for the mode index but now with the criteria for a solution to exist in the interval between $n_{\rm m1}$ and $n_{\rm m2}$ that

$$|\text{phase}\{f(n_{\rm m1})\} - \text{phase}\{f(n_{\rm m2})\}| > \pi/2. \tag{F.7}$$

By using the jump in phase and bisection, the mode index of guided modes versus waveguide thickness d has been obtained in Fig. F.2(c). The fields $E(y)$ of guided modes in the case of $d = 400$ nm are shown in Fig. F.2(d). Note that here the mode index must be in the interval $n_3 < n_{\rm m} < n_2$ since $n_3 > n_1$.

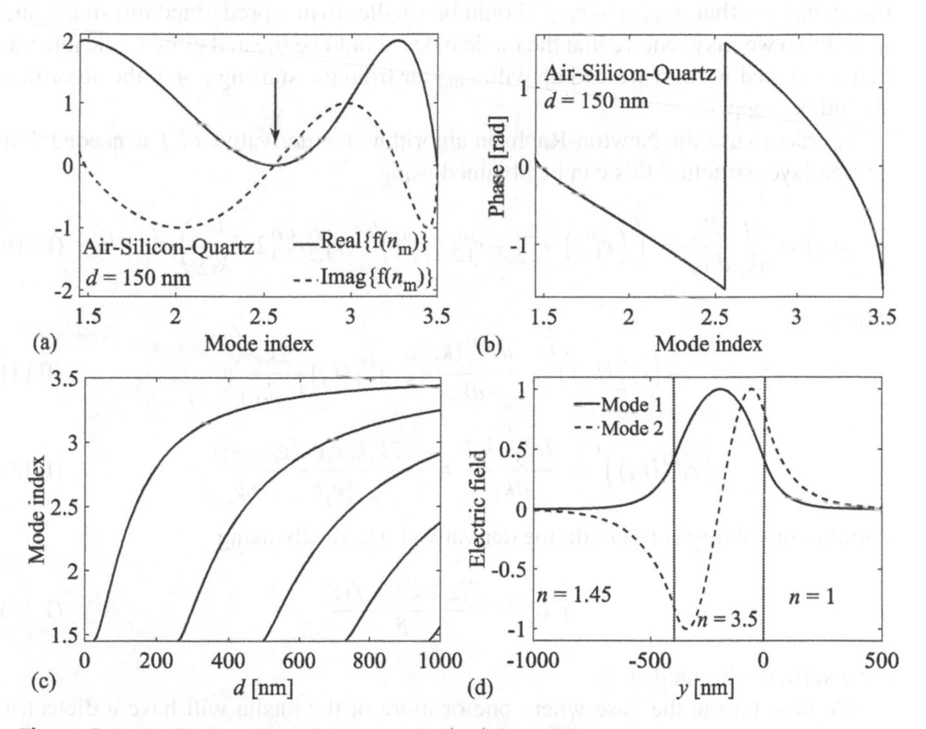

Figure F.2 (a) Real and imaginary part of $f(n_{\rm m})$ for a dielectric waveguide with $n_1 = 1$, $n_2 = 3.5$, $n_3 = 1.45$, waveguide thickness $d = 150$ nm, wavelength 1500 nm, and s polarization. (b) Corresponding phase of $f(n_{\rm m})$. (c) Mode indices of guided modes versus waveguide thickness d. (d) $E(y)$ for the two guided modes corresponding to $d = 400$ nm.

An alternative method of finding solutions, which is much faster but also less robust, is to use the *Newton-Raphson* algorithm. The first step with calculating $f(n_{\rm m})$ for a suitable number of $n_{\rm m}$ in the interval from n_3 to n_2 is the same, and sign changes or phase jumps of $f(n_{\rm m})$ between neighbor points $n_{\rm m,a}$ and $n_{\rm m,b}$ are used to obtain values of $n_{\rm m}$ that are close to a solution. We may now use $n_{\rm m1} = (n_{\rm m,a} + n_{\rm m,b})/2$ as the first estimate of the solution. By assuming that f is approximately a linear function in a region close to $n_{\rm m1}$, i.e.,

$$f(n_{\rm m}) \approx f(n_{\rm m1}) + (n_{\rm m} - n_{\rm m1}) f'(n_{\rm m1}), \tag{F.8}$$

where $f'(x) = df/dx$, the next estimate for the solution to $f(n_{\rm m}) = 0$ becomes

$$n_{{\rm m},i+1} = n_{{\rm m},i} - \frac{f(n_{{\rm m},i})}{f'(n_{{\rm m},i})}. \tag{F.9}$$

The series in (F.9) may very quickly converge to the mode index if the starting point $n_{\rm m1}$ is close enough to the solution. Two criteria can be used for having found a mode, namely that $|n_{m,i+1} - n_{m,i}|$ should be smaller than a predefined threshold, and in addition we may require that the mode index should be located close to the interval that we started with. If we end up with $n_{{\rm m},i}$ far from the starting point, the algorithm should be stopped.

In order to use the Newton-Raphson algorithm, the derivative of f is needed. For a three-layer structure this can be obtained using

$$f'(n_{\rm m}) = \frac{df}{dk_x}\frac{dk_x}{dn_{\rm m}} = \left(\left(r_{1,2}^{(u)} \right)' r_{2,3}^{(u)} + r_{1,2}^{(u)} \left(r_{2,3}^{(u)} \right)' - r_{1,2}^{(u)} r_{2,3}^{(u)} 2id \frac{k_x}{k_{y2}} \right) e^{2ik_{y2}d} k_0, \tag{F.10}$$

where

$$\left(r_{i,j}^{(s)}(k_x) \right)' \equiv \frac{dr_{i,j}^{(s)}(k_x)}{dk_x} = r_{i,j}^{(s)}(k_x) \frac{2k_x}{k_{y,i}k_{y,j}}, \tag{F.11}$$

$$\left(r_{i,j}^{(p)}(k_x) \right)' \equiv \frac{dr_{i,j}^{(p)}(k_x)}{dk_x} = \frac{2k_x k_0^2 \varepsilon_i \varepsilon_j (\varepsilon_i - \varepsilon_j)}{k_{y,i}k_{y,j} (\varepsilon_j k_{y,i} + \varepsilon_i k_{y,j})^2}. \tag{F.12}$$

Another possibility is to obtain the derivative numerically using

$$f'(z) \approx \frac{f(z+\delta) - f(z)}{\delta} \tag{F.13}$$

for a sufficiently small δ.

We now turn to the case where one or more of the media will have a dielectric constant with an imaginary part, i.e., lossy media, where the mode index will no longer be a purely real number, and we must search for solutions $n_{\rm m}$ to Eq. (F.2) in the complex plane. As an example we will consider a gold-air-gold waveguide, p polarization, and wavelength 800 nm. The refractive index of gold at this wavelength can be set to $n_1 = n_3 = 0.1532 + 4.8984i$ [40]. Even for ultra-small air gaps of thickness d between the metal surfaces this type of waveguide supports a type of guided mode known as a gap-plasmon-polariton.

The Newton-Raphson algorithm still applies in the complex case, and with the same expressions, as long as n_m is considered in regions where the square roots involved in expressions for $k_{y,i}$ are analytic, which means that branch-cuts where these functions jump must not be crossed. The branch-cuts are governed by the requirement $\text{Imag}\{k_{y,i}\} \geq 0$. If we can determine a rough estimate of the solution n_{m1} the Newton-Raphson algorithm can be used to find the exact value. For analytic functions the approach (F.13) can also be applied for any non-zero complex number δ of sufficiently small magnitude, as long as z is sufficiently farther away from a branch-cut than the magnitude of δ.

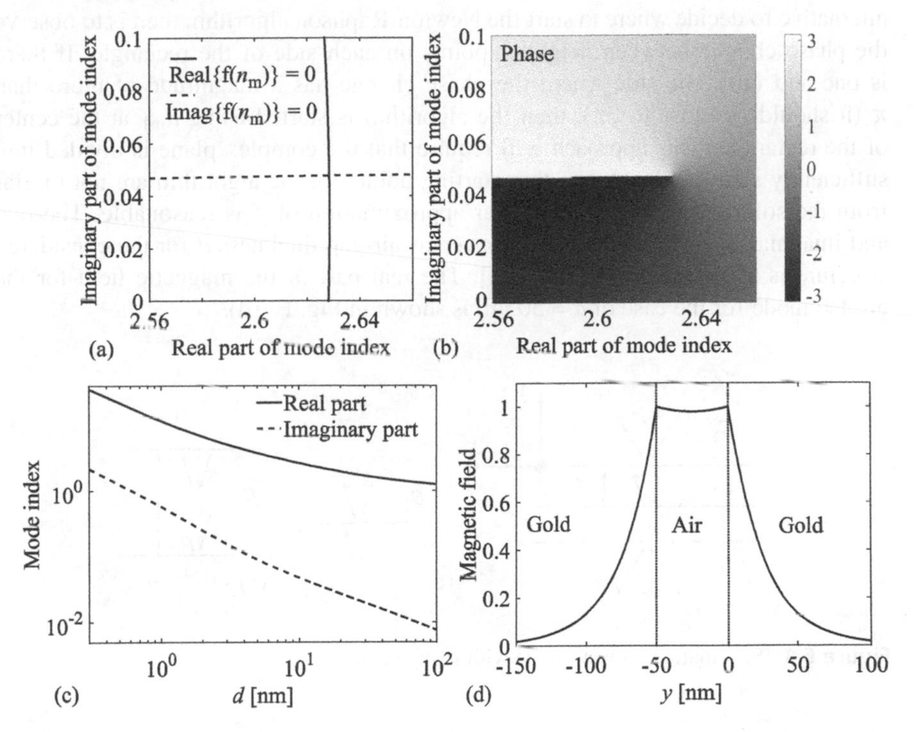

Figure F.3 (a) Lines where either the real part or the imaginary part of $f(n_m)$ is zero for a gold-air-gold dielectric waveguide with air gap $d = 150$ nm, p polarization, and wavelength 800 nm. (b) Corresponding phase of $f(n_m)$. (c) Mode index of the guided mode versus waveguide thickness d. (d) $H(y)$ for the guided mode corresponding to $d = 50$ nm.

In order to roughly estimate the solutions for the mode index we may now divide the part of the complex plane, where we wish to search for solutions, into rectangular regions, and then calculate the function $f(n_m)$ at the four corners of each rectangle. A solution may now possibly exist in the rectangular region if both the real and imaginary part change sign across at least one side in the rectangle. This is shown in Fig. F.3(a) for the gold-air-gold waveguide with gap $d = 10$ nm. The lines show where either the real part or the imaginary part of f is zero, and the solution is at the crossing point. The curves are exactly perpendicular at the crossing point, which is a feature of analytic functions. Thus, by looking for sign changes of real and imaginary parts across the sides of the rectangle we can decide if the rectangle represents a region where it is worth applying the Newton-Raphson algorithm for further search of solutions.

The phase of the complex function $f(n_{\mathrm{m}})$ is shown in Fig. F.3(b) for the same situation as in Fig. F.3(a). It can be seen that the root of f is a phase singularity. The phase is continuously increasing in one direction as the position is moved around the singularity, except for one place where the phase (modulo 2π) jumps by 2π. An alternative to decide where to start the Newton-Raphson algorithm then is to observe the phase change between neighbor points on each side of the rectangle. If there is one and only one side where the phase change has a magnitude of more than π (it should be close to 2π), then the algorithm is started using $n_{\mathrm{m}1}$ at the center of the rectangle. This approach will require that the complex plane is divided into sufficiently small rectangles so that starting points for the algorithm are not too far from the solution, such that the linear approximation of f is reasonable. The real and imaginary part of the mode index versus air-gap thickness d for the considered structure is shown in Fig. F.3(c) [72]. The real part of the magnetic field for the guided mode for the case of d = 50 nm is shown in Fig. F.3(d).

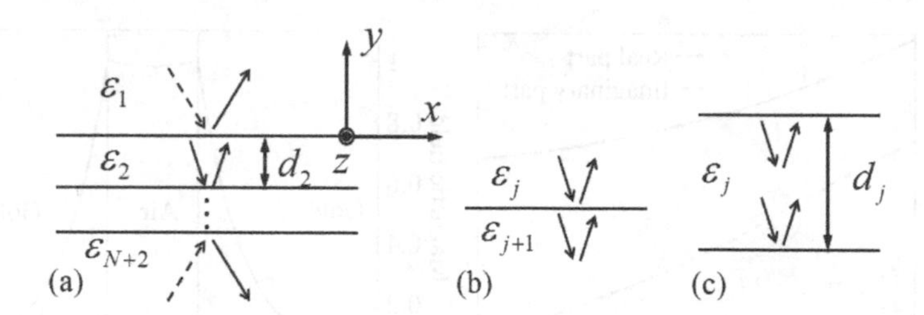

Figure F.4 Schematic of a waveguide with many layers.

In the last part of this appendix the case of waveguides with many layers, as illustrated in Fig. F.4(a), will be considered. The field for $y > 0$ and $y < -d$, where d is the total thickness of layers between medium 1 and $N+2$, can be expressed as

$$H(\mathbf{r}) = e^{ik_0 n_{\mathrm{m}} x} \begin{cases} H_1^+ e^{-ik_{y1}y} + H_1^- e^{ik_{y1}y}, & y > 0 \\ H_{N+2}^+ e^{-ik_{y,N+2}(y+d)} + H_{N+2}^- e^{ik_{y,N+2}(y+d)}, & y < -d \end{cases}. \tag{F.14}$$

The boundary condition for a guided mode is that the coefficients $H_1^+ = H_{N+2}^- = 0$.

Now consider the fields in two neighbor layers j and $j+1$ with boundary between layers at $y = y_{\mathrm{ref}}$ as illustrated in Fig. F.4(b). The fields in these layers can be expressed as

$$H(\mathbf{r}) = e^{ik_0 n_m x} \begin{cases} H_j^+ e^{-ik_{y,j}(y-y_{ref})} + H_j^- e^{ik_{y,j}(y-y_{ref})}, & d_j > y - y_{ref} > 0 \\ H_{j+1}^+ e^{-ik_{y,j+1}(y-y_{ref})} + H_{j+1}^- e^{ik_{y,j+1}(y-y_{ref})}, & -d_{j+1} < y - y_{ref} < 0 \end{cases}. \tag{F.15}$$

The electromagnetics boundary conditions can be used to obtain the following matrix relation between coefficients governing fields on either side of the interface [38]

$$\begin{bmatrix} H_j^+ \\ H_j^- \end{bmatrix} = \frac{1}{t_{j,j+1}} \begin{bmatrix} 1 & r_{j,j+1} \\ r_{j,j+1} & 1 \end{bmatrix} \begin{bmatrix} H_{j+1}^+ \\ H_{j+1}^- \end{bmatrix} = \overline{\overline{H}}_{j,j+1} \begin{bmatrix} H_{j+1}^+ \\ H_{j+1}^- \end{bmatrix}. \tag{F.16}$$

Here $r_{j,j+1}(k_0 n_m)$ and $t_{j,j+1}(k_0 n_m)$ are the two-layer Fresnel reflection and transmission coefficients between layer j and $j+1$ similar to Eq. (2.50) for p polarization.

The fields inside a layer j of thickness d_j illustrated in Fig. F.4(c) can be expressed relative to the top of the layer at $y = y_{ref}$ as follows

$$H(\mathbf{r}) = e^{ik_0 n_m x} \left(H_j^+ e^{-ik_{y,j}(y-y_{ref})} + H_j^- e^{ik_{y,j}(y-y_{ref})} \right), \quad -d_j < y - y_{ref} < 0, \tag{F.17}$$

or relative to the bottom of the layer at $y = y_{ref,2} = y_{ref} - d_j$ as

$$H(\mathbf{r}) = e^{ik_0 n_m x} \left(\tilde{H}_j^+ e^{-ik_{y,j}(y-y_{ref,2})} + \tilde{H}_j^- e^{ik_{y,j}(y-y_{ref,2})} \right), \quad d_j > y - y_{ref,2} > 0. \tag{F.18}$$

Both expressions are representations of the same field but based on different reference positions. The coefficients are related by the matrix relation [38]

$$\begin{bmatrix} H_j^+ \\ H_j^- \end{bmatrix} = \begin{bmatrix} e^{-ik_{y,j} d_j} & 0 \\ 0 & e^{ik_{y,j} d_j} \end{bmatrix} \begin{bmatrix} \tilde{H}_j^+ \\ \tilde{H}_j^- \end{bmatrix} = \overline{\overline{L}}_j \begin{bmatrix} \tilde{H}_j^+ \\ \tilde{H}_j^- \end{bmatrix}. \tag{F.19}$$

By applying the matrices in Eqs. (F.16) and (F.19) across each interface and across each layer, the following matrix relation for the coefficients of outer layers can be obtained

$$\begin{bmatrix} H_1^+ \\ H_1^- \end{bmatrix} = \overline{\overline{H}}_{1,2} \overline{\overline{L}}_2 \overline{\overline{H}}_{2,3} \overline{\overline{L}}_3 \ldots \overline{\overline{H}}_{N+1,N+2} \begin{bmatrix} H_{N+2}^+ \\ H_{N+2}^- \end{bmatrix} = \begin{bmatrix} M_{11} & M_{12} \\ M_{21} & M_{22} \end{bmatrix} \begin{bmatrix} H_{N+2}^+ \\ H_{N+2}^- \end{bmatrix}. \tag{F.20}$$

The boundary condition $H_1^+ = H_{N+2}^- = 0$ now leads to the requirement $M_{11} = 0$. The matrix element M_{11} is, on the other hand, a function of the mode index n_m, and the mode indices for a waveguide with many layers can thus be obtained by requiring $f(n_m) = M_{11}(n_m) = 0$ and using the same techniques that have already been discussed.

F.1 EXERCISES

1. Prove the expressions (F.11) and (F.12).

2. Construct a numerical program for finding the mode index of an asymmetric dielectric waveguide with $n_2 > n_3 > n_1$ (no absorption). 1. Define n_1, n_2, n_3, d, λ, and the desired accuracy. 2. Divide the interval from n_3 to n_2 in N steps, and check, for each step, if the phase of $f(n_\mathrm{m})$ jumps. Apply the *bisection* method if it does. Store the obtained mode indices in a vector. 3. Plot the fields using (2.70)-(2.72) or equivalent expressions for s polarization. 4. Construct a loop where λ or d is varied, and where the mode indices for each step are stored in a matrix. 4. Plot the mode indices versus λ or d.

3. Construct a numerical program for finding the mode index of a waveguide with absorption losses. The steps are similar to Problem 2 except that in Step 2 the complex plane is divided into small rectangles, and it is checked for each rectangle if there is a phase singularity inside. In that case the *Newton-Raphson* algorithm is applied.

G Plane-wave-expansion theory

This appendix is concerned with the plane-wave-expansion method for calculating band diagrams of 2D-periodic photonic structures. We shall only consider in-plane propagation of light, in which case it is possible to either consider *s*- or *p*-polarized light, where either the electric field or the magnetic field is purely polarized along the *z*-axis. In the literature these polarizations are also sometimes referred to instead as transverse-electric (TE) and transverse-magnetic (TM). The structures of interest are periodic in the *xy*-plane with a period, or unit cell, being repeated on a rectangular lattice with lattice constants Λ_x and Λ_y along the *x* and *y* directions, respectively.

The unit cell for a simple photonic crystal with a single dielectric rod being repeated periodically is shown with the dashed line in Fig. G.1(a). The unit cell is being repeated on a square lattice ($\Lambda_x = \Lambda_y = \Lambda$). A rectangular unit cell that can be used for modeling of photonic crystal waveguides is shown in Fig. G.1(b). This unit cell contains 6 rods. The structure describes a photonic crystal with a line defect along the *y*-axis which can be used as a waveguide. Note that the waveguide will also be repeated periodically along the *x* axis, and the distance between waveguides must be chosen large enough that coupling between waveguides can be ignored.

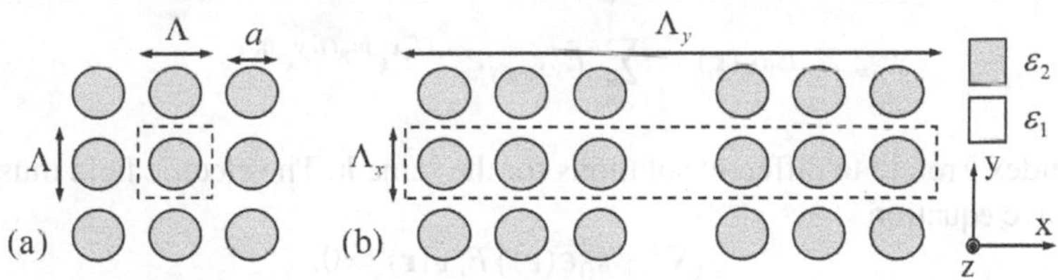

Figure G.1 (a) Square shaped unit cell with a single dielectric cylinder being repeated on a square lattice. (b) Rectangular unit cell with 6 dielectric cylinders being repeated on a rectangular lattice.

The dielectric constant of a periodic structure such as those in Fig. G.1 can be expressed as a Fourier series of the form

$$\varepsilon(\mathbf{r}) = \sum_{n_x,n_y} \tilde{\varepsilon}_{n_x,n_y} e^{in_xG_xx} e^{in_yG_yx}, \tag{G.1}$$

where $G_x = 2\pi/\Lambda_x$ and $G_y = 2\pi/\Lambda_y$, and n_x and n_y assume integer values. The Fourier coefficients are straightforwardly given by

$$\tilde{\varepsilon}_{n_x,n_y} = \frac{1}{\Lambda_x\Lambda_y} \int_{x=0}^{\Lambda_x} \int_{y=0}^{\Lambda_y} \varepsilon(\mathbf{r}) e^{-in_xG_xx} e^{-in_yG_yy} dxdy. \tag{G.2}$$

By evaluating this expression for the simple unit cell in Fig. G.1(a) with a single dielectric cylinder with radius *a* on a square lattice with period Λ, the following

Fourier coefficients are obtained

$$\tilde{\varepsilon}_{n_x,n_y} = \begin{cases} \varepsilon_1\left(1-\frac{\pi a^2}{\Lambda^2}\right)+\varepsilon_2\frac{\pi a^2}{\Lambda^2}, & n_x=n_y=0 \\ (\varepsilon_2-\varepsilon_1)\frac{a}{\Lambda}\frac{2\pi}{|\mathbf{G}_{n_x,n_y}|\Lambda}J_1(|\mathbf{G}_{n_x,n_y}|a), & \text{otherwise} \end{cases}, \tag{G.3}$$

where $|\mathbf{G}_{n_x,n_y}| = G\sqrt{n_x^2+n_y^2}$.

In the case of the rectangular unit cell as in Fig. G.1(b) with N circular rods within the unit cell it can be shown that the Fourier coefficients are given by

$$\tilde{\varepsilon}_{n_x,n_y} = \begin{cases} \varepsilon_1\left(1-\frac{N\pi a^2}{\Lambda_x\Lambda_y}\right)+\varepsilon_2\frac{N\pi a^2}{\Lambda_x\Lambda_y}, & n_x=n_y=0 \\ (\varepsilon_2-\varepsilon_1)\frac{a}{\Lambda_x\Lambda_y}\frac{2\pi}{|\mathbf{G}_{n_x,n_y}|}J_1(|\mathbf{G}_{n_x,n_y}|a)\sum_{j=1}^{N}e^{in_xG_xx_j}e^{in_yG_yy_j}, & \text{otherwise} \end{cases}, \tag{G.4}$$

where now $|\mathbf{G}_{n_x,n_y}| = \sqrt{n_x^2G_x^2+n_y^2G_y^2}$, and (x_j,y_j) represents the center of rod j.

Considering first s polarization, the z-component of the electric field can be expanded in Bloch waves in the form

$$E_{\nu,\mathbf{k}}(\mathbf{r}) = U_{\nu,\mathbf{k}}(\mathbf{r})e^{i\mathbf{k}\cdot\mathbf{r}}, \tag{G.5}$$

where $\mathbf{k} = \hat{\mathbf{x}}k_x+\hat{\mathbf{y}}k_y$ is the Bloch-wave vector, and $U_{\nu,\mathbf{k}}(\mathbf{r})$ is a periodic function which is periodic with the same period as the structure. This function can also be expanded in a Fourier series leading to the electric field

$$E_{\nu,\mathbf{k}}(\mathbf{r}) = \sum_{n_x,n_y}\tilde{E}_{\nu,n_x,n_y}e^{in_xG_xx}e^{in_yG_yy}e^{i\mathbf{k}\cdot\mathbf{r}}. \tag{G.6}$$

The index ν refers to different solutions for the same $\mathbf{k}$. The electric field must satisfy the wave equation

$$\left(\nabla^2+k_0^2\varepsilon(\mathbf{r})\right)E_{\nu,\mathbf{k}}(\mathbf{r}) = 0. \tag{G.7}$$

By inserting the Fourier series for the electric field and the dielectric constant, this leads to

$$\sum_{n_x,n_y}\left((n_xG_x+k_x)^2+(n_yG_y+k_y)^2\right)E_{\nu,n_x,n_y}e^{in_xG_xx}e^{in_yG_yy}e^{i\mathbf{k}\cdot\mathbf{r}} =$$

$$k_0^2\sum_{n_x'',n_y''}\sum_{n_x',n_y'}\tilde{\varepsilon}_{n_x'',n_y''}\tilde{E}_{\nu,n_x',n_y'}e^{i(n_x''+n_x')G_xx}e^{i(n_y''+n_y')G_yy}e^{i\mathbf{k}\cdot\mathbf{r}}. \tag{G.8}$$

By requiring that this should be satisfied for all values of x and y, the expression reduces to

$$\left((n_xG_x+k_x)^2+(n_yG_y+k_y)^2\right)E_{\nu,n_x,n_y} = k_0^2\sum_{n_x',n_y'}\tilde{\varepsilon}_{n_x-n_x',n_y-n_y'}\tilde{E}_{\nu,n_x',n_y'}. \tag{G.9}$$

Thus, one equation results for each (n_x,n_y). It is necessary to truncate the Fourier-series and only consider a finite number of n_x and n_y, e.g. $-N_x \le n_x \le +N_x$ and $-N_y \le n_y \le +N_y$.

The resulting equations can be organized in the matrix form

$$\overline{\overline{A}}\,\overline{E} = k_0^2 \overline{\overline{B}}\,\overline{E}, \tag{G.10}$$

which is a generalized matrix eigenvalue problem. This can also be rewritten as a standard matrix eigenvalue problem

$$\overline{\overline{C}}\,\overline{E} = k_0^2 \overline{E}. \tag{G.11}$$

The eigenvalue problems can be solved by standard solvers. In MATLAB®, for example, the matrices can be used as input in the routines *eig* or *eigs*, which will give as output the eigenvectors and corresponding eigenvalues. Here, the index v is used to distinguish between the different eigenvectors (eigenvalues). The index v is also referred to as the band number. The eigenvalues k_0^2 will depend on the choice of Bloch-wave vector.

It is usually sufficient to obtain the eigenvalues for Bloch-wave vectors within the irreducible 1st Brillouin zone in order to investigate if a periodic structure possesses a photonic bandgap [12]. The irreducible Brillouin zone is obtained from symmetry considerations. The reciprocal lattice for the photonic crystal in Fig. G.1(a) is shown in Fig. G.2(a). The Brillouin zone is shown with the dashed line. Any Bloch mode can be obtained by considering Bloch-wave vectors within the Brillouin zone. In addition it is possible by using symmetry considerations to further obtain any Bloch mode by considering only the Bloch-wave vectors within the gray region, which is the triangle with corners Γ, X and M. This region is known as the irreducible Brillouin zone. Band diagrams are usually obtained by considering only the boundary of

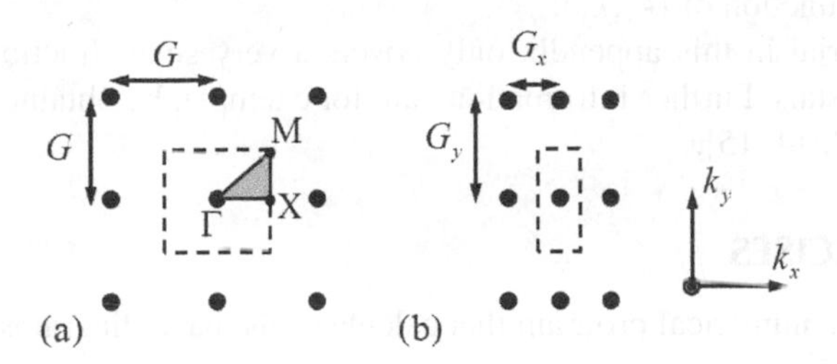

Figure G.2 (a) Reciprocal lattice and irreducible 1. Brillouin zone for a periodic structure with a square-shaped unit cell. (b) Reciprocal lattice and Brillouin zone for a rectangular-shaped unit cell.

the irreducible B.Z. ($\Gamma \rightarrow X \rightarrow M \rightarrow \Gamma$) as maxima and minima of eigenvalues are expected to occur on this boundary. The corners are defined by $\Gamma : (k_x, k_y) = (0, 0)$; $X : (k_x, k_y) = (G/2, 0)$; $M : (k_x, k_y) = (G/2, G/2)$. An example of a band diagram for the case of rods with dielectric constant $\varepsilon_2 = 8.9$ in air and with $a = 0.2\ \Lambda$ is shown in Fig. 4.63.

In the case of the photonic crystal waveguide in Fig. G.1(b), the Brillouin zone collapses along the x direction as $G_x \rightarrow 0$ for $\Lambda_x \rightarrow \infty$. For a large Λ_x it may often be sufficient to only consider values of $\mathbf{k}$ with $k_x = 0$ and $0 \leq k_y \leq G_y/2$. An example of a band diagram for this situation is shown in Fig. 4.65.

In the case of p polarization with the magnetic field along the z-axis, the wave equation can be expressed as the following eigenvalue problem

$$-\frac{\partial}{\partial x}\left(\eta(\mathbf{r})\frac{\partial H(\mathbf{r})}{\partial x}\right) - \frac{\partial}{\partial y}\left(\eta(\mathbf{r})\frac{\partial H(\mathbf{r})}{\partial y}\right) = k_0^2 H(\mathbf{r}), \tag{G.12}$$

where

$$\eta(\mathbf{r}) \equiv \frac{1}{\varepsilon(\mathbf{r})} = \sum_{n_x,n_y} \tilde{\eta}_{n_x,n_y} e^{in_x G_x x} e^{in_y G_y y}. \tag{G.13}$$

The Fourier expansion of the magnetic field can be expressed as

$$H_{\nu,\mathbf{k}}(\mathbf{r}) = \sum_{n_x,n_y} \tilde{H}_{\nu,n_x,n_y} e^{in_x G_x x} e^{in_y G_y y} e^{ik_x x} e^{ik_y y}, \tag{G.14}$$

and by inserting the respective Fourier expansions in the wave equation (G.12) the following equation for the Fourier coefficients may be obtained

$$\sum_{n'_x,n'_y} \mathbf{K}_{n_x,n_y} \cdot \mathbf{K}_{n'_x,n'_y} \tilde{\eta}_{n_x-n'_x,n_y-n'_y} \tilde{H}_{\nu,n'_x,n'_y} = k_0^2 \tilde{H}_{\nu,n_x,n_y}, \tag{G.15}$$

where

$$\mathbf{K}_{n_x,n_y} = \hat{\mathbf{x}}(n_x G_x + k_x) + \hat{\mathbf{y}}(n_y G_y + k_y). \tag{G.16}$$

Again, this may be formulated as a matrix eigenvalue problem, and the eigenvalues k_0^2 will be a function of (k_x, k_y).

The material in this appendix only covers a very small fraction of the field of photonic crystals. Further information can, for example, be obtained from the references [12, 13, 14, 15].

G.1 EXERCISES

1. Construct a numerical program that calculates the band diagrams in Fig. 4.63 and Fig. 4.65.

2. Derive the expressions (G.3) and (G.4) for the Fourier coefficients of the array of dielectric rods on a square lattice and the photonic crystal waveguide.

References

1. P. M. Morse and H. Feshbach, *Methods of Theoretical Physics*. New York: McGraw-Hill, 1953.
2. C.-T. Tai, *Dyadic Green's Functions in Electromagnetic Theory*. London: Intext Educational Publishers, 1971.
3. E. Economou, *Green's Functions in Quantum Physics*, 1st edn. Berlin, Germany: Springer-Verlag, 1979.
4. J. Jin, *The Finite Element Method in Electromagnetics*, 2nd edn. New York: Wiley, 2002.
5. J. L. Volakis, A. Chatterjee, and L. C. Kempel, *Finite Element Method for Electromagnetics: Antennas, Microwave Circuits, and Scattering Applications*. New York: IEEE press, 1998.
6. A. C. Polycarpou, *Introduction to the Finite Element Method in Electromagnetics*. San Rafael: Morgan & Claypool Publishers, 2006.
7. M. G. Larson and F. Bengzon, *The Finite Element Method: Theory, Implementation, and Applications*. Berlin: Springer Berlin Heidelberg, 2013.
8. M. G. Moharam, and T. K. Gaylord, Rigorous coupled-wave analysis of planar-grating diffraction, *J. Opt. Soc. Am.* **71**, 811–818 (1981).
9. L. Li, *Fourier Modal Method* in *Gratings: Theory and Numeric Applications, 2nd edn.* Institut Fresnel: www.fresnel.fr/numerical-grating-book-2 (2014).
10. R.-B. Hwang, *Periodic Structures - Mode-Matching Approach and Applications in Electromagnetic Engineering*. Singapore: IEEE Press and John Wiley & Sons, 2013.
11. A. V. Lavrinenko, J. Lægsgaard, N. Gregersen, F. Schmidt, and T. Søndergaard, *Numerical Methods in Photonics*. Boca Raton: CRC Press, 2014.
12. J. D. Joannopoulos, S. G. Johnson, J. N. Winn, and R. D. Meade, *Photonic Crystals: Molding the Flow of Light*, 2nd edn. Princeton, N. J.: Princeton University Press, 2008.
13. M. Skorobogatiy and J. Yang, *Fundamentals of Photonic Crystal Guiding*. Cambridge: Cambridge University Press, 2009.
14. K. Sakoda, *Optical Properties of Photonic Crystals*, 2nd edn. Berlin: Springer-Verlag Berlin Heidelberg, 2005.
15. W. L. Dahl (editor), *Photonic Crystals: Optical Properties, Fabrication and Applications*. New York: Nova Science Publishers, 2011.
16. H. Raether, *Surface Plasmons on Smooth and Rough Surfaces and on Gratings*. New York: Springer-Verlag, 1986.
17. S. A. Maier, *Plasmonics: Fundamentals and Applications*. Bath: Springer, 2007.
18. L. Novotny and B. Hecht, *Principles of Nano-Optics, Second Edition*. Cambridge: Cambridge University Press, 2012.
19. W. Cai, and V. Shalaev, *Optical Metamaterials*. New York: Springer New York, 2010.
20. W. C. Chew, *Waves and Fields in Inhomogeneous Media*. New York, IEEE Press, 1995.
21. W. C. Chew, M. S. Tong, and B. Hu, *Integral Equation Methods for Electromagnetic and Elastic Waves*. San Rafael: Morgan & Claypool Publishers, 2009.
22. W. C. Gibson, *The Method of Moments in Electromagnetics*, 2nd edn. Boca Raton: CRC Press, 2015.
23. A. F. Peterson, S. L. Ray, and R. Mittra, *Computational Methods for Electromagnetics*. New York: IEEE Press, 1998.

24. R. Mittra, ed., *Computer Techniques for Electromagnetics*. Oxford: Pergamon Press, 1973.
25. R. F. Harrington, *Field Computation by Moment Methods*. New York: Macmillan, 1968.
26. N. Morita, N. Kumagai, and J. R. Mautz, *Integral Equation Methods for Electromagnetics*. London: Artech House, 1990.
27. Y. Ozaki, K. Kneipp, and R. Aroca, *Frontiers of Surface-Enhanced Raman Scattering: Single Nanoparticles and Single Cells*. Chichester, England: Wiley, 2014.
28. K. Kneipp, M. Moskovits, and H. Kneipp (eds.), *Surface-Enhanced Raman Scattering: Physics and Applications*. Berlin, Heidelberg: Springer-Verlag Berlin Heidelberg, 2006.
29. L. M. Liz-Marzán, Nanometal: Formation and color, *Materials Today*, February, 26–31 (2004).
30. R. D. Averitt, S. L. Westcott, and N. J. Halas, Linear optical properties of gold nanoshells, *J. Opt. Soc. Am. B* **16**, 1824–1832 (1999).
31. E. Purcell, Spontaneous emission probabilities at radio frequencies, *Proc. Am. Phys. Soc.* **69**, 681 (1946)
32. P. Milonni, *The Quantum Vacuum: An Introduction to Quantum Electrodynamics*. Boston: Academic Press, 1994.
33. H. Yokoyama and K. Ujihara, *Spontaneous Emission and Laser Oscillations in Microcavities*. New York: CRC Press, 1995.
34. R. Chang and A. Campillo, *Optical Processes in Microcavities*. Singapore: World Scientific, 1996.
35. J. D. Jackson, *Classical Electrodynamics*, 3rd edn., New York: John Wiley & Sons Inc, 1999.
36. J. A. Stratton, *Electromagnetic Theory*. New Jersey: IEEE Press (An IEEE press classic reissue), 2007.
37. J. R. Reitz, F. J. Milford, R. W. Christy, *Foundations of Electromagnetic Theory*, 4th edn., New York: Addison-Wesley Publishing Company, 1992.
38. M. V. Klein, and T. E. Furtak, *Optics*, 2nd edn., New Jersey: John Wiley & Sons, 1986.
39. H. Angus Macleod, *Thin-Film Optical Filters*, 4th edn. Boca Raton: CRC Press, 2010.
40. P. B. Johnson and R. W. Christy, Optical constants of the noble metals, *Phys. Rev. B* **6**, 4370 (1972).
41. A. E. Siegmann, Lasers without photons - or should it be lasers with too many photons?, *Appl. Phys. B: Lasers Opt.* **60**, 247–257 (1995).
42. O. J. F. Martin, A. Dereux, and C. Girard, Iterative scheme for computing exactly the total field propagating in dielectric structures of arbitrary shape, *J. Opt. Soc. Am. A* **11**, 1073–1080 (1994).
43. J. Van Bladel, *Singular Electromagnetic Fields and Sources*. New Jersey: IEEE press, 1991.
44. E. N. Economou, Surface plasmons in thin films, *Phys. Rev.* **182**, 539–554 (1969).
45. J. J. Burke, G. I. Stegeman, and T. Tamir, Surface-polariton-like waves guided by thin, lossy metal films, *Phys. Rev. B* **33**, 5186–5201 (1986).
46. T. Søndergaard, J. Jung, S. I. Bozhevolnyi, and G. Della Valle, Theoretical analysis of gold nano-strip gap plasmon resonators, *New J. Phys.* **10**, 105008 (2008).
47. R. Gordon, Light in a subwavelength slit in a metal: Propagation and reflection, *Phys. Rev. B* **73**, 153405 (2006).
48. Y. Kurokawa and H. T. Miyazaki, Metal-insulator-metal plasmon nanocavities: Analysis of optical properties, *Phys. Rev. B* **75**, 035411 (2007).
49. A. Chandran, E. S. Barnard, J. S. White, and M. L. Brongersma, Metal-dielectric-metal

surface plasmon-polariton resonators, *Phys. Rev. B* **85**, 085416 (2012).
50. T. Søndergaard and S. I. Bozhevolnyi, Strip and gap plasmon polariton optical resonators, *Phys. Stat. Sol. (b)* **245**, 9–19 (2008).
51. T. Søndergaard, Modeling of plasmonic nanostructures: Green's function integral equation methods, *Phys. Stat. Sol. (b)* **244**, 3448–3462 (2007).
52. T. Søndergaard and S. I. Bozhevolnyi, Slow-plasmon resonant nanostructures: Scattering and field enhancements, *Phys. Rev. B* **75**, 073402 (2007).
53. P. Mühlschlegel, H.-J. Eisler, O. J. F. Martin, B. Hecht, and D. W. Pohl, Resonant optical antennas, *Science* **308**, 1607 (2005).
54. D. W. Prather, M. S. Mirotznik, and J. N. Mait, Boundary integral methods applied to the analysis of diffractive optical elements , *J. Opt. Soc. Am. A* **14**, 34–43 (1997).
55. J. Jung, T. Søndergaard, J. Beermann, A. Boltasseva, and S.I. Bozhevolnyi, Theoretical analysis and experimental demonstration of resonant light scattering from metal nanostrips on quartz, *J. Opt. Soc. Am. B* **26**, 121–124 (2009).
56. J. Jung, T. Søndergaard, and S. I. Bozhevolnyi, Gap plasmon-polariton nanoresonators: Scattering enhancement and launching of surface plasmon polaritons, *Phys. Rev. B* **79**, 035401-1-8 (2009).
57. V. Siahpoush, T. Søndergaard, and J. Jung, Green's function approach to investigate the excitation of surface plasmon polaritons in a nanometer-thin metal film, *Phys. Rev. B* **85**, 075305 (2012).
58. T. Søndergaard, V. Siahpoush, and J. Jung, Coupling light into and out from the surface plasmon polaritons of a nanometer-thin metal film with a metal nanostrip, *Phys. Rev. B* **86**, 085455 (2012).
59. J. Jung, and T. Søndergaard, Green's function surface integral equation method for theoretical analysis of scatterers close to a metal interface, *Phys. Rev. B* **77**, 245310-1-12 (2008).
60. J.-J. Greffet, R. Carminati, K. Joulain, J.-P. Mulet, S. Mainguy, and Y. Chen, Coherent emission of light by thermal sources, *Nature* **416**, 61 (2002).
61. H. T. Miyazaki, K. Ikeda, T. Kasaya, K. Yamamoto, Y. Inoue, K. Fujimara, T. Kanakugi, M. Okada, K. Hatade, and S. Kitagawa, Thermal emission of two-color polarized infrared waves from integrated plasmon cavities, *Appl. Phys. Lett.* **92**, 141114 (2008).
62. J. A. Mason, D. C. Adams, Z. Johnson, S. Smith, A. W. Davis, and D. Wasserman, Selective thermal emission from patterned steel, *Opt. Express* **18**, 25192 (2010).
63. R. Stanley, Plasmonics in the mid-infrared, *Nat. Photon.* **6**, 409 (2012).
64. T. Bauer, *Thermophotovoltaics: Basic Principles and Critical Aspects of System Design*. Berlin, Heidelberg : Springer-Verlag Berlin Heidelberg, 2011.
65. J. Le Perchec, P. Quémerais, A. Barbara, and T. López-Ríos, Why metallic surfaces with grooves a few nanometers deep and wide may strongly absorb visible light, *Phys. Rev. Lett.* **100**, 066408 (2008).
66. F. Pardo, P. Bouchon, R. Haïdar, and J.-L. Pelouard, Light funneling mechanism explained by magnetoelectric interference, *Phys. Rev. Lett.* **107**, 093902 (2011).
67. X. Liu, T. Tyler, T. Starr, A. F. Starr, N. M. Jokerst, and W. J. Padilla, Taming the blackbody with infrared metamaterials as selective thermal emitters, *Phys. Rev. Lett.* **107**, 045901 (2011).
68. K. Aydin, V. E. Ferry, R. M. Briggs, and H. Atwater, Broadband polarization-independent resonant light absorption using ultrathin plasmonic super absorbers, *Nat. Commun.* 2:517 doi:10.1038/ncomms1528 (2011).
69. S. I. Bozhevolnyi, V. S. Volkov, E. Devaux, J.-Y. Laluet, and T. W. Ebbesen, Channel

plasmon subwavelength waveguide components including interferometers and ring resonators, *Nature* **440**, 508 (2006).

70. C. L. C. Smith, N. Stenger, A. Kristensen, N. A. Mortensen, and S. I. Bozhevolnyi, Gap and channeled plasmons in tapered grooves: A review, *Nanoscale* **7**, 9355 (2015).
71. T. Søndergaard, S. M. Novikov, T. Holmgaard, R. L. Eriksen, J. Beermann, Z. Han, K. Pedersen, and S. I. Bozhevolnyi, Plasmonic black gold by adiabatic nanofocusing and absorption of light in ultra-sharp convex grooves, *Nat. Commun.* 3:969 doi: 10.1038/ncomms1976 (2012).
72. T. Søndergaard, and S. I. Bozhevolnyi, Theoretical analysis of plasmonic black gold: Periodic arrays of ultra-sharp grooves, *New J. Phys.* **15**, 0134034 (2013).
73. J. Beermann, R. L. Eriksen, T. Søndergaard, T. Holmgaard, K. Pedersen, and S. I. Bozhevolnyi, Plasmonic black metals by broadband light absorption in ultra-sharp convex grooves, *New J. Phys.* **15**, 073007 (2013).
74. M. Odgaard, M. G. Laursen, and T. Søndergaard, Modeling the reflectivity of plasmonic ultrasharp groove arrays: General direction of light incidence, *J. Opt. Soc. Am. B* **31**, 1853–1860 (2014).
75. M. Bora, E. M. Behymer, D. A. Dehlinger, J. A. Britten, C. C. Larson, A. S. P. Chang, K. Munechika, H. T. Nguyen, and T. C. Bond, Plasmonic black metals in resonant nanocavities, *Appl. Phys. Lett.* **102**, 251105 (2013).
76. A. S. Roberts, T. Søndergaard, M. Chirumamilla, A. Pors, and J. Beermann, Light extinction and scattering from individual and arrayed high-aspect-ratio trenches in metal, *Phys. Rev. B* **93**, 075413 (2016).
77. T. Søndergaard, and S. I. Bozhevolnyi, Optics of a single ultrasharp groove in metal, *Opt. Lett.* **41**, 2903 (2016).
78. T. Søndergaard, Y.-C. Tsao, P. K. Kristensen, T. G. Pedersen, and K. Pedersen, Light trapping in guided modes of thin-film silicon-on-silver waveguides by scattering from a nanostrip, *J. Opt. Soc. Am. B* **31**, 2036 (2014).
79. H. A. Atwater and A. Polman, Plasmonics for improved photovoltaic devices, *Nat. Mater.* **9**, 205 (2010).
80. P. Spinelli, V. E. Ferry, H. van de Groep, M. van Lare, M. A. Verschuuren, R. E. I. Schropp, H. A. Atwater, and A. Polman, Plasmonics light-trapping in thin-film Si solar cells, *J. Opt.* **14**, 024002 (2012).
81. H. R. Stuart and D. G. Hall, Absorption enhancement in silicon-on-insulator waveguides using metal island films, *Appl. Phys. Lett.* **69**, 2327 (1996).
82. H. R. Stuart and D. G. Hall, Island size effects in nanoparticle-enhanced photodetectors, *Appl. Phys. Lett.* **73**, 3815 (1998).
83. D. M. Schaadt, B. Feng, and E. T. Yu, Enhanced semiconductor optical absorption via surface plasmon excitation in metal nanoparticles, *Appl. Phys. Lett.* **86**, 063106 (2005).
84. S. Pillai, K. R. Catchpole, T. Trupke, and M. A. Green, Surface plasmon enhanced silicon solar cells, *J. Appl. Phys.* **101**, 093105 (2007).
85. K. R. Catchpole and A. Polman, Design principles for particle plasmon enhanced solar cells, *Appl. Phys. Lett.* **93**, 191113 (2008).
86. S. Mokkapati, F. J. Beck, R. de Waele, A. Polman, and K. R. Catchpole, Resonant nanoantennas for light-trapping in plasmonic solar cells, *J. Phys. D* **44**, 185101 (2011).
87. M. Brincker, P. Karlsen, E. Skovsen, and T. Søndergaard, Microstructured gradient-index lenses for THz photoconductive antennas, *AIP Advances* **6**, 025015 (2016).
88. S.-G. Park, K. Lee, D. Han, and K.-H. Jeong, Subwavelength silicon through-hole arrays as an all-dielectric broadband terahertz gradient index material, *Appl. Phys. Lett.* **105**,

091101 (2014).

89. A. I. Hernandez-Serrano, M. Weidenbach, S. F. Busch, M. Koch, and E. Castro-Camus, Fabrication of GRIN lenses for terahertz applications by three-dimensional modeling, *J. Opt. Soc. Am. B* **33**, 928 (2016).
90. T. Søndergaard, J. Gadegaard, P. K. Kristensen, T. K. Jensen, T. G. Pedersen, and K. Pedersen, Guidelines for 1D-periodic surface microstructures for antireflective lenses, *Opt. Express* **18**, 26245 (2010).
91. J. Zhu, Z. Yu, S. Fan, and Y. Cui, Nanostructured photon management for high performance solar cells, *Mat. Sci. Eng. R* **70**, 330-340 (2010).
92. A. Gombert et. al., Subwavelength-structured antireflective surfaces on glass, *Thin Solid Films* **351**, 73-78 (1999).
93. M. Niggemann, M. Glatthaar, A. Gombert, A. Hinsch, and V. Wittwer, Diffraction gratings and buried nano-electrodes - architectures for organic solar cells, *Thin Solid Films* **451-452**, 619-623 (2004).
94. C. David et. al., Nano-structured anti-reflective surfaces replicated by hot embossing, *Microelectron Eng.* **61-62**, 435-440 (2002).
95. J. Zhu et. al., Optical absorption enhancement in amorphous silicon nanowire and nanocone arrays, *Nano Lett.* **9**, 279-282 (2009).
96. Y. Wang, F. Hu, Y. Kanamori, T. Wu, and K. Hane, Large area freestanding GaN nanocolumn membrane with bottom subwavelength nanostructure, *Opt. Express* **18**, 5504-5511 (2010).
97. C.-H. Sun, W.-L. Min, N. C. Lin, P. Jiang, and B. Jiang, Templated fabrication of large area subwavelength antireflection gratings on silicon, *Appl. Phys. Lett.* **91**, 231105 (2007).
98. H. L. Chen, K. T. Huang, C. H. Lin, W. Y. Yang, and W. Fan, Fabrication of subwavelength antireflective structures in solar cells by utilizing modified illumination and defocus techniques in optical lithography, *Microelectron. Eng.* **84**, 750-754 (2007).
99. D. G. Stavenga, S. Foletti, G. Palasantzas, and K. Arikawa, Light on the moth-eye corneal nipple array of butterflies, *Proc. Biol. Sci.* **273** (1587), 661-667 (2006).
100. D. L. Brundrett, E. N. Glytsis, and T. K. Gaylord, Homogeneous layer models for high-spatial-frequency dielectric surface-relief gratings: Conical diffraction and antireflection designs, *Appl. Opt.* **33**, 2695-2706 (1994).
101. T. K. Gaylord, W. E. Baird, and M. G. Moharam, Zero-reflectivity high-spatial-frequency rectangular-groove dielectric surface-relief gratings, *Appl. Opt.* **25**, 4562-4567 (1986).
102. D. E. Aspnes, Local-field effects and effective-medium theory: A microscopic perspective, *Am. J. Phys.* **50**, 704-709 (1982).
103. P. Lalanne, Waveguiding in blazed-binary diffractive elements, *J. Opt. Soc. Am. A* **16**, 2517-2520 (1999).
104. E. Skovsen, T. Søndergaard, C. Lemke, T. Holmgaard, T. Leissner, R. L. Eriksen, J. Beermann, M. Bauer, K. Pedersen, and S. I. Bozhevolnyi, Plasmonic black gold based broadband polarizers for ultra-short laser pulses, *Appl. Phys. Lett.* **103**, 211102 (2013).
105. T. Søndergaard, E. Skovsen, C. Lemke, T. H. Stær, T. Leissner, R. L. Eriksen, J. Beermann, M. Bauer, K. Pedersen, and S. Bozhevolnyi, Plasmonic black metal polarizers for ultra-short laser pulses, *Proceedings of SPIE 9163*, paper 916308 (2014).
106. E. J. Skjølstrup, T. Søndergaard, and T. G. Pedersen, Quantum spill-out in few-nanometer metal gaps: Effect on gap plasmons and reflectance from ultrasharp groove arrays, *Phys. Rev. B* **97**, 115429 (2018).

107. J. Nachamkin, Integrating the dyadic Green's function near sources, *IEEE Trans. Antennas. Propag.* **38**, 919 (1990).
108. A. D. Yaghjian, Electric dyadic Green's functions in the source region, *Proc. IEEE* **68**, 248 (1980).
109. J. P. Kottmann, O. J. F. Martin, D. R. Smith, and S. Schultz, Spectral response of plasmon resonant nanoparticles with non-regular shape, *Opt. Express* **6**, 213–219 (2000).
110. J. P. Kottmann and O. J. F. Martin, Accurate solution of the volume integral equation for high-permittivity scatterers, *IEEE Trans. Ant. Prop.* **48**, 1719–1726 (2000).
111. Y. Saad and M. H. Schultz, GMRES: A generalized minimal residual algorithm for solving nonsymmetric linear systems, *SIAM J. Sci. Stat. Comput.* **7**, 856-869 (1986).
112. R. Barrett, M. Berry, T. F. Chan, J. Demmel, J. Donato, J. Dongarra, V. Eijkhout, R. Pozo, C. Romine, and H. Van der Vorst, *Templates for the Solution of Linear Systems: Building Blocks for Iterative Methods*, 2nd edn., Philadelphia: SIAM, 1994.
113. P. Goy, J. M. Raimond, M. Gross, and S. Haroche, Observation of cavity-enhanced single-atom spontaneous emission, *Phys. Rev. Lett.* **50**, 1903-1906 (1983).
114. R. G. Hulet, E. S. Hilfer, and D. Kleppner, Inhibited spontaneous emission by a Rydberg atom, *Phys. Rev. Lett.* **55**, 2137-2140 (1985)
115. G. Gabrielse and H. Dehmelt, Observation of inhibited spontaneous emission, *Phys. Rev. Lett.* **55**, 67-70 (1985).
116. J. M. Gérard, B. Sermage, B. Gayral, B. Legrand, E. Costard, and V. Thierry-Mieg, Enhanced spontaneous emission by quantum boxes in a monolithic optical microcavity, *Phys. Rev. Lett.* **81**, 1110-1113 (1998).
117. K. H. Drexhage, Influence of a dielectric interface on fluorescence decay time, *J. Luminescence* **1,2**, 693-701 (1970).
118. M. R. Philpott, Fluorescence from molecules between mirrors, *Chem. Phys. Lett.* **19**, 435-439 (1973).
119. W. Lukosz, Theory of optical-environment-dependent spontaneous-emission rates for emitters in thin layers, *Phys. Rev. B* **22**, 3030-3038 (1980).
120. W. L. Barnes, Fluorescence near interfaces: The role of photonic mode density, *J. Mod. Opt.* **45**, 661-699 (1998).
121. E. Snoeks, A. Lagendijk, and A. Polman, Measuring and modifying the spontaneous emission rate of Erbium near an interface, *Phys. Rev. Lett.* **74**, 2459-2462 (1995).
122. C. Hooijer, G.-X. Li, K. Allaart, and D. Lenstra, Spontaneous emission in multilayer semiconductor structures, *IEEE J. Quantum. Electron.* **37**, 1161-1170 (2001).
123. G. Björk, S. Machida, Y. Yamamoto, and K. Igeta, Modification of spontaneous emission rate in planar dielectric microcavity structures, *Phys. Rev. A* **44**, 669-681 (1991).
124. I. Abram, I. Robert, and R. Kuszelewicz, Spontaneous emission control in semiconductor microcavities with metallic or Bragg mirrors, *IEEE J. Quantum. Electron.* **34**, 71-76 (1998).
125. J. Barthes, G. Colas des Francs, A. Bouhelier, J.-C. Weeber, and A. Dereux, Purcell factor for a point-like emitter coupled to a two-dimensional plasmonic waveguide, *Phys. Rev. B* **84**, 073403 (2011).
126. J. N. Farahani, D. W. Pohl, H.-J. Eisler, and B. Hecht, Single quantum dot coupled to a scanning optical antenna: A tunable superemitter, *Phys. Rev. Lett.* **95**, 017402 (2005).
127. S. Kühn, U. Hkanson, L. Rogobete, and V. Sandoghdar, Enhancement of single-molecule fluorescence using a gold nanoparticle as an optical nanoantenna, *Phys. Rev. Lett.* **97**, 017402 (2006).
128. M. Frimmer and A. F. Koenderink, Spontaneous emission control in a tunable hybrid

photonic system, *Phys. Rev. Lett.* **110**, 217405 (2013).

129. T. Søndergaard and B. Tromborg, General theory for spontaneous emission in active dielectric microstructures: Example of a fiber amplifier, *Phys. Rev. A* **64**, 033812 (2001).
130. Y. Chen, T. R. Nielsen, N. Gregersen, P. Lodahl, and J. Mørk, Finite-element modeling of spontaneous emission of a quantum emitter at nanoscale proximity to plasmonic waveguides, *Phys. Rev. B* **81**, 125431 (2010).
131. E. Yablonovitch, Inhibited spontaneous emission in solid-state physics and electronics, *Phys. Rev. Lett.* **58**, 2059-2062 (1987).
132. S. Noda, M. Fujita, and T. Asano, Spontaneous-emission control by photonic crystals and nanocavities, *Nature photonics* **1**, 449-458 (2007).
133. J. R. de Lasson, P. T. Kristensen, J. Mørk, and N. Gregersen, Semianalytical quasinormal mode theory for the local density of states in coupled photonic crystal cavity-waveguide structures, *Opt. Lett.* **40**, 5790-5793 (2015)
134. O. Painter, J. Vuckovic, and A. Scherer, Defect modes of a two-dimensional photonic crystal in an optically thin dielectric slab, *J. Opt. Soc. Am. B* **16**, 275-285 (1999).
135. T. Søndergaard, Spontaneous emission in two-dimensional photonic crystal microcavities, *IEEE J. Quantum Electron.* **36**, 450-457 (2000).
136. P. T. Kristensen, J. Mørk, P. Lodahl, and S. Hughes, Decay dynamics of radiatively coupled quantum dots in photonic crystal slabs, *Phys. Rev. B* **83**, 075305 (2011).
137. K. Pedersen, M. Schiek, J. Rafaelsen, and H.-G. Rubahn, Second-harmonic generation spectroscopy on organic nanofibers, *Appl. Phys. B* **96**, 821–826 (2009).
138. J. Brewer, M. Schiek, and H.-G. Rubahn, Nonlinear optical properties of CNHP4 nanofibers: Molecular dipole orientations and two photon absorption cross sections, *Opt. Commun.* **283**, 1514–1518 (2010).
139. P. Simesen, T. Søndergaard, E. Skovsen, J. Fiutowski, H.-G. Rubahn, S. I. Bozhevolnyi, and K. Pedersen, Surface plasmon polariton excitation by second harmonic generation in single organic nanofibers, *Opt. Express* **23**, 16356–16362 (2015).
140. I. Gryczynski, J. Malicka, Z. Gryczynski, and J. R. Lakowich, Surface plasmon-coupled emission with gold films, *J. Phys. Chem. B* **108**, 12568–12574 (2004).
141. M. Trnavsky, J. Enderlein, T. Ruckstuhl, C. McDonagh, and B. D. MacCraith, Experimental and theoretical evaluation of surface-plasmon-coupled emission for sensitive fluorescence detection, *J. Biomed. Opt.* **13**, 054021 (2008).
142. D. S. Smith, Y. Kostov, and G. Rao, SPCE-based sensors: Ultra-fast oxygen sensing using surface plasmon-coupled emission from ruthenium probes, *Sens. Actuators B* **127**, 432–440 (2007).
143. A. Drezet, A. Hohenau, D. Koller, A. Stepanov, H. Ditlbacher, B. Steinberger, F. R. Aussenegg, A. Leitner, and J. R. Krenn, Leakage radiation microscopy of surface plasmon polaritons, *Mater. Sci. Eng. B* **149**, 220–229 (2008).
144. D. Zhang, X. Yuan, and A. Bouhelier, Direct image of surface-plasmon-coupled emission by leakage radiation microscopy, *Appl. Opt.* **49**, 875–879 (2010).
145. S. Massenot, J. Grandidier, A. Bouhelier, G. Colas des Francs, L. Markey, J.-C. Weeber, A. Dereux, J. Renger, M. U. González, and R. Quidant, Polymer-metal waveguides characterization by Fourier plane leakage radiation microscopy, *Appl. Phys. Lett.* **91**, 243102 (2007).
146. M. Böhmler, N. Hartmann, C. Georgi, F. Heinrich, A. A. Green, M. C. Hersam, and A. Hartschuh, Enhancing and redirecting carbon nanotube photoluminescence by an optical antenna, *Opt. Express* **18**, 16443–16451 (2010).
147. B. T. Draine, The discrete-dipole approximation and its application to interstellar

graphite grains, *Astrophys. J.* **333**, 848-72 (1988).
148. E. M. Purcell and C. R. Pennypacker, Scattering and absorption of light by nonspherical dielectric grains, *Astrophys. J.* **186**, 705–714 (1973).
149. H. Weyl, Ausbreitung elektromagnetischer Wellen über einem ebenen Leiter, *Ann. Phys.* **365**, 481-500 (1919).
150. T. Søndergaard and S. I. Bozhevolnyi, Surface plasmon polariton scattering by a small particle placed near a metal surface: An analytical study, *Phys. Rev. B* **69**, 045422 (2004).
151. L. Novotny, B. Hecht, and D. W. Pohl, Interference of locally excited surface plasmons, *J. Appl. Phys.* **81**, 1798–1806 (1997).
152. G. Kobidze, B. Shanker, and D. P. Nyquist, Efficient integral-equation-based method for accurate analysis of scattering from periodically arranged nanostructures, *Phys. Rev. E* **72**, 056702 (2005).
153. B. Gallinet, A. M. Kern, and O. J. F. Martin, Accurate and versatile modeling of electromagnetic scattering on periodic nanostructures with a surface integral approach, *J. Opt. Soc. Am. A* **27**, 2261–2271 (2010).
154. T. Søndergaard, D. Lenstra, and B. Tromborg, Near fields and far fields generated by sources in the presence of dielectric structures with cylindrical symmetry, *Opt. Lett.* **26**, 1705–1707 (2001).
155. T. Søndergaard and B. Tromborg, Lippmann-Schwinger integral equation approach to the emission of radiation by sources located inside finite-sized dielectric structures, *Phys. Rev. B* **66**, 155309 (2002).
156. W. Lukosz and R. E. Kunz, Light emission by magnetic and electric dipoles close to a plane interface. I. Total radiated power, *J. Opt. Soc. Am.* **67**, 1607–1615 (1977).
157. W. Lukosz and R. E. Kunz, Light emission by magnetic and electric dipoles close to a plane dielectric interface. II. Radiation patterns of perpendicular oriented dipoles, *J. Opt. Soc. Am.* **67**, 1615–1619 (1977).
158. J. Mertz, Radiative absorption, fluorescence, and scattering, of a classical dipole near a lossless interface: A unified description, *J. Opt. Soc. Am. B* **17**, 1906–1913 (2000).
159. S. R. J. Brueck, Radiation from a dipole embedded in a dielectric slab, *IEEE J. Sel. Top. Quantum Electron.* **6**, 899–910 (2000).
160. A. F. Peterson, *Mapped Vector Basis Functions for Electromagnetic Integral Equations*. San Rafael: Morgan & Claypool publishers, 2005.
161. A. M. Kern, and O. J. F. Martin, Surface integral formulation for 3D simulations of plasmonic and high permittivity nanostructrues, *J. Opt. Soc. Am. B* **26**, 732-40 (2009).
162. U. Hohenester and J. Krenn, Surface plasmon resonances of single and coupled metallic nanoparticles: A boundary integral method approach, *Phys. Rev. B* **72**, 195429 (2005).
163. M. Petravic and G. Kuo-Petravic, An ILUCG algorithm which minimizes in the euclidean norm, *J. Comput. Phys.* **32**, 263 (1979).
164. M. R. Spiegel, *Mathematical Handbook of Formulas and Tables*, Schaum's Outline Series. New York: McGraw-Hill, Inc., 1993.
165. M. Abramovitz and I. A. Stegun, *Handbook of Mathematical Functions*. New York: Dover Publications, Inc., 1965.
166. C. F. Bohren and D. R. Craig, *Absorption and Scattering of Light by Small Particles*. Weinheim: Wiley-VCH Verlag GmbH, 2004.
167. A. L. Aden and M. Kerker, Scattering of electromagnetic waves from two concentric spheres, *J. Appl. Phys.* **22**, 1242–1246 (1951).

Index